Wild Crop Relatives:
Genomic and Breeding Resources

Chittaranjan Kole
Editor

Wild Crop Relatives: Genomic and Breeding Resources

Vegetables

Springer

Editor
Prof. Chittaranjan Kole
Director of Research
Institute of Nutraceutical Research
Clemson University
109 Jordan Hall
Clemson, SC 29634
CKOLE@clemson.edu

ISBN 978-3-642-20449-4 e-ISBN 978-3-642-20450-0
DOI 10.1007/978-3-642-20450-0
Springer Heidelberg Dordrecht London New York

Library of Congress Control Number: 2011922649

Cover design: deblik, Berlin

Printed on acid-free paper

Springer is part of Springer Science+Business Media (www.springer.com)

Dedication

Dr. Norman Ernest Borlaug,[1] the Father of Green Revolution, is well respected for his contributions to science and society. There was or is not and never will be a single person on this Earth whose single-handed service to science could save millions of people from death due to starvation over a period of over four decades like Dr. Borlaug's. Even the Nobel Peace Prize he received in 1970 does not do such a great and noble person as Dr. Borlaug justice. His life and contributions are well known and will remain in the pages of history of science. I wish here only to share some facets of this elegant and ideal personality I had been blessed to observe during my personal interactions with him.

It was early 2007 while I was at the Clemson University as a visiting scientist one of my lab colleagues told me that "somebody wants to talk to you; he appears to be an old man". I took the telephone receiver casually and said hello. The response from the other side was – "I am Norman Borlaug; am I talking to Chitta?" Even a million words would be insufficient to define and depict the exact feelings and thrills I experienced at that moment!

[1]The photo of Dr. Borlaug was kindly provided by Julie Borlaug (Norman Borlaug Institute for International Agriculture, Texas A&M Agriculture) the granddaughter of Dr. Borlaug.

I had seen Dr. Borlaug only once, way back in 1983, when he came to New Delhi, India to deliver the Coromandal Lecture organized by Prof. M.S. Swaminathan on the occasion of the 15th International Genetic Congress. However, my real interaction with him began in 2004 when I had been formulating a 7-volume book series entitled *Genome Mapping and Molecular Breeding in Plants*. Initially, I was neither confident of my ability as a series/book editor nor of the quality of the contents of the book volumes. I sent an email to Dr. Borlaug attaching the table of contents and the tentative outline of the chapters along with manuscripts of only a few sample chapters, including one authored by me and others, to learn about his views as a source of inspiration (or caution!) I was almost sure that a person of his stature would have no time and purpose to get back to a small science worker like me. To my utter (and pleasant) surprise I received an email from him that read: "May all Ph.D.'s, future scientists, and students that are devoted to agriculture get an inspiration as it refers to your work or future work from the pages of this important book. My wholehearted wishes for a success on your important job". I got a shot in my arm (and in mind for sure)! Rest is a pleasant experience – the seven volumes were published by Springer in 2006 and 2007, and were welcome and liked by students, scientists and their societies, libraries, and industries. As a token of my humble regards and gratitude, I sent Dr. Borlaug the Volume I on *Cereals and Millets* that was published in 2006. And here started my discovery of the simplest person on Earth who solved the most complex and critical problem of people on it – hunger and death.

Just one month after receiving the volume, Dr. Borlaug called me one day and said, "Chitta, you know I cannot read a lot now-a-days, but I have gone through only on the chapters on wheat, maize and rice. Please excuse me. Other chapters of this and other volumes of the series will be equally excellent, I believe". He was highly excited to know that many other Nobel Laureates including Profs. Arthur Kornberg, Werner Arber, Phillip Sharp, Günter Blobel, and Lee Hartwell also expressed generous comments regarding the utility and impact of the book series on science and the academic society. While we were discussing many other textbooks and review book series that I was editing at that time, again in my night hours for the benefit of students, scientists, and industries, he became emotional and said to me, "Chitta, forget about your original contributions to basic and applied sciences, you deserved Nobel Prize for Peace like me for providing academic foods to millions of starving students and scientists over the world particularly in the developing countries. I will recommend your name for the World Food Prize, but it will not do enough justice to the sacrifice you are doing for science and society in your sleepless nights over so many years. Take some rest Chitta and give time to Phullara, Sourav and Devleena" (he was so particular to ask about my wife and our kids during most of our conversations). I felt honored but really very ashamed as I am aware of my almost insignificant contribution in comparison to his monumental contribution and thousands of scientists over the world are doing at least hundred-times better jobs than me as scientist or author/editor of books! So, I was unable to utter any words for a couple of minutes but realized later that he must been too affectionate to me and his huge affection is the best award for a small science worker as me!

In another occasion he wanted some documents from me. I told him that I will send them as attachments in emails. Immediately he shouted and told me: "You know, Julie (his granddaughter) is not at home now and I cannot check email myself. Julie does this for me. I can type myself in type writer but I am not good in computer. You know what, I have a xerox machine and it receives fax also. Send me the documents by fax". Here was the ever-present child in him. Julie emailed me later to send the documents as attachment to her as the 'xerox machine' of Dr. Borlaug ran out of ink!

Another occasion is when I was talking with him in a low voice, and he immediately chided me: "You know that I cannot hear well now-a-days; I don't know where Julie has kept the hearing apparatus, can't you speak louder?" Here was the fatherly figure who was eager to hear each of my words!

I still shed tears when I remember during one of our telephone conversations he asked: "You know I have never seen you, can you come to Dallas in the near future by chance?" I remember we were going through a financial paucity at that time and I could not make a visit to Dallas (Texas) to see him, though it would have been a great honor.

In late 2007, whenever I tried to talk to Dr. Borlaug, he used to beckon Julie to bring the telephone to him, and in course of time Julie used to keep alive all the communications between us when he slowly succumbed to his health problems.

The remaining volumes of the *Genome Mapping and Molecular Breeding in Plants* series were published in 2007, and I sent him all the seven volumes. I wished to learn about his views. During this period he could not speak and write well. Julie prepared a letter based on his words to her that read: "Dear Chitta, I have reviewed the seven volumes of the series on *Genome Mapping and Molecular Breeding in Plants*, which you have authored. You have brought together genetic linkage maps based on molecular markers for the most important crop species that will be a valuable guide and tool to further molecular crop improvements. Congratulations for a job well done".

During one of our conversations in mid-2007, he asked me what other book projects I was planning for Ph.D. students and scientists (who had always been his all-time beloved folks). I told him that the wealth of wild species already utilized and to be utilized for genetic analysis and improvement of domesticated crop species have not been deliberated in any book project. He was very excited and told me to take up the book project as soon as possible. But during that period I had a huge commitment to editing a number of book volumes and could not start the series he was so interested about.

His sudden demise in September 2009 kept me so morose for a number of months that I could not even communicate my personal loss to Julie. But in the meantime, I formulated a 10-volume series on *Wild Crop Relatives: Genomic and Breeding Resources* for Springer. And whom else to dedicate this series to other than Dr. Borlaug!

I wrote to Julie for her formal permission and she immediately wrote me: "Chitta, Thank you for contacting me and yes I think my grandfather would be honored with the dedication of the series. I remember him talking of you and this undertaking quite often. Congratulations on all that you have accomplished!" This helped me a lot as I could at least feel consoled that I could do a job he wanted me to do and I will always remain grateful to Julie for this help and also for taking care of Dr. Borlaug, not only as his granddaughter but also as the representative of millions of poor people from around the world and hundreds of plant and agricultural scientists who try to follow his philosophy and worship him as a father figure.

It is another sad experience of growing older in life that we walk alone and miss the affectionate shadows, inspirations, encouragements, and blessings from the fatherly figures in our professional and personal lives. How I wish I could treat my next generations in the same way as personalities like Mother Teresa and Dr. Norman Borlaug and many other great people from around the world treated me!

During most of our conversations he used to emphasize on the immediate impact of research on the society and its people. A couple of times he even told me that my works on molecular genetics and biotechnology, particularly of 1980s and 1990s, have high fundamental importance, but I should also do some works that will benefit people immediately. This advice elicited a change in my thoughts and workplans and since then I have been devotedly endeavoring to develop crop varieties enriched with phytomedicines and nutraceuticals. Borlaug influenced both my personal and professional life, particularly my approach to science, and I dedicate this series to him in remembrance of his great contribution to science and society and for all his personal affection, love and blessings for me.

I emailed the above draft of the dedication page to Julie for her views and I wish to complete my humble dedication with great satisfaction with the words of Julie who served as the living ladder for me to reach and stay closer to such as great human being as Dr. Borlaug and express my deep regards and gratitude to her. Julie's email read: "Chitta, Thank you for sending me the draft dedication page. I really enjoyed reading it and I think you captured my grandfather's spirit wonderfully. . . . So thank you very much for your beautiful words. I know he would be and is honored."

Clemson, USA Chittaranjan Kole

Preface

Wild crop relatives have been playing enormously important roles both in the depiction of plant genomes and the genetic improvement of their cultivated counterparts. They have contributed immensely to resolving several fundamental questions, particularly those related to the origin, evolution, phylogenetic relationship, cytological status and inheritance of genes of an array of crop plants; provided several desirable donor genes for the genetic improvement of their domesticated counterparts; and facilitated the innovation of many novel concepts and technologies while working on them directly or while using their resources. More recently, they have even been used for the verification of their potential threats of gene flow from genetically modified plants and invasive habits. Above all, some of them are contributing enormously as model plant species to the elucidation and amelioration of the genomes of crop plant species.

As a matter of fact, as a student, a teacher, and a humble science worker I was, still am and surely will remain fascinated by the wild allies of crop plants for their invaluable wealth for genetics, genomics and breeding in crop plants and as such share a deep concern for their conservation and comprehensive characterization for future utilization. It is by now a well established fact that wild crop relatives deserve serious attention for domestication, especially for the utilization of their phytomedicines and nutraceuticals, bioenergy production, soil reclamation, and the phytoremediation of ecology and environment. While these vastly positive impacts of wild crop relatives on the development and deployment of new varieties for various purposes in the major crop plants of the world agriculture, along with a few negative potential concerns, are envisaged the need for reference books with comprehensive deliberations on the wild relatives of all the major field and plantation crops and fruit and forest trees is indeed imperative. This was the driving force behind the inception and publication of this series.

Unlike the previous six book projects I have edited alone or with co-editors, this time it was very difficult to formulate uniform outlines for the chapters of this book series for several obvious reasons. Firstly, the status of the crop relatives is highly diverse. Some of them are completely wild, some are sporadically cultivated and some are at the initial stage of domestication for specific breeding objectives recently deemed essential. Secondly, the status of their conservation varies widely: some have been conserved, characterized and utilized; some have been eroded completely except for their presence in their center(s) of origin; some are at-risk or endangered due to genetic erosion, and some of them have yet to be explored. The third constraint is the variation in their relative worth, e.g. as academic model, breeding resource, and/or potential as "new crops."

The most perplexing problem for me was to assign the chapters each on a particular genus to different volumes dedicated to crop relatives of diverse crops grouped based on their utility. This can be exemplified with *Arabidopsis*, which has primarily benefited the Brassicaceae crops but also facilitated genetic analyses and improvement in crop plants in other distant families; or with many wild relatives of forage crops that paved the way for the genetic analyses and breeding of some major cereal and millet crops. The same is true for wild crop relatives such as *Medicago truncatula*, which has paved the way for in-depth research on two crop groups of diverse use: oilseed and pulse crops belonging to the Fabaceae family. The list is too long to enumerate. I had no other choice but to compromise and assign the genera of crop relatives in a volume on the crop group to which they are taxonomically the closest and to which they have relatively greater contributions. For example, I placed the chapter on genus *Arabidopsis* in the volume on oilseeds, which deals with the wild relatives of Brassicaceae crops amongst others.

However, we have tried to include deliberations pertinent to the individual genera of the wild crop relatives to which the chapters are devoted. Descriptions of the geographical locations of origin and genetic diversity, geographical distribution, karyotype and genome size, morphology, etc. have been included for most of them. Their current utility status – whether recognized as model species, weeds, invasive species or potentially cultivable taxa – is also delineated. The academic, agricultural, medicinal, ecological, environmental and industrial potential of both the cultivated and/or wild allied taxa are discussed.

The conservation of wild crop relatives is a much discussed yet equally neglected issue albeit the in situ and ex situ conservations of some luckier species were initiated earlier or are being initiated now. We have included discussions on what has happened and what is happening with regard to the conservation of the crop relatives, thanks to the national and international endeavors, in most of the chapters and also included what should happen for the wild relatives of the so-called new, minor, orphan or future crops.

The botanical origin, evolutionary pathway and phylogenetic relationship of crop plants have always attracted the attention of plant scientists. For these studies morphological attributes, cytological features and biochemical parameters were used individually or in combinations at different periods based on the availability of the required tools and techniques. Access to different molecular markers based on nuclear and especially cytoplasmic DNAs that emerged after 1980 refined the strategies required for precise and unequivocal conclusions regarding these aspects. Illustrations of these classical and recent tools have been included in the chapters.

Positioning genes and defining gene functions required in many cases different cytogenetic stocks, including substitution lines, addition lines, haploids, monoploids and aneuploids, particularly in polyploid crops. These aspects have been dealt in the relevant chapters. Employment of colchiploidy, fluorescent or genomic in situ hybridization and Southern hybridization have reinforced the theoretical and applied studies on these stocks. Chapters on relevant genera/species include details on these cytogenetic stocks.

Wild crop relatives, particularly wild allied species and subspecies, have been used since the birth of genetics in the twentieth century in several instances such as studies of inheritance, linkage, function, transmission and evolution of genes. They have been frequently used in genetic studies since the advent of molecular markers. Their involvement in molecular mapping has facilitated the development of mapping populations with optimum polymorphism to construct saturated maps and also illuminating the organization, reorganization and functional aspects of genes and genomes. Many phenomena such as genomic duplication, genome reorganization, self-incompatibility, segregation distortion, transgressive segregation and defining genes and their phenotypes have in many cases been made possible due to the utilization of wild species or subspecies. Most of the chapters contain detailed elucidations on these aspects.

The richness of crop relatives with biotic and abiotic stress resistance genes was well recognized and documented with the transfer of several alien genes into their cultivated counterparts through wide or distant hybridization with or without employing embryo-rescue and mutagenesis. However, the amazing revelation that the wild relatives are also a source of yield-related genes is a development of the molecular era. Apomictic genes are another asset of many crop relatives that deserve mention. All of these past and the present factors have led to the realization that the so-called inferior species are highly superior in conserving desirable genes and can serve as a goldmine for breeding elite plant varieties. This is particularly true at a point when natural genetic variability has been depleted or exhausted in most of the major crop species, particularly due to growing and promoting only a handful of so-called high-yielding varieties while disregarding the traditional cultivars and landraces. In the era of molecular breeding, we can map desirable genes and polygenes, identify their donors and utilize tightly linked markers for gene introgression, mitigating the constraint of linkage drag, and even pyramid genes from multiple sources, cultivated or wild taxa. The evaluation of primary, secondary and tertiary gene pools and utilization of their novel genes is one of the leading strategies in present-day plant breeding. It is obvious that many wide hybridizations will never be easy and involve near-impossible constraints such as complete or partial sterility. In such cases gene cloning and gene discovery, complemented by intransgenic breeding, will hopefully pave the way for success. The utilization of wild relatives through traditional and molecular breeding has been thoroughly enumerated over the chapters throughout this series.

Enormous genomic resources have been developed in the model crop relatives, for example *Arabidopsis thaliana* and *Medicago truncatula*. BAC, cDNA and EST libraries have also been developed in some other crop relatives. Transcriptomes and metabolomes have also been dissected in some of them. However, similar genomic resources are yet to be constructed in many crop relatives. Hence this section has been included only in chapters on the relevant genera.

In this book series, we have included a section on recommendations for future steps to create awareness about the wealth of wild crop relatives in society at large and also for concerns for their alarmingly rapid decrease due to genetic erosion. The authors of the chapters have also emphasized on the imperative requirement of their conservation, envisaging the importance of biodiversity. The importance of intellectual property rights and also farmers' rights as owners of local landraces, botanical varieties, wild species and subspecies has also been dealt in many of the chapters.

I feel satisfied that the authors of the chapters in this series have deliberated on all the crucial aspects relevant to a particular genus in their chapters.

I am also very pleased to present many chapters in this series authored by a large number of globally reputed leading scientists, many of whom have contributed to the development of novel concepts, strategies and tools of genetics, genomics and breeding and/or pioneered the elucidation and improvement of particular plant genomes using both traditional and molecular tools. Many of them have already retired or will be retiring soon, leaving behind their legacies and philosophies for us to follow and practice. I am saddened that a few of them have passed away during preparation of the manuscripts for this series. At the same time, I feel blessed that all of these stalwarts shared equally with me the wealth of crop relatives and contributed to their recognition and promotion through this endeavor.

I would also like to be candid with regard to my own limitations. Initially I planned for about 150 chapters devoted to the essential genera of wild crop relatives. However, I had to exclude some of them either due to insignificant progress made on them during the preparation of this series, my failure to identify interested authors willing to produce acceptable manuscripts in time or authors' backing out in the last minute, leaving no time to find replacements. I console myself for this lapse with the rationale that it is simply too large a series to achieve complete satisfaction on the contents. Still I was able to arrange about 125 chapters in the ten volumes, contributed by nearly 400 authors from over 40 countries of the world. I extend my heartfelt thanks to all these scientists, who have cooperated with me since the inception of this series not only with their contributions, but also in some cases by suggesting suitable authors for chapters on other genera. As happens with a mega-series, a few authors had delays for personal or professional reasons, and in a few cases, for no reason at all. This caused delays in the publication of some of the volumes and forced the remaining authors to update their manuscripts and wait too long to see their manuscripts in published form. I do shoulder all the responsibilities for this myself and tender my sincere apologies.

Another unique feature of this series is that the authors of chapters dedicated to some genera have dedicated their chapters to scientists who pioneered the exploration, description and utilization of the wild species of those genera. We have duly honored their sincere decision with equal respect for the scientists they rightly reminded us to commemorate.

Editing this series was, to be honest, very taxing and painstaking, as my own expertise is limited to a few cereal, oilseed, pulse, vegetable, and fruit crops, and some medicinal and aromatic plants. I spent innumerable nights studying to attain the minimum eligibility to edit the manuscripts authored by experts with even life-time contributions on the concerned genera or species. However, this indirectly awakened the "student-for-life" within me and enriched my arsenal with so many new concepts, strategies, tools, techniques and even new terminologies! Above all, this helped me to realize that individually we know almost nothing about the plants on this planet! And this realization strikingly reminded me of the affectionate and sincere advice of Dr. Norman Borlaug to keep abreast with what is happening in the crop sciences, which he used to do himself even when he had been advised to strictly limit himself to bed rest. He was always enthusiastic about this series and inspired me to take up this huge task. This is one of the personal and professional reasons I dedicated this book series to him with a hope that the present and future generations of plant scientists will share the similar feelings of love and respect for all plants around us for the sake of meeting our never-ending needs for food, shelter, clothing, medicines, and all other items used for our basic requirements and comfort. I am also grateful to his granddaughter, Julie Borlaug, for kindly extending her permission to dedicate this series to him.

I started editing books with the 7-volume series on Genome Mapping and Molecular Breeding in Plants with Springer way back in 2005, and I have since edited many other book series with Springer. I always feel proud and satisfied to be a member of the Springer family, particularly because of my warm and enriching working relationship with Dr. Sabine Schwarz and Dr. Jutta Lindenborn, with whom I have been working all along. My special thanks go out to them for publishing this "dream series" in an elegant form and also for appreciating my difficulties and accommodating many of my last-minute changes and updates.

I would be remiss in my duties if I failed to mention the contributions of Phullara – my wife, friend, philosopher and guide – who has always shared with me a love of the collection, conservation, evaluation, and utilization of wild crop relatives and has enormously supported me in the translation of these priorities in my own research endeavors – for her assistance in formulating the contents of this series, for monitoring its progress and above all for taking care of all the domestic and personal responsibilities I am supposed to shoulder. I feel myself alien to the digital world that is the sine qua non today for maintaining constant communication and ensuring the preparation of manuscripts in a desirable format. Our son Sourav and daughter Devleena made my life easier by balancing out my limitations and also by willingly sacrificing the spare amount of time I ought to spend with them. Editing of this series would not be possible without their unwavering support.

I take the responsibility for any lapses in content, format and approach of the series and individual volumes and also for any other errors, either scientific or linguistic, and will look forward to receiving readers' corrections or suggestions for improvement.

As I mentioned earlier this series consists of ten volumes. These volumes are dedicated to wild relatives of Cereals, Millets and Grasses, Oilseeds, Legume Crops and Forages, Vegetables, Temperate Fruits, Tropical and Subtropical Fruits, Industrial Crops, Plantation and Ornamental Crops, and Forest Trees.

This volume "Wild Crop Relatives: Genomic and Breeding Resources – Vegetables" includes 13 chapters dedicated to *Allium*, *Amaranthus*, *Asparagus*, *Capsicum*, *Citrulus*, *Cucumis*, *Daucus*, *Lactuca*, *Lycopersicon*, *Momordica*, *Raphanus*, *Solanum*, and *Spinacea*. The chapters of this volume were authored by 58 scientists from 12 countries of the world, namely Argentina, Australia, China, Denmark, India, Italy, Japan, Poland, South Africa, Thailand, UK and the USA.

It is my sincere hope that this volume and the series as a whole will serve the requirements of students, scientists and industries involved in studies, teaching, research and the extension of vegetable crops with an intention of serving science and society.

Clemson, USA Chittaranjan Kole

Contents

Abbreviations

2-TD	2-Tridecanone
3G	3-Glucoside
6-PGD	6-Phosphogluconate dehydrogenase
7G	7-Glucoside
aadA	Aldehyde–alcohol dehydrogenase gene
AAL	*Alternaria alternata* f. sp. *Lycopersici*
AB	Advanced backcross
ABF3	Gene encoding a transcription factor for expression of abscisic acid response genes
ABL	Advanced breeding line
accD	Acetyl-coenzyme D-subunit gene
ADH	Alcohol dehydrogenase
AFLP	Amplified fragment length polymorphism
AG	*AGAMOUS* gene
AIDS	Acquired immune deficiency syndrome
ALS	Acetolactate synthase
AMV	Alfalfa mosaic virus
Ap	Apigenin
ARS	Agriculture Research Service (of USDA)
AU	*Aurora* locus
AVGRIS	The AVRDC Vegetable Genetic Resources Information System
Avr	Avirulence gene
AVRDC	Asian Vegetable Research and Development Center (presently The World Vegetable Center), Taiwan
BAC	Bacterial artificial chromosome
BC	Backcross
BC_1	Backcross (first generation)
BCRIL	Backcross recombinant inbred line
BIL	Backcross inbred line
BWYV	Beet western yellows virus
CAAS	Chinese Academy of Agricultural Science
CAPS	Cleaved amplified polymorphic sequence
CATIE	Centro Agronómico Tropical de Investigación y Enseñanza (Costa Rica)
CC	Coiled
cDNA	Complementary-DNA
CG*AA*	Candidate genes associated with ascorbic acid biosynthesis

CG*C*	Candidate carotenoid gene
CG*FL*	Candidate gene for flowering
CG*FSC*	Candidate gene for fruit size and composition
CGIAR	Consultative Group on International Agricultural Research
Chr	Chrysoeriol
CHS	Chalcone synthase
CIP	International Center for the Potato (Peru)
cM	CentiMorgan
Cm-ERS1	Mutated melon ethylene receptor gene
Cmm	*Clavibacter michiganensis* ssp. *michiganensis*
CMS	Cytoplasmic male sterility
CMV	Cucumber mosaic virus
COS	Conserved ortholog set
COSII	Conserved ortholog set II
CPC	Commonwealth Potato Collection (SCRI)
cpDNA	Chloroplast-DNA
CR-EST	Candidate resistance/defense-response EST
cv.	Cultivar
CVMV	Chili veinal mottle virus
DEF	*DEFICIENS* gene
Dm3	Downy mildew resistance gene
EB	Early blight
ECPGR	European Cooperative Program for Plant Genetic Resources
ELISA	Enzyme-linked immunosorbent assay
EPSPS	5-Enolypyruvyl-shikimate-3-phosphate synthase
EST	Expressed sequence tag
ETS	External transcribed spacer
EU	European Union
F_1	First filial generation
FAO	Food and Agriculture Organization (of the United Nations)
FAOSTAT	FAO Statistics
FISH	Fluorescence/Fluorescent in situ hybridization
FW	Fresh weight
G(M1)/G(m1)	Ganglioside
GBSSI	Granule-bound starch synthase
GDH	Glutamate dehydrogenase
GFP	Green fluorescent protein
GISH	Genomic in situ hybridization
GLO	*GLOBOSA* gene
GLOase	L-Gulono-gamma-lactone oxidase
GM	Genetically modified/modification
GM	Tomato gray mold
GRIN	Germplasm Resources Information Network (of USDA)
gus	β-Glucuronidase gene
HIV	Human immunodeficiency virus
HPT	Homogentisate phytyl transferase
HR	Hypersensitive response
IAA	Indole acetic acid

IBA	Indole butyric acid
IBC	Inbred backcross population
IBPGR	International Board for Plant Genetic Resources
IDH	Isocitrate dehydrogenase
IL	Introgression line
ILD	Incongruence length difference test
ILH	Introgression line hybrid
INIBAP	International Network for Improvement of Banana and Plantain
IPGRI	International Plant Genetic Resources Institute
ISO	Isozyme
ISSR	Intersimple sequence repeat
ITAG	The International Tomato Annotation Group
ITS	Internal transcribed spacer
IUCN	International Union for Conservation of Nature
Ka	Kaempferol
KNAT1	Knotted 1 gene from *Arabidopsis thaliana*
LA	Linoleic acid
LAP	Leucine amino peptidase
LB	Late blight
LBVaV	Lettuce big vein associated virus
LMV	Lettuce mosaic virus
LRR	Leucine rich repeat
LTB	Labile enterotoxin B
Lu	Luteolin
MAAL	Monosomic alien addition line
MAS	Marker-assisted selection
MDH	Malate dehydrogenase
ME-4aN4	Late embryogenesis abundant protein gene from *Brassica napus*
MH	Maleic hydrazide
MHa	Million hectare(s)
MI	First metaphase
MiBASE	Micro-Tom Database
MLBVV	Mirafiori lettuce big vein virus
MO	Morphological marker
Mpb	Million base pairs
MSAP	Methylation-sensitive amplified polymorphism
mtDNA	Mitochondrial-DNA
MW	*Milky Way* locus
NATP	National Agricultural Technology Project (India)
NBN	National Biodiversity Network
NBS	Nucleotide binding site
NCBI	National Center for Biotechnology Information
NCGRP	National Center for Genetic Resources Preservation
NIL	Near isogenic line
NOR	Nucleolus organizing region
NPGS	National Plant Germplasm System
*npt*II	Neomycin phosphotransferase gene
nrDNA	Nuclear ribosomal DNA

Nv	Non-virion gene
ODO	Overdominant
OR	*Orion* locus
OxdC	Decarboxylase gene
PA	Petroselinic acid
PBMC	Peripheral blood mononuclear cells
PCN	Potato cyst nematode
PCR	Polymerase chain reaction
PEG	Polyethylene glycol
PepMoV	Pepper mottle virus
PGI	Phosphoglucose isomerase
PGM	Phosphoglucomutase
PGRU	Plant Genetic Resources Unit
PM	Powdery mildew
pMAC	Hybrid promoter of cauliflower mosaic virus 35S protein and the Ti plasmid mannopine synthase gene promoter
PMC	Pollen mother cell
PPO	Protoporphyrinogen oxidase
PPX2	Allelic form of protoporphyrinogen oxidase
PRV	Papaya ringspot virus
PSII	Photosystem II
PVY	Potato virus Y
QAL	Queen Anne's lace
QTL	Quantitative trait loci
Qu	Quercetin
RAPD	Random amplified polymorphic DNA
RDA	Recommended dietary allowances
rDNA	Ribosomal DNA
RFLP	Restriction fragment length polymorphism
RG	Resistance gene
RGA	Resistance gene analog
RIL	Recombinant inbred line
RNAi	RNA interference
RS	Reproductive stage
RT-PCR	Reverse transcriptase-PCR
SAMPL	Selectively amplified microsatellite polygenic loci
SC	Self-compatibility/Self-compatible
SCAR	Sequence characterized amplified region
SCRI	Scottish Crop Research Institute
sCTB	Synthetic cholera toxin B subunit
SG	Seed germination
SGe	Selective genotyping
SGN	SOL Genomics Network
SI	Self-Incompatibility or Self-Incompatible
SINGER	System-wide Information Exchange Network (of CGIAR)
SNP	Single nucleotide polymorphism
SOL	International Solanaceae Genome Project
SSC	Soluble solids content

SSR	Simple sequence repeat
ST	Salt tolerance
STS	Sequence tagged site
TC	Test cross
TC	Tocopherol cyclase
TED	Tomato Expression Database
TEI	Tomato EST-derived intronic polymorphism
TES	Tomato EST-derived SSR
TEV	Tobacco etch virus
TGRC	Tomato Genetics Resource Center
TGS	Tomato genome-derived SSR
TILLING	Targeting induced local lesions in genomes
TIR	Tol interleukin receptor homology
TL	Tocopherol cyclase
TMRD	Tomato Mapping Resource Database
TMV/ToMV	Tobacco mosaic/mottle virus
t-NBS-LRR	Tomato-nucleotide binding site-leucine rich repeat
TPI	Triose phosphate isomerase
TRAP	Target region amplification polymorphism
TSED	Tomato Stress EST Database
TSWV	Tomato spotted wilt virus
TYLCV	Tomato yellow leaf curl virus
UI	Unilateral incongruity
UNU	United Nations University
UPGMA	Unweighted pair group method
USDA	United States Department of Agriculture
USFWS	United States Fish and Wildlife Service
VG	Vegetative growth
VIGS	Virus induced gene silencing
WHO	World Health Organization
WMV	Watermelon mosaic virus
WUE	Water use efficiency
YAC	Yeast artificial chromosome
ZYMV	Zucchini yellow mosaic virus

Contributors

Charlotte Allender Warwick Crop Centre, School of Life Sciences, The University of Warwick, Wellesbourne Campus, Warwick, CV35 9EF, UK, charlotte.allender@warwick.ac.uk

Sven B. Andersen Plant and Soil Science section, Department of Agriculture and Ecology, Faculty of Life Sciences, University of Copenhagen, Copenhagen, Denmark, sba@life.ku.dk

Paul Anthony Plant and Crop Sciences Division, School of Biosciences, University of Nottingham, Sutton Bonington Campus, Loughborough LE12 5RD, UK, paul.anthony@nottingham.ac.uk

Sang Woo Bang Laboratory of Plant Breeding, Faculty of Agriculture Organization, Utsunomiya University, 350 Minemachi, Utsunomiya 321-8505, Japan, bang@cc.utsunomiya-u.ac.jp

Rafal Baranski Department of Genetics, Plant Breeding and Seed Science, University of Agriculture in Krakow, Al. 29 Listopada 54, 31-425, Kraków, Poland, baranski@ogr.ar.krakow.pl

T.K. Behera Division of Vegetable Science, Indian Agricultural Research Institute, New Delhi 110012, India, tusar@iari.res.in

L.K. Bharathi Division of Vegetable Science, Indian Agricultural Research Institute, New Delhi 110012, India, alkb@rediffmail.com

Glenn Bryan The James Hutton Institute, Invergowrie, Dundee DD2 5DA, UK, glenn.bryan@hutton.ac.uk

Maria Cammareri CNR – Institute of Plant Genetics, Res. Div. Portici, National Research Council, Via Università 133, 80055, Portici, NA, Italy, cammarer@unina.it

Jin-Feng Chen State Key Laboratory of Crop Genetics and Germplasm Enhancement, Nanjing Agricultural University, No.6 Tongwei Road, Nanjing 210095, China, jfchen@njau.edu.cn

Roger Chetelat C. M. Rick Tomato Genetics Resource Center, Department of Plant Sciences, University of California, 1 Shields Avenue, Davis, CA 95616, USA, trchetelat@ucdavis.edu

Maria Luisa Chiusano Department of Soil, Plant, Environmental and Animal Production Sciences, University of Naples Federico II, Via Università 100, 80055, Portici, Italy, chiusano@unina.it

Michael R. Davey Plant and Crop Sciences Division, School of Biosciences, University of Nottingham, Sutton Bonington Campus, Loughborough LE12 5RD, UK, mike.davey@nottingham.ac.uk

Maria Raffaella Ercolano Department of Soil, Plant, Environmental and Animal Production Sciences, University of Naples Federico II, Via Università 100, 80055, Portici, Italy, ercolano@unina.it

Luigi Frusciante Department of Soil, Plant, Environmental and Animal Production Sciences, University of Naples Federico II, Via Università 100, 80055, Portici, Italy, fruscian@unina.it

Silvana Grandillo CNR – Institute of Plant Genetics, Res. Div. Portici, National Research Council, Via Università 133, 80055, Portici, NA, Italy, grandill@unina.it

Dariusz Grzebelus Department of Genetics, Plant Breeding and Seed Science, University of Agriculture in Krakow, Al. 29 Listopada 54, 31-425, Kraków, Poland, dgrzebel@ogr.ar.krakow.pl

Nurul Islam-Faridi Forest Tree Molecular Cytogenetics Laboratory, Southern Institute of Forest Genetics, Southern Research Station, US Forest Service, Department of Ecosystem Science and Management, Texas A&M University, College Station, TX 77843, USA, nfaridi@tamu.edu

K. Joseph John National Bureau of Plant Genetic Resources (ICAR, KAU (P.O.), Thrissur Dt, Kerala 680656, India, josephjohnk@rediffmail.com

Yukio Kaneko Laboratory of Plant Breeding, Faculty of Agriculture, Utsunomiya University, 350 Minemachi, Utsunomiya 321-8505, Japan, kaneko@cc.utsunomiya-u.ac.jp

Akira Kanno Graduate School of Life Sciences, Tohoku University, 2-1-1, Katahira, Aoba-ku, Sendai 980-8577, Japan, kanno@ige.tohoku.ac.jp

R. Karuppaiyan ICAR Research Complex for NEH Region, Sikkim Centre, Tadong Post, Gangtok, Sikkim 737102, India, kanno@ige.tohoku.ac.jp

Sandra Knapp Department of Botany, The Natural History Museum, Cromwell Road, London SW7 5BD, UK, s.knapp@nhm.ac.uk

Amnon Levi US Vegetable Laboratory, USDA-ARS, 2875 Savannah Highway, Charleston, SC 29414, USA, Amnon.Levi@ARS.USDA.GOV

Frank Lutz Douglass Land-grant Institute, West Virginia State University, Institute, WV 25112, USA, lutz@wvstateu.edu

Yasuo Matsuzawa Laboratory of Plant Breeding, Faculty of Agriculture Organization, Utsunomiya University, 350 Minemachi, Utsunomiya 321-8505, Japan, aht60my@orion.ocn.ne.jp

Orarat Mongkolporn Department of Horticulture, Faculty of Agriculture Kamphaeng Saen, Kasetsart University, Kamphaeng Saen Campus, Nakhon Pathom 73140, Thailand, orarat.m@ku.ac.th

Luigi Monti Department of Soil, Plant, Environmental and Animal Production Sciences, University of Naples Federico II, Via Università 100, 80055, Portici, Italy, lmonti@unina.it

Satya S. Narina Department of Biology, Virginia State University, 1 Hayden Drive, Petersburg, VA 23806, USA; Agricultural Research Station, Virginia State University, Petersburg, VA 23806, snarina@vsu.edu

Padma Nimmakayala Douglass Land-grant Institute, West Virginia State University, Institute, Kanawha, WV 25112, USA, padma@wvstateu.edu

Damaris Odeny Agricultural Research Council – Vegetable and Ornamental Plants Institute, Private Bag X293, Pretoria 0001, South Africa, DOdeny@arc.agric.za

Iris Peralta Department of Agronomy, National University of Cuyo, Almirante Brown 500, 5505 Chacras de Coria, Luján, Mendoza, Argentina; IADIZA CCT Mendoza CONICET, C.C. 507, 5500 Mendoza, Argentina, iperalta@fca.uncu.edu.ar

Olga Perez Scuola Superiore Sant'Anna, International Doctoral Programme on Agrobiodiversity – Plant Genetic Resources, ENEA-Cr. Casaccia, Rome, Italy, olgayperez@gmail.com

Domenico Pignone CNR – Institute of Plant Genetics, National Research Council, Via Amendola 165/A, 70126 Bari, Italy, domenico.pignone@igv.cnr.it

Gavin Ramsay The James Hutton Institute, Invergowrie, Dundee DD2 5DA, UK, gavin.ramsay@hutton.ac.uk

Umesh K. Reddy Douglass Land-grant Institute, West Virginia State University, Institute, WV 25112, USA, ureddy@wvstateu.edu

Philipp W. Simon Department of Horticulture, University of Wisconsin, 1575 Linden Drive, Madison, WI 53706, USA, psimon@wisc.edu

Krzysztof Spalik Department of Plant Systematics and Geography, Institute of Botany, University of Warsaw, Aleje Ujazdowskie 4, 00-478 Warszawa, Poland, spalik@biol.uw.edu.pl

David Spooner Vegetable Crops Research Unit, USDA-ARS, Department of Horticulture, University of Wisconsin, 1575 Linden Drive, Madison, WI 53706-1590, USA, David.Spooner@ARS.USDA.GOV

Paul W. J. Taylor Center for Plant Health/BioMarka, Department of Agriculture and Food Systems, Melbourne School of Land and Environment, The University of Melbourne, Melbourne, VIC 3010, Australia, paulwjt@unimelb.edu.au

Pasquale Termolino CNR – Institute of Plant Genetics, Res. Div. Portici, National Research Council, Via Università 133, 80055, Portici, NA, Italy, termolin@unina.it

Yan R. Tomason Douglass Land-grant Institute, West Virginia State University, Institute, Kanawha, WV 25112, USA, yan@wvstateu.edu

Anna Maria Torp Plant and Soil Science section, Department of Agriculture and Ecology, Faculty of Life Sciences, University of Copenhagen, Copenhagen, Denmark, amt@life.ku.dk

Patrick J. Tranel Department of Crop Sciences, University of Illinois, Urbana, IL 61801, USA, tranel@illinois.edu

Pasquale Tripodi CNR – Institute of Plant Genetics, Res. Div. Portici, National Research Council, Via Università 133, 80055 Portici, NA, Italy, ptripodi@unina.it

Federico Trucco Instituto de Agrobiotecnología Rosario, CP2000 Rosario, Santa Fe, Argentina, trucco@indear.com

Jun Yokoyama Department of Biology, Faculty of Science, Yamagata University, Kojirakawa 1-4-12, Yamagata-shi, Yamagata 990-8560, Japan, jyokoyam@sci.kj.yamagata-u.ac.jp

Xiao-Hui Zhou State Key Laboratory of Crop Genetics and Germplasm Enhancement, Nanjing Agricultural University, No.6 Tongwei Road, Nanjing 210095, China, xhzhou1984@sina.com

Chapter 1
Allium

Damaris A. Odeny and Satya S. Narina

1.1 Introduction

The word "*Allium*" comes from the Greek word "aloe," which means to avoid, and was given to the genus because of the characteristic offensive smell of its members. The genus *Allium* is one of the largest plant genera and includes about 780 species. Cultivation of *Allium* species is reportedly very old and as extensive as civilization itself (Block 2010). The wealth of *Allium* species was mentioned in the ancient civilizations both as flavorful foods and healing herbs.

Allium is a genus within the family Alliaceae and belongs to the order Asparagales of the monocot division. Asparagales and the Poales (which includes the grasses) are two well-supported monophyletic orders within the monocots (Rudall et al. 1997). The genus *Allium* is mainly restricted to the regions that are seasonally dry, with centers of diversity in Soutwest/Central Asia, eastern Asia, and in North America. Members of the genus *Allium* include many economically important crops such as onions (*Allium cepa*), shallots (*Allium oschaninii*), leeks (*Allium ampeloprasum* var. *porrum*), scallions (*Allium ascalonicum*), garlic (*Allium sativum*), chives (*Allium schoenoprasum*), Japanese bunching onions (*A. fistulosum*), rakkyo (*A. chinense*) and Chinese chives (*A. tuberosum*). Others such as *A. karataviense* and *A. christophii* are cultivated as ornamentals (Huxley et al. 1992).

Different communities use several wild *Allium* species for different purposes while breeders use them as sources of economically important traits for improving the cultivated species. Many wild *Allium* species are believed to have significant potential contribution to the *Allium* breeding community and to the global food and health needs. Despite the importance of these species, there are very few studies and research investments on them. The taxonomy is not extensive and conservation efforts have been limited. The limited extent of genomic tools within the genus as a whole has further limited research focus to cultivated species. We provide here a brief review on wild *Allium* species including basic botany, conservation initiatives, contribution toward the development of cytogenetic stocks, genetic tools, and improvement of the cultivated species. The potential for more domestication and commercialization of wild *Allium* species as a source of income, nutrition, and medicinal remedy are also mentioned.

1.2 Basic Botany of the Species

The taxonomy of *Allium* has been described as complicated with a great number of synonyms and intrageneric groupings (Klaas 1998). The genus comprises either short- or long-lived perennials (Brewster 1998) with characteristic storage organs (Fritsch and Friesen 2002). The typical storage organs are rhizomes, roots, or bulbs.

The bulbs are generally enclosed in membranous tunics, free or almost free tepals, and often in a subgynobasic style (Friesen et al. 2006). The bulbs can be large and single, or small, forming clusters (Le Guen-

D.A. Odeny (✉)
Agricultural Research Council, Vegetable and Ornamental Plants Institute, Private Bag X293, Pretoria 0001, South Africa

Present Address
ARC-OVI (Biotechnology Platform), Private Bag X5, Onderstepoort 0110, South Africa
e-mail: dodeny@arc.agric.za

C. Kole (ed.), *Wild Crop Relatives: Genomic and Breeding Resources, Vegetables*,
DOI 10.1007/978-3-642-20450-0_1, © Springer-Verlag Berlin Heidelberg 2011

Le Saos et al. 2002) and range in size from 2–3 mm to 8–10 cm in diameter.

The leaves are tubular (as in onions) or flat arising from the underground stem with long sheathing base, which can give the appearance of a stem (Block 2010). Many *Allium* species have basal leaves that wither away from the tips downward before or while the plant is flowering (http://www.absoluteastronomy.com/topics/Allium). The inflorescence can be fasciculate, umbel or head like, few to many, and loose to dense flowers have been reported (Fritsch and Friesen 2002). Flowering time varies with various *Allium* species. Flowering has been recorded in spring, summer, or autumn (Brewster 1998). A majority of *Allium* species are outbreeders, but the occurrence of both outbreeding and inbreeding forms in the same species has been reported (Brat 1965).

Allium species contain nectaries, which are located between the base of the ovary and the flattened widened base of the inner whorl of stamen and filaments (Peumans et al. 1997). The ovaries are largely trilocular and the styles are single and slender (Fritsch and Friesen 2002). The seeds can be angular or globular (Fritsch and Friesen 2002). Polyembryony (the development of multiple seedlings from a single seed) has been observed in several *Allium* species within the subgenus *Rhizirideum* (Specht et al. 2001).

1.3 Conservation Initiatives

As in many other crop species, there is evidence of genetic erosion in *Allium* species (Mingochi and Swai 1994). Goodding's onion (*Allium gooddingii*), a delicate perennial with reddish-purple flowers and a pungent onion aroma, was listed as a candidate for federal endangered/threatened status (http://www.fws.gov/southwest/es/Arizona/Goodings.htm). *Allium roseum* var. *odoratissimum*, which is harvested from roadsides during spring in Tunisia, is marketed very cheaply creating the risk of genetic erosion. *Allium roylei*, one of the most important sources of disease resistance genes in the genus *Allium* (Kofoet et al. 1990), is also a threatened species (Araki et al. 2010) and an urgent plea for its conservation has been made (Kohli and Gohil 2009). The need for protection of *A. gooddingii* was recognized in 1998 by the Forest Service and the Environmental Protection Agency of the United States of America, which helped considerably toward the protection of this species against erosion (USFWS 2000). Ghrabi (2010) describes the need to protect and domesticate *A. roseum* as an urgent necessity.

Global networks have, therefore, been formed to coordinate the collection, conservation, and utilization of the genetic resources within the genus *Allium*. Examples include the European Cooperative Program for Plant Genetic Resources (ECPGR) *Allium* Working Group, the United Kingdom Natural Resources Institute Onion Newsletter for the Tropics network, and the *Allium* Improvement Newsletter (Astley 1994). The *Allium* working group (http://www.ecpgr.cgiar.org/Workgroups/Allium/Allium.htm) established in 1982 in Europe is one of the original six working groups constituted during the first phase of ECPGR. The work of the ECPGR *Allium* working group benefited significantly from the work program of the European Genetic Resources *Allium* project (ended 31 Mar 2000), and the research projects FAIR (1996–2001) and "Garlic and Health" (1998–2003). In the framework of the European Genetic Resources 20 Project, over 200 accessions of wild taxa belonging to 35 species were collected and preserved in Greek gene Bank, Greece (Samaras 2001).

In India, the National Bureau of Plant Genetic Resources (NBPGR) conducted extensive plant exploration in different *Allium*-growing states resulting in the collection of over 2,200 accessions of *Allium* species including wild relatives such as *A. ampeloprasum, A. auriculatum, A. ascalonicum, A. carolinianum, A. chinensis, A. wallachi, A. tuberosum,* and *A. rubellum* (Singh and Rana 1994). NBPGR also introduced over 1,100 accessions of *Allium* germplasm from over 40 countries (Singh and Rana 1994). In Tunisia, the Medenine Institut des Regions Arides has recently developed a research program to protect, conserve, and domesticate *A. roseum* (http://www.uicnmed.org/nabp/database/HTM/PDF/p5.pdf).

Given the outbreeding nature of most *Allium* species, cryopreservation has been suggested as a viable method of conservation (Volk et al. 2004) but this will need to be combined with traditional field maintenance, pollen storage, and in vitro culture (Working Group on Allium 2009). Cryopreservation should be done carefully due to the high risk of contamination and virus infection.

1.4 Role in Elucidation of Origin and Evolution of Cultivated *Allium* Species

The genus *Allium* is one of the largest genera of perennial bulbous plants on earth. The use of internal transcribed spacer (ITS) region of ribosomal DNA recently classified the genus into 780 species, 56 sections, and 15 subgenera (Friesen et al. 2006). Earlier molecular comparisons recognized 67 sections and 14 subgenera (Fritsch and Friesen 2002). The fact that so many species exist suggests that a lot of evolutionary differentiation has occurred (Stevenson et al. 1999).

Both morphological and molecular data have been used to study evolution of the genus *Allium*. Wendelbo (1969) initially suggested that all other groups in *Allium* might be derived from rhizomatous species without bulbs. Rhizomes are a distinct feature within the subgenus *Rhizirideum*. More recent reports indicate that rhizomes have evolved from an ancestral bulbous life form that was subsequently lost at least twice independently (Ricroch et al. 2005). Other *Allium* species (*A. caeruleum, A. proliferium, A. vineale, A. carinatum,* and *A. scorodoprasum*) have been grouped as viviparous due to the development of topsets instead of flowers or intermingle with flowers in the florescence (Kamenetsky and Rabinowitch 2001).

More molecular studies of the genus have given insight into evolution of the cultivated *Allium* species. Amplified fragment length polymorphism (AFLP) studies in garlic clustered garlic clones very closely with *A. longicuspis* suggesting that the two species are not genetically distinct (Ipek and Simon 2001). *A. longicuspis* and *A. tuncelianum* are genetically identical to garlic and therefore, were earlier suggested as possible ancestors of garlic (Block 2010). However, DNA analysis indicated that neither *A. longicuspis* nor *A. tuncelianum* are ancestor species of garlic (Block 2010).

In the subgenus *Cepa*, section *Cepa*, the closest wild relative of *A. cepa* and *A. fistulosum* has been identified as *A. vavilovii* and *A. attaicium*, respectively (Klaas and Friesen 2002). *A. oschaninii,* on the other hand, appears to be a sister group to *A. cepa*/*A. vavilovii* evolutionary line (Friesen and Klaas 1998). *A. ampeloprasum* is thought to be the ancestor species of leek and kurrat (*A. ampeloprasum* var. *kurrat*) (Block 2010).

The use of nuclear ribosomal DNA [ITS and external transcribed spacer (ETS)] has been suggested to provide sufficient resolution for investigating evolutionary relationships within *Allium* (Nguyen et al. 2008). Nguyen et al. (2008) used combined sequences from 39 native Californian *Allium* species with 154 ITS sequences available on GenBank to develop a global *Allium* phylogeny with the simultaneous goals of investigating the evolutionary history of *Allium* in the Californian center of diversity and exploring patterns of adaptation to serpentine soils (Nguyen et al. 2008). The ITS region alone was sufficient to resolve the deeper relationships in North American species.

1.5 Role in Development of Cytogenetic Stocks and Their Utility

Allium species show variation in several cytogenetic characters such as basic chromosome number, ploidy level, and genome size. *Allium* species are identified by their symmetrical and uniform karyotypes, which can often make chromosome identification difficult (Stevenson et al. 1999). The somatic chromosome number ranges from $2n = 16$–40 with basic chromosome numbers of $x = 7$, 8, and 9 (Karpaviciene 2007).

The ploidy level varies from $2x$ to $16x$ (De Sarker et al. 1997; Klaas 1998; Bennett et al. 2000) while the 2C DNA amounts per genome ranges from 16.93 to 63.57 pg (Ricrocha and Brown 1997). In a study of 25 *Allium* species, Jones and Rees (1968) found considerable differences among 2C-values measured by Feulgen densitometry. Ohri et al. (1998) confirmed this in a survey of 86 *Allium* species (representing all six subgenera), measured in 4C nuclei by Feulgen densitometry. The same conclusions were drawn from the study of genome size in 28 *Allium* species (Baranyi and Greilhuber 1999).

To study chromosome organization within *Allium* species, an integrated map of *Allium cepa* × (*A. roylei* × *A. fistulosum*) was used (Khrustaleva et al. 2005), which showed that *Allium* species recombination predominantly occurs in the proximal half of chromosome arms. Cytological analysis in populations of *A. roylei* has reportedly revealed presence of complex chromosomal configurations during male meiosis and chromosomal heteromorphicity in somatic metaphase spreads (Kohli and Gohil 2009). Unstable

B-chromosomes have also been reported in some European populations of *A. schoenoprasum* (Stevens and Bougourd 1994).

Overall, very little cytogenetic work has been done in wild *Allium* species. The most notable is the use of *A. galanthum* cytoplasm to develop cytoplasmic male sterile (CMS) substitution lines of cultivated *Allium* species (Yamashita 2005). More studies are required in this area to improve our understanding of these complex genomes. Significant utilization of wild *Allium* cytogenetic stocks in the breeding of cultivated species will depend a lot on our deep understanding of the specific genomes involved.

1.6 Role in Classical and Molecular Genetics

Allium genetics is poorly understood mainly due to the large genomes of the *Allium* species. There have been almost no classical genetic studies and the use of molecular tools has only been recently initiated. The few genetic studies in *Allium* species have focused on specific traits of economic importance such as bulb color, flavor, and disease and pest resistance.

Lack of genetic variability within some cultivated *Allium* species necessitates the use of closely related germplasm for the development of linkage maps. In the section *Cepa*, for example, *A. roylei* has been crossed with *A. cepa* and *A. fistulosum* to develop a linkage map. *A. roylei* shows a unique position in the taxonomy of the genus *Allium* (Fritsch and Friesen 2002). The nuclear DNA profile of *A. roylei* is related to the species of *Cepa* and *Phyllodolon* though its chloroplast DNA profile is related to the section *Schoenoprasum* (van Raamsdonk et al. 2000). *A. roylei* crosses easily with *A. cepa* and *A. fistulosum.* Crosses between *A. roylei* and *A. cepa* yield fertile hybrids.

The first *Allium* linkage map based on an F_2 population of an interspecific cross between *A. cepa* and *A. roylei* used AFLP markers (van Heusden et al. 2000a). This map was later used to locate a resistance locus for downy mildew (*Peronospora destructor*) to the distal end of chromosome 3 (van Heusden et al. 2000b). Downy mildew is a common and serious disease in onions, which results in major yield losses. Kofoet et al. (1990) reported that resistance to downy mildew in A. *roylei* was controlled by a single dominant gene named *Pd1*. The location of the resistance gene was confirmed using genomic in situ hybridization (GISH) and this resistance was successfully introduced into a bulb onion cultivar by the use of conventional back-crossing (Scholten et al. 2007).

Studies of resistance to anthracnose (*Colletotrichum gloesporioides* Penz) among *A. cepa* and *A. roylei* (Galvan et al. 1997) revealed that the resistance of *A. roylei* to the Brazilian anthracnose isolate is dominantly inherited. The same study also suggested that the resistance was most likely determined by more than one gene (Galvan et al. 1997).

Another important trait in *Allium* species is bulb color, which has been widely studied. The pigments responsible for bulb color are flavonoids. Flavonoids are common plant secondary metabolites that have been shown to function as antioxidant agents (Sengupta et al. 2004). Flavonoids are involved in UV protection, plant–microbe interactions, fertility, and pigmentation (Shirley 1996). Inheritance of bulb color has been shown to be very complex (Masuzaki et al. 2006), involving five major genes (Davis and El-Shafie 1967). *I* and *C* are the genes concerned with the expression of the pigmentation. *C* of the basic color factor is likely a regulatory gene controlling chalcone synthase (CHS) gene transcription (Kim et al. 2005). *I* is also presumed to be a regulatory gene as it inhibits pigment formation regardless of the other dominant genes (Kim et al. 2004). CHS is presumably located on chromosome 5A (Masuzaki et al. 2006) of *A. cepa*.

Although these genes have been studied in cultivated *Allium* species, it is expected that the flavonoid synthesis pathway in wild *Allium* species would contain similar genes making such reports applicable in the wild species.

1.7 Role of Wild *Allium* Species in Crop Improvement

Wild *Allium* species act as a reservoir of important traits that can be used to broaden the genetic base of the cultivated *Allium* species. However, the relatively long juvenile phase in most cultivated species and severe inbreeding depression have further hindered the use of wild *Allium* species for the improvement of cultivated *Allium* species. There are a few reports of interspecific gene introgression through classical breeding most of which have been done within the section *Cepa*.

Genes for resistance to downy mildew and Botrytis leaf blight (*Botrytis squamosa*) have been transferred from *A. roylei* (Kofoet et al. 1990; de Vries and Wietsma 1992) to *A. cepa*. *A. roylei* has also been used as a bridging species to stably introgress disease resistance genes from *A. fistulosum* into *A. cepa*. *A. fistulosum* harbors a number of genes for resistance to diseases and pests (Rabinowitch 1997). Although *A. fistulosum* and *A. cepa* are grouped under the same section, *Cepa*, the genome of *A. cepa* has a 28% higher DNA content than *A. fistulosum* (Labani and Elkington 1987) making crosses between the two species non-viable. Using the bridge-cross approach, *A. fistulosum* was stably integrated into *A. cepa* genome and unique populations were developed in which the important resistance genes segregated (Khrustaleva and Kik 2000).

Breeding for increased flavonoid content has become a major objective within *Allium* as people pay more attention to health-promoting compounds. In onions, flavonoid content has been the subject of many studies (Masuzaki et al. 2006; Kim et al. 2009a). Given the importance of flavonoids for human consumption, it is important to find ways of developing *Allium* cultivars that accumulate flavonoids to a larger extent (Masuzaki et al. 2006) and the wild germplasm presents a viable option. *Allium ursinum* (wild garlic) and *A. victorialis* have been reported to contain novel flavonoids (Andersen and Fossen 1995; Wu et al. 2009) that could be introgressed into garlic (*A. sativum*). Although garlic breeding in the past was limited to clonal selection of wild varieties or spontaneous mutants, routine seed production has been developed (Simon and Jenderek 2004) making it possible to introgress genes from closely related species using conventional means.

Higher fructan content is another trait of interest in *Allium* that could be improved using wild species. Fructans are a significant source of soluble dietary fiber (Kleessen et al. 1997). Fructan consumption has been correlated with lower rates of colorectal cancers (Roberfroid and Delzenne 1998). Analyses have reported that accumulation of fructans is associated with greater thiosulfinate concentrations (Havey et al. 2004). Thiosulfinates have incredible health benefits and have been shown to be abundant especially in *A. tuberosum* L. (Kim et al. 2008). More screening of the wild germplasm using traditional and molecular means could reveal more novel sources for the improvement of fructan content in the cultivated species.

Wild *Allium* species have played a role in creating and restoring CMS in cultivated *Allium* species. CMS is a maternally inherited condition in which a plant is unable to produce functional pollen (Schnable and Wise 1998). Male sterile plants are essential to exempt breeders from the difficulty of emasculation and results in the production of large numbers of F_1 seeds. Compared with genic male sterility, which is controlled only by nuclear genes, the CMS system allows easy propagation by using the appropriate maintainer line (Yamashita et al. 2010). The cytoplasm of *A. galanthum* Kar. et Kir., a wild species in section *Cepa*, has been used to develop a male sterile line for shallot and bunching onion (Yamashita et al. 1999) thus making it possible to produce hybrid seeds in these species at a commercial scale. *A. roylei* restores CMS (type T) in *A. cepa* (de Vries and Wietsma 1992).

Embryo rescue has been employed where distant crosses do not result in viable offspring. The procedure of embryo culture defines the techniques used to promote development of an immature or weak embryo into a viable plant. Depending on the organ used, embryo rescue is referred to as embryo, ovule, or ovary culture. Interspecific hybrids between *A. fistulosum* L. and *A. macrostemon* Bunge have been developed through ovary culture (Umehara et al. 2006a). *A. macrostemon* is a perennial herb with medicinal properties that propagates vegetatively and grows wild in China, Korea, and Japan (Fritsch and Friesen 2002). Embryo rescue has also been done to improve the aroma profile of onions (Keusgen et al. 2002). Embryo rescue of *Allium* species is usually done on a phytohormone-free medium (Umehara et al. 2006b).

Tsukazaki et al. (2006) proposed a "simple sequence repeats (SSR)-tagged breeding" scheme to enhance the rapidity, ease, and accuracy of variety identification and F_1 purity test in bunching onion. Their breeding scheme could be extended to other *Allium* species and utilized in backcrossing programs. With the current progress in *Allium* transformation (Eady et al. 2005; Kenel et al. 2010), it will be possible to transfer genes of interest from distant wild relatives into cultivated *Allium* species. Despite the few successes in introgression of resistance genes within the section *Cepa*, there are many diseases and pests such as neck rot (*Botrytis allii* Munn.), basal rot (*Fusarium* spp.), black mold (*Aspergillus niger* Tiegheim) for which sources of resistance are yet to be identified (Singh and Rana 1994). There is potential in identifying these genes from the wild germplasm. As domestication of more *Allium* species is still going on, future research will need to focus on optimization of screening, transformation protocols, and development of more molecular markers for the newly domesticated species.

1.8 Genomics Resources

Molecular tools for the improvement of *Allium* species have been lacking and the huge nuclear genomes are the major constraint. The first isolation of SSRs in the genus *Allium* was from bulb onion (*A. cepa*) (Fischer and Bachman 2000) only recently. In wild *Allium* species, the AFLP linkage map developed using F_2 population of an interspecific cross between *A. roylei* and *A. cepa* was the first *Allium* genetic map based on an interspecific cross (van Heusden et al. 2000a, b). However, this effort resulted in two maps; one map based on *A. cepa* markers and the second map based on *A. roylei* markers. The two maps were not integrated and 25% of the markers remained unlinked (van Heusden et al. 2000a, b). More recently, SSR markers have been developed and evaluated in both cultivated (*A. fistulosum, A. cepa*) and wild *Allium* species (*A. roylei, A. vavilovii, A. galanthum, A. altaicum*) (Araki et al. 2010).

A number of expressed sequence tags (ESTs) have been developed for cultivated *Allium* species (Kuhl et al. 2004; Kim et al. 2009b), even though none was detected from the National Center for Biotechnology Information (NCBI) database for wild *Allium* species. ESTs are a useful resource for identifying full-length genes and could be used in wild *Allium* species to identify novel genes. SSRs and single nucleotide polymorphisms (SNPs) derived from ESTs are likely to be transferred to the wild species as they are derived from less variable regions of the genome. There are currently EST-derived SSRs developed from an analysis of a bulb onion cDNA library (Kuhl et al. 2004) that could be tested for transferability in wild *Allium* species of interest. A garlic EST database (Kim et al. 2009b) is also available for the development and testing of more markers in interesting closely related wild species.

There are no known reports of bacterial artificial chromosomes (BACs) developed for the wild *Allium* species but those developed for onion (Suzuki et al. 2002) and garlic (Lee et al. 2003) will be a starting point toward the identification of candidate genes in wild *Allium* species. Given the close phylogenetic relationship between the Poales (includes the grasses) and Asparagales (Chase et al. 2000), it is highly expected that the advanced genomic resources available within the Poales (complete genomic sequence in *Oryza sativa* L., assembled BAC contigs, comprehensive EST databases) could be used for genetic improvement of Asparagales. However, only scant colinearity has been observed so far at the recombinational level between onion and rice (*O. sativa* L.) (Martin et al. 2005). More analysis within the wild *Allium* species may reveal more significant colinearity and provide insights into the evolution of the *Allium* species.

With the advances recently made in sequencing technologies, more genomic resources will be developed at a faster and cheaper rate than was possible a decade ago. These resources will make it possible to exploit the useful genetic variation available within the wild *Allium* species.

1.9 Scope for Domestication and Commercialization

The genus *Allium* comprises about 700 species of which less than 10% are cultivated but reports show evidence that domestication is still going on (Fritsch and Friesen 2002). There are several species of minor importance grown as semi-domesticated types or as wild economic species (Pandey et al. 2005) in India and other parts of the world. Domestication has therefore been done for different purposes.

A. komarovianum Vved was recently introduced into cultivation in North Korea as a vegetable (Hanelt 2001). Other wild *Allium* species used as vegetable in different parts of the world include *A. fasciculatum* in Bhutan (Pandey et al. 2008) and *A. roylei* in India (Kohli and Gohil 2009). Najjah et al. (2009) demonstrated the potential of *A. roseum* as an antimicrobial agent against food poisoning bacteria. *A. oschaninii*, *A. pskemense*, and *A. praemixtum* are used in Tajikistan and Uzbekistan as spices. In the western Himalayas, dried leaves of *A. stracheyi* are used for seasoning vegetable curries (Negi and Gaur 1991). *Allium vineale*, a perennial from bulblets emits a strong garlic or onion smell when crushed and has potential for use as spice.

The use of curative members of the genus *Allium* has a long tradition in several Asian populations with deep historical roots (Keusgen et al. 2006). *A. pskemense* and *A. praemixtum* are used in Tajikistan and Uzbekistan for medicinal purposes. In India, *A. wallichii*, *A. ampeloprasum*, *A. griffithianum*, and *A. tuberosum* are used as medicine (Pandey et al. 2008). *A. roylei* is used in India to relieve headache and fed to horses to relieve colic (Kohli and Gohil 2009).

The potential of domestication of *A. mongolicum* in China has been explored for its high nutritious, medicinal, and ecological value (Zi-Zhu et al. 2006). Wild *Allium* species have also received attention as ornamentals due to a wide range of attractive colors and persistence of floral or long vegetative cycle (Pandey et al. 2008). *Allium* species are used in rock gardens, herbaceous beds, perennial borders, as pot plants, as decorative items, and in dry arrangements (Kamenetsky and Fritsch 2002). Processed products of wild species of *A. auriculatum, A. carolinianum, A. griffithianum, A. humile, A. roylei,* and *A. wallichii* are reportedly in great demand (Pandey et al. 2008).

Wild garlic capable of producing seed has played a major role in generation of sexually producing garlic enabling commercial seed production in garlic. As important traits continue to be selected by end users of various wild *Allium* species, more domestication of these important species will need to be done in order to protect them from overexploitation.

1.10 Some Dark Sides and Recommendations for Future Actions

The major problem faced with some wild *Allium* species is their weediness. The most studied weedy *Allium* is *A. vineale* (wild garlic), a perennial bulb flower, native to Europe, North Africa, and western Asia. *A. vineale* has a highly developed system of reproduction (Ronsheim and Bever 2000), which allows it to survive a wide range of environmental conditions as well as various chemical and control measures (Leys and Slife 1986). This species was introduced in Australia and North America, where it has become an invasive species. *A. vineale* is especially a problem among cultivated grasses (Ferguson et al. 1992). Wild garlic is resistant to herbicides due to the structure of its leaves, being vertical, smooth, and waxy (Davies 1992; Block 2010). The use of herbicides is not effective especially because unsprouted underground bulbs may remain dormant and gradually sprout over a 4–5 year period. More knowledge of this species at DNA level would benefit the development of novel control approaches.

The other problem is the strong odor of most species within the genus *Allium* normally produced upon damage of vegetative parts. Although evidence shows that the substances responsible for the odor have health benefits (Rose et al. 2005), stronger and more offensive odors reduce their successful utilization in breeding and consumption. Allelopathic effects have been reported in several *Allium* species both cultivated and wild (Djurdjevic et al. 2003). These plants release allelochemicals by exudation, decomposition, leaching, and vaporization. Such chemicals may be toxic within the rhizosphere (Yu 1999) and have been reported to inhibit germination of some plant species (Djurdjevic et al. 2003).

The ongoing activities relating to collection and conservation of genetic resources of *Allium* will require much more support in the future, if exhaustive collection of all species is to be achieved and genetic erosion avoided. The use of more modern techniques in germplasm enhancement through characterization and conservation will be needed. More genomic resources such as ESTs and molecular markers, as well as transcriptome and/or genome sequence data will be necessary in these species for more efficient utilization of their resources. Breeders will also need to come up with more long-term pre-breeding programs that will be necessary for the transfer of useful traits from the wild to the domesticated species. More investment in biotechnology-related research projects in future will go along way toward the improvement of these species.

References

Andersen OM, Fossen T (1995) Anthocyanins with an unusual acylation pattern from stem of *Allium victorialis*. Phytochemistry 40:1809–1812

Araki N, Masuzaki SI, Tsukazaki H, Yaguchi S, Wako T, Tashiro Y, Yamauchi N, Shigyo M (2010) Development of microsatellite markers in cultivated and wild species of sections Cepa and Phyllodolon in Allium. Euphytica 173:321–328

Astley D (1994) A network approach to the conservation of Allium genetic resources. Int Symp on Alliums for the Tropics, ISHS. Acta Hortic 358:135–142

Baranyi M, Greilhuber J (1999) Genome size in Allium: in quest of reproducible data. Ann Bot 83:687–695

Bennett MD, Bhandol P, Leitch IJ (2000) Nuclear DNA amounts in angiosperms and their modern uses – 807 new estimates. Ann Bot 86:859–909

Block E (2010) Garlic and other Alliums: the lore and the science. Royal Society of Chemistry, Cambridge, UK. ISBN 978-0-85404-190-9

Brat SV (1965) Genetic systems in Allium III. Meiosis and breeding systems. Heredity 20:325–339

Brewster JL (1998) Onions and other Alliums. CABI, Wallingford, UK. ISBN 978-1-84593-399-9

Chase MW, Soltis DE, Soltis PS, Rudall PJ, Fay MF, Hahn WH, Sullivan S, Joseph J, Molvray M, Kores PJ, Givnish TJ, Sytsma KJ, Pires JC (2000) Higher-level systematics of the monocotyledons: an assessment of current knowledge and a new classification. In: Wilson K, Morrison S (eds) Monocots: systematics and evolution, vol 1. CSIRO, Melbourne, Australia, pp 3–16

Davies D (1992) Alliums: the ornamental onions. Timber, Portland, OR, USA. ISBN 0-88192-241-2

Davis GN, El-Shafie MW (1967) Inheritance of bulb color in the onion (*Allium cepa* L.). Hilgardia 38:607–622

De Sarker D, Johnson MAT, Reynolds A, Brandham PE (1997) Cytology of the highly polyploid disjunct species, *Allium dregeanum* (Alliaceae), and of some Eurasian relatives. Bot J Linn Soc 124:361–373

De Vries JN, Wietsma WA (1992) *Allium roylei* Stearn restores cytoplasmic male sterility of Rijnsburger onion (*A. cepa* L.). J Genet Breed 46:379–382

Djurdjevic L, Dinic A, Pavlovic P, Mitrovic M, Karadzic B, Tesevic V (2003) Allelopathic potential of *Allium ursinum* L. Biochem Syst Ecol 32:533–544

Eady C, Davis S, Catanach A, Kenel F, Hunger S (2005) *Agrobacterium tumefaciens*-mediated transformation of leek (*Allium porrum*) and garlic (*Allium sativum*). Plant Cell Rep 24:209–215

Ferguson GP, Coats EG, Wilson GB, Shaw DR (1992) Postemergence control of wild garlic (*Allium vineale*) in turfgrass. Weed Technol 6:144–148

Fischer D, Bachman K (2000) Onion microsatellites for germplasm analysis and their use in assessing intra- and interspecific relatedness within the subgenus Rhizirideum. Theor Appl Genet 101:153–164

Friesen N, Klaas M (1998) Origin of some minor vegetatively propagated Allium crops studied with RAPD and GISH. Genet Resour Crop Evol 45:511–523

Friesen N, Fritsch RM, Blattner FR (2006) Phylogeny and new intrageneric classifion of *Allium* L. (Alliaceae) based on nuclear ribosomal DNA ITS sequences. Aliso 22:372–395

Fritsch RM, Friesen N (2002) Evolution, domestication and taxonomy. In: Rabinowitch HD, Currah L (eds) Allium crop science: recent advances. CABI, Wallingford, UK, pp 5–30

Galvan GA, Wietsma WA, Putrasemedja S, Permadi AH, Kik C (1997) Screening for resistance to anthracnose (Colletotrichum gloeosporioides Penz.) in *Allium cepa* and its wild relatives. Euphytica 95:173–178

Ghrabi Z (2010) *Allium roseum* var. odoratissimum: a guide to medicinal plants in North Africa, pp 23–24. http://www.uicnmed.org/nabp/database/HTM/PDF/p5.pdf. Accessed 5 July 2010

Hanelt P (2001) Alliaceae. In: Hanelt P and Institute of Plant Genetics and Crop Plant Research (eds) Mansfelds encyclopedia of agricultural and horticultural crops, vol 4, 3rd edn. Springer, Berlin, Germany, pp 2250–2269

Havey MJ, Galmarini CR, Gökçe AF, Henson C (2004) QTL affecting soluble carbohydrate concentrations in stored onion bulbs and their association with flavor and health-enhancing attributes. Genome 47:463–468

Huxley A, Griffiths M, Levy M (1992) The New Royal Horticultural Society dictionary of gardening. Macmillan, Stockton, London, UK. ISBN 978-0-333-47494-5

Ipek M, Simon P (2001) Genetic diversity in garlic (*Allium sativum* L.) as assessed by AFLPs and isozymes. In: 98th Annual conference & exhibition, American Society of Horticultural Science, P3, 22–25 July 2001

Jones RN, Rees H (1968) Nuclear DNA variation in Allium. Heredity 23:591–605

Kamenetsky R, Fritsch R (2002) Ornamental Alliums. In: Rabinowitch HD, Currah L (eds) Allium crop science: recent advances. CABI, Wallington, UK, pp 459–492

Kamenetsky R, Rabinowitch HD (2001) Floral development in bolting garlic. Sex Plant Reprod 13:235–241

Karpaviciene B (2007) Chromosome numbers of Allium from Lithuania. Ann Bot Fennici 44:345–352

Kenel F, Eady C, Brinch S (2010) Efficient *Agrobacterium tumefaciens*-mediated transformation and regeneration of garlic (*Allium sativum*) immature leaf tissue. Plant Cell Rep 29:223–230

Keusgen M, Schulz H, Glodek J, Krest I, Krüger H, Herchert N, Keller J (2002) Characterization of some Allium hybrids by aroma precursors, aroma profiles, and allinase activity. J Agric Food Chem 10:2884–2890

Keusgen M, Fritsch RM, Hisoriev H, Kurbonova PA, Khassanov FO (2006) Wild Allium species (Alliaceae) used in folk medicine of Tajikistan and Uzbekistan. J Ethnobiol Ethnomed 2:18

Khrustaleva LI, Kik C (2000) Introgression of *Allium fistulosum* into *A. cepa* mediated by *A. roylei*. Theor Appl Genet 100:17–26

Khrustaleva LI, de Melo PE, van Heusden AW, Kik C (2005) The integration of recombination and physical maps in a large-genome monocot using haploid genome analysis in a trihybrid Allium population. Genetics 169:1673–1685

Kim S, Binzel ML, Park S, Yoo K, Pike LM (2004) Inactivation of DFR (Dihydroflavonol 4-reductase) gene transcription results in blockage of anthocyanin production in yellow onions (*Allium cepa*). Mol Breed 14:253–256

Kim S, Yoo K, Pike LM (2005) The basic color factor, the C locus, encodes a regulatory gene controlling transcription of chalcone synthase genes in onions (*Allium cepa*). Euphytica 142:273–282

Kim SY, Park KW, Kim JY, Shon MY, Yee ST, Kim KH, Rhim JS, Yamada K, Seo KI (2008) Induction of apoptosis by thiosulfinates in primary human prostrate cancer cells. Int J Oncol 32:869–875

Kim S, Baek D, Cho DY, Lee ET, Yoon MK (2009a) Identification of two novel inactive DFR-A alleles responsible for failure to produce anthocyanin and development of a simple PCR-based molecular marker for bulb color selection in onion (*Allium cepa* L.). Theor Appl Genet 118:1391–1399

Kim DW, Jung TS, Nam SH, Kwon HR, Kim A, Chae SH, Choi SH, Kim DW, Kim RN, Park HS (2009b) GarlicESTdb: an online database and mining tool for garlic EST sequences. BMC Plant Biol 9:61

Klaas M (1998) Applications and impact of molecular markers on evolutionary and diversity studies in the genus *Allium*. Plant Breed 117:297–308

Klaas M, Friesen N (2002) Molecular markers in Allium. In: Rabinowitch HD, Currah L (eds) Allium crop science – recent advances. CABI, Wallingford, UK, pp 159–185

Kleessen B, Sykura B, Zunft H, Blaut M (1997) Effects of inulin and lactose on fecal microflora, microbial activity, and bowel habit in elderly constipated persons. Am J Clin Nutr 65:1397–1402

Kofoet A, Kik C, Wietsma WA, de Vries JN (1990) Inheritance of resistance to downy mildew (Peronospora destructor (Berk.) Casp.) from *Allium roylei* Stearn in the backcross *Allium cepa* L. (A. roy- lei x A. cepa). Plant Breed 105:144–149

Kohli B, Gohil RN (2009) Need to conserve *Allium roylei* Stearn: a potential gene reservoir. Genet Resour Crop Evol 56:891–893

Kuhl JC, Cheung F, Yuan Q, Martin W, Zewdie Y, McCallum J, Catanach A, Rutherford P, Sink KC, Jenderek M, Prince JP, Town CD, Havey MJ (2004) A unique set of 11,008 onion expressed sequence tags reveals expressed sequence and genomic differences between the monocot orders Asparagales and Poales. Plant Cell 16:114–125

Labani R, Elkington T (1987) Nuclear DNA variation in the genus *Allium* L. (Liliaceae). Heredity 59:119–128

Le Guen-Le Saos F, Hourmant A, Esnault F, Chauvin JE (2002) In vitro bulb development in shallot (*Allium cepa* L. aggregatum group): effects of anti-gibberellins, sucrose and light. Ann Bot 89:419–425

Lee HR, Eon EM, Lim YP, Bang JW, Lee DH (2003) Construction of a garlic BAC library and the chromosomal assignment of BAC clones using the FISH technique. Genome 46:514–520

Leys A, Slife FW (1986) The response of wild garlic (*Allium vineale*) to the timing of spray applications of chlorsulfuron. Weed Sci 34:718–723

Martin WJ, McCallum J, Shigyo M, Jakse J, Kuhl JC, Yamane N, Pither-Joyce M, Gokce AF, Sink KC, Town CD, Havey MJ (2005) Genetic mapping of expressed sequences in onion and in silico comparisons with rice show scant colinearity. Mol Genet Genom 274:197–204

Masuzaki S, Shigyo M, Yamauchi N (2006) Direct comparison between genomic constitution and flavonoid contents in Allium multiple alien addition lines reveals chromosomal locations of genes related to biosynthesis from dihydrokaempferol to quercetin glucosides in scaly leaf of shallot (*Allium cepa* L.). Theor Appl Genet 112:607–617

Mingochi DS, Swai REA (1994) Role of the Southern Africa development community in the conservation of Allium genetic resources. In: International symposium on Alliums for the tropics, ISHS. Acta Hortic 358:161–164. http://www.actahort.org/books/358/358_25.htm

Najjah H, Ammar E, Neffati M (2009) Antimicrobial activities of *Allium roseum* L., a wild edible species in North Africa. J Food Agric Environ 7:150–154

Negi KS, Gaur RD (1991) Little known endemic wild Allium species in the Uttar Pradesh hills. Mt Res Dev 11:162–164

Nguyen NH, Driscolla HE, Specht CD (2008) A molecular phylogeny of the wild onions (Allium; Alliaceae) with a focus on the western North American center of diversity. Mol Phylogenet Evol 47:1157–1172

Ohri D, Fritsch RM, Hanelt P (1998) Evolution of genome size in Allium (Alliaceae). Plant Syst Evol 210:57–86

Pandey A, Pandey R, Negi KS (2005) Wild Allium species in India: biodiversity distribution and systematic studies. In: National Conference on Allium (Abstr), Banaras Hindu University, Varanasi, UP, India, 24–25 Feb 2005, 44 p

Pandey A, Pandey R, Negi KS, Radhamani J (2008) Realizing value of genetic resources of Allium in India. Genet Resour Crop Evol 55:985–994

Peumans WJ, Smeets K, Van Nerum K, Van Leuven F, Van Damme EJM (1997) Lectin and alliinase are the predominant proteins in nectar from leek (*Allium porrum* L.) flowers. Planta 201:298–302

Rabinowitch HD (1997) Breeding alliaceous crops for pest resistance. Acta Hortic 433:223–246

Ricroch A, Yockteng R, Brown SC, Nadot S (2005) Evolution of genome size across some cultivated Allium species. Genome 48:511–520

Ricrocha A, Brown SC (1997) Junk DNA: the role and the evolution of non-coding sequences – DNA base composition of Allium genomes with different chromosome numbers. Gene 205:255–260

Roberfroid MB, Delzenne NM (1998) Dietary fructans. Annu Rev Nutr 18:117–143

Ronsheim ML, Bever JD (2000) Genetic variation and evolutionary trade-offs for sexual and asexual reproductive modes in *Allium vineale* (liliaceae). Am J Bot 87:1769–1777

Rose P, Whiteman M, Mooreb PK, Zhun Zhu Y (2005) Bioactive S-alk(en)yl cysteine sulfoxide metabolites in the genus Allium: the chemistry of potential therapeutic agents. Nat Prod Rep 22:351–368

Rudall P, Furness C, Chase M, Fay M (1997) Microsporogeneisis and pollen sulcus type in Asparagales (Lilianae). Can J Bot 75:408–430

Samaras S (2001) European collections of vegetatively propoagated Allium. A report on the current status of the greek Allium wild taxa, ECPGR, 20–21 May 2001, 44 p

Schnable PS, Wise RP (1998) The molecular basis of cytoplasmic male sterility and fertility restoration. Trends Plant Sci 3:175–180

Scholten OE, van Heusden AW, Khrustaleva LI, Burger-Meijer K, Mank RA, Antonise RGC, Harrewijn JL, Van Haecke W, Oost EH, Peters RJ, Kik C (2007) The long and winding road leading to the successful introgression of downy mildew resistance into onion. Euphytica 156:345–353

Sengupta A, Ghosh S, Bhattacharjee S (2004) Allium vegetables in cancer prevention: An Overview. Asian Pac J Cancer Prevent 5:237–245

Shirley BW (1996) Flavonoid biosynthesis: 'new' functions for an 'old' pathway. Trends Plant Sci 1:377–382

Simon PW, Jenderek MM (2004) Flowering, seed production and the genesis of garlic breeding. Plant Breed Rev 23:211–244

Singh BP, Rana RS (1994) Collection and conservation of Allium genetic resources: an Indian perspective. Acta Hortic 358:181–190

Specht CE, Meister A, Keller ERJ, Korzun L, Börner A (2001) Polyembryony in species of the genus *Allium*. Euphytica 121:37–44

Stevens JP, Bougourd SM (1994) Unstable B-chromosomes in a European population of Allium schoenoprasum L. (Liliaceae). Biol J Linn Soc 52:357–363

Stevenson M, Armstrong SJ, Jones GH, Ford-Lloyd BV (1999) Distribution of a 375 bp repeat sequence in Allium (Alliaceae) as revealed by FISH. Plant Syst Evol 217:31–42

Suzuki G, Do GS, Mukai Y (2002) Efficient storage and screening system for onion BAC clones. Breed Sci 52:157–159

Tsukazaki H, Fukuoka H, Song YS, Yamashita KI, Wako T, Kujima A (2006) Considerable heterogeneity in commercial F1 varieties of bunching onion (*Allium fistulosum*) and proposal of breeding scheme for conferring variety traceability using SSR markers. Breed Sci 56:321–326

Umehara M, Sueyoshi T, Shimomura K, Nakaha T (2006a) Production of interspecific hybrids between *Allium fistulosum* L. and *A. macrostemon* Bunge through ovary culture. Plant Cell Tiss Org Cult 87:297–304

Umehara M, Sueyoshi T, Shimomura K, Iwai M, Shigyo M, Hirashima K, Nakahara T (2006b) Interspecific hybrids between *Allium fistulosum* and *Allium schoenoprasum* reveal carotene-rich phenotype. Euphytica 148:295–301

USFWS (2000) Notice of reclassification of nine candidate taxa. Fed Regist 64(204):63044–63047

van Heusden AW, van Ooijen JW, van Vrielink Ginkel R, Verbeek WHJ, Wietsma WA, Kik C (2000a) A genetic map of an interspecific cross in Allium based on amplified fragment length polymorphism (AFLP) markers. Theor Appl Genet 100:118–126

van Heusden AW, Shigyo M, Tashiro Y, van Vrielink Ginkel R, Kik C (2000b) AFLP linkage group assignment to the chromosomes of *Allium cepa* L. via monosomic addition lines. Theor Appl Genet 100:480–486

van Raamsdonk LWD, van Vrielink Ginkel M, Kik C (2000) Phylogeny reconstruction and hybrid analysis in Allium subgenus Rhizirideum. Theor Appl Genet 100: 1000–1009

Volk GM, Maness N, Rotindo K (2004) Cryopreservation of garlic (*Allium sativum* L.) using plant vitrification solution 2. Cryoletters 25:219–226

Wendelbo P (1969) New subgenera, sections and species of Allium. Bot Notiser 122:25

Working Group on Allium (2009) In: Astley D, Bas N, Branca F, Daunay MC, Díez MJ, Keller J, van Dooijeweert W, van Treuren R, Maggioni L, Lipman E (eds) Report of a vegetables network. Reports by the Working Groups' rapporteurs on issues discussed during the parallel meetings. 2nd Meeting, 26–28 June 2007, Olomouc, Czech Republic, Bioversity International, Rome, Italy. ISBN 978-92-9043-792-5

Wu H, Dushenkov S, Ho C-T, Sang S (2009) Novel acetylated flavonoid glycosides from the leaves of Allium Ursinum. Food Chem 115:592–595

Yamashita K (2005) Breeding of Allium cultivated species using cytoplasm engineering technique. Rearing of male sterile line using the cytoplasm of wild species. Agric Hortic 80:7–14

Yamashita K, Arita H, Tashiro Y (1999) Cytoplasm of a wild species, *Allium galanthum* Kar. et Kir., is useful for developing the male sterile line of *A. fistulosum* L. J Jpn Soc Hortic Sci 68:788–797

Yamashita K, Tsukazaki H, Kojima A, Ohara T, Wako T (2010) Inheritance mode of male sterility in bunching onion (*Allium fistulosum* L.) accessions. Euphytica 173:357–367

Yu JQ (1999) Allelopathic suppression of Pseudomonas solanacearum infection of tomato (Lycopersicon esculentum) in a tomato-Chinese Chive (*Allium tuberosum*) intercropping System. J Chem Ecol 25:2409–2417

Zi-zhu Y, Shi-zeng L, Ai-de L, Shu-juan SA (2006) Tentative exploring to the developing value of wild vegetable resources – *Allium mongolicum*. J Gansu For Sci Technol. doi:cnki:ISSN:1006-0960.0.20

Chapter 2
Amaranthus

Federico Trucco and Patrick J. Tranel

2.1 Basic Botany of the Species

The *Amaranthus* genus (Magnoliophyta: Caryophyllidae) comprises 70 species grouped into three subgenera (Mosyakin and Robertson 2003). The most economically important is the subgenus *Amaranthus* proper, which includes the three species domesticated for grain production: *Amaranthus hypochondriacus*, *Amaranthus cruentus*, and *Amaranthus caudatus*. Other species of amaranths have been domesticated as leaf-vegetables, for fodder, as potherbs, or as ornamentals; among these species, *A. tricolor*, from South Asia, is probably the most important (Sauer 1967). This chapter, however, will focus on the wild relatives of the grain crops, particularly on species of the *Amaranthus hybridus* aggregate (*A. hybridus* proper, *A. retroflexus*, and *A. powellii*), from which the "pseudo-cereals" are believed to be domesticated. Part of the discussion however – especially that dealing with the development of genomic resources and hybridization with potential for future breeding programs – will refer to wild species of the subgenus *Acnida*, where increasingly studied dioecious (unisexual) weeds are enlisted.

2.1.1 Subgenus Amaranthus

The subgenus *Amaranthus* consists of 20 species of annual herbs that are monoecious (Mosyakin and Robertson 2003), that is, have separate male and female flowers. The species are native to the Americas, with the exception of only one species of possible European origin (Mosyakin and Robertson 2003). Monoecious amaranths are primarily self-pollinated, as female and male flowers are arranged in close proximity (Murray 1940). Stems are usually erect and both axillary and terminal inflorescences are arranged in cylindrical spikes or panicles (Mosyakin and Robertson 2003). Much of the difficulty in taxonomic discrimination of species within the group can be attributed to attempts at recognizing taxa based on pigmentation or growth forms, which are extremely variable within amaranths (Sauer 1967). However, examination of floral parts can result in constant characters from which discontinuities can be used to define well-established taxa. In this sense, tepal (petals and sepals are combined in a single floral whorl) number and morphology are commonly used in taxonomic keys.

A. hybridus is a basal species in the crop subgenus and conforms an interbreeding complex with two other *Amaranthus* weeds: *A. retroflexus* and *A. powellii*. As presented by Sauer (1967) *A. hybridus* originated as a riverbank pioneer of eastern North America, with earlier range expanding throughout milder and moister regions to Mexico, Central America, and northern South America. The earliest European records of the species date back approximately 300 years, with spread in Europe taking place primarily in the Mediterranean region. Spread of *A. hybridus* has been slower than that of other *Amaranthus* weeds, especially when compared to *A. retroflexus*. Presence of the species in western North America, eastern Asia, Australia, and South Africa has been reported as of early to mid 1900s. Today, *A. hybridus* is a worldwide distributed weed of agricultural fields and other

P.J. Tranel (✉)
Department of Crop Sciences, University of Illinois, Urbana, IL 61801, USA
e-mail: tranel@illinois.edu

C. Kole (ed.), *Wild Crop Relatives: Genomic and Breeding Resources, Vegetables*,
DOI 10.1007/978-3-642-20450-0_2,

disturbed habitats, and it ranks among the 18 most serious weeds in the world (Holm et al. 1991).

A. retroflexus, like *A. hybridus* and many other amaranths, is a riverbank pioneer. Its earliest distribution expanded from the central-eastern United States to adjacent Canada and Mexico. Sauer (1967) mentions that Linnaeus is blamed for introducing the weed to Europe, where the species quickly spread. By early 1800s, it became a common weed in the temperate regions of the Old World, reaching the Near East and northern Africa soon after. Today, the species is introduced or naturalized nearly worldwide, ranking among the most widely distributed weeds of the world (Holm et al. 1997).

A. powellii's initial distribution included canyons, desert washes, and other open habitats west of the Cordilleran system of America, with wide gaps in wetter regions of Central America. The earliest European record of this species is found in German herbarium specimens from the late 1800s, and later introductions can be interpreted from samples of southern India and South Africa. Expansion of *A. powellii* to eastern North America occurred only during the last century.

Partially fertile hybrid swarms between these species can be found in the United States, in areas where their distributions overlap, and in Europe, where all three species are recent immigrants. The amaranth grain crop is derived of ancient domestications of these species or their hybrids, or from their South American close relative, *Amaranthus quitensis*.

A. hypochondriacus, one of the three grain amaranths, is cultivated as an alternative crop in North America and Asia. Although initially thought to have Asian origin, it is believed that this distribution is secondary and that the species derives from an *A. powellii* domestication in North America. Hybridization has had a significant role in the evolution of *A. hypochondriacus*, with several hybrid races cultivated by American aborigines. Sauer (1967) identified stable hybrid cultivars derived from crosses presumably between *A. hypochondriacus* and local admixtures of *A. cruentus* – an *A. hybridus* domesticated form originating in southern Mexico or Guatemala – and its progenitor. For instance, in the region of Reyes (Michoacan), a cultivar grown to make special "dark" tamales was a putative hybrid between *A. hypochondriacus* and *A. hybridus*. Likewise, a Warihio Indian crop from Rancho Trigo (Chihuahua) was classified as a hybrid between *A. hybridus* and *A. powellii*. Another putative hybrid between *A. cruentus* and *A. hypochondriacus* is cultivated in the region of Oaxaca (southern Mexico) and is the same crop found in small gardens in Madras, India.

A. caudatus, the grain amaranth of South America, is thought to originate from a domestication of *A. quitensis* in the Andean region (Sauer 1967). *A. quitensis* is a weedy member of the *A. hybridus* aggregate, with original distribution as a riverbank pioneer of South America, in mountains in the northwest and at lower elevations in the temperate south. Cultivation of *A. quitensis* forms with incipient domestication is observed from Ecuador to northern Argentina, mainly for the production of pigments needed for coloring of chicha and other maize dishes. Although some cultivated forms of *A. caudatus–A. quitensis* are suspected to be the result of interbreeding with *A. cruentus*, the South American amaranths are not thought to readily hybridize with the North American members of this cluster.

2.1.2 *Subgenus* Acnida

The subgenus *Acnida* includes nine dioecious species – that is, taxa with separate male and female plants – which are native to North America and have no immediate evolutionary relationship with the amaranth crop. However, recent studies (Trucco et al. 2005a) show that gene exchange may occur between *Amaranthus tuberculatus*, an infamous member of *Acnida,* and *A. hybridus* – as discussed previously, a crop progenitor with residual compatibility with some domesticated forms. In fact, studies of *A. tuberculatus* and *A. hybridus* gene exchange reveal interesting insights as to how the genetic diversity of the dioecious taxon may be available for crop improvement. In addition, as *A. tuberculatus* is increasingly accepted as a model organism for the study of weeds (Tranel and Trucco 2009), a wealth of genomic resources are being developed that may be of use for programs dedicated to the crops. Since these matters will be discussed in more detail in later sections of this chapter, we feel it is pertinent to introduce the basic botany of the species herein.

A. tuberculatus is an annual herb flowering during the summer or fall. It has erect stems, which are usually branched and have terminal inflorescences in the forms of linear spikes to panicles. Female flowers usually have no tepals, although one or two rudimentary tepals may be observed at times. Sauer (1972) separated *A. rudis* (formerly *A. tamariscina*) as distinct from *A. tuberculatus*, primarily based on utricle dehiscence and absence of female tepals.

Sauer's *A. rudis* was first described in Oklahoma in the 1830s and since has shown continuous northward and eastward accretion into midwestern states, overlapping with *A. tuberculatus*, of static range, in sandy and muddy streambanks, lakeshores, and pond margins, along the Missouri, Mississippi and Ohio River systems (Sauer 1957, 1972). Where both *A. tuberculatus* and *A. rudis* coexisted, the record of the former was on average 40 years prior to that of the latter. Many of the samples collected in these areas were classified as putative *A. tuberculatus* by *A. rudis* hybrids, with a higher ratio of hybrids to non-hybrids in artificial habitats compared to natural settings. In Sauer's assessment of dioecious amaranths (1957), *A. tuberculatus* by *A. rudis* hybrids are the most abundant hybrid combination. The author also notes that actual hybridization among these species may be underestimated due to the nature of morphological determinations based on character intermediacy, which is often diluted after a few generations of backcrossing with the predominant genotype. More recent work using molecular and morphological markers suggested both species to be one and the same (Pratt and Clark 2001), and a single polymorphic species, *A. tuberculatus*, is presently recognized (Mosyakin and Robertson 2003). Costea and Tardif (2003), however, encouraged recognition of the two entities at the variety level: *A. tuberculatus* var. *rudis* having more weedy tendencies than *A. tuberculatus* var. *tuberculatus*.

Over the last 20 years, *A. tuberculatus* has gone from virtual anonymity to becoming the most significant weed problem in the midwestern United States (Steckel 2007), one of the world's premier agricultural regions. Success as a weed is attributed, among other things, to its remarkable ability to evolve resistance to herbicides. Herbicide resistance studies with *A. tuberculatus* are discussed in detail in a later section. Although of great concern from a weed management perspective, the ability of this species to respond to selection and the diversity of adaptations identified thus far may be of potential profit to less orthodox crop-breeding initiatives.

2.2 Conservation Initiatives

Wild species of amaranths, particularly those closely related to the grain crops, are successful agricultural weeds and found abundantly in areas from which they are native. In situ conservation initiatives of wild *Amaranthus* species are not known, though genetic erosion problems are of concern (Grubben and van Sloten 1981), and materials of interest are actively collected for ex situ conservation. Ex situ conservation of *Amaranthus* germplasm is not very laborious, since seeds are small and long-lived, and efficient protocols for seed regeneration and conservation of genetic diversity exist (Brenner and Widrlechner 1998).

The most significant efforts at ex situ conservation of amaranth germplasm were initiated during the late 1970s, mainly as a result of the amaranth breeding initiative by scientists at the Rodale Research Center in Pennsylvania (Kauffman 1992). At its peak, the Rodale collection contained approximately 1,400 accessions, which were donated in 1990 to the North Central Regional Plant Introduction Station, a part of the USDA National Plant Germplasm System (Brenner et al. 2000). The USDA collection is by far the most comprehensive, including 3,200 accessions, with close to 80% of the accessions representing domesticated species. The *A. hybridus* aggregate (including *A. quitensis*) is represented by less than 300 entries, and 51 entries are listed for *A. tuberculatus*.

Other germplasm collections are held by at least 60 different groups or institutions, although most of these tend to have fewer than 100 entries (Brenner et al. 2000). In general, non-domesticated germplasm is poorly represented in these collections. However, at the University of Illinois, for instance, seed collected from several weedy populations of the midwestern USA are conserved for herbicide resistance research and genetic diversity studies.

2.3 Studies Using Molecular Markers

As discussed previously, grain amaranths are associated to three putative progenitors: *A. powellii*, *A. hybridus*, and *A. quitensis*. Evolutionary associations are based on morphology, distribution, and some degree of sexual compatibility among species. More recently, molecular marker analyses have contributed to the elucidation of origin and evolution of cultivated amaranths, and allied wild species have been critical in these analyses. Hauptli and Jain (1984) were among the first to use molecular markers to address evolutionary relationships among the grain amaranths. They observed that with the exception of the *A. caudatus* – *A. quitensis* pair, grain amaranths are more closely related to each other than either is to their putative wild progenitor. This work was based on isozyme polymorphisms and several authors have since expanded molecular diversity studies in the genus.

In a study including both isozyme and random amplified polymorphic DNA (RAPD) markers, Chan and Sun (1997) generated molecular phylogenies of cultivated and wild amaranths. These authors evaluated 23 different species, including the three cultivated for grain as well as accessions of all species in the *A. hybridus* aggregate. For the crop species, they obtained 15 and 240 polymorphic isozymes and RAPD markers, respectively. The level of polymorphic markers increased slightly when considering accessions from putative wild progenitors. Up to 70% of all evaluated RAPD markers (600 in total) were polymorphic when all 23 species were included in the analysis. Both RAPD and isozyme data sets supported a monophyletic origin for grain amaranths, with *A. hybridus* as the common ancestor.

However, molecular studies do not show unanimous consensus regarding the evolutionary origin and proximity of crop–wild allies. Classical studies dealing with hybrid fertility and chromosome numbers tend to support the hypothesis of independent domestication, with *A. hypochondriacus* and *A. caudatus* as the most related crop species (Pal and Khoshoo 1972, 1973). Some molecular data supported a similar conclusion (Transue et al. 1994; Kirkpatrick 1995). However, studies based on restriction site variations in nuclear and cytoplasmic DNA found that *A. caudatus* and *A. cruentus* are more closely related to each other and to their supposed progenitors than either is to *A. hypochondriacus* (Lanoue et al. 1996). Isozyme and RAPD markers were used by other authors and findings tended to agree with the different evolutionary hypothesis presented herein (Ranade et al. 1997; Zheleznov et al. 1997).

The assembly of different phylogenies with different evolutionary implications may result from the intraspecific genetic variation found across amaranth populations, from the residual cross breeding among sympatric species, and from the choice of accessions selected by researchers for each experiment. For instance, Brenner et al. (2000) note that the number of accessions surveyed by Chan and Sun (1997) was limited, with approximately five accessions representing each crop species and fewer for most wild taxa. Additionally, wild taxa surveyed were not from the area of origin of domesticated material, so they could not represent adequately the diversity within the putative progenitors. Other factors contributing to the ambiguity of molecular phylogenies may be related to the DNA-marker system employed. Many of the early molecular studies used RAPD markers, which are known to provide inconsistent results. Current studies are applying microsatellite markers and genomic sequencing to address evolutionary questions, and these are discussed in more detail in Sect. 2.5.

Some molecular studies have been strictly dedicated to weed species. Wetzel et al. (1999) generated ribosomal ITS restriction-site-based PCR markers to identify common amaranth weeds, which are difficult to identify based on morphological evaluations with a casual eye. Pratt and Clark (2001) used isozymes to address whether *A. rudis* and *A. tuberculatus* should be considered a single species or two. And Wassom and Tranel (2005) used amplified fragment length polymorphism (AFLP)-based markers to assemble a phylogeny of both dioecious and monoecious *Amaranthus* weeds. In this last study, eight weedy species were considered, represented by 141 individuals from 98 different accessions. Interestingly, the dioecious weeds *A. palmeri* and *A. tuberculatus* did not group together, perhaps indicating independent evolutionary paths.

2.4 Interspecies Hybridization Studies

Hybridization studies have been very important in establishing evolutionary relations and gene pools accessible for conventional breeding programs. Merritt Murray (1940) was one of the first to systematically assess interspecies hybridization within the genus, in a study to elucidate the mechanisms involved with sex determination in Amaranthaceae.

Murray classified monoecious species according to the arrangement pattern shown by male flowers in inflorescences. He identified two types of species, with type I plants having male flowers interspersed with female flowers, whereas type II plants have male flowers clustered at the terminal ends of inflorescences. Murray performed a number of different crosses between and among type I monoecious species (including *A. caudatus*, *A. hybridus*, *A. retroflexus*, and *A. powellii*), type II monoecious species (*A. spinosus*), and dioecious taxa. Crosses between monoecious species produced hybrids with different ease, with type I by type II crosses showing the most difficulty at hybrid production. Hybrids were readily obtained among species of the type I floral arrangement and between type I species and dioecious taxa, suggesting evolutionary proximity between these species. *A. hybridus* by *A. caudatus* crosses were among the most prolific, consistent with the weak pre-zygotic isolation expected of closely related taxa. Interestingly, similarly prolific were crosses between *A. hybridus* and *A. caudatus* with *A. tuberculatus* (referred to as *Acnida tamariscina* in Murray's work), insinuating an evolutionary relationship that is closer than is morphologically apparent.

2.4.1 Hybridization Within Subgenus Amaranthus

Grant (1959) has reviewed reports by different authors on the occurrence of spontaneous *Amaranthus* hybrids, validating in nature part of Murray's greenhouse results. In the studies cited by Grant, natural hybrids were identified by character intermediacy between *A. caudatus* or *A. cruentus* and species of the *A. hybridus* complex. In some instances, hybrid morphology suggested three-way hybridizations among these species (Tucker and Sauer 1958), and the cytogenetic data produced by Grant were consistent with this notion. The fact that hybrid forms may be observed in nature indicates that first generation hybrids are fertile enough to advance to more stable generations. In this sense, speculations regarding the possible hybrid origin of some domesticated forms appear reasonable – see discussion from Sect. 2.1.1.

A somatic chromosome number of 32 or 34 was observed for all 30 species analyzed by Grant, with the exception of the polyploid *A. dubius*, with 64 chromosomes. Khoshoo and Pal (1972) used *A. hypochondriacus* as the male parent in crosses with *A. hybridus* and *A. caudatus*, all with 32 chromosomes. Hybrids from these crosses showed the formation of 16 bivalent chromosomal associations. However, hybrids produced with *A. hybridus* showed much greater pollen fertility than hybrids produced with *A. caudatus*. Interestingly, *A. hybridus* by *A. caudatus* hybrids here were seedling lethal, a phenomenon not reported previously by Murray for this cross. Hybrid fertility in this study is in consonance with the notion that amaranth domestication occurred independently in the crop species and that *A. hybridus* may be evolutionary closer to *A. hypochondriacus* than to *A. caudatus*.

The two basic chromosome numbers are observed among the grain crops and their putative progenitors. While *A. caudatus*, *A. hypochondriacus*, *A. hybridus* and *A. quitensis* have 32 chromosomes ($n = 16$), *A. cruentus* and *A. powellii* both have 34 ($n = 17$) (Grant 1959). Pal et al. (1982) have explored the evolutionary relationship between the two basic numbers in the grain group by performing a dibasic cross between *A. hypochondriacus* and an African race of *A. hybridus* with 34 chromosomes. At metaphase I, the majority of meiotic cells from the interspecific F_1 showed 15 bivalent and 1 trivalent chromosomal associations. Hybrid progeny (F_2) showed 1:2:1 segregation for 32, 33 and 34 somatic chromosomes, respectively. The observance of this meiotic configuration in the dibasic hybrid suggested that $n = 17$ arose through aneuploidy, perhaps involving a reciprocal translocation resulting in a decrease in chromosome number from $n = 17$ to $n = 16$. Other authors also have analyzed meiotic behavior in crop–wild hybrids, and much of what we know about gene pool accessibility and phylogenetic relations is derived from these studies. Greizerstein and Poggio (1995) studied the meiotic configuration of 13 different crop–wild and wild–wild spontaneous hybrids and this information was used to configure the first set of genomic formulae for these species (Brenner et al. 2000).

Hybridization studies have been of great value for breeders interested in crop improvement through traditional means. Hybrids with wild species have been produced to address all major breeding objectives, including raising yield, improving pest tolerance, and improving grain harvestability. For instance, one of the most widely utilized grain varieties in the USA, *A. hypochondriacus* var. Plainsman, is derived from a cross with a Pakistani *A. hybridus* accession used as a source for earliness (Baltensperger et al. 1992).

Interspecific crosses with *A. hybridus* have been used to measure biomass heterosis and combining ability with domesticated species, in a first step to exploit heterosis in the development of cultivars improved for forage, energy feedstock, or as vegetables (Lehmann et al. 1991).

Brenner et al. (2000) report that crop–wild hybrids have been produced to transfer *A. powellii* nondehiscence to *A. cruentus* and *A. hypochondriacus* breeding lines, in efforts to reduce grain shattering. The authors also propose that hybridization with *A. cannabinus*, a wild dioecious species, may be useful to obtain germplasm with greater seed size. In an ongoing breeding program, weedy *A. hybridus* with evolved herbicide resistance is being used to introduce herbicide selectivity to *A. hypochondriacus* and *A. cruentus* elite breeding lines (Federico Trucco unpublished data). The introgression of herbicide resistance from wild species and the potential implications of herbicide resistant cultivars is discussed in more detail in Sect. 2.6.

2.4.2 Hybridization Between A. hybridus and A. tuberculatus, A Species from Subgenus Acnida

A recent aspect of research regarding hybridization among wild crop relatives focused in the study of gene flow between two problematic weeds that have been already introduced, namely *A. hybridus* and *A. tuberculatus* (Trucco et al. 2005b). Although this research has been conducted from a weed science perspective, the fact that *A. hybridus* is a common progenitor to the domesticated species makes the findings of these studies of value for breeders interested in exploiting the diversity of the dioecious taxon. Previous experiments indicated that hybrids (F_1s) between *A. tuberculatus* and *A. hybridus* could be produced but failed to quantify the extent to which this could occur (Murray 1940). Also, subsequent introgression was thought to be compromised by severe sterility in the F_1 (Sauer 1957), and the only viable BC_1 progeny were thought to be those derived from unreduced gametes from the hybrid parent (Murray 1940), resulting in triploidy. First generation backcross progeny would have a full complement of the recurrent species' genome and only a haploid complement of the non-recurrent parent, and exhibit sterility due to abnormal chromosome pairing. These observations suggested little if any chance for homoploid gene exchange between *A. tuberculatus* and *A. hybridus* – that is, gene exchange without changes in ploidy.

Tranel et al. (2002) were able to transfer a herbicide-resistance allele of acetolactate synthase (ALS) from *A. hybridus* to an advanced hybrid population (BC_2) – where *A. tuberculatus* was recurrently used paternally – and with the use of DNA content data suggested that introgression could occur in a homoploid background. However, these authors did not address directly the fertility and genome structure of introgressants. Were heterozygous BC_2s more fertile than heterozygous BC_1s, or F_1s? Was the genomic constitution of these introgressants recombinant (on average 12.5% *A. hybridus* and 87.5% *A. tuberculatus*, or a reconstitution of the F_1)? What about introgression in the reciprocal direction? From a crop-breeding perspective, all these questions need to be addressed in order to establish the potential for conventional (sex-mediated) gene pool exploitation.

Later experiments showed that hybrids can be produced at relatively high frequencies under field conditions (Trucco et al. 2005a, b). In the case where the monoecious parent was used maternally, the maximum hybridization frequency obtained accounted for close to 50% of the believed intraspecific outcrossing potential of the species. In the reciprocal case, more than 200,000 hybrids could be obtained from a single *A. tuberculatus* plant. These data indicated that little if any gametic incompatibility exists between the studied species and that F_1 production is unlikely to constitute a significant bottleneck for gene introgression.

Although hybrid sterility was well documented by Murray and Sauer, to the extent hybrids were quoted to "run into a blind alley of sterility" (Sauer 1957), a detailed assessment of hybrid fertility showed that as many as 800 seeds could be recovered from a single F_1 (Trucco et al. 2006b). This number is substantial enough to allow expectations of successful gene introgression. However, successful introgression is dependent on recombination between the donor and recipient genomes and this could be unveiled adequately with the use of cytogenetic and molecular markers. Using these tools to profile hybrid progeny from backcrosses to the "pure" species, the following observations were made by Trucco et al. (2005c):

- Most BC_1s (98%) were homoploid ($2n = 32$), and triploidy was not necessarily the product of unreduced hybrid gametes. This is in agreement with Tranel et al. (2002) and in contrast to Murray (1940). Production of unreduced gametes may vary among populations and this may explain the discrepancy observed in triploidy occurrence in Murray's work (100%) versus Trucco et al. (2%).
- Fertility restitution was not a strict function of reconstitution of the parental species' genomes; in fact, hybrid sterility could be explained by as few as five independently assorting loci. In which case, advantageous alleles unlinked to these loci may be introgressed quickly. The introgression of linked alleles (genes linked to post-zygotic reproductive barriers) may depend on the selection coefficient and population size.

These authors also examined introgression of a herbicide resistance allele of *ALS* from *A. tuberculatus* to *A. hybridus* (a reciprocal of that evaluated by Tranel et al. 2002), and they observed that the *A. tuberculatus* allele could not be introduced into *A. hybridus* monoecious background. Allele introgression was limited to a small number of non-monoecious individuals exhibiting high sterility. The main speculation then was that lack of introgression resulted from linkage of *ALS* to a hybrid sterility locus associated with sex determination, a taxonomically discriminating character for these species.

In a subsequent study by the same authors (Trucco et al. 2009), 192 homoploid BC_1s were evaluated at 197 AFLP loci, as well as at *ALS* and *PPO* (the gene for protoporphyrinogen oxidase, the enzyme targeted by a second family of herbicides). The parental populations used were polymorphic at the herbicide target genes, and just as observed in the prior study, *A. tuberculatus*' *ALS* and *PPO* alleles could not be transferred to *A. hybridus* monoecious background. This indicated that gene exchange likely is limited by a phenomenon beyond circumstantial linkage.

Indeed, Trucco et al. (2009) were unable to transfer most of 133 AFLP markers from *A. tuberculatus* to *A. hybridus*, with the exception of introgression in a smaller group of non-monoecious BC_1s, characterized by anomalous phenotypes and high sterility. This observation is striking as the authors uncovered a very different scenario in the reciprocal exchange. They were able to transfer from *A. hybridus* to *A. tuberculatus* not only *ALS* and *PPO* alleles but also most of the *A. hybridus*-specific AFLP markers. Although introgression at some loci appeared to be disfavored (i.e., showed negative segregation distortion or a fecundity penalty), *ALS* and *PPO* alleles as well as many of the AFLP-markers showed Mendelian segregation in backcross progeny with *A. tuberculatus* and no association with BC_1 reproductive output (measured as pollen viability or seed production). This was not observed in progeny from reciprocal backcrosses, where almost all *A. hybridus* markers showed strong negative distortion and increased introgression was associated with reduced reproductive viability.

Taking monoecism and dioecism as the taxonomic distinguishing characters for *A. hybridus* and *A. tuberculatus*, respectively, we may say that gene exchange between these species is unidirectional. Even if we considered non-monoecious *A. hybridus* backcross progeny as being more "*A. hybridus*" than "*A. tuberculatus*," the use of these individuals in crop-breeding programs seems restricted by the dramatic fitness disadvantage at which they stand compared to their reciprocals. In fact, there is no association between the level of introgression measured in *A. tuberculatus* BC_1s and their relative seed output.

The data produced by these experiments indicate that *A. tuberculatus* adaptations, some of which may be of great value to the *Amaranthus* crops, may not be transferred to *A. hybridus*, a species from which the crops have evolved. Yet, crop adaptations may be equally transferable to *A. tuberculatus* as those from *A. hybridus*, and this may have alternative implications for crop breeders. First, crop traits such as disease resistance or herbicide tolerance may end up in a highly problematic weed, although crop erosion with *A. tuberculatus* alleles should be of little concern. Secondly, given that *Amaranthus* breeding is at an early stage, the development of an *A. tuberculatus* crop by transferring crop adaptations to this wild relative may be a bold but profitable proposition. Considering that the most distinctive crop adaptations show simple inheritance (Brenner et al. 2000) and that gene transfer may be accomplished readily, the development of a dioecious crop seems within reach. Such possibility could provide a unique opportunity for the exploitation of heterosis in amaranths.

2.5 Genomics Resources Developed for *Amaranthus*

Within the last 2 years, genomics resources have been developed for *Amaranthus* species. Two research groups independently developed microsatellite markers from *A. hypochondriacus* (Lee et al. 2008; Mallory et al. 2008). In both cases, the markers were demonstrated to be transferable to other cultivated as well as weedy *Amaranthus* species. A preliminary phylogenetic analysis using some of these markers placed *A. hybridus* within multiple grain amaranth clades, suggesting multiple domestication events from *A. hybridus* (Mallory et al. 2008). Additional *Amaranthus* microsatellite markers were obtained recently from *A. tuberculatus* (Lee et al. 2009). Collectively, these microsatellite markers will be valuable for more detailed phylogenetic studies, as well as for various genetic studies and breeding efforts (e.g., population genetics, construction of genetic maps, map based cloning, and marker-assisted selection).

A second genomics resource for *Amaranthus* is a bacterial artificial chromosome (BAC) library from *A. hypochondriacus* (Maughan et al. 2008). This library contains over 35,000 clones averaging 147 kb, or about a 10-fold coverage of the genome. Utility of the library was demonstrated by using it to obtain the full-length sequences of the *ALS* and *PPO* genes, both of which encode important herbicide target sites (Maughan et al. 2008). In addition to serving as a source for candidate gene isolation and sequencing, this BAC library could be used further to develop a physical map of the *Amaranthus* genome, and could serve as a scaffold for whole-genome sequencing.

A third *Amaranthus* genomics resource recently obtained is shotgun sequence data from *A. tuberculatus* (Lee et al. 2009). Using next-generation sequencing technology, over 40 Mbp of sequence was obtained from *A. tuberculatus*. Included in the dataset was a nearly complete sequence of the chloroplast genome and partial sequences of most currently known herbicide target-site genes. The dataset also provided leads for microsatellite markers, mentioned above. Although the dataset contains only partial sequences for nuclear genes, such information serves as a starting point for candidate gene isolation insofar as designing primers for PCR-based approaches. The same technology has been used to sequence the *A. tuberculatus* transcriptome (P. Tranel unpublished data). Both the genomic and transcriptomic datasets are being made publicly available via the National Center for Biotechnology Information (http://www.ncbi.nlm.nih.gov/).

Finally, a forth resource in development is a collection of recombinant inbred lines (RILs) derived from an initial crop–wild cross between *A. hypochondriacus* and *A. hybridus* (P. Tranel unpublished data). The initial F_1 plant was selected based on herbicide resistance (which was present in the *A. hybridus* parent, used paternally in the cross) and over 200 lines derived from selfing the F_1 are being propagated via single-seed decent. Given the expected high genetic diversity among the RILs, they should provide an ideal population for development of an *Amaranthus* genetic map.

A current need is the development of a facile genetic transformation system for *Amaranthus*. Only modest success has been reported in regenerating plants from *Amaranthus* callus tissue (Brenner et al. 2000). There is, however, one report of successful transformation of *A. hypochondriacus* by inoculation of mature embryo explants with *Agrobacterium* (Jofre-Garfias et al. 1997). It is surprising that there are not more reports of *Amaranthus* transformation, and we do not know if this is due to lack of effort or technical challenges. Possibly the weedy species are more amenable to genetic transformation and could serve as model systems for developing and optimizing protocols that could then be adapted to the cultivated species.

2.6 Herbicide Resistances in *Amaranthus* Weeds

A primary characteristic contributing to the infamy of *Amaranthus* species as weeds of modern agriculture is their demonstrated ability to evolve herbicide resistance. *Amaranthus* weeds comprise over 5% of worldwide cases of herbicide-resistant weeds and have evolved resistances to diverse herbicide modes of action (Heap 2010; Tranel and Trucco 2009). For example, *A. tuberculatus* has evolved resistance to herbicides that inhibit photosystem II (PSII), ALS, PPO, and 5-enolypyruvyl-shikimate-3-phosphate synthase (EPSPS) (Patzoldt et al. 2005; Legleiter and Bradley 2008). In some cases, resistances to more than one of these herbicide groups is present within a single population (or even within a single plant), making control of *A. tuberculatus* a significant practical problem. In the southeastern US, resistance to glyphosate (which inhibits EPSPS) has become widespread in *A. palmeri* in recent years and is posing a very significant weed management challenge (Culpepper et al. 2006, 2008; Norsworthy et al. 2008).

The frequent occurrence of herbicide resistance in *Amaranthus* weeds suggests it should be possible to select the same traits in cultivated *Amaranthus* crops. Alternatively, it should be possible to transfer the resistance traits from the weeds to the crops via hybridization (although see Sect. 2.4.2). For example, it should be straightforward to cross grain amaranth with *A. hybridus* containing resistance to ALS inhibitors (Trucco et al. 2006a), and then obtain the herbicide-resistant crop by recurrent backcrossing along with selection for the resistance. Unfortunately, however, that these herbicide resistances are widespread in many of the *Amaranthus* weeds would limit their utility in the crop. Nevertheless, the only *Amaranthus* species thus far to have evolved resistance to PPO inhibitors is *A. tuberculatus*, and to EPSPS inhibitors are *A. tuberculatus* and *A. palmeri*. Thus, resistance to one or both of these herbicides in cultivated amaranth may have value, particularly in regions where these two weeds are not present.

The mechanism conferring resistance to PPO inhibitors in *A. tuberculatus* was determined to be a deletion of a glycine residue in a conserved region of the *PPX2* gene (Patzoldt et al. 2006). The gene was predicted to encode both mitochondria- and chloroplast-targeted PPO, thereby resulting in herbicide-insensitive enzymes in both organelles. Through genetic transformation, one could insert the *A. tuberculatus* herbicide resistant *PPX2* into the crop species. The homologous gene was obtained from *A. hypochondriacus* and also shown to contain the dual-targeting signal sequences (Maughan et al. 2008). Site-directed mutagenesis of the native *A. hypochondriacus PPX2* to obtain the glycine codon deletion followed by transformation would be another route to obtain resistance to PPO inhibitors. This latter approach might be met with greater public acceptance since the crop would not be carrying a gene from a weed species (although the encoded proteins from the two genes are over 97% identical; Tranel and Trucco 2009).

A major challenge beyond the development of a herbicide-resistant amaranth crop would be maintaining the utility of the trait by preventing its escape into coexisting *Amaranthus* weeds. In this regard, the body of work on interspecific hybridization (reviewed previously herein) should provide the framework for the development of adequate protocols for technology stewardship.

2.7 Recommendations for Future Actions

Amaranths have been a staple crop of pre-Columbian cultures, and they have received interest in the last two to three decades as an alternative crop. Much of the recent interest in amaranths is based on the exceptional nutritional profile of the grain proteins, which are rich in amino acids that are usually deficient in other crops (Bejosano and Corke 1998). Additional interest is generated by the oil and carbohydrate profiles of amaranth seeds, which present opportunities for different industrial applications, from the use of amaranth squalene as a cosmetic oil (Budin et al. 1996) to that of micro-sized starch in the formulation of foods (Uriyapongson and Rayas-Duarte 1994). Numerous studies have been conducted to develop and optimize technologies aimed at exploiting these amaranth properties (see works in Paredes-López 1994 for greater detail).

From an agronomic perspective, drought tolerance and environmental plasticity are attractive traits promoting amaranth adoption in areas where traditional crops face greater challenges (Brenner et al. 2000). Yet, modern amaranth cultivars still face several difficulties, which have been overcome in most major crops. Recent breeding efforts to try to solve some of these difficulties have been modest, and very few cultivars have been registered over the last decade. In an unusual contrast, *Amaranthus* weeds have been the subject of leading weed science research over the same timeframe (see Tranel and Trucco 2009 for a revision on the subject). In fact, weedy amaranths have been proposed as a model system for the study of plant weediness (Basu et al. 2004), and valuable genomic resources are being generated with these species as discussed in Sect. 2.5.

This chapter attempts to bridge the research conducted by the weed science and the crop-breeding communities, realizing that perhaps the path to improving important crop traits may be realized through the judicious exploitation of the wealth found in weedy amaranth resources. The discussions regarding the patterns of gene exchange among the different taxonomic groups constitute a first and incomplete attempt at drafting a roadmap for the exchange of adaptations among species. Similarly, the discussion dealing with herbicide resistance covers a number of possibilities yet to be explored by amaranth breeders. It is important to note that the great success of amaranths as weeds is not found in any one adaptation but in their ability to adapt quickly to changing weed management practices. Infamy due to the evolution of numerous herbicide resistant populations is a reflection of their adaptability, or from a different perspective a reflection of their ability to successfully respond to selection. Interestingly, what constitutes a threat to farm economies at one level may be the most valuable asset for the development of competitive cultivars at another. It is up to us to transform this serious challenge into a beneficial force.

References

Baltensperger DD, Weber LE, Nelson LA (1992) Registration of 'Plainsman' grain amaranth. Crop Sci 32:1510–1511

Basu C, Halfhill MD, Mueller TC, Stewart NC (2004) Weed genomics: new tools to understand weed biology. Trends Plant Sci 9:391–398

Bejosano FP, Corke H (1998) Protein quality evaluation of *Amaranthus* wholemeal flours and protein concentrates. J Sci Food Agric 76:100–106

Brenner DM, Widrlechner MP (1998) *Amaranthus* seed regeneration in plastic tents in greenhouses. FAO Plant Genet Resour Newsl 116:1–4

Brenner DM, Baltensperger DD, Kulakow PA, Lehmann JW, Myers RL, Slabbert MM, Sleugh BB (2000) Genetic resources and breeding in *Amaranthus*. In: Janick J (ed) Plant breeding reviews, vol 19. Wiley, New York, USA, pp 227–285

Budin JT, Breene WM, Putnam DH (1996) Some compositional properties of seeds and oils of eight *Amaranthus* species. J Am Chem Soc 73:475–481

Chan KF, Sun M (1997) Genetic diversity and relationships detected by isozyme and RAPD analysis of crop and wild species of *Amaranthus*. Theor Appl Genet 95:865–873

Costea M, Tardif FJ (2003) Conspectus and notes on the genus *Amaranthus* (Amaranthaceae) in Canada. Rhodora 105:260–281

Culpepper AS, Grey TL, Vencill WK, Kichler JM, Webster TM, Brown SM, York AC, Davis JW, Hanna WW (2006) Glyphosate-resistant Palmer amaranth (*Amaranthus palmeri*) confirmed in Georgia. Weed Sci 54:620–626

Culpepper AS, Whitaker JR, MacRae AW, York AC (2008) Distribution of glyphosate-resistant Palmer amaranth (*Amaranthus palmeri*) in Georgia and North Carolina during 2005–2006. J Cotton Sci 12:306–310

Grant WF (1959) Cytogenetic studies in *Amaranthus*III. Chromosome numbers and phylogenetic aspects. Can J Genet Cytol 1:313–328

Greizerstein EJ, Poggio L (1995) Meiotic studies of spontaneous hybrids of *Amaranthus*: genome analysis. Plant Breed 114:448–450

Grubben GJ, van Sloten DH (1981) Genetic resources of amaranths: a global plan of action. International Board of Plant Genetic Resources, Rome, Italy

Hauptli H, Jain S (1984) Genetic structure of landrace populations of the New World amaranths. Euphytica 33: 875–884

Heap I (2010) The international survey of herbicide resistant weeds. http://www.weedscience.com. Accessed 29 June 2010

Holm LG, Plucknett DL, Pancho JV, Herberger JP (1991) The world's worst weeds: distribution and biology. Krieger, Malabar, FL, USA

Holm L, Doll J, Holm E, Pancho J, Herberger J (1997) World weeds: natural histories and distribution. Wiley, New York, NY, USA

Jofre-Garfias AE, Villegas-Sepúlveda N, Cabrera-Ponce JL, Adame-Alvarez RM, Herrera-Estrella L, Simpson J (1997) *Agrobacterium*-mediated transformation of *Amaranthus hypochondriacus*: light- and tissue-specific expression of a pea chlorophyll a/b-binding protein promoter. Plant Cell Rep 16:847–852

Kauffman CS (1992) Realizing the potential of grain amaranth. Food Rev Int 8:5–21

Khoshoo TN, Pal M (1972) Cytogenetic patterns in *Amaranthus*. Chrom Today 3:259–267

Kirkpatrick BA (1995) Interspecies relationships within the genus *Amaranthus* (Amaranthaceae). PhD Dissertation, Texas A&M University, College Station, TX, USA

Lanoue KZ, Wolf PG, Browning S, Hood EE (1996) Phylogenetic analysis of restriction site variation in wild and cultivated *Amaranthus* species (Amaranthaceae). Theor Appl Genet 93:722–732

Lee JR, Hong GY, Dixit A, Chung JW, Ma KH, Lee JH, Kang HK, Cho YH, Gwag JG, Park YJ (2008) Characterization of microsatellite loci developed for *Amaranthus hypochondriacus* and their cross-amplifications in wild species. Conserv Genet 9:243–246

Lee RM, Thimmapuram J, Thinglum KA, Gong G, Hernandez AG, Wright CL, Kim RW, Mikel MA, Tranel PJ (2009) Sampling the waterhemp (*Amaranthus tuberculatus*) genome using pyrosequencing technology. Weed Sci 57:463–469

Legleiter TR, Bradley KW (2008) Glyphosate and multiple herbicide resistance in waterhemp (*Amaranthus rudis*) populations from Missouri. Weed Sci 56:582–587

Lehmann JW, Clark RL, Frey KJ (1991) Biomass heterosis and combining ability in interspecific and intraspecific matings of grain amaranths. Crop Sci 31:1111–1116

Mallory MA, Hall RV, Mcnabb AR, Pratt DB, Jellen EN, Maughan PJ (2008) Development and characterization of microsatellite markers for the grain amaranths. Crop Sci 48:1098–1106

Maughan PJ, Sisneros N, Luo M, Kudrna D, Ammiraju JSS, Wing RA (2008) Construction of an *Amaranthus hypochondriacus* bacterial artificial chromosome library and genomic sequencing of herbicide target genes. Crop Sci 48:S85–S94

Mosyakin SL, Robertson KR (2003) Amaranthus. In: Flora of North America. North of Mexico. Oxford University Press, New York, USA

Murray MJ (1940) The genetics of sex determination in the family *Amaranthaceae*. Genetics 25:409–431

Norsworthy JK, Griffith GM, Scott RC, Smith KL, Oliver LR (2008) Confirmation and control of glyphosate-resistant Palmer amaranth (*Amaranthus palmeri*) in Arkansas. Weed Technol 22:108–113

Pal M, Khoshoo TN (1972) Evolution and improvement of cultivated amaranths. V. Inviability, weakness, and sterility in hybrids. J Hered 73:467

Pal M, Khoshoo TN (1973) Evolution and improvement of cultivated amaranths VII. Cytogenetic relationships in vegetable amaranths. Theor Appl Genet 43:343–350

Pal M, Pandey RM, Khoshoo TN (1982) Evolution and improvement of cultivated amaranths. J Hered 73:353–356

Paredes-López O (1994) Amaranth: biology, chemistry, and technology. CRC, Boca Raton, FL, USA

Patzoldt WL, Tranel PJ, Hager AG (2005) A waterhemp (*Amaranthus tuberculatus*) biotype with multiple resistance across three herbicide sites of action. Weed Sci 53:30–36

Patzoldt WL, Hager AG, McCormic JS, Tranel PJ (2006) A codon deletion confers resistance to herbicides inhibiting protoporphyrinogen oxidase. Proc Natl Acad Sci USA 103: 12329–12334

Pratt DB, Clark LG (2001) *Amaranthus rudis* and *A. tuberculatus* – one species or two? J Torr Bot Soc 128:282–296

Ranade SA, Kumar A, Goswami M, Farooqui N, Sane PV (1997) Genome analysis of amaranths: determination of inter- and intra-species variations. J Biosci 22:457–464

Sauer JD (1957) Recent migration and evolution of the dioecious amaranths. Evolution 11:11–31

Sauer JD (1967) The grain amaranths and their relatives: a revised taxonomic and geographic survey. Ann MO Bot Gard 54:102–137

Sauer JD (1972) The dioecious amaranths: a new species name and major range extensions. Madroño 21:425–434

Steckel LE (2007) The dioecious *Amaranthus* spp.: here to stay. Weed Technol 21:567–570

Tranel PJ, Trucco F (2009) 21st century weed science: a call for *Amaranthus* genomics. In: Stewart CN Jr (ed) Weedy and invasive plant genomics. Blackwell, Ames, IA, USA, pp 53–81

Tranel PJ, Wassom JJ, Jeschke MR, Rayburn AL (2002) Transmission of herbicide resistance from a monoecious to a dioecious weedy *Amaranthus* species. Theor Appl Genet 105:674–679

Transue DK, Fairbanks DJ, Robison LR, Andersen WR (1994) Species identification by RAPD analysis of grain amaranth genetic resources. Crop Sci 34:1385–1389

Trucco F, Jeschke MR, Rayburn AL, Tranel PJ (2005a) *Amaranthus hybridus* can be pollinated frequently by *A. tuberculatus* under field conditions. Heredity 94:64–70

Trucco F, Jeschke MR, Rayburn AL, Tranel PJ (2005b) Promiscuity in weedy amaranths: high frequency of female tall waterhemp (*Amaranthus tuberculatus*) x smooth pigweed (*A. hybridus*) hybridization under field conditions. Weed Sci 53:46–54

Trucco F, Tatum T, Rayburn AL, Tranel PJ (2005c) Fertility, segregation at a herbicide resistance locus, and genome structure in BC_1 hybrids between two important weedy *Amaranthus* species. Mol Ecol 14:2717–2728

Trucco F, Hager AG, Tranel PJ (2006a) Acetolactate synthase mutation conferring imidazolinone-specific herbicide resistance in *Amaranthus hybridus*. J Plant Physiol 163:475–479

Trucco F, Tatum T, Robertson KR, Rayburn AL, Tranel PJ (2006b) Morphological, reproductive, and cytogenetic characterization of *Amaranthus tuberculatus* × *A. hybridus* F_1 hybrids. Weed Technol 20:14–22

Trucco F, Tatum T, Rayburn AL, Tranel PJ (2009) Out of the swamp: Unidirectional hybridization with weedy species may explain *Amaranthus tuberculatus*' prevalence as a weed. New Phytol 184:819–827

Tucker JM, Sauer JD (1958) Aberrant *Amaranthus* populations of the Sacramento-San Joaquin Delta, California. Madroño 14:252–261

Uriyapongson J, Rayas-Duarte P (1994) Comparison of yield and properties of amaranth starches using wet and dry-wet milling processes. Cereal Chem 71:571–577

Wassom JJ, Tranel PJ (2005) Amplified fragment length polymorphism-based genetic relationships among weedy *Amaranthus* species. J Hered 96:410–416

Wetzel DK, Horak MJ, Skinner DZ (1999) Use of PCR-based molecular markers to identify weedy *Amaranthus* species. Weed Sci 47:518–523

Zheleznov AV, Solonenko LP, Zheleznova NB (1997) Seed protein of the wild and cultivated *Amaranthus* species. Euphytica 97:177–182

Chapter 3
Asparagus

Akira Kanno and Jun Yokoyama

3.1 Introduction

Asparagus is known because of the important vegetable plant, *A. officinalis,* which is cultivated throughout the world. The cultivated area of *A. officinalis* is more than 200,000 ha, mainly in China, Europe, and Central and South America. Except for this species, the utilization of other *Asparagus* species by humans is limited. Some species are used as ornamental plants (e.g., *A. asparagoides*, *A. scandens*, *A. plumosus*, and *A. falcatus*), and some others are used for medicinal purposes (e.g., *A. cochinchinensis* and *A. racemosus* in Asia). However, this genus comprises more than 200 species and is widely distributed in the Old World continents. Thus, a large amount of untapped genetic resources exists within this genus. Here, we provide an overview of the species diversity and evolutionary relationships of genus *Asparagus*, the developmental genetics that control its floral structures, and the generation of hybrids with cultivated *Asparagus* and its wild relatives for future breeding purposes.

A. Kanno (✉)
Graduate School of Life Sciences, Tohoku University, 2-1-1, Katahira, Aoba-ku, Sendai 980–8577, Japan
e-mail: kanno@ige.tohoku.ac.jp

3.2 Species Diversity and Phylogenetic Relationships of *Asparagus*

3.2.1 *Asparagaceae and Related Families*

The genus *Asparagus* was recognized based on *A. officinalis* L. by Linnaeus in "Species Plantarum" (1753). The genus was once placed within the huge monocot family Liliaceae, but the family was extensively reconstructed with other monocot families based on morphological and molecular phylogenetic studies. *Asparagus* is now the main genus of the Asparagaceae, belonging to Asparagales (Dahlgren et al. 1985; Angiosperm Phylogeny Group 2003; Janssen and Bremer 2004; Stevens 2008).

Asparagales, one of the two major orders of lilioid monocots, contains Asparagaceae and other 23 families (Stevens 2008). Dividing lilioid monocots into two large groups was first proposed by Dahlgren et al. (1985) and was supported by various studies of molecular phylogenies with slight modifications of the original concept (Chase et al. 2000a, b; Janssen and Bremer 2004). For Asparagales, molecular phylogenetic studies indicated that Orchidaceae was a sister family of all other families in Asparagales and Iridaceae is involved in Asparagales, although these two families have both been treated as members of the Liliales since the publication of Dahlgren et al. (1985).

Asparagaceae belongs to the monophyletic clade with Laxmanniaceae and Ruscaceae (Fay et al. 2000). These three families share apomorphic characteristics such as steroidal saponins, articulated pedicels, and helobial, thick-walled, pitted endosperm (Stevens 2008). Two closely related families of Asparagaceae

C. Kole (ed.), *Wild Crop Relatives: Genomic and Breeding Resources, Vegetables*,
DOI 10.1007/978-3-642-20450-0_3,

are distributed almost allopatrically in the world: Laxmanniaceae is predominantly distributed in Australia and adjacent areas, whereas Ruscaceae is mainly distributed in the northern hemisphere and Africa. Laxmanniaceae comprises 15 genera and more than 170 species that are divided into three groups: *Lomandra* (five genera), *Laxmannia* (eight genera), and *Cordyline* (two genera). *Cordyline* is cultivated as ornamental plants. Phylogenetic analyses have indicated that Ruscaceae is the sister group of Asparagaceae, although statistical support for this idea is not high (Fay et al. 2000). Ruscaceae is a large family that contains 26 genera and more than 450 species. Some of the represented genera are well known as ornamental plants, such as *Convallaria*, *Dracaena*, *Ophiopogon*, and *Ruscus*. *Ruscus*, the nominal genus of the family, draws particular interest because this genus also has a "phylloclade" as *Asparagus* species (Cooney-Sovetts and Sattler 1987). Although the resemblance does not reflect a direct phylogenetic relationship, the morphological characteristics may have a similar genetic background.

Asparagaceae contains two genera, *Asparagus* and *Hemiphylacus*. The latter genus was once treated as a member of the Asphodelaceae, but molecular phylogenetic studies clearly indicated that the genus is not a part of this family (Chase et al. 2000a). This genus was monotypic for a long time [only the type species *H. latifolius* has been known for more than 100 years (Watson 1883)], but now comprises five species, all distributed only in Mexico (Hernandez 1995). The five species show quite restricted distribution ranges and occur on limestone or rhyolitic areas. Although the ranges are limited, the New World distribution of this sister genus of *Asparagus* is in contrast to the Old World distribution of *Asparagus*.

3.2.2 Species Diversity of Asparagus

Until now, about 400 taxa have been described under the genus *Asparagus* since Linnaeus's time. Among them, about 210 taxa are considered to be accepted as species (Tutin et al. 1980; Chen and Tamanian 2000; African Flowering Plants Database 2008; Appendix 1). The latest infrageneric classification divides *Asparagus* into three subgenera: *Asparagus*, *Myrsiphyllum*, and *Protasparagus* (Clifford and Conran 1987). The African species belonging to the latter two subgenera are sometimes treated as distinct from *Asparagus* at the genus level (Obermeyer 1983). The subgenus *Asparagus* comprises all dioecious species with unisexual flowers, whereas the subgenus *Myrsiphyllum* consists of hermaphroditic, southern African species with shortly united tepals and annual aerial stems. The remaining species with bisexual flowers, separated tepals, and perennial stems are classified within the subgenus *Protasparagus* (Clifford and Conran 1987). Of these subgenera, *Protasparagus* is considered to retain the primitive characteristics of the genus, such as multiple ovules per locule and globular seeds (Obermeyer 1983).

The center of species diversity is Africa, especially South Africa and adjacent regions. More than half of the total species are distributed there, and still undescribed species are also found there (e.g., Burrows and Burrows 2008). Most of the South African *Asparagus* species are shrubby, and some are climbers. The plant sizes vary extensively among species, from less than 20 cm to more than 5 m. Madagascar, the large island adjacent to continental Africa, also has a considerable amount of *Asparagus* species (A Catalog of the Vascular Plants of Madagascar, http://www.efloras.org/flora_page.aspx?flora_id=12). Although only seven species are currently listed as accepted names, another 13 species are considered to be undescribed endemic species of the island. Including those undescribed species, 95% of *Asparagus* species are endemic to Madagascar, suggesting a characteristic floral composition, as mentioned in other previous studies (e.g., Grubb 2003; Janssen et al. 2008). The Macaronesian Islands, offshore of the northwestern coast of Africa, also have some characteristic *Asparagus* species; they harbor six species of *Asparagus*, five of which are endemic to the islands. Two species, *A. nesiotes* and *A. umbellatus*, have further diversified into allopatric subspecies in distant island clusters (Canary and Madeire).

In the Eurasian continent, Europe (including northern Africa) and China have a relatively rich flora of *Asparagus* species. In China, 29 native species are listed, of which more than half (15 species) are endemic (Chen and Tamanian 2000). Most of the Chinese endemic species are regional and are thought to be neo-endemic species that diversified relatively recently in central to southern China.

Some narrow endemic species now face extinction in the wild. For example, only 2–5 populations of *A. calcicolus*, endemic to Madagascar, are left in wild conditions. *Asparagus mairei* is only known in the Kunming city of Yunnan, China. Although its ranked category is listed as "Data Deficient" and its detailed situation in the wild is unknown, an undescribed species from Socorta, Yemen ("*Asparagus* sp. nov. A") is listed in the IUCN Red List.

3.2.3 Phylogenetic Relationships of Asparagus

Phylogenetic relationships of *Asparagus* species have been poorly understood because only a few studies have been conducted so far. Lee et al. (1997) suggested that dioecious species in *Asparagus* (subgenus *Asparagus*) comprise a monophyletic group based on restriction fragment length polymorphism (RFLP) analysis of chloroplast DNA (cpDNA) using ten species of the genus. Stajner et al. (2002b) conducted a phylogenetic study based on an RFLP analysis of the nuclear DNA internal transcribed spacer (ITS) region in ten *Asparagus* species and revealed that European and African species comprise distinct monophyletic groups. The most comprehensive study of *Asparagus* phylogeny was conducted by Fukuda et al. (2005) based on nucleotide sequences of the intron of *petB* and the intergenic region between *petD* and *rpoA* in the cpDNA. In this study, 24 taxa from all three subgenera were analyzed, revealing the following results.

(1) The subgenera *Protasparagus* and *Myrsiphyllum* do not consist of monophyletic groups. (2) The subgenus *Asparagus* is monophyletic, as previously mentioned. (3) Phylloclades evolved from leaf-like forms to branch-like forms. (4) South Africa is considered to be the ancestral area of *Asparagus*, and Eurasian species, except *A. racemosus* (the most widely distributed *Asparagus* species), evolved once from the African ancestor.

Even this analysis, however, only reveals an outline of the evolutionary history of *Asparagus* because only small amounts of useful phylogenetic information were obtained from molecular data. This problem is largely due to the diverse history of *Asparagus* itself. Lee et al. (1996) and Kanno et al. (1997) suggested that most species of this genus are very closely related on the basis of their cpDNA physical map. Previously conducted RFLP studies also faced challenges. A nucleotide sequence study by Fukuda et al. (2005) failed to detect any variation in the *trnL-F* intergenic region or the *trnL* intron in the cpDNA, which are frequently used for investigating intrageneric relationships. These facts indicate that the degree of genetic differentiation among *Asparagus* species is extremely small, even though these species are morphologically diverse, such as in their shoot architectures, phylloclade morphologies, and floral colors. Therefore, *Asparagus* has undergone a recent, rapid radiation of species. Recent studies have shown that plant groups that dominate in the Cape Floral Province have experienced relatively rapid radiations, and their evolutionary patterns may contribute to the extremely isolated and specialized floral composition in this region (Richardson et al. 2001; Goldblatt et al. 2002; Forest et al. 2007). Furthermore, the evolution of extreme xeromorphic habits following phylloclade formation may have been a key innovation that facilitated the radiation of *Asparagus* into arid to subarid regions.

3.3 Chromosome Numbers and Ploidy Evolution of *Asparagus*

The chromosome numbers of 38 *Asparagus* species are known (Index to Plant Chromosome Numbers, http://mobot.mobot.org/W3T/Search/ipcn.html). The basic number of chromosome in *Asparagus* is $x = 10$, and diploid records have been made from more than 70% of the 27 recorded species (including species with an intraspecific polyploidy series, see below). Others are tetraploids (12 species) or hexaploids (five species). Some irregular counts are known for Chinese species ($2n = 18$ for *A. dauricus*, Lu and Li 1999; $2n = 16, 18$ for *A. filicinus*, Ge et al. 1988; Shang et al. 1992). The presence of B chromosomes has been reported in eight species (Sheidai and

Inamdar 1993). Polyploid species are well known in African species, and ploidy-level evolution also contributes in part to the diversification of *Asparagus* on the African continent. Intraspecies ploidy variations have been recorded from eight species, including *A. officinalis*. Tetraploids of *A. officinalis* seem to have originated at least twice independently from diploid races (Moreno et al. 2008). In the case of *A. racemosus*, the nominal variety (*A. r.* var. *racemosus*) is diploid, whereas another examined variety, *A. r.* var. *subacerosa*, is tetraploid, and yet another variety, *A. r.* var. *javanica*, contains both diploid and tetraploid individuals (Sheidai and Inamdar 1997).

Karyotypic analysis was conducted for 11 taxa by Sheidai and Inamdar (1997). All the species were found to have submetacentric chromosomes with gradual decreases in size from the largest to the smallest chromosomes. Secondary constrictions were observed in one to seven chromosomes, and large chromosomes tended to have the constrictions. Interestingly, two ploidy-level variations of *A. racemosus* var. *javanica* showed different numbers of chromosomes with secondary constrictions (diploid, 7; tetraploid, 2). This may indicate that chromosomal rearrangements or genomic reductions occurred following polyploid formation.

In contrast, chromosome information on *Hemiphylacus*, another genus in the Asparagaceae, is limited; the only known number is $2n = 112$ (*H. alatostylus*: Rudall et al. 1998). This species is considered $8x$ or $16x$, and thus the basic number of the genus is $x = 7$ or 14.

3.4 Sex Expression and Floral Homeotic Genes in the Genus *Asparagus*

3.4.1 *Floral Morphology in the Genus* Asparagus

The genus *Asparagus* contains dioecious and hermaphrodite species. The dioecious species are clustered into the subgenus *Asparagus*, and hermaphrodite species into the subgenera *Protasparagus* and *Myrsiphyllum*. Phylogenetic analysis using ITS sequences showed that dioecious species evolved from hermaphrodite species (Fukuda et al. 2005). Hermaphrodite species have bisexual flowers that have stamens and pistils in one flower, whereas dioecious species have male and female flowers that have fertile stamens and sterile pistils or sterile stamens and fertile pistils in one flower (Fig. 3.1). A morphological analysis of male and female floral development has been conducted in *A. officinalis* (Caporali et al. 1994). Male and female reproductive organs develop normally at the beginning of floral development. After that, a difference between male and female flowers becomes visible. At stage −1, the style begins to develop in the female flower, but not in the male flower. The degeneration of stamens in female flowers becomes visible at stages 3 and 4 (Caporali et al. 1994). No detailed analysis of floral development has been reported in any other *Asparagus* species. As the morphological differences between male and female flowers are similar among some dioecious *Asparagus* species, such as *A. officinalis*, *A. maritimus*, and *A. schoberioides* (Fig. 3.1), morphological events such as the inhibition of style development in male flowers and the degeneration of stamens in female flowers are similar among these species.

3.4.2 *Floral Homeotic Genes in the Genus* Asparagus

Molecular analyses of floral development have been carried out in many kinds of plants. Dicot flowers consist of four floral organs: sepals, petals, stamens, and carpels. To explain the floral organ identity of dicot flowers, the floral ABC model was proposed by Coen and Meyerowitz (1991; Fig. 3.2a). According to this model, class A genes specify sepals, class A and class B, petals, class B and class C, stamens, and class C, carpels. In contrast to dicot flowers, many monocot flowers such as those of lily and tulip consist of two whorls of the perianths, stamens, and carpels. To explain this monocot floral morphology, the modified ABC model has been proposed (van Tunen et al. 1993; Fig. 3.2b). According to this model, two whorls of petaloid perianth are caused by the extended expression of the class B genes. Class B genes have been isolated from tulip, *Agapanthus praecox*, and *Alstroemeria ligtu*, and the expression pattern of these genes is consistent with the modified ABC model (Kanno et al. 2003; Nakamura et al. 2005;

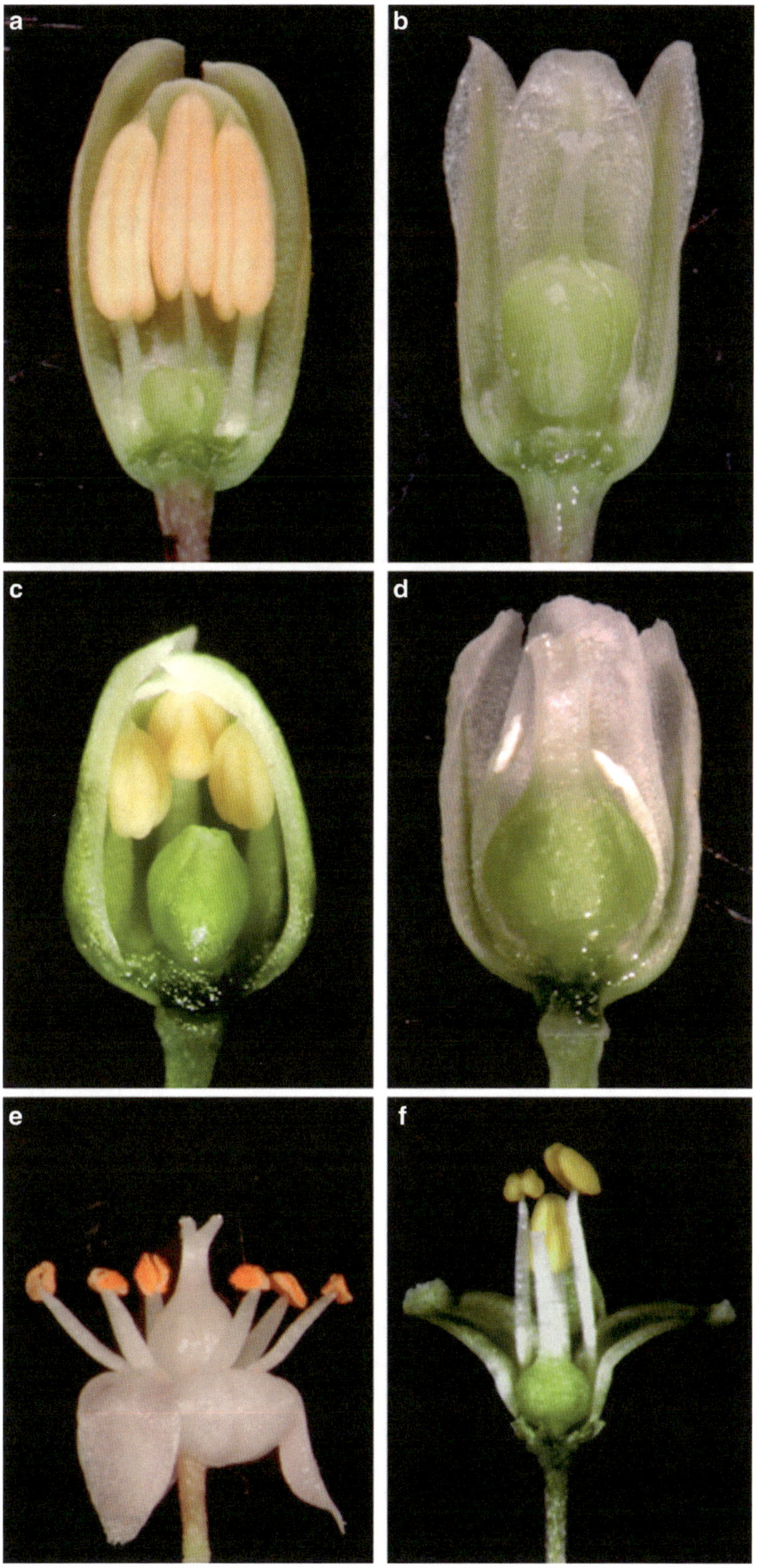

Fig. 3.1 Flower of dioecious and hermaphrodite *Asparagus* species. (**a**) Male flower of *Asparagus maritimus*. (**b**) Female flower of *A. maritimus*. (**c**) Male flower of *A. schoberioides*. (**d**) Female flower of *A. schoberioides*. (**e**) *A. densiflorus* "Sprengeri." (**f**) *A. virgatus*. Bar = 1 mm

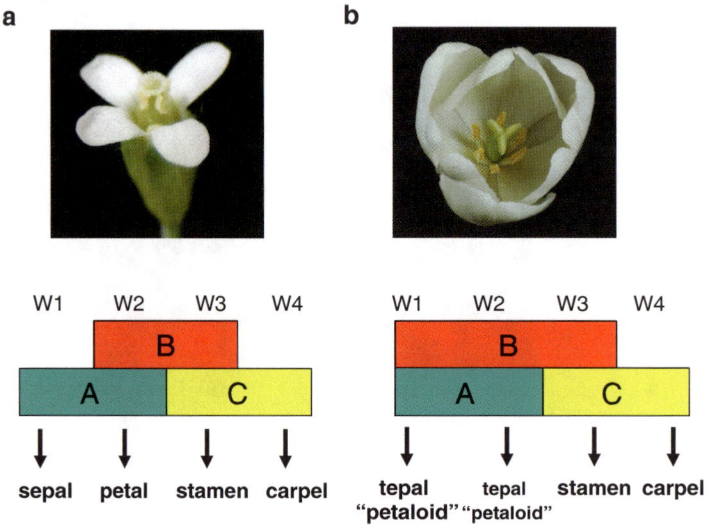

Fig. 3.2 Classical (**a**, *Arabidopsis thaliana*) and modified (**b**, *Tulipa gesneriana*) ABC model

Hirai et al. 2007). Three class B genes have been isolated from *Asparagus officinalis*, one *DEFICIENS*-like (*AODEF*) and two *GLOBOSA*-like (*AOGLOA* and *AOGLOB*) genes (Park et al. 2003, 2004). In contrast to the case in tulip and *Agapanthus*, in *A. officinalis*, the class B genes are expressed in whorls 2 and 3 and not in whorl 1 (Park et al. 2003, 2004). This expression pattern is not consistent with the modified ABC model, although *A. officinalis* has two petaloid perianths. However, one *A. officinalis* floral homeotic mutant has been reported; the perianth of this mutant has a leaf-like structure in whorls 1 and 2, and stamens are replaced with carpels (Asada et al. 2006) that appear very similar to those of the putative B mutant in tulip (van Tunen et al. 1993). Thus, the modified ABC model is basically applicable in *A. officinalis* (Kanno et al. 2004).

As mentioned above, two *GLO*-like genes have been isolated from *A. officinalis* (Park et al. 2004). Phylogenetic analysis showed that these *GLO*-like genes belong to a different group, and *GLO*-like genes from *Muscari armeniacum* and *Crocus sativus* have been classified into each cluster (Kanno et al. 2007; Fig. 3.3), indicating that duplication of *GLO*-like genes found in *A. officinalis*, *Muscari*, and *Crocus* occurred before the diversification of these species. Like tulips, *Muscari* and *Crocus* have two whorls of petaloid perianths, and *GLO*-like and *DEF*-like genes isolated from these plants are expressed in the outer two whorls of perianths (Kanno et al. 2003; Nakada et al. 2006; Tsaftaris et al. 2006; Kanno et al. 2007), even though *GLO*-like genes from *A. officinalis* are expressed only in whorls 2 and 3, not in whorl 1 (Park et al. 2004). Thus, the expression of two *GLO*-like genes in whorl 1 was lost during the evolution of *Asparagus* species. Within the genus *Asparagus*, three class B genes were isolated from *A. virgatus*: *AVDEF*, *AVGLOA*, and *AVGLOB* (Ito 2006; Fig. 3.3). *AVGLOA* and *AVGLOB* are closely related to *AOGLOA* and *AOGLOB*, respectively, and belong to different groups, like *AOGLOA* and *AOGLOB* (Ito 2006; Fig. 3.3). Given that *A. officinalis* and *A. virgatus* belong to different subgenera, *Asparagus* and *Protasparagus*, respectively, many *Asparagus* species likely have duplicated *GLO*-like genes. Although the expression patterns of *GLO*-like and *DEF*-like genes have been analyzed only in *A. officinalis*, it would be very interesting to analyze and compare the expression pattern of class B genes among *Asparagus* species. This would reveal when the expression of class B genes in whorl 1 was lost during the specification of genus *Asparagus*.

Another type of floral homeotic gene, the *AGAMOUS*(*AG*)-like genes, was isolated from the dioecious *A. officinalis* (*AOAG1* and *AOAG2*; Kanno et al. 2002) and the hermaphrodite *A. virgatus* (*AVAG1* and *AVAG2*; Yun et al. 2004a, b). *AG*-like genes are very important for the development of male and

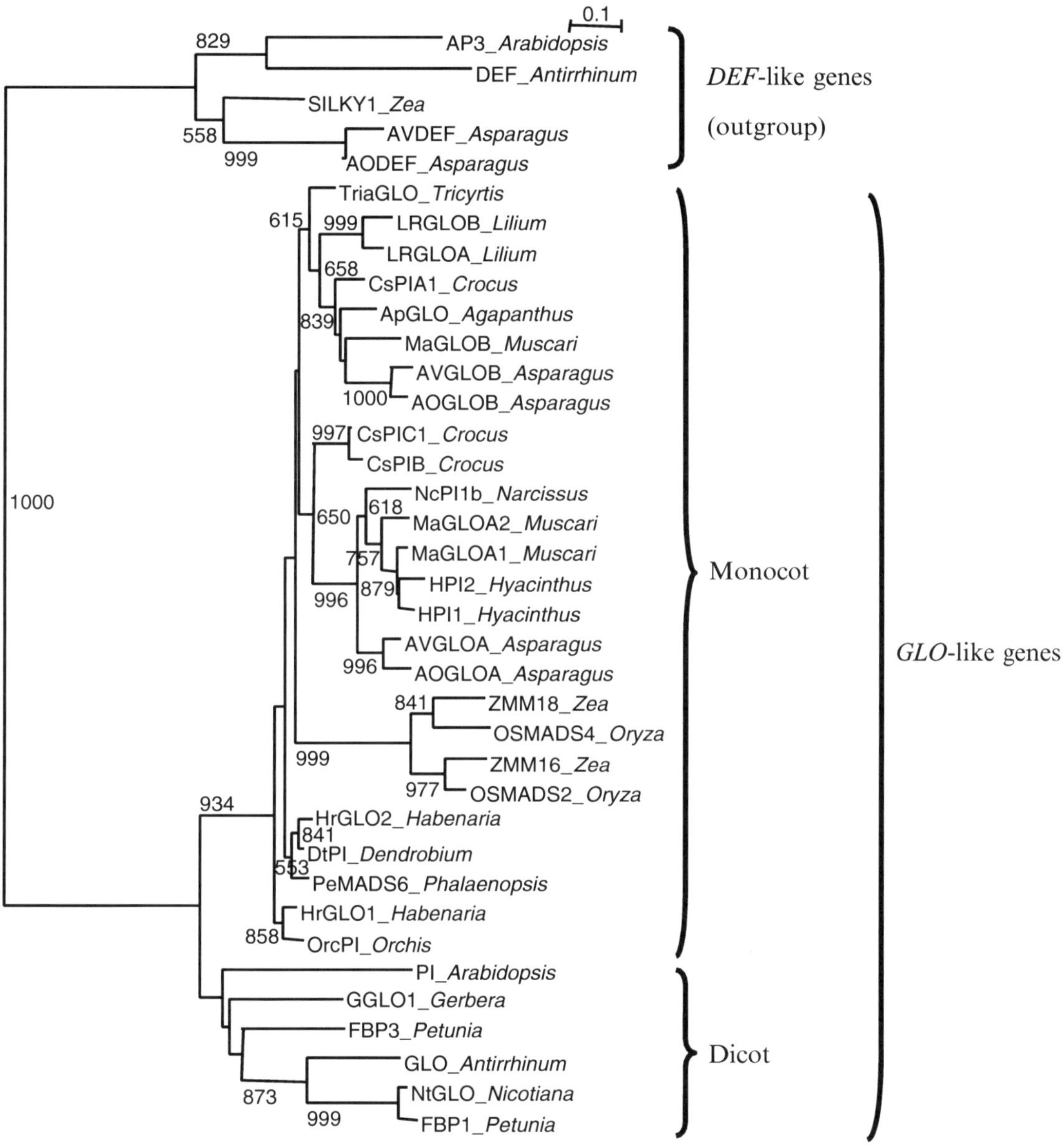

Fig. 3.3 Phylogenetic tree of *GLO*-like genes. *DEF*-like genes were used as outgroup. The *neighbor joining tree* was generated by Clustal W. The *numbers* next to the nodes give bootstrap values from 1,000 replicates, and only values above 500 are shown

female reproductive organs (Yanofsky et al. 1990; Colombo et al. 1995). In situ hybridization analysis showed that the *AVAG1* and *AVAG2* genes are expressed in floral primordia, stamen and carpel primordia, and ovules (Yun et al. 2004a, b). In *Silene latifolia* and *Rumex acetosa*, which are dioecious species, the expression of the *AG*-like genes differs in male and female flowers, and *AG*-like gene expression is lower in depressed organs of male and female plants (Hardenack et al. 1994; Ainsworth et al. 1995; Sather et al. 2005). Although the detailed expression patterns of *AG*-like genes from *A. officinalis* have not been determined, comparative analyses of the expression of *AG*-like genes in dioecious and hermaphrodite

Asparagus species would reveal the relationship between sex differentiation and the *AG*-like gene expression in the genus *Asparagus*.

3.5 Functional Substances Found in *Asparagus*

Most *Asparagus* species are poor sources of biologically active substances. They contain no cyanogenic substrates, alkaloids, proanthocyanidins, ellagic acid, or arbutin (Watson and Dallwitz 1992). Instead, steroidal saponins/saponidins are present in most of the species.

As mentioned previously, *Asparagus* species other than *A. officinalis* are rarely used by humans, but some species have traditionally been used as foods or medicines. For example, tuberous roots of *A. kansuensis* are edible and are used by local peoples (Chen and Tamanian 2000). Tuberous roots of *A. cochinchinensis* and *A. racemosus* are used as medicines, the former in traditional Chinese medicine, and the latter as the medicinal plant for the Indian *Ayurveda* as "*satavari*." A similar usage has been reported for *A. lycopodineus* in Myanmar (H. Yamaji personal communication). These medicinal usages are based on the presence of saponins/saponidins.

Interestingly, some species of *Asparagus* contain phytoecdysteroids, plant compounds that are analogs of insect steroid hormones. Dinan et al. (2001) reported that phytoecdysteroids occur in seven out of 16 species of *Asparagus*. The contents varied from detectable to high concentrations depending on the species. Phytoecdysteroids can act as insect deterrents and may be useful for resistance to insect herbivores.

3.6 Introgression of Agricultural Traits from Wild Asparagus into *Asparagus officinalis*

3.6.1 *Agricultural Traits of Wild* Asparagus

The genus *Asparagus* contains more than 200 species and is distributed in Europe, Asia, and Africa. *A. officinalis* is mainly cultivated all over the world and is the most economically important species in this genus. However, the genetic diversity of this species is relatively low (Stajner et al. 2002b). Wild relative species in the genus *Asparagus* have many agricultural traits that *A. officinalis* does not have, such as salt tolerance, drought tolerance, acid soil tolerance, and disease resistance (Venezia et al. 1993). This is not surprising, as the genus *Asparagus* contains numerous species that are spread all over the world and are adapted against various environments. *A. maritimus* (= *A. scaber*) grows on sandy soils, mainly near the coast (Tutin et al. 1980), and is known as a salt-tolerant species (Venezia et al. 1993; Stajner et al. 2002a). *A. kiusianus* is distributed in Kyushu, Japan, and grows on sandy soil near the seashore like *A. maritimus* (Fig. 3.4). Although no reports have described the analysis of salt tolerance in *A. kiusianus*, this species appears to be highly resistant to salt. Drought-tolerant species include *A. acutifolius*, *A. aphyllus*, *A. albus*, *A. stipuralis*, and *A. scoparius*, and *A. tenuifolius* is an acid soil-resistant species (Bozzini 1959; Venezia et al. 1993; Gonzalez Castanon and Falavigna 2008). These characteristics are likely related to their habitats.

Among the agricultural traits, disease resistance has been extensively analyzed in many *Asparagus* species. *A. officinalis* production is reduced by infection with many diseases. Thus, the introgression of disease resistance into *A. officinalis* is an very important objective in asparagus breeding, and a trial of introgression of disease resistance from wild species was reported as early as 1913 (Norton 1913). Well-known diseases in *A. officinalis* are crown and root rot caused by *Fusarium*, rust caused by *Puccinia asparagi*, purple spot caused by *Stemphylium*, and stem blight caused by *Phomopsis asparagi*. Disease-resistant species of *Asparagus* have been reported by many researchers and are listed in Table 3.1. As far as we know, the first report of a disease-resistant species is on *A. virgatus*, which is resistant to rust caused by *Puccinia asparagi* (Norton 1913). As shown in Table 3.1, many species have resistance characteristics. *A. stipuralis* is resistant to rust caused by *Puccinia asparagi*, but is susceptible to *Fusarium* crown and root rot (Gonzalez Castanon and Falavigna 2008; Stephen and Elmer 1988). *A. setaceus* (= *A. plumosus*) is resistant to *Stemphyllum* leaf spot and rust caused by *Puccinia asparagi*, but is susceptible to *Fusarium* crown and root rot (Kahn et al. 1952; Bansal et al. 1986; Stephen

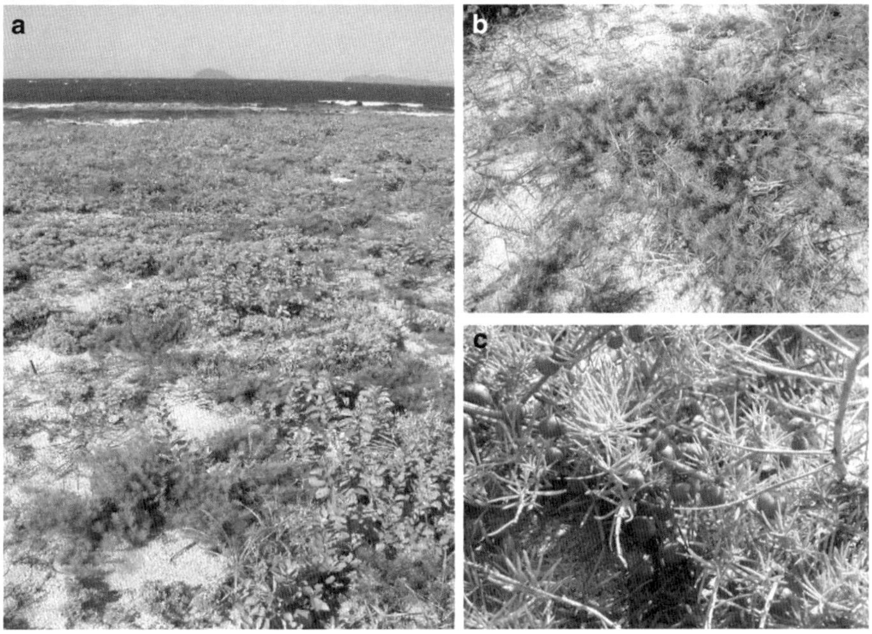

Fig. 3.4 *Asparagus kiusianus*. (**a**) *A. kiusianus* grow on sandy soil near seashore in Kyushu, Japan. (**b**) A plant of *A. kiusianus*. (**c**) Red fruits of *A. kiusianus*

and Elmer 1988). *A. virgatus* is resistant to *Stemphyllum* leaf spot, rust caused by *Puccinia asparagi*, and stem blight caused by *Phomopsis asparagi* (Norton 1913; Kahn et al. 1952; Bansal et al. 1986; Sonoda et al. 2001). *A. densiflorus* is noteworthy because this species is resistant to all four diseases listed in Table 3.1 (Kahn et al. 1952; Thompson and Hepler 1956; Bansal et al. 1986; Stephen et al. 1989; Sonoda et al. 2001). To date, more than ten *Asparagus* species have been identified as disease resistant, but most *Asparagus* species have not yet been analyzed. It is likely that more *Asparagus* species show disease resistance because the genus contains more than 200 species.

3.6.2 Interspecific Hybridization Among Asparagus Species

To introgress agricultural traits of wild *Asparagus* into *A. officinalis*, interspecific hybridization was carried out by hand-pollinated crossings, as listed in Table 3.2. To our knowledge, the first interspecific hybridization between *Asparagus* species was reported by Norton (1913). In this report on the rust resistance of *A. officinalis*, an interspecific hybridization between *A. officinalis* and a rust-resistant species, *A. virgatus*, was tried, although this hybridization was not successful (Norton 1913). Another hybrid between *A. officinalis* and *A. davuricus* appeared to be successful because the progeny showed hybrid characteristics, but the rust-resistance status of *A. davuricus* was not known, and no detailed analysis of this hybrid has been reported (Norton 1913) (Table 3.2).

Later, some research groups reported interspecific crossings among *Asparagus* species, mainly with *A. officinalis*, such as *A. officinalis* × *A. tenuifolius*, *A. officinalis* × *A. brachyphyllus*, *A. officinalis* × *A. maritimus*, *A. officinalis* × *A. acutifolius*, and *A. officinalis* × *A. prostratus* (Bozzini 1962a, b; Ito and Currence 1965; Thevenin 1974; McCollum 1988a). McCollum (1988b) reported interspecific hybridization mainly with *A. densiflorus*, which is disease resistant, as mentioned above. Many of the crosses, such as *A. acutifolius* × *A. densiflorus*, *A. falcatus* × *A. densiflorus*, and *A. virgatus* × *A. densiflorus*, were not successful. However, cross-pollination between *A. officinalis* and *A. oligoclonos* was successful, because they used this hybrid as bridge crossing between *A. officinalis* and *A. densiflorus*

Table 3.1 Disease resistance found in the genus *Asparagus*

Agricultural trait of interest	Species	References
Resistance to *Fusarium*	*A. densiflorus*	Stephen et al. (1989)
	A. myersii (syn. *A. densiflorus*)	Stephen and Elmer (1988)
	A. sprengeri (syn. *A. densiflorus*)	Stephen and Elmer (1988)
Resistance to *Stemphyllum* leaf spot	*A. densiflorus*	Bansal et al. (1986)
	A. acutifolius	Falavigna et al. (2008), Gonzalez Castanon and Falavigna (2008)
	A. aphyllus	Falavigna et al. (2008), Gonzalez Castanon and Falavigna (2008)
	A. albus	Falavigna et al. (2008), Gonzalez Castanon and Falavigna (2008)
	A. stipuralis	Falavigna et al. (2008), Gonzalez Castanon and Falavigna (2008)
	A. asparagoides	Highly resistant, Bansal et al. (1986)
	A. compactus	Highly resistant, Bansal et al. (1986)
	A. larcinus	Highly resistant, Bansal et al. (1986)
	A. verticillatus	Highly resistant, Bansal et al. (1986)
	A. virgatus	Highly resistant, Bansal et al. (1986)
	A. retrofractus	Low levels of infection, Bansal et al. (1986)
	A. macowanii	Low levels of infection, Bansal et al. (1986)
	A. setaceus	Low levels of infection, Bansal et al. (1986)
Resistance to rust (*Puccinia asparagi*)	*A. densiflorus*	Kahn et al. 1952; Thompson and Helper (1956)
	A. maritimus	Falavigna et al. (2008), Gonzalez Castanon and Falavigna (2008)
	A. acutifolius	Gonzalez Castanon and Falavigna (2008)
	A. aphyllus	Gonzalez Castanon and Falavigna (2008)
	A. albus	Gonzalez Castanon and Falavigna (2008)
	A. stipuralis	Gonzalez Castanon and Falavigna (2008)
	A. virgatus	Norton (1913), Kahn et al. (1952)
	A. plumosus (syn. *A. setaceus*)	Kahn et al. (1952)
	A. scandens	Kahn et al. (1952)
Resistance to *Phomopsis asparagi*	*A. densiflorus*	Sonoda et al. (2001)
	A. asparagoides	Sonoda et al. (2001)
	A. virgatus	Sonoda et al. (2001)
	A. macowanii	Sonoda et al. (2001)

(McCollum 1988b). Recently, an Italian group and our group reported that *A. maritimus*, *A. schoberioides*, and *A. kiusianus* were able to be crossed with *A. officinalis* (Ochiai et al. 2002; Ito et al. 2007, 2008), and the crossing of *A. acutifolius* and the hybrid between *A. maritimus* and *A. officinalis* was successful (Falavigna et al. 2008). All of the successful crossings were among dioecious *Asparagus* species, not between dioecious and hermaphrodite species (Table 3.2). In one study, about 33,000 cross-pollinations between *A. officinalis* (dioecious) and *A. densiflorus* (hermaphrodite) were carried out, but these crossings were unsuccessful (McCollum 1988b). Because dioecious and hermaphrodite species of *Asparagus* are not closely related, interspecific hybridization between dioecious and hermaphrodite species is likely to be difficult (Ito et al. 2008).

The production of somatic hybrids by electrofusion between the dioecious *A. officinalis* and the hermaphrodite *A. macowanii* has been reported by Kunitake et al. (1996). Because *A. macowanii* is resistant to *Stemphylium* leaf spot (Bansal et al. 1986) and *Phomopsis asparagi* stem blight (Sonoda et al. 2001), these resistance traits might be incorporated into *A. officinalis* by interspecific hybridization. However, the obtained somatic embryos showed abnormal germination, and it was difficult to transfer them to soil (Kunitake et al. 1996; Kunitake and Mii 1998). The interspecific somatic hybrids had a chromosome number of $2n = 50$, although both of the parents had

Table 3.2 List of successful and unsuccessful hand-pollinated crosses among *Asparagus* species

Successful cross	Unsuccessful cross	References
A. officinalis × *A. davuricus*	*A. officinalis* × *A. virgatus*	Norton (1913)
	A. officinalis × *A. densiflorus* (=*A. sprengeri*) *A. officinalis* × *A. plumosus* (=*A. setaceus*) *A. officinalis* × *A. virgatus* *A. officinalis* × *A. scandens*	Kahn et al. (1952)
A. officinalis × *A. tenuifolius*		Bozzini (1962a)
A. tenuifolius × *A. officinalis* *A. davuricus* × *A. scaber* (=*A. maritimus*) *A. acutifolius* × *A. aphyllus*	*A. officinalis* × *A. retroflactus* *A. virgatus* × *A. officinalis* *A. densiflorus* × *A. officinalis* *A. tenuifolius* × *A. scaber* (=*A. maritimus*)	Bozzini (1962b)
A. officinalis × *A. brachyphyllus*		Ito and Currence (1965)
A. officinalis × *A. scaber* (=*A. maritimus*) *A. tenuifolius* × *A. officinalis* *A. acutifolius* × *A. officinallis*		Thevenin (1974)
A. officinalis × *A. prostratus*		McCollum (1988a)
A. officinalis × *A. oligoclonos*	*A. officinalis* × *A. densiflorus* *A. prostratus* × *A. densiflorus* (*A. officinalis* × *A. prostratus*) × *A. densiflorus* *A. acutifolius* × *A. densiflorus* *A. aphyllus* × *A. densiflorus* *A. arborescens* × *A. densiflorus* *A. falcatus* × *A. densiflorus* *A. laricinus* × *A. densiflorus* *A. oligoclonos* × *A. densiflorus* (*A. officinalis* × *A. oligoclonos*) × *A. densiflorus* *A. pastorianus* × *A. densiflorus* *A. setaceus* × *A. densiflorus* *A. virgatus* × *A. densiflorus*	McCollum (1988b)
	A. officinalis × *A. densiflorus*	Marcellán and Camadro (1996, 1999)
A. officinalis × *A. schoberioides*		Ochiai et al. (2002)
A. officinalis × *A. schoberioides*		Ito et al. (2007)
A. officinalis × *A. maritimus* (*A. officinalis* × *A. maritimus*) × *A. acutifolius*	*A. officinalis* × *A. acutifolius* *A. maritimus* × *A. acutifolius* *A. albus* × *A. officinalis* *A. officinalis* × *A. stipularis*	Falavigna et al. (2008)
A. offcinalis × *A. kiusianus*	*A. officinalis* × *A. densiflorus* *A. officinalis* × *A. virgatus* *A. officinalis* × *A. asparagoides* *A. officinalis* × *A. cochinchinensis* *A. officinalis* × *A. verticillatus*	Ito et al. (2008)

Fig. 3.5 Interspecific hybrid between *Asparagus schoberioides* and *A. officinalis*. (**a**) *A. schoberioides*. (**b**) Interspecific hybrid. (**c**) *A. officinalis*

$2n = 20$, suggesting that the abnormal growth of the somatic hybrid may be partly due to an aberrant chromosome number (Kunitake et al. 1996; Kunitake and Mii 1998).

3.6.3 Edible Wild Asparagus

Asparagus is one of the most important vegetables grown in the world. Edible asparagus is mainly of the species *A. officinalis*. In addition to this species, several *Asparagus* species have been reported to be edible. Young shoots of *A. acutifolius* are eaten in omelets, stewed, or boiled and consumed alone or with scrambled eggs and fresh cheese in Italy, Spain, Cyprus, and Turkey (Bonnet and Valles 2002; Ertug 2004; Pieroni et al. 2005; Tardio et al. 2006; Della et al. 2006). *A. acutifolius* has been introduced as an interesting niche crop for marginal areas in Europe, and a cultivation technique has been reported (Benincasa et al. 2007). The young shoots of *A. aphyllus*, *A. albus*, and *A. stipuralis* are also harvested and eaten in omelets or with scrambled eggs in Spain, although the geographical distributions of these species are limited (Tardio et al. 2006). Outside Europe, shoots and roots of *A. flagellaris* and *A. africanus*, shoots of *A. suaveolens*, and fruits of *A. racemosus* are eaten in sub-Saharan Africa (Peters et al. 1992), and young *A. wildemanii* shoots arising below the soil surface (like white asparagus) are boiled and served with butter or used in soups, omelets, and other dishes in Zimbabwe (Tredgold et al. 1986). The parts of these *Asparagus* species that are eaten are mainly young shoots, and various species are eaten in diverse areas. It is likely that many other wild species of *Asparagus* are eaten around the world.

3.6.4 Possibility of Asparagus Breeding Using Wild Species

Asparagus (*A. officinalis*) is cultivated throughout the world. Although this species includes many cultivars, the gene pool of the species is relatively limited (Stajner et al. 2002b). Nevertheless, the genus *Asparagus* contains more than 200 species with the many agricultural traits mentioned in Sects. 3.5 and 3.6.1. One of these species, *A. maritimus*, shows resistance to *P. asparagi*, and this resistance trait has been inherited by an interspecific hybrid between *A. maritimus* and *A. officinalis* (Falavigna et al. 2008). It is expected that the agricultural traits of wild asparagus will be introgressed into *A. officinalis* by continuing interspecific crossing using hand pollination. Meanwhile, the molecular mechanisms of the disease resistance traits will be clarified, and the gene(s) responsible for disease resistance will be transferred into *A. officinalis* by genetic transformation. In addition, the young shoots of several wild asparagus species are edible, such as those of *A. acutifolius* and *A. stipuralis*. Although a comparative analysis of taste among *Asparagus* species has not been reported, asparagus breeding using these wild *Asparagus* species may produce interspecific cultivars with new tastes.

Appendix: A List of *Asparagus* Species in the World

Species	Distribution	Chromosome nos.	Remarks
Asparagus acerosus Thunb. ex Schult. & Schult. f.	South Africa		Status uncertain
Asparagus acicularis F.T. Wang & S.C. Chen	China		
Asparagus acocksii Jessop	South Africa		*Protasparagus acocksii* (Jessop) Oberm.
Asparagus acutifolius L.	Europe, North Africa, West Asia	$2n = 20$	
Asparagus adscendens Roxb.	India, Pakistan	$n = 10, 2n = 20$	
Asparagus aethiopicus L.	Tropical Africa, South Africa		*Protasparagus aethiopicus* (L.) Oberm.
Asparagus africanus Lam.	Africa, Arabian Peninsula	$2n = 20$	
Asparagus aggregatus (Oberm.) Fellingham & N.L. Mey.	South Africa		
Asparagus albus L.	Europe, North Africa	$2n = 20$	
Asparagus alopecurus (Schltr. Ex Oberm.) S.T. Malcomber & Sebsebe Demissew	South Africa		*Myrsiphyllum alopecurum* Schltr. Ex Oberm.
Asparagus altiscandens Engl. & Gilg	Tropical Africa		
Asparagus altissimus Munby	Africa		
Asparagus angulofractus Iljin	China (SW Xinjiang), Kazakhstan		
Asparagus angusticladus (Jessop) J.-P. Lebrun & Stork	Tropical Africa, South Africa		*Protasparagus angusticladus* (Jessop) Oberm.
Asparagus aphyllus L.	Europe, North Africa, West Asia	$2n = 20$	
Asparagus arborescens Willd.	Canary	$2n = 20$	
Asparagus aridicola Sebsebe	Tropical Africa		
Asparagus asiaticus L.	India		
Asparagus asparagoides (L.) Druce	Africa	$2n = 20$	
Asparagus aspergillus Jessop	Tropical Africa, South Africa		*Protasparagus aspergillus* (Jessop) Oberm.
Asparagus baguirmiensis A. Chev.	Tropical Africa		nom. nud.
Asparagus baumii Engl. & Gilg	Tropical Africa		
Asparagus bayeri (Oberm.) Fellingham & N.L. Mey.	South Africa		*Protasparagus bayeri* Oberm
Asparagus bechuanicus Baker	South Africa		
Asparagus benguellensis Baker	Tropical Africa		
Asparagus bequaertii De Wild.	Tropical Africa		
Asparagus biflorus (Oberm.) Fellingham & N.L. Mey.	South Africa		*Protasparagus biflorus* Oberm
Asparagus botschantzevii Vlassova	Middle Asia		
Asparagus brachiatus Thulin	Tropical Africa		
Asparagus brachyphyllus Turcz.	China, Kazakhstan, Korea, Mongolia, Tajikistan, Turkmenistan, Uzbekistan		
Asparagus breslerianus Schult. f.	China, Kazakhstan, Mongolia, Russia, Turkmenistan, Uzbekistan; SW Asia	$2n = 20$	
Asparagus buchananii Baker	Tropical Africa, South Africa		*Protasparagus buchananii* (Baker) Oberm.
Asparagus bucharicus Iljin	Tajikistan		
Asparagus burchellii Baker	South Africa		*Protasparagus burchellii* (Baker) Oberm.
Asparagus burjaticus Peshkova	Russia (Siberia)		
Asparagus buruensis Engl.	Tropical Africa		
Asparagus calcicolus H. Perrier	Madagascar		
Asparagus capensis L.	South Africa		*Protasparagus capensis* (L.) Oberm.
Asparagus caspius Schult. & Schult. f.			
Asparagus clareae (Oberm.) Fellingham & N.L. Mey.	South Africa		*Protasparagus clareae* Oberm.
Asparagus cochinchinensis (Lour.) Merr.	China, Japan, Korea, Laos, Vietnam, Philippines	$2n = 20$	

(*continued*)

Species	Distribution	Chromosome nos.	Remarks
Asparagus coddii (Oberm.) Fellingham & N.L. Mey.	South Africa		*Protasparagus coddii* Oberm.
Asparagus concinnus Kies	South Africa		*Protasparagus concinnus* (Baker) Oberm. & Immelman
Asparagus confertus K. Krause	South Africa		*Protasparagus confertus* (K.Krause) Oberm.
Asparagus consanguineus (Kunth) Baker	Madagascar		
Asparagus cooperi Baker	South Africa	$2n = 40$	*Protasparagus cooperi* (Baker) Oberm.
Asparagus courtetii A. Chev.	Tropical Africa		nom. nud.
Asparagus crassicladus Jessop	South Africa		*Protasparagus crassicladus* (Jessop) Oberm.
Asparagus dauricus Link	China, Korea, Mongolia, Russia (Far East, Siberia)	$2n = 18$	
Asparagus dauricus var. elongatus Pamp.	China (Hubei)		
Asparagus debilis A. Chev.	South Africa		nom. nud.
Asparagus declinatus L.	South Africa		*Myrsiphyllum declinatum* (L.) Oberm.
Asparagus deflexus Baker	Tropical Africa		
Asparagus densiflorus (Kunth) Jessop	Tropical Africa, South Africa	$n = 20$, $2n = 40$	*Protasparagus densiflorus* (Kunth) Oberm.
Asparagus denudatus (Kunth) Baker	South Africa		*Protasparagus denudatus* (Kunth) Oberm.
Asparagus denudatus subsp. nudicaulis Sebsebe	Tropical Africa		
Asparagus devenishii (Oberm.) Fellingham & N.L. Mey.	South Africa		*Protasparagus devenishii* Oberm.
Asparagus divaricatus (Oberm.) Fellingham & N.L. Mey.	South Africa		*Protasparagus divaricatus* Oberm.
Asparagus drepanophyllus Welw. & Baker	Tropical Africa		
Asparagus duchesnei L. Linden	Tropical Africa		
Asparagus dumosus Baker	Pakistan (Sindh)	$n = 30$	
Asparagus edulis (Oberm.) J.-P. Lebrun & Stork	Tropical Africa, South Africa		*Protasparagus edulis* Oberm.
Asparagus elephantinus S.M. Burrows	South Africa		
Asparagus equisetoides Welw. ex Baker	Tropical Africa		
Asparagus exsertus (Oberm.) Fellingham & N.L. Mey.	South Africa		*Protasparagus exsertus* Oberm.
Asparagus exuvialis Burch.	South Africa, Namibia		*Protasparagus exuvialis* (Burch.) Oberm.
Asparagus falcatus L.	Tropical Africa, South Africa, Comoros, Madagascar, Yemen, Sri Lanka	$2n = 20, 40$	*Protasparagus falcatus* (L.) Oberm.
Asparagus fallax Svent.	Canary	$2n = 20$	
Asparagus fasciculatus Thunb.	South Africa		*Myrsiphyllum fasciculatum* (Thunb.) Oberm.
Asparagus faulkneri Sebsebe	Tropical Africa		
Asparagus ferganensis Vved.	Russia		
Asparagus filicinus Buch.-Ham. ex D. Don	China, Bhutan, Nepal, India, Myanmar, Thailand	$2n = 16,18,20$	
Asparagus filicladus (Oberm.) Fellingham & N.L. Mey.	South Africa		*Protasparagus filicladus* Oberm.
Asparagus flagellaris (Kunth) Baker	Tropical Africa		
Asparagus flavicaulis (Oberm.) Fellingham & N.L. Mey.	South Africa		
Asparagus fourei (Oberm.) Fellingham & N.L. Mey.	South Africa		*Protasparagus fouriei* Oberm.
Asparagus fractiflexus (Oberm.) Fellingham & N.L. Mey.	South Africa		*Protasparagus fractiflexus* Oberm.
Asparagus glaucus Kies	South Africa		*Protasparagus glaucus* (Kies) Oberm.

(*continued*)

Species	Distribution	Chromosome nos.	Remarks
Asparagus gobicus Ivan. ex Grubov	China, Mongolia		
Asparagus gonocladus Baker	India, Sri Lanka	$n = 10{,}30$, $2n = 60$	
Asparagus graniticus (Oberm.) Fellingham & N.L. Mey.	South Africa		*Protasparagus graniticus* Oberm.
Asparagus greveanus H. Perrier	Madagascar		
Asparagus griffithii Baker	Afghanistan		
Asparagus gypsaceus Vved.	Central Asia		
Asparagus hirsutus S.M. Burrows	South Africa		
Asparagus homblei De Wild.	Tropical Africa		
Asparagus humilis Engl.	Tropical Africa		
Asparagus intricatus (Oberm.) Fellingham & N.L. Mey.	South Africa		*Protasparagus intricatus* Oberm.
Asparagus juniperoides Engl.	South Africa		*Myrsiphyllum juniperoides* (Engl.) Oberm.
Asparagus kaessneri De Wild.	Tropical Africa		
Asparagus kansuensis F.T. Wang & T. Tang	China (S Gansu)		
Asparagus kasakstanicus Iljin	Kazakhstan		
Asparagus katangensis De Wild. & T. Durand	Tropical Africa		
Asparagus kraussianus J.F. Macbr.	South Africa		*Myrsiphyllum kraussianum* Kunth
Asparagus krebsianus (Kunth) Jessop	Tropical Africa, South Africa		*Protasparagus krebsianus* (Kunth) Oberm.
Asparagus laevissimus Steud.	India	$n = 10{,}20$, $2n = 40$	
Asparagus laricinus Burch.	Tropical Africa, South Africa	$2n = 20$	*Protasparagus laricinus* (Burch.) Oberm.
Asparagus lecardii De Wild.	Tropical Africa		
Asparagus ledebourii Mishchenko	South Caucasus		
Asparagus leptocladodius Chiov.	Tropical Africa		
Asparagus leptophyllus Schischk.	Armenia		
Asparagus levinae Klokov	Ukraine		
Asparagus lignosus Burm. f.	South Africa		
Asparagus litoralis Steven	Ukraine, Kazakhstan		
Asparagus longicladus N.E. Br.	Tropical Africa, South Africa		*Protasparagus longicladus* (N.E.Br.) B. Mathew
Asparagus longiflorus Franch.	China		
Asparagus longipes Baker	Tropical Africa		
Asparagus lujae De Wild.	Tropical Africa		
Asparagus lycopodineus (Baker) F.T. Wang & T. Tang	China, Bhutan, India, Myanmar		
Asparagus lynetteae (Oberm.) Fellingham & N.L. Mey.	South Africa		*Protasparagus lynetteae* Oberm.
Asparagus macowanii Baker	Tropical Africa, South Africa, Madagascar		*Protasparagus macowanii* (Baker) Oberm.
Asparagus madecassus H. Perrier	Madagascar		
Asparagus mahafalensis H. Perrier	Madagascar		
Asparagus mairei H. Lev.	China (Yunnan (Kunming Shi))		
Asparagus mariae (Oberm.) Fellingham & N.L. Mey.	South Africa		*Protasparagus mariae* Oberm.
Asparagus maritimus (L.) Mill.	Europe, North Africa		
Asparagus martretii A. Chev.	Tropical Africa		nom. nud.
Asparagus meioclados H. Lev.	China (Guizhou, Sichuan, Yunnan)		
Asparagus merkeri K. Krause	Tropical Africa		
Asparagus microraphis (Kunth) Baker	South Africa		*Protasparagus microrhaphis* (Kunth) Oberm.
Asparagus migeodii Sebsebe	Tropical Africa		
Asparagus minutiflorus (Kunth) Baker	Tropical Africa, South Africa		*Protasparagus minutiflorus* (Kunth) Oberm.
Asparagus mollis (Oberm.) Fellingham & N.L. Mey.	South Africa		*Protasparagus mollis* Oberm.

(*continued*)

Species	Distribution	Chromosome nos.	Remarks
Asparagus monophyllus Baker	Pakistan, Afghanistan, Iran		
Asparagus mozambicus Kunth	Tropical Africa		
Asparagus mucronatus Jessop	South Africa		*Protasparagus mucronatus* (Jessop) Oberm.
Asparagus multiflorus Baker	South Africa		*Protasparagus multiflorus* (Baker) Oberm.
Asparagus multituberosus R.A. Dyer	South Africa		*Myrsiphyllum multituberosum* (R.A. Dyer) Oberm.
Asparagus munitus F.T. Wang & S.C. Chen	China (SW Sichuan, Yunnan)		
Asparagus myriacanthus F.T. Wang & S.C. Chen	China (SE Xizang, NW Yunnan)		
Asparagus natalensis (Baker) J.-P. Lebrun & Stork	Tropical Africa, South Africa		*Protasparagus natalensis* (Baker) Oberm. 3.
Asparagus ndelleensis A. Chev.	Tropical Africa		nom. nud.
Asparagus neglectus Kar. & Kir.	China (N Xinjiang), Afghanistan, Kazakhstan, Mongolia, Pakistan, Russia (E Siberia), Tajikistan, Turkmenistan, Uzbekistan		May be synonym of Asparagus trichophyllus
Asparagus nelsii Schinz	Tropical Africa, South Africa		*Protasparagus nelsii* (Schinz) Oberm.
Asparagus nesiotes Svent. Subsp. nesiotes	Madeire	$n = 30$	
Asparagus nesiotes subsp. purpuriensis A.Marrero & A.Ramos	Canary	$2n = 60$	
Asparagus nodulosus (Oberm.) J.-P. Lebrun & Stork	Tropical Africa, South Africa		*Protasparagus nodulosus* Oberm.
Asparagus officinalis L.	Europe, North Africa, West Asia, Russia, Kazakhstan, Mongolia	$n = 10$, $2n = 20,40$	
Asparagus oligoclonos Maxim.	China, Japan, Korea, Mongolia, Russia (Far East, Siberia)	$2n = 40$	
Asparagus oligophyllus Baker	Asia?		
Asparagus oliveri Fellingham & N.L. Mey.	South Africa		*Protasparagus oliveri* Oberm.
Asparagus ovatus T.M. Salter	South Africa		*Myrsiphyllum ovatum* (Salter) Oberm.
Asparagus oxyacanthus Baker	South Africa		*Protasparagus oxyacanthus* (Baker) Oberm.
Asparagus pachyrrhizus Ivanova ex Vlassova	Turkmenistan		
Asparagus parviflorus Turcz.	Russia (Siberia)		
Asparagus pastorianus Webb & Berthel.	North Africa, Macaronesia	$2n = 40$	
Asparagus pearsonii Kies	South Africa		*Protasparagus pearsonii* (Kies) Oberm.
Asparagus pendulus (Oberm.) J.-P. Lebrun & Stork	South Africa		*Protasparagus pendulus* Oberm.
Asparagus persicus Baker	Russia	$2n = 20,40,60$	
Asparagus petersianus Kunth	Tropical Africa		
Asparagus planiusculus Burm. f.	South Africa		
Asparagus plocamoides Webb ex Svent.	Canary	$2n = 20$	
Asparagus poissonii H. Perrier	Madagascar		
Asparagus popovii Iljin	Turkmenistan		
Asparagus przewalskyi N.A. Ivanova ex Grubov & T.V. Egorova	China (Qinghai)		
Asparagus pseudoscaber Grec.	Europe		
Asparagus psilurus Welw. ex Baker	Tropical Africa		
Asparagus racemosus Willd.	China (S Xizang), Bhutan, India, Malaysia, Myanmar, Nepal, Pakistan, Sikkim; Africa, Madagascar, Australia	$n = 10$, $2n = 20$	
Asparagus ramosissimus Baker	South Africa, Madagascar		*Myrsiphyllum ramosissimum* (Baker) Oberm.
Asparagus recurvispinus (Oberm.) Fellingham & N.L. Mey.	South Africa		*Protasparagus recurvispinus* Oberm.

(*continued*)

Species	Distribution	Chromosome nos.	Remarks
Asparagus retrofractus L.	South Africa		*Protasparagus retrofractus* (L.) Oberm.
Asparagus rigidus Jessop	South Africa		*Protasparagus rigidus* (Jessop) Oberm.
Asparagus ritschardii De Wild.	Tropical Africa		
Asparagus rogersii R.E. Fr.	Tropical Africa		
Asparagus rubicundus P.J. Bergius	South Africa		*Protasparagus rubicundus* (P.J. Bergius) Oberm.
Asparagus sapinii De Wild.	Tropical Africa		
Asparagus scaberulus A. Rich.	Tropical Africa		
Asparagus scandens Thunb.	South Africa		*Myrsiphyllum scandens* (Thunb.) Oberm.
Asparagus schoberioides Kunth	China, Japan, Korea, Mongolia, Russia (Far East, Kurile Islands, Sakhalin, Siberia)	$2n = 20$	
Asparagus schroederi Engl.	Tropical Africa, South Africa		*Protasparagus schroederi* (Engl.) Oberm.
Asparagus schumanianus Schltr. ex H. Perrier	Africa, Madagascar		nom. nud.
Asparagus scoparius Lowe	North Africa (Morocco)	$n = 10$, $2n = 20$	
Asparagus sekukuniensis (Oberm.) Fellingham & N.L. Mey.	South Africa		*Protasparagus sekukuniensis* Oberm.
Asparagus setaceus (Kunth) Jessop	Africa		
Asparagus setiformis Kryl.	Russia (Siberia)		
Asparagus sichuanicus S.C. Chen & D.Q. Liu	China (Sichuan, Xizang)		
Asparagus simulans Baker	Madagascar		
Asparagus spinescens Steud. ex Roem. & Schult.	South Africa		*Protasparagus spinescens* (Steud. ex Roem. & Schult.) Oberm.
Asparagus squarrosus J.A. Schmidt	Tropical Africa		
Asparagus stellatus Baker	South Africa		*Protasparagus stellatus* (Baker) Oberm.
Asparagus stipulaceus Lam.	Tropical Africa, South Africa		*Protasparagus stipulaceus* (Lam.) Oberm.
Asparagus stipularis Forssk.	Europe, Libya, Saudi Arabia, North Africa	$2n = 20$	
Asparagus striatus (L.f.) Thunb.	South Africa		*Protasparagus striatus* (L. f.) Oberm.
Asparagus striatus var. linearifolius Baker	South Africa		
Asparagus suaveolens Burch.	Tropical Africa, South Africa		*Protasparagus suaveolens* (Burch.) Oberm.
Asparagus subfalcatus De Wild.	Tropical Africa		
Asparagus subscandens F.T. Wang & S.C. Chen	China (S Yunnan)		
Asparagus subulatus Thunb.	Tropical Africa, South Africa		*Protasparagus subulatus* (Thunb.) Oberm.
Asparagus sylvicola S.M. Burrows	Tropical Africa		
Asparagus taliensis F.T. Wang & T. Tang ex S.C. Chen	China (Yunnan)		
Asparagus tamariscinus Ivanova ex Grub.	Central Asia		
Asparagus ternifolius Hook. f.	Europe		
Asparagus tibeticus F.T. Wang & S.C. Chen	China (Xizang)		
Asparagus transvaalensis (Oberm.) Fellingham & N.L. Mey.	South Africa		*Protasparagus transvaalensis* Oberm.
Asparagus trichoclados (F.T. Wang & T. Tang) F.T. Wang & S.C. Chen	China (C Yunnan)		
Asparagus trichophyllus Bunge	China, Mongolia, Russia (E Siberia)	$2n = 20$	
Asparagus turkestanicus Popov	Turkmenistan		
Asparagus uhligii K. Krause	Tropical Africa		
Asparagus umbellatus Link subsp. umbellatus	Canary		
Asparagus umbellatus Link subsp. lowei (Kunth) Valdes	Madeire		
Asparagus umbellulatus Sieber ex Baker	Mauritius	$2n = 20$	
Asparagus undulatus (L.f.) Thunb.	South Africa		*Myrsiphyllum undulatum* Oberm.

(*continued*)

Species	Distribution	Chromosome nos.	Remarks
Asparagus usambarensis Sebsebe	Tropical Africa		
Asparagus vaginellatus Bojer ex Baker	Madagascar		
Asparagus verticillatus L.	Europe, North Africa	$2n = 20$	
Asparagus virgatus Baker	Tropical Africa, South Africa	$n = 20, 2n = 40$	*Protasparagus virgatus* (Baker) Oberm.
Asparagus volubilis Thunb.	South Africa		*Myrsiphyllum volubile* (Thunb.) Oberm.
Asparagus vvedenskyi Botsch.	Central Asia		
Asparagus warneckei (Engl.) Hutch.	Tropical Africa		
Asparagus yanbianensis S.C. Chen	China (SW Sichuan)		
Asparagus yanyuanensis S.C. Chen	China (SW Sichuan)		
Asparagus zeylanicus (Baker) Hook. f.	Sri Lanka		

Acknowledgments We wish to thank Dr. A. Uragami and Dr. T. Sonoda for their useful advices and information. We also thank Mr. I. Konno for the pictures of the *Asparagus* flowers.

References

African Flowering Plants Database (2008) Conservatoire et Jardin botaniques de la Villle de Genève and South African National Biodiversity Institute, Pretoria. http://www.ville-ge.ch/musinfo/bd/cjb/africa/. Accessed 31 Oct 2008

Ainsworth C, Crossley S, Buchanan Wollaston V, Thangavelu M, Parker J (1995) Male and female flowers of the dioecious plant sorrel show different patterns of MADS box gene expression. Plant Cell 7:1583–1598

Angiosperm Phylogeny Group (2003) An update of the Angiosperm Phylogeny Group classification for the orders and families of flowering plants: APG II. Bot J Linn Soc 141: 399–436

Asada Y, Kasai N, Adachi Y, Kanno A, Ito N, Yun PY, Masuda K (2006) A vegetative line of asparagus (*Asparagus officinalis*) with a homeotic change in flower development is correlated with a functional deficiency in class-B MADS-box genes. J Hortic Sci Biotechnol 81:874–882

Bansal RK, Menzies SA, Broadhurst PG (1986) Screening of *Asparagus* species for resistance to *Stemphylium* leaf spot. NZ J Agric Res 29:539–545

Benincasa P, Tei F, Rosati A (2007) Plant density and genotype effects on wild asparagus (*Asparagus acutifolius* L.) spear yield and quality. Hortic Sci 42:1163–1166

Bonnet MA, Valles J (2002) Use of non-crop food vascular plants in Montseny biosphere reserve (Catalonia, Iberian Peninsula). Int J Food Sci Nutr 53:225–248

Bozzini A (1959) Revisione cito-sistematica del genere *Asparagus* I: Le specie di Asparagus della flora Italiana e chiave analítica per la loro determinazione. Caryologia 12: 199–264

Bozzini A (1962a) Interspecific hybridization and experimental mutagenesis in the genetic improvement of asparagus. Genet Agrar 16:212–218

Bozzini A (1962b) Genetic affinity between Asparagus spp. (Affinità genetica tra specie di Asparago). Atti Assoc Genet Ital 7:277–278

Burrows SM, Burrows JE (2008) Three new species of *Asparagus* (Asparagaceae) from South Africa, with notes on other taxa. Bothalia 38:23–29

Caporali E, Carboni A, Galli MG, Rossi G, Spada A, Marziani Longo GP (1994) Development of male and female flower in *Asparagus officinalis* Search for point of transition from hermaphroditic to unisexual developmental pathway. Sex Plant Reprod 7:239–249

Chase MW, Soltis DE, Soltis PS, Rudall PJ, Fay MF, Hahn WH, Sullivan S, Joseph J, Molvray M, Kores PJ, Givnish TJ, Sytsma KJ, Pires JC (2000a) Higher-level systematics of the monocotyledons: an assessment of current knowledge and a new classification. In: Wilson KL, Morrison DA (eds) Monocots: systematics and evolution. CSIRO, Collingwood, Australia, pp 3–16

Chase MW, de Bruijn AY, Cox AV, Reeves G, Rudall PJ, Johnson MAT, Eguiarte LE (2000b) Phylogenetics of Asphodelaceae (Asparagales): an analysis of plastid *rbc*L and *trn*L-F DNA sequences. Ann Bot 86:935–951

Chen XQ, Tamanian KG (2000) Asparagus. In: Wu ZY, Raven PH (eds) Flora of China, vol 24 (Flagellariaceae through Marantaceae). Science and MO Botanical Garden, Beijing, St. Louis, MO, USA, pp 209–216

Clifford HT, Conran JG (1987) Asparagaceae. In: George AS (ed) Flora of Australia. Australian Government of Public Service, Canberra, Australia, pp 159–164

Coen ES, Meyerowitz EM (1991) The war of the whorl: genetic interactions controlling flower development. Nature 353: 31–37

Colombo L, Franken J, Koetje E, van Went J, Dons HJ, Angenent GC, van Tunen AJ (1995) The petunia MADS-box gene *FBP11* determines ovule identity. Plant Cell 7: 1859–1868

Cooney-Sovetts C, Sattler R (1987) Phylloclade development in the Asparagaceae: an example of homeosis. Bot J Linn Soc 94:327–371

Della A, Paraskeva-Hadjichambi D, Hadjichambis AC (2006) An ethnobotanical survey of wild edible plants of Paphos and Larnaca countryside of Cyprus. J Ethnobiol Ethnomed 2:34

Dinan L, Savchenko T, Whiting P (2001) Phytoecdysteroids in the genus *Asparagus* (Asparagaceae). Phytochemistry 56: 569–576

Dahlgren RMT, Clifford HT, Yeo PF (1985) The Families of the Monocotyledons. Springer, Berlin

Ertug F (2004) Wild edible plants of the Bodrum area (Mugla, Turkey). Turk J Bot 28:161–174

Falavigna A, Alberti P, Casali PE, Toppino L, Huaisong W, Mennella G (2008) Interspecific hybridization for asparagus breeding in Italy. Acta Hortic 776:291–297

Fay MF, Rudall PJ, Sullivan S, Stobart KL, de Bruijn AY, Reeves G, Qamaruz-Zaman F, Hong WP, Joseph J, Hahn WJ, Conran JG, Chase MW (2000) Phylogenetic studies of Asparagales based on four plastid DNA regions. In: Wilson KL, Morrison DA (eds) Monocots: systematics and evolution. CSIRO, Collingwood, Australia, pp 360–371

Forest F, Grenyer R, Rouget M, Jonathan Davies T, Cowling RM, Faith DP, Balmford A, Manning JC, Proches S, van der Bank M, Reeves G, Hedderson TAJ, Savolainen V (2007) Preserving the evolutionary potential of floras in biodiversity hotspots. Nature 445:757–760

Fukuda T, Ashizawa H, Suzuki R, Ochiai T, Nakamura T, Kanno A, Kameya T, Yokoyama J (2005) Molecular phylogeny of the genus *Asparagus* (Asparagaceae) inferred from plastid *petB* intron and *petD-rpoA* intergenic spacer sequences. Plant Species Biol 20:121–132

Ge CJ, Li YK, Wan P, Li YX, Jiang FH (1988) Observations on the chromosome numbers of medicinal plants from Shandong Province (V). J Shandong Coll Tradition Chin Med 12:55–57

Goldblatt P, Savolainen V, Porteous O, Sostaric I, Powell M, Reeves G, Manning JC, Barraclough TG, Chase MW (2002) Radiation in the Cape flora and the phylogeny of peacock irises *Moraea* (Iridaceae) based on four plastid DNA regions. Mol Phylogenet Evol 25:341–360

Gonzalez Castanon ML, Falavigna A (2008) Asparagus germplasm and interspecific hybridization. Acta Hortic 776: 319–326

Grubb PJ (2003) Interpreting some outstanding features of the flora and vegetation of Madgascar. Perspect Plant Ecol 6: 125–146

Hardenack S, Ye D, Saedler H, Grant S (1994) Comparison of MADS box gene expression in developing male and female flowers of the dioecious plant white campion. Plant Cell 6: 1775–1787

Hernandez SL (1995) Taxonomic study of the Mexican genus *Hemiphylacus* (Hyacinthaceae). Syst Bot 20:546–554

Hirai M, Kamimura T, Kanno A (2007) The expression patterns of three class B genes in distinctive two whorls of petaloid tepals in *Alstroemeria ligtu*. Plant Cell Physiol 48:310–321

Ito T (2006) Genetic variation in genus Asparagus: Cross compatibility and flower development. Ph.D thesis, Tohoku University.

Ito PJ, Currence TM (1965) Inbreeding and heterosis in asparagus. Am Soc Hortic Sci 86:338–346

Ito T, Ochiai T, Ashizawa H, Shimodate T, Sonoda T, Fukuda T, Yokoyama J, Kameya T, Kanno A (2007) Production and analysis of reciprocal hybrids between *Asparagus officinalis* L. and *A. schoberioides* Kunth. Genet Resour Crop Evol 54: 1063–1071

Ito T, Ochiai T, Fukuda T, Ashizawa H, Sonoda T, Kameya T, Kanno A (2008) Potential of interspecific hybrids in Asparagaceae. Acta Hortic 776:279–284

Janssen T, Bremer K (2004) The age of major monocot groups inferred from 800+ *rbcL* sequences. Bot J Linn Soc 146:385–398

Janssen T, Bystriakova N, Rakotondrainibe F, Coomes D, Labat JN, Schneider H (2008) Neoendemism in Madagascan scaly tree ferns results from recent, coincident diversification bursts. Evolution 62:1876–1889

Kahn RP, Anderson HW, Hepler PR, Linn MB (1952) An investigation of asparagus rust in Illinois. University of Illinois Agricultual Experiment Station, Bull 559, IL, USA

Kanno A, Lee YO, Kameya T (1997) The structure of the chloroplast genome in members of the genus *Asparagus*. Theor Appl Genet 95:1196–1202

Kanno A, Park JH, Yun PY, Choi HM, Yoshida R, Kameya T (2002) Isolation and characterization of floral organ identity genes from Asparagus officinalis L. Acta Hortic 589:267–272

Kanno A, Saeki H, Kameya T, Saedler H, Theissen G (2003) Heterotopic expression of class B floral homeotic genes supports a modified ABC model for tulip (*Tulipa gesneriana*). Plant Mol Biol 52:831–841

Kanno A, Park JH, Ochiai T, Kameya T (2004) Floral organ identity genes involved in tepal development in asparagus. Flower Newsl 38:10–18

Kanno A, Nakada M, Akita Y, Hirai M (2007) Class B gene expression and the modified ABC model in nongrass monocots. TSW Development and Embryology 2:17–28

Kunitake H, Mii M (1998) Somatic embryogenesis and its application for breeding and micropropagation in asparagus (*Asparagus officinalis* L.). Plant Biotechnol 15:51–61

Kunitake H, Nakashima T, Mori K, Tanaka M, Saito A, Mii M (1996) Production of interspecific somatic hybrid plants between *Asparagus officinalis* and *A. macowanii* throuth electrofusion. Plant Sci 116:213–222

Lee YO, Kanno A, Kameya T (1996) The physical map of the chloroplast DNA from *Asparagus officinalis* L. Theor Appl Genet 92:10–14

Lee YO, Kanno A, Kameya T (1997) Phylogenetic relationships in the genus *Asparagus* based on the restriction enzyme analysis of the chloroplast DNA. Breed Sci 47: 375–378

Lu YQ, Li FZ (1999) A study of chromosomes of *Asparagus dauricus*. J Shandong Normal Univ Nat Sci Edn 14:67–68

McCollum GD (1988a) Asp 8271 and Asp 8284 asparagus germplasm. Hortic Sci 23:641

Marcellán ON, Camadro EL (1996) Self- and cross-incompatibility in Asparagus officinalis and Asparagus densiflorus cv. Sprengeri. Can J Bot 74:1621–1625

Marcellán ON, Camadro EL (1999) Formation and development of embryo and endosperm in intra- and inter-specific cross of Asparagus officinalis and A. densiflorus cv. Sprengeri. Sci Hort 81:1–11

McCollum GD (1988b) Asparagus densiflorus cultivars Sprengeri and Myers cross-pollinations with A. officinalis and other species. Asparagus Newsl 6:1–10

Moreno R, Espejo JA, Cabrera A, Gil J (2008) Origin of tetraploid cutivated asparagus landraces inferred from nuclear ribosomal DNA internal transcribed spacers' polymorphisms. Ann Appl Biol 153:233–241

Nakada M, Komatsu M, Ochiai T, Ohtsu K, Nakazono M, Nishizawa NK, Nitta K, Nishiyama R, Kameya T, Kanno A (2006) Isolation of MaDEF from Muscari armeniacum and analysis of its expression using laser microdissection. Plant Sci 170:143–150

Nakamura T, Fukuda T, Nakano M, Hasebe M, Kameya T, Kanno A (2005) The modified ABC model explains the development of the petaloid perianth of *Agapanthus praecox* ssp. *orientalis* (Agapanthaceae) flowers. Plant Mol Biol 58: 435–445

Norton JB (1913) Methods used in breeding asparagus for rust resistance. USDA, Bureau of Plant Industry, Bull no 263

Obermeyer AA (1983) *Protasparagus* Oberm, nom nov. new combinations. S Afr J Bot 2:243–244

Ochiai T, Sonoda T, Kanno A, Kameya T (2002) Interspecific hybrids between *Asparagus schoberioides* Kunth and *A. officinalis* L. Acta Hortic 589:225–229

Park JH, Ishikawa Y, Yoshida R, Kanno A, Kameya T (2003) Expression of *AODEF*, a B-functional MADS-box gene, in stamens and inner tepals of dioecious species *Asparagus officinalis* L. Plant Mol Biol 51:867–875

Park JH, Ishikawa Y, Ochiai T, Kanno A, Kameya T (2004) Two *GLOBOSA*-like genes are expressed in second and third whorls of homochlamydeous flowers in *Asparagus officinalis* L. Plant Cell Physiol 45:325–332

Peters CR, O'Brien EM, Drummond RB (1992) Asparagaceae. In: Edible wild plants of Subsaharan Africa. Royal Botanic Gardens, Kew, UK, p 13

Pieroni A, Nebel S, Santoro RF, Heinrich M (2005) Food for two seasons: culinary uses of non-cultivated local vegetables and mushrooms in a south Italian. Int J Food Sci Nutr 56: 245–272

Richardson JE, Weitz FM, Fay MF, Cronk QCB, Peter Linder H, Reeves G, Chase MW (2001) Rapid and recent origin of species richness in the cape flora of South Africa. Nature 412:181–183

Rudall PJ, Engleman EM, Hanson L, Chase MW (1998) Embryology, cytology and systematics of *Hemiphylacus*, *Asparagus* and *Anemarrhena* (*Asparagales*). Plant Syst Evol 211:181–199

Sather DN, York A, Pobursky KJ, Golenberg EM (2005) Sequence evolution and sex-specific expression patterns of the C class floral identity gene, *SpAGAMOUS*, in dioecious *Spinacia oleracea* L. Planta 222:284–292

Shang ZY, Li RJ, Cui TC (1992) Report on karyotypes of 10 species of Liliaceae (s.l.) from Qinling Range. Acta Phytotax Sin 30:438–449

Sheidai M, Inamdar AC (1993) B-chromosomes in *Asparagus* L. Nucleus 36:141–144

Sheidai M, Inamdar AC (1997) Cytomorphology of *Asparagus* taxa using multivariate statistical analysis. Nucleus 40:7–12

Sonoda T, Uragami A, Itoh K, Kohmura H, Ohwada M, Kaji K (2001) Evaluation of *Asparagus* species and comparison between sexes in *A. officinalis* cultivars for resistance to stem blight. J Jpn Soc Hortic Sci 70:244–250

Stajner N, Bohance B, Jakse M (2002a) In vitro propagation of Asparagus maritimus – a rare Mediterranean salt-resistant species. Plant Cell Tiss Org Cult 70:269–274

Stajner N, Bohance B, Javornik B (2002b) Genetic variability of economically important *Asparagus* species as revealed by genome size analysis and rDNA ITS polymorphisms. Plant Sci 162:931–937

Stephen CT, Elmer WH (1988) An in vitro assay to evaluate sources of resistance in *Asparagus* spp. to *Fusarium* crown and root rot. Plant Dis 72:334–337

Stephen CT, De Vries RM, Sink KC (1989) Evaluation of *Asparagus* species for resistance to *Fusarium oxysporum* f. sp. *asparagi* and *F. moniliforme*. Hortic Sci 24:365–368

Stevens PF (2008) Angiosperm phylogeny website. Version 9, June 2008 [and more or less continuously updated since]. http://www.mobot.org/MOBOT/research/APweb/

Tardio J, Padro-de-Santayana M, Morales R (2006) Ethnobotanical review of wild edible plants in Spain. Bot J Linn Soc 152:27–71

Thevenin L (1974) Asperge III. Croisements interspecifiques. In: Rapport d'activité 1972–1974. Sta Génét d'Amélior Pl, CNRA, Versailles, France, p H8-9

Thompson AE, Hepler AR (1956) A summary of resistance and susceptibility to *Puccinia asparagi* DC within the genus *Asparagus*. Plant Dis Rep 40:133–137

Tredgold MH, Biegel HM, Mavi S, Ashton H (1986) Asparagus wildemanii. In: Food plants of Zimbabwe. Mambo, Zimbabwe, pp 64–65

Tsaftaris AS, Polidoros AN, Pasentsis K, Kalivas A (2006) Tepal formation and expression pattern of Bclass paleoAP3 like MADSbox genes in crocus (*Crocus sativus* L.). Plant Sci 170: 238–246

Tutin TG, Heywood VH, Burges N, Moore DM, Valentine DH, Walter S, Webb DA (1980) Asparagus L. In: Flora Europaea, vol 5. Alismataceae to Orchidaceae (Monocotyledones). Cambridge University Press, Cambridge, UK, pp 71–73

van Tunen AJ, Eikeboom W, Angenent GC (1993) Floral organogenesis in Tulipa. Flower Newsl 16:33–38

Venezia A, Soressi GP, Falavigna A (1993) Aspects related to utilisation of wild *Asparagus* species in Italy. Agric Ric 141: 41–48

Watson S (1883) List of plants from southwestern Texas and northern Mexico, collected chiefly by Dr. E. Palmer in 1879–80. Proc Am Acad Arts Sci 18:164–165

Watson L, Dallwitz M (1992) The Families of Flowering Plants: Descriptions, Illustrations, Identification, and Information Retrieval. http://www.biologie.uni-hamburg.de/b-online/delta/angio/. Accessed 31 Oct 2008

Yanofsky MF, Ma H, Bowman JL, Drews GN, Feldmann KA, Meyerowitz EM (1990) The protein encoded by the Arabidopsis homeotic gene agamous resembles transcription factors. Nature 346:35–39

Yun PY, Ito T, Kim SY, Kanno A, Kameya T (2004a) *AVAG1* gene is involved in the development of redproductive organs in ornamental asparagus, *Asparagus virgatus*. Sex Plant Reprod 17:1–8

Yun PY, Kim SY, Ochiai T, Fukuda T, Ito T, Kanno A, Kameya T (2004b) *AVAG2* is a putative D-class gene from an ornamental asparagus. Sex Plant Reprod 17:107–116

Chapter 4
Capsicum

Orarat Mongkolporn and Paul W.J. Taylor

4.1 Introduction

Capsicum is an important crop grown worldwide as a vegetable and spice crop. *Capsicum* is ranked among the world's one of the most important vegetables. The world average annual production from 2000 to 2007 was approximately 23.9 million tons for fresh produce and 2.5 million tons for dry chili from an average world harvested area of 3.5 MHa (FAOSTAT 2008a). Production has increased by approximately 22% since 2000. Figure 4.1 shows total production of chili (fresh and dry forms) from different parts of the world in the years 2005 and 2006, with Asia as the largest producer. The world trade values of both fresh and dry *Capsicum* have been fast growing from 1.7 in 2000 to 2.8 billion USD in 2005 for fresh produce, and from 405 to 655 million USD for dry products (FAOSTAT 2008b). In addition, *Capsicum* is importantly used in food industries for flavoring and coloring, and is valued as medicine and ornamentals.

4.2 Basic Botany, Origin, and Evolution of *Capsicum*

The *Capsicum* species is a member of the Solanaceae family (tribe Solaneae, subtribe Capsicinae) that includes tomato, potato, tobacco, and petunia. *Capsicum* is native to the tropics and subtropics of America. Four locations including southern USA and Mexico to western South America, northeastern Brazil and coastal Venezuela, eastern coastal Brazil, and central Bolivia and Paraguay to northern and central Argentina are suggested by Hunziker (2001) to be the centers of distribution of *Capsicum*. To date, 31 species have been identified as explained by Moscone et al. (2007). The list of the 31 *Capsicum* spp. with some taxonomic traits is provided in Table 4.1. Predominantly, *Capsicums* are perennial shrubs, while some grow biennially and a few are trees.

Capsicum has various names, including pepper, chile, chili, chilli, aji, and paprika. Among the 31 species, five are domesticated including *C. annuum* L., *C. baccatum* L., *C. chinense* Jacq., *C. frutescens* L., and *C. pubescens* Ruiz & Pavón. (Heiser and Pickersgill 1969; IBPGR 1983). An up to date knowledge of *Capsicum* evolution based on cytogenetic study was summarized by Moscone et al. (2007). The *Capsicum* species were distinguished into two groups based on chromosome numbers, $2n = 24$ and $2n = 26$. Only one species, *C. annuum* var. *glabriusculum*, is tetrapoid ($2n = 4x = 48$) (Pickersgill 1977). *Capsicum chacoense* appeared to be the most primitive taxa, and the species with chromosome number 26 were likely to be more advanced than the species with $2n = 24$. The *Capsicum* species, more advance in evolution, exhibited more DNA content (1C or haploid DNA content 3.35–5.77 pg) and heterochromatin amount (1.8–38.9% of the karyotype length) (Pickersgill 1977).

A possible evolutionary relationship among the *Capsicum* species based on karyotype features proposed by Moscone et al. (2007) is displayed in Fig. 4.2. All *Capsicum* species shared a common ancestor, which was a diploid with $x = 12$. *C. chacoense* differentiated

O. Mongkolporn (✉)
Department of Horticulture, Kasetsart University, Kamphaeng Saen Campus, Nakhon Pathom 73140, Thailand
e-mail: orarat.m@ku.ac.th

C. Kole (ed.), *Wild Crop Relatives: Genomic and Breeding Resources, Vegetables*,
DOI 10.1007/978-3-642-20450-0_4,

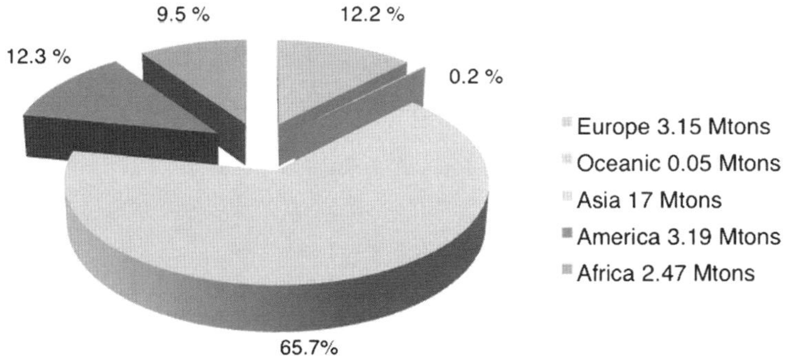

Fig. 4.1 Total world production of chili in 2005–2006 (FAOSTAT 2008a)

Table 4.1 List of the 31 identified *Capsicum* species with some characteristics and geographic distribution (Moscone et al. 2007)

Species and variety	Growth form	Corolla shape and color	Fruit shape and color	Seed color	$2n^{a}$	Geographic distribution
C. annuum L. var. *annuum*	Herb or subshrub (1–2 m)	Stellate; white or cream (exceptionally violet)	Highly variable shape; violet, red, orange, yellow or green	Yellowish	$24^{1-3,10-13}$ 48^{2}	Cultivated worldwide
var. *glabriusculum* (Dunal) Heiser & Pickersgill [syn. = var. minimum (Mill.) Heiser]	Herb or subshrub (1–2 m)	Stellate; white or cream	Ovoid or spherical; red	Yellowish	24^{1-3}	Southern USA, Mexico, Antilles, Belize, Honduras, El Salvador, Panama, Costa Rica, Guatemala, Surinam, Venezuela, Colombia, Ecuador, Peru, northern and northeastern Brazil
C. *baccatum* L. var. *baccatum*	Shrub (0.6–3.5 m)	Stellate; white with greenish spots in the throat	Ovoid or elliptic; red	Yellowish	$24^{2-3,13-14}$	Colombia, Peru, Bolivia, Paraguay, southern and southeastern Brazil, northern Argentina
var. *pendulum* (Willd.) Eshbaugh	Shrub (0.6–2 m)	Stellate; white with greenish spots in the throat	Elliptic or fusiform; red or yellow	Yellowish	$24^{2-3,10-13}$	Cultivated in USA, Mexico, Costa Rica, Colombia, Ecuador, Peru, Brazil, Bolivia, Paraguay, Chile, Argentina, India
var. *umbilicatum* (Vellozo) Hunz. & Barboza	Herb or shrub (1.6–2 m)	Stellate; white with greenish spots in the throat	Umbonate umbilicate; red	Yellowish	$24^{9,13}$	Cultivated in USA, Mexico, Jamaica, Peru, Brazil, Bolivia, Paraguay, Argentina
C. caballeroi M. Nee	Herb, shrub or tree (1–7 m)	Campanulate; lemon yellow	Spherical; red	Yellowish	$?^{19}$	Bolivia
C. campylopodium Sendtn.	Shrub (0.7–1.1 m)	Stellate; white with golden spots in the throat	Spherical compressed; yellowish green	Blackish	$26^{10-11,13,17}$	Brazil
C. cardenasii Heiser & Smith	Shrub (1 m)	Campanulate; violet lobules with azure throat	Spherical; red	brownish	$24^{2-3,15}$	Bolivia
C. ceratocalyx M. Nee	Shrub (1.5 m)	Rotate; yellow with green spots in the throat	Spherical; red	?	$?^{19}$	Bolivia

(*continued*)

Table 4.1 (continued)

Species and variety	Growth form	Corolla shape and color	Fruit shape and color	Seed color	$2n^{a}$	Geographic distribution
C. chacoense Hunz.	Shrub (0.4–0.8 m)	Stellate; white	Ovoid or elliptic; red	Yellowish	24[2–3, 6, 10–11,13]	Southern Bolivia, Paraguay, northern and central Argentina
C. chinense Jacq.	Herb or shrub (0.5–2 m)	Stellate; white or cream	Spherical or conical; red, orange, yellow or white	Yellowish	24[2–3,11–13, 17]	Cultivated in USA, Mexico, Central America, Ecuador, Peru, Bolivia, Brazil, Argentina, China, Japan, Thailand[b]
C. coccineum (Rusby) Hunz.	Herb or climbing subshrub (1.5–3 m)	Stellate; yellowish white with purplish spots in the throat	Spherical; orange or red	Brownish	?	Peru, Bolivia
C. cornutum (Hiern.) Hunz.	Shrub (1.2–1.8 m)	Rotate; white with violet or brownish spots in the throat, green in the tube	Spherical depressed; yellowish green	Blackish	26[17]	Brazil
C. dimorphum (Miers) Kuntze	Shrub (1.5–2 m)	Stellate; yellow, sometimes with violet spots in the throat	Spherical; orange or red	Brownish	?	Colombia, Ecuador
C. eximium Hunz.	Herb, shrub or tree (0.6–4 m)	Stellate; white with violet lobules, greenish in the tube	Spherical; red	Brownish	24[2–3,13]	Southern Bolivia, northern Argentina
C. flexuosum Sendtn.	Shrub (0.5–2 m)	Stellate; white with greenish spots in the throat	Spherical depressed; red	Blackish	24[7,17]	Paraguay, southern and southeastern Brazil, northeastern Argentina
C. friburgense Bianchetti & Barboza	Shrub (0.8–2.5 m)	Campanulate urceolate; pink or lilac	Spherical depressed; yellowish green	Blackish	26[17]	Brazil
C. frutescens L.	Herb or shrub (1–2 m)	Stellate; white or cream	Elongate; red	Yellowish	24[2–3,11–13]	Cultivated in USA, Mexico, Central and South America, Africa, India, China, Japan, Thailand[b]
C. galapagoense Hunz.	Shrub (1–4 m)	Stellate; white	Spherical; red	Yellowish	24[2–3]	Ecuador
C. geminifolium (Dammer) Hunz.	Shrub (0.7–4 m)	Rotate; white or yellowish with violet spots in the throat	Spherical; red	Brownish	?	Colombia, Ecuador, Peru
C. hookerianum (Miers) Kuntze	Shrub (1–3 m)	Stellate; ocher	Spherical; color unknown	Brownish	?	Southern Ecuador, northern Peru
C. hunzikerianum Barboza & Bianchetti	Shrub (1–3 m)	Stellate; white with purple spots in lobules and throat, yellowish in the tube	Spherical; yellowish green	Blackish	?	Brazil

(*continued*)

Table 4.1 (continued)

Species and variety	Growth form	Corolla shape and color	Fruit shape and color	Seed color	$2n$[a]	Geographic distribution
C. lanceolum (Greenm.) Morton & Standley	Herb or shrub (1–5 m)	Stellate campanulate; white or yellowish	Spherical; pale orange or red	Brownish	26[16]	Mexico, Guatemala
C. mirabile Mart. (syn. = *C. buforum* Hunz.)	Herb or shrub (0.5–3 m)	Stellate; white with purple spots in the lobules, greenish in the throat and tube	Spherical; yellowish green	Blackish	26[11,17]	Brazil
C. parvifolium Sendtn.	Shrub or tree (1.5–5 m)	Rotate; white with purple spots in the lobules, greenish in the throat and tube	Spherical; orange or red	Brownish	24[8, 10–11,13, 17]	Colombia, Venezuela, northeastern Brazil
C. pereirae Barboza & Bianchetti	Shrub (0.5–3 m)	Stellate; white with purple spots in the lobules, yellowish in the throat and tube	Spherical; yellowish green	Blackish	26[5,17]	Brazil
C. praetermissum Heiser & Smith [syn. = *C. baccatum* var. *praetermissum* (Heiser & Smith) Hunz.]	Herb or shrub (0.8–1.8 m)	Rotate; white with purple lobule margins and greenish spots in the throat	Spherical or elliptic; orange or red	Yellowish	24[2–3,14,17]	Central and southeastern Brazil, Paraguay
C. pubescens Ruiz & Pav.	Shrub (0.8–2 m)	Rotate; purple or violet in the lobules, white or yellowish in the tube	Turban shaped, spherical or elongate; red, orange or yellow	Blackish	24[2–3,10–13]	Cultivated in Mexico, Central and South America
C. recurvatum Witas.	Herb or shrub (0.5–3 m)	Stellate; white with greenish spots in the throat	Spherical; yellowish green	Blackish	26[18]	Brazil
C. rhomboideum (Dunal) Kuntze [syn. = *C. cilatum* (Kunth) Kuntze]	Shrub or small tree (0.8–4 m)	Rotate; yellow	Spherical; red	Brownish	26[2–3]	Mexico, Guatemala, Honduras, Colombia, Venezuela, Ecuador, Peru
C. schottianum Sendtn.	Shrub (1.2–3 m)	Stellate; white with violet or brownish spots in the throat, greenish in the tube	Spherical; yellowish green	Blackish	26[5,17]	Brazil
C. scolnikianum Hunz.	Shrub (1.5 m)	Campanulate; yellowish white	Spherical depressed; red	Brownish	?	Ecuador
C. tovarii Eshbaugh, Smith & Nickrent	Shrub (1 m)	Stellate; variable color (purple or cream with greenish spots in the lobules)	Spherical; red	Brownish	24[2–4,15,20]	Peru

(*continued*)

Table 4.1 (continued)

Species and variety	Growth form	Corolla shape and color	Fruit shape and color	Seed color	$2n$[a]	Geographic distribution
C. villosum Sendtn.	Subshrub or shrub (1–3 m)	Stellate; white with violet or brownish spots in the throat, greenish in the tube	Spherical; yellowish green	Blackish	26[17]	Brazil

[a]Main chromosome report references: 1 = Pickersgill (1971), 2 = Pickersgill (1977), 3 = Pickersgill (1991), 4 = Eshbaugh et al. (1983), 5 = Moscone (1989), 6 = Moscone (1990), 7 = Moscone (1992), Moscone (1993), Moscone (1999), 10 = Moscone et al. (1993), 11 = Moscone et al. (1995), 12 = Moscone et al. (1996), 13 = Moscone et al. (2003), 14 = Cecchini and Moscone (2002), 15 = Cecchini et al. (2003), 16 = Tong and Bosland (1997), 17 = Pozzobon et al. (2006), 18 = Moscone et al. (2007), 19 = Nee et al. (2006), 20 = Tong and Bosland (1999)
[b]Newly added in this article

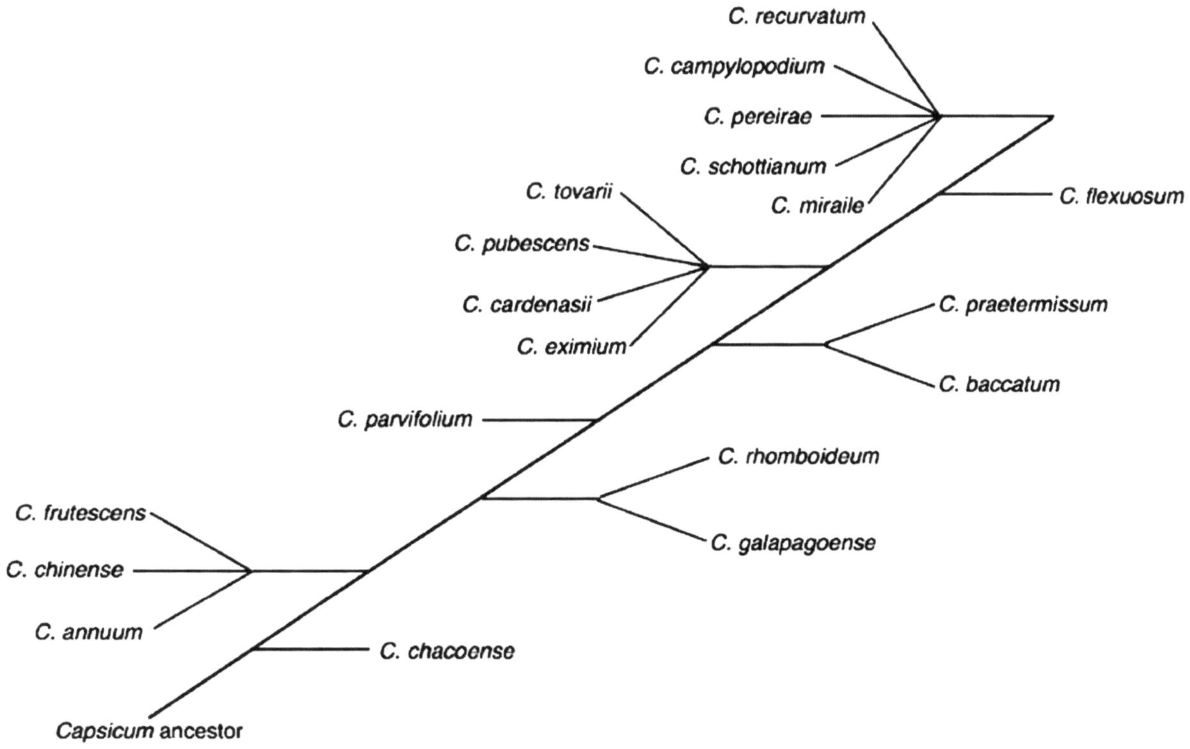

Fig. 4.2 Diagram showing possible evolutionary relationships between the *Capsicum* species based on karyotype studies (adapted from Moscone et al. 2007)

at the earliest stage of the evolution line, and soon after evolved the *C. annuum* complex (*C. annuum*, *C. chinense*, and *C. frutescens*). The primitive group, which is represented by species with white flowers, comprises *C. galapagoense*, *C. rhomboideum*, and *C. parvifolium*. The recent species group started with *C. baccatum* and *C. praetermissum*, and three most advanced groups were beyond this point. The first was the group with purple flowered species (*C. eximium*, *C. cardenasii*, *C. pubescens*, and *C. tovarii*), the second was *C. flexuosum*, and the third was the most advanced (majority were $2n = 26$) group with *C. mirabile* as the core species.

The following taxonomic key to identify wild *Capsicum* species was developed by Barboza and Bianchetti (2005).

1. Style cylindrical, equal in width from the base to the apex. Fruit red, generally elliptical, ovoid, or sometimes globose. Seeds yellowish-brown, the episperm smooth. Corolla 4–7.5 mm long.
   2. Corolla stellate, unspotted, white or cream-colored, the lobes generally oblong, and more or less the same length as the limb and tube. Filaments as long as or shorter than the anthers. Northern and northeastern Brazil (Acre, Amazonas, Maranhão, Rondonia, Roraima) . . . . *C. annuum* var. *glabriusculum* (Dunal) Heiser & Pickersgill.
   2. Corolla rotate, white with greenish-yellow spots on the lobes and limb inside, the lobes broader than long and markedly shorter than the limb and tube. Filaments generally 1.5 times or more longer than the anthers.
      3. Corolla with the inside margin white. South and southeastern Brazil (Espírito Santo, Mato Grosso do Sul, Minas Gerais, Paraná, Rio de Janeiro, Rio Grande do Sul, Santa Catarina, São Paulo) . . . . *C. baccatum* L. var. *baccatum*.
      3. Corolla with the inside margin lilac or violet. Southeastern and west-central Brazil (Goiás, Minas Gerais, Paraná, Santa Catarina, São Paulo) . . . . *C. baccatum* var. *praetermissum* (Heiser & Smith) Hunz.
1. Style clavate, widening from a moderately narrow base to a gradually broadened apex. Fruit generally yellow or yellowish-green at maturity, rarely red-colored, globose or globose-depressed, or globose-compressed. Seeds generally brownish or blackish (yellowish-brown only in *C. parvifolium*), the episperm foveolate with spine-like projections. Corolla (5.5) 6–15 (16) mm long.
   4. Corolla entirely pink or lilac, clearly campanulate to urceolate, tube (5.5) 7–9 (11) mm. Leaves generally ovate. Eastern Brazil (Rio de Janeiro) . . . . *C. friburgense* Bianchetti & Barboza.
   4. Corolla white with yellowish-green and sometimes also purple spots inside, stellate or rotate, never campanulate–urceolate, tube (2) 2.6–6 (8) mm. Leaves ovate, elliptical, or narrowly elliptical.
      5. Pedicels non-geniculate at anthesis, the flowers pendant.
         6. Shrubs or trees up to 4-m tall or more. Fascicles 5–20 flowered. Calyx 5-toothed. Anthers as long as or longer than the filaments. Seeds yellowish-brown. Northeastern Brazil (Bahía, Ceará, Paraiba, Pernambuco, Piauí, Rio Grande do Norte) . . . . *C. parvifolium* Sendtn.
         6. Shrubs 0.5–2 (3) m tall. Flowers solitary or the fascicles 2–3 flowered. Calyx toothless or with 5 minuscule teeth. Anthers clearly shorter than the filaments. Seeds brownish or blackish.
            7. Corolla white with yellowish-green spots in the lobes and limb inside, 5.5–6 mm long. Leaves membranaceous, ovate, 2–3 (3.5) times longer than broad, glabrescent to pubescent. Fruits red at maturity. South and southeastern Brazil (Minas Gerais, Paraná, Rio Grande do Sul, São Paulo, Santa Catarina) . . . . *C. flexuosum* Sendtn.
            7. Corolla white with purple spots followed by an interrupted yellowish-green zone in the lobes and limb, 9–10 mm long. Leaves coriaceous, elliptical to narrowly elliptical, 3–5.5 (10) times longer than broad, glabrate. Fruits yellowish-green at maturity. Southeastern Brazil (Espírito Santo, Minas Gerais) . . . . *C. pereirae* Barboza & Bianchetti.
      5. Pedicels geniculate at anthesis, the flowers twisted 90°.
         8. Corolla lacking purple spots inside.
            9. Calyx toothless. Corolla with yellow or golden spots in lobes and limb. Ovules 2 per locule. Androecium heterodynamous with 3 short stamens and 2 long stamens. Fruits globose-compressed. Southeastern Brazil (Espírito Santo, Minas Gerais, Rio de Janeiro) . . . . *C. campylopodium* Sendtn.
            9. Calyx with 5, or 6–9 horizontal or recurved teeth. Corolla with greenish spots inside. Ovules 5–8 per locule. Androecium homodynamous with all stamens equal in length. Fruits globose-depressed. South and southeastern Brazil (Paraná, Rio de Janeiro, Santa Catarina, São Paulo) . . . . *C. recurvatum* Witas.
         8. Corolla with purple or brownish or violaceous spots followed by yellowish-green zones inside.
            10. Calyx toothless or sometimes with 5 tiny teeth. Southeastern Brazil (Minas Gerais, Rio de Janeiro, São Paulo) . . . . *C. schottianum* Sendtn.
            10. Calyx 5–10 toothed.
               11. Calyx with only 5 short teeth (0.5–3 mm long).
                  12. Plants glabrescent, the hairs antrorse. Leaves elliptical to narrowly elliptical. Southeastern Brazil (Minas Gerais, Rio de Janeiro, São Paulo) . . . . *C. mirabile* Mart.
                  12. Plants densely hairy, the hairs flexuous and patent on stems, petioles, pedicels, and sometimes also on the leaf nerves beneath. Leaves ovate. Southeastern Brazil (Minas Gerais, Rio de Janeiro, São Paulo) . . . . *C. villosum* Sendtn.
               11. Calyx up to 6–10 long teeth (3.2–6 mm long).
                  13. Shrubs 1.2–1.8 m tall, densely hairy. Corolla (8) 9–12 (14) mm long. Leaves membranaceous, ovate to broadly ovate. Southeastern Brazil (Rio de Janeiro, São Paulo) . . . . . *C. cornutum* (Hiern) Hunz.
                  13. Shrubs up to 3-m tall, glabrate. Corolla 10–14 (16) mm long. Leaves coriaceous, slightly ovate to elliptical. Southeastern Brazil (São Paulo) . . . . *C. hunzikerianum* Barboza & Bianchetti.

Historically, *Capsicum* was domesticated about 7000 BC (Andrews 1984). Among the domesticated *Capsicum*, *C. annuum* is the most important and widely grown around the world. *C. annuum* is closely related to *C. chinense* and *C. frutescens*, which are known as the *C. annuum* complex (DeWitt and Bosland 1996). These three species are more widely grown than *C. baccatum* and *C. pubescens* (the latter two are limited only to South America). A taxonomic key to identify domesticated *Capsicum* was provided by the International Board for Plant Genetic Resources (IBPGR 1983).

1. Seed dark, corolla purple .... *C. pubescens*
1. Seed straw-colored, corolla white or greenish white (rarely purple) 2
   2. Corolla with diffuse yellow spots at bases of lobes .... *C. baccatum*
   2. Corolla without diffuse yellow spots at bases of lobes ... 3
      3. Corolla purple 4
         4. Flowers solitary .... *C. annuum*
         4. Flowers two or more at each node .... *C. chinense*
      3. Corolla white or greenish white 5
         5. Calyx of mature fruit with annular constriction at junction with pedicel .... *C. chinense*
         5. Calyx of mature fruit without annular constriction at junction with pedicel.................... 6
            6. Flowers solitary 7
               7. Corolla milky white, lobes usually straight, pedicels often declining at anthesis .... *C. annuum*
               7. Corolla greenish white, lobes usually slightly revolute, pedicels erect at anthesis .... *C. frutescens*
            6. Flowers two or more at each node 8
               8. Corolla milky white .... *C. annuum*
               8. Corolla greenish white 9
                  9. Pedicels erect at anthesis, corolla lobes usually slightly revolute ..... *C. frutescens*
                  9. Pedicels declining at anthesis, corolla lobes straight .... *C. chinense*

4.3 Conservation Initiatives

The IBPGR was established in 1974 under the aegis of the Consultative Group on International Agricultural Research (CGIAR) to promote an international network of genetic resources centers to further the collection, conservation, documentation, evaluation, and use of plant germplasm, and thereby contribute to raising the standard of living and welfare of people throughout the world. IBPGR was renamed to IPGRI (International Plant Genetic Resources Institute) in 1991 and recently was combined with the International Network for Improvement of Banana and Plantain (INIBAP) to a new operating organization Bioversity International, or Bioversity for short.

With the concern for genetic erosion of crops especially grown in the tropics, *Capsicum* was one of the high global priorities to be conserved. IBPGR and the Centro Agronómico Tropical de Investigación y Enseñanza (CATIE) in Costa Rica formed a global plan of action for *Capsicum* genetic resources in 1979 and decided on three priority areas for germplasm exploration and collection. The first was in the South American regions where *Capsicum* was thought to originate and thus should be rich in wild species. The second was in Central and South Asia, and the Mediterranean, and the third was Southeast Asia, China, and Africa (IBPGR 1983). IBPGR also provided official descriptors for characterization of the *Capsicum* germplasm.

Major collections are operated by several key institutes in USA, South America, Europe, and Asia. A list of the *Capsicum* germplasm collections was provided in a directory by IBPGR (1990), and recently by Djian-Caporalino et al. (2007). Three key networks for *Capsicum* germplasm collections, Germplasm Resources Information Network (GRIN), The World Vegetable Center (formerly Asian Vegetable Research and Development Center, located in Taiwan – AVRDC), and CATIE, provide accessible computerized germplasm databases covering passport data, characterization, evaluation, inventory, and distribution data important for the effective management. GRIN is operated under the United States Department of Agriculture (USDA) and has 3,210 accessions in the collection as of December 2008 (http://www.ars-grin.gov/cgi-bin/npgs/html/taxon.pl?8904). AVRDC is now part of the System-wide Information Network for Genetic Resources (SINGER), which is the germplasm information exchange network of the Consultative Group on International Agricultural Research (CGIAR) and its partners. AVRDC holds the largest collection of *Capsicum* germplasm, the number of accessions is 7,514 (as of December 2008: http://singer.grinfo.net/index.jsp). CATIE has 14 regular members of the South American countries.

4.4 Role in Classical and Molecular Genetic Studies

Genetic maps of *Capsicum* have been constructed using both intra- and interspecific populations. The latest *Capsicum* map was published by Paran et al. (2004) as an integrated map from six distinct populations with 2,262 markers – 1,528 amplified fragment length polymorphisms (AFLP), 440 restriction fragment length polymorphisms (RFLP), 288 random amplified polymorphic DNAs (RAPD), several known genes, isozymes, and morphological traits. Only 320 common markers from two maps were used for the map integration. The map contained 13 linkage groups covering 1,832 cM with average marker density of 1 marker per 0.8 cM, which was much improved compared to the most dense individual previous maps (1 marker per 2.1 cM). The small linkage groups obtained from the map integration indicated that the map was not saturated, as the arrangement of the linkage groups was still not corresponding to the 12 *Capsicum* chromosomes (named as P1 to P12). In addition, chromosomes P1 and P8 were not congruent with the intraspecific map of Lefebvre et al. (2002). The six populations contributing to the map were derived from four intraspcific *C. annuum* and two interspecific *C. annuum* and *C. chinense* crosses. Three wild accessions PI 152225 and PI 159234 of *C. chinense* and *C. annuum* CM334 made up three of the six populations.

The very first linkage map of *Capsicum* was constructed using interspecific populations of *C. annuum* cv. NuMex RNaky and *C. chinense* PI 159234 (Tanksley 1984) to map enzyme-coding genes. However, a wide genome coverage map (containing 85 loci) was constructed by Tanksley et al. (1988) using a different *C. annuum* parent crossed with the same *C. chinense*. These populations have been used further using tomato-derived markers as probes to comparatively study the genome structure of *Capsicum* and tomato (Prince et al. 1993; Livingstone et al. 1999). *Capsicum* contains fourfolds of the DNA amount greater than tomato; however, it shares the same chromosome number. Tanksley et al. (1988) reported that there was high conservation of the gene repertoire of the two species, but the gene order on their chromosomes was greatly modified. Livingstone et al. (1999) suggested that there were 18 homeologous linkage blocks covering 95% of the *Capsicum* and 98% of the tomato genomes. Rearrangements of DNA, i.e., translocations, inversions, dis- or associations, and duplications or deletions differentiated these genomes. In addition, two quantitative trait loci (QTL) involving the flower number per node were originally mapped from this cross (Prince et al. 1993).

Several intraspecific *C. annuum* populations were produced to map disease-resistant genes, most of which were doubled haploid progenies, and one was a cross derived from the wild *C. annuum* CM 334 (Lefebvre et al. 1995). The map covered 59% of the total *Capsicum* genome and located the gene positions of resistance to tobacco mosaic virus (TMV), upright fruit, and pungency. A more saturated map was reported by Lefebvre et al. (2002), which was also derived from integrated intraspecific *C. annuum* populations. The consensus map contained 12 linkage groups corresponding to the basic chromosome number of *Capsicum*, 100 known-function gene markers and added five more disease-resistant gene loci to the map, i.e., *pvr2*, *Pvr4*, *Tsw*, *Me3* (nematode resistance), and *Bs3* (resistance to *Xanthomonas*). Also the *y* locus (yellow fruit color) was mapped. Details of *Capsicum* molecular maps on parental species, population types, types and number of markers with locations of important genes, and QTLs were summarized by Djian-Caporalino et al. (2007).

C. frutescens played a small role in mapping, while most interspecific maps were derived from *C. chinense* due to higher compatibility and fertility of the hybrid progeny. Rao et al. (2003) used the wild *C. frutescens* BG2816 to cross with *C. annuum* cv. Maor to study yield-related traits. Maor is a common large-fruited blocky cultivar, whereas BG2816 is a wild accession with a small, more elongated, oval fruit. Ten yield-related traits (i.e., fruit characters, fruit number, yield, flowering, maturity, and seed weight) were mapped, and 58 QTLs relating to the traits were identified on the map.

4.5 Role in Crop Improvement Through Traditional and Advanced Tools

C. annuum is the most important species of the genus as it is widely grown at commercial scale. Therefore, most breeding programs around the world are focused toward *C. annuum* improvement. Some interspecific crosses among wild species have also been attempted, however, only for taxonomical study purposes.

There are wild forms of all five domesticated species, except *C. pubescens* (DeWitt and Bosland 1996). Wild species and wild accessions of cultivated species have played an important role in crop improvement, especially for pest resistance. In *Capsicum* substantial utilization of pest resistance from wild species introgressed into elite cultivars to improve disease resistance has been documented. It is important to consider the crossability between species. Within each species there are wild, semi-domesticated, and domesticated accessions from the primary gene pool, which are reciprocally crossable and produce fertile hybrids.

For domesticated *Capsicum* species, the secondary gene pool includes wild and domesticated species within the same genetic complex. In *Capsicum* there are three complexes including the *C. annuum* (*C. chacoense*, *C. chinense*, *C. frutescens*, and *C. galapagoense*), *C. baccatum* (*C. praetermissum* and *C. tovarii*), and *C. pubescens* (*C. eximium* and *C. cardenasii*) complexes (Fig. 4.3; Djian-Caporalino et al. 2007). Most interspecific crosses within the same complex produce partially fertile hybrids without aids such as embryo rescue. Crossability between two species is not always reciprocally successful. Djian-Caporalino et al. (2007) summarized more or less successful interspecific crosses within the secondary gene pool in Fig. 4.3. When crossing *C. annuum* with

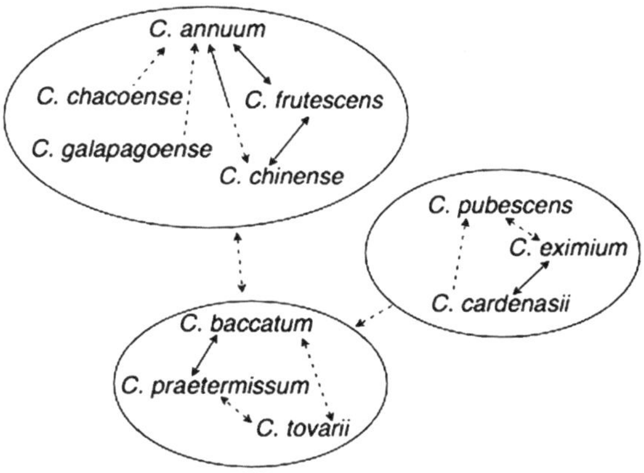

Fig. 4.3 Genetic exchanges between and within gene pools in the *Capsicum* genus. *Arrows* point female parent. *Solid lines* indicate relatively successful fertile hybrids, *broken lines* indicate very few hybrids with very low fertility. Adapted from Djian-Caporalino et al. (2007)

C. chacoense or *C. galapagoense*, or crossing *C. pubescens* with *C. cardenasii*, it is recommended that *C. annuum* and *C. pubescens* should be the female parents to gain more success.

Genetic exchanges with the tertiary gene pool are restricted because of genetic incompatibility. However, with the aid of embryo rescue, fertile hybrids have been obtained. By using *C. annuum* as the female parent to hybridize with *C. baccatum*, embryo development after pollination was observed and rescued (Yoon et al. 2006). Embryo rescue technique overcomes embryo abortion (post-fertilization barrier); however, there are cases where post-fertilization embryo abortion occurred before the globular stage. To overcome this type of genetic barrier, trispecies bridge crosses can be an alternative. *C. chinense* was used as the bridge parent to accomplish the hybridization between *C. annuum* and *C. baccatum* (Yoon and Park 2005). Wild *C. chacoense*, wild accessions of *C. chinense*, and *C. baccatum* have been reported as resistance sources of several diseases.

4.5.1 Resistance to Anthracnose

Anthracnose, caused by a complex of *Colletotrichum* species, is considered a serious problem to chili production in the tropics and subtropics worldwide, and thus cause fruit yield losses to over 80% (Mahasuk et al. 2009a). Typical anthracnose symptoms are sunken necrotic tissues, with concentric rings of acervuli that are often moist (Than et al. 2008; Montri et al. 2009).

Breeding for resistance to anthracnose began in the early 1990s (Park et al. 1990a, b); however, to date there is still no available commercial resistant variety. The key factor that hinders the success of anthracnose resistance breeding is the complexity of the causal pathogen and also of the host–pathogen interaction.

As *C. annuum* lacks anthracnose resistance, interspecific hybridization is essential. A few wild varieties from *C. baccatum* "PBC80" and "PBC81" and *C. chinense* "PBC932" were identified to be highly resistant (immune) to anthracnose (AVRDC 1998), and recently *C. baccatum* "PI594137" has been newly identified (Kim et al. 2008b). *C. chinense* "PBC932", small round fruited and pungent chili, was easily crossable with some *C. annuum* varieties without any aids.

The resistance derived from PBC932 differentially expressed at seedling and fruit stages. Interestingly, three recessive genes *co1*, *co2*, and *co3* were identified as being responsible for the resistance at mature green fruit (Pakdeevaraporn et al. 2005), ripe fruit, and seedling (Mahasuk et al. 2009a), respectively. The *co1* and *co2* genes were linked (approximately 25% recombination), while *co3* was independent to the others. Since anthracnose mainly infects chili fruit, the finding of seedling resistance being independent to the fruit resistance suggested that the selection process for fruit resistance cannot be executed at seedling. In addition, the selection for resistance should be assessed at different fruit stages to obtain more durable resistance. The recessive resistance from PBC932 was also confirmed by a different study using different *Colletotrichum* species (Kim et al. 2008a). Advanced breeding of *C. annuum* lines with high resistance derived from PBC932 is going on at AVRDC, Kasetsart University (Thailand), and Seoul National University (Korea) (Kim et al. 2007).

C. baccatum is another excellent source of anthracnose resistance, containing the broadest spectrum of resistance to the key *Colletotrichum* species (i.e., *C. truncatum* (formerly *capsici*), *C. gloeosporioides*, and *C. acutatum*; Montri et al. 2009; Mahasuk et al. 2009b; Mongkolporn et al. 2010); however, the species is not easy to cross with *C. annuum*. The introgression of the resistance from PBC81 into *C. annuum* was managed via embryo rescue technique, and then to achieve advanced resistance, breeding was conducted by backcrossing (Yoon et al. 2006). PBC80 was introgressed into *C. annuum* via trispecies cross by using *C. chinense* as a bridge (Yoon and Park 2005). The resistance genes derived from *C. baccatum* accessions PBC80 (Yoon and Park 2005) and PI594137 (Kim et al. 2008b) appeared to be of single dominant gene action. However, similar to *C. chinense* PBC932, the resistance genes differentially expressed at different fruit stages. In an intraspecific cross derived from *C. baccatum* PBC80 and PBC1422, the resistance at mature green and ripe red fruit stages appeared to be controlled by a single recessive and single dominant gene, respectively (Mahasuk et al. 2009b).

Nevertheless, pathotypes of each of the major *Colletotrichum* species causing anthracnose in chili have been identified (Montri et al. 2009; Mongkolporn et al. 2010). Three pathotypes were identified for *C. truncatum* isolates on differential genotypes of *C. chinense* – PBC932 and C04714. Three pathotypes were also identified for *C. acutatum* on *C. baccatum* – PBC1422. For *C. gloeosporioides* six pathotypes were identified at the green and five at the red fruit stages on differential genotypes of *C. chinense* – C04714, *C. annuum* – Bangchang, 83–168, Jinda and *C. frutescens* – Kee Noo Suan and Karen. This result will have profound effect on chili breeding programs where novel sources of resistance genes from the related species are being incorporated into the commercial *C. annuum* varieties to enhance resistance to anthracnose.

4.5.2 Virus Resistance

Viruses are also serious problems of chili and in some cases whole fields of chili have been abandoned before harvest (Green and Kim 1994). Around 70 viruses are known to infect chili (Pernezny et al. 2009). Virus infection can interfere with plant chlorophyll synthesis, thus causing chlorosis and mottling of the foliage producing the mosaic symptoms. However, viral symptoms vary dramatically in both severity and forms, with some disorders causing mottle, leaf puckering, leaf distortion, shoe stringing, and plant stunting. Since there are a high number of viral pathogens infecting chili, screening chili germplasm for resistance to key viruses in different countries has been performed by several breeders. Good resistance was found in both cultivated and wild varieties of *Capsicum* (Greenleaf 1986; Green and Kim 1994; Bosland 2000; Hernández-Verdugo et al. 2001; Suzuki et al. 2003).

Potyviridae is the largest and most economically devastating family of plant viruses (Kyle and Palloix 1997). Typically several potyviruses regularly coinfect a crop. Three potyviruses, potato virus Y (PVY), tobacco etch virus (TEV), and pepper mottle virus (PepMoV), predominate in Europe and North America, while the Chili veinal mottle virus (CVMV) is important in Africa and Asia. Kyle and Palloix (1997) revised the nomenclature for the resistance genes to potyviruses in *Capsicum*. Five gene loci were identified (*pvr1* to *pvr5*, reviewed by Kyle and Palloix 1997). The *pvr1*, *pvr2*, *Pvr4*, and *pvr5* were derived from wild accessions of either *C. annuum* or *C. chinense*, although some cultivars possessed some resistance *pvr* genes, such as Yolo RP10 (*pvr2*^{1}), Yolo Y (*pvr2*^{1}), Florida VR2 (*pvr2*^{2}), and Avelar (*pvr3*). The *pvr1* derived from *C. chinense* PI 159236 and PI 152225 conferred the resistances to TEV-C, TEV-F, PepMoV, PVY. The *pvr2*^{2} derived from *C. annuum* PI 264281 and SC46252 conferred the resistances to PVY (pathotypes 0 and 1) and TEV. The *Pvr4* was derived from *C. annuum* CM334 and resistant to PVY (pathotypes 0, 1, and 2) and PepMoV. The *pvr5* was also derived from CM334 that is resistant to PVY (pathotype 0). Newly identified *Pvr7* was also derived from PI 159236 that is resistant to PepMoV (Grube et al. 2000). Comparative mapping conducted by Grube et al. (2000) indicated a cluster of *Pvr4*, *Pvr7*, and *Tsw* (resistance gene for Tomato spotted wilt virus) on the *Capsicum* chromosome 10.

There have been several other screening trials for more viral resistant sources in *Capsicum* wild

Table 4.2 List of wild accessions of *Capsicum* species and their disease-resistant sources

Capsicum species	Accession	Disease resistances/gene loci	References
C. annuum	CM334 (Serrano Criollo de Morelos-334)	*Phytophthora capsici*	Gil Ortega et al. (1991)
		Pvr4 (PVY pathotypes 0, 1, and 2; and PepMoV)	Dogimont et al. (1996), Kyle and Palloix (1997)
		pvr5 (common PVY strains)	Dogimont et al. (1996), Kyle and Palloix (1997)
		Me (*Meloidogyne* spp.)	Djian-Caporalino et al. (1999)
	PI264281	*pvr2* (PVY pathotypes 0 and 1; TEV)	Kyle and Palloix (1997)
	SC46252	*pvr2* (PVY pathotypes 0 and 1; TEV)	Kyle and Palloix (1997)
	PM687 (inbred PI322719 – a local Indian population)	Bacterial wilt (*Ralstonia solanacearum*)	Lafortune et al. (2005)
		Meloidogyne spp.	Djian-Caporalino et al. (1999)
	PM217 (PI201234)	*Meloidogyne* spp.	Djian-Caporalino et al. (1999)
	AC2258 (PI201234)	*Phytophthora capsici*	Ares et al. (2005)
C. baccatum	PBC80	*Colletotrichum* spp.	AVRDC (1998), Montri et al. (2009), Mongkolporn et al. (2010)
	PBC81	*Colletotrichum* spp.	AVRDC (1998), Montri et al. (2009), Mongkolporn et al. (2010)
	C-153	TSWV	Rosellol et al. (1996)
C. chacoense	PI260435	*Bs2* (*Xanthomonas campestris* pv. *vesicatoria*)	Cook and Guevara (1984)
C. chinense	7204	*Tsw* (tomato spotted wilt tospovirus)	Moury et al. (1997)
	CNPH275	*Tsw* (tomato spotted wilt tospovirus)	Boiteux et al. (1993)
	ECU-973	*Tsw* (tomato spotted wilt tospovirus)	Cebolla-Cornejo et al. (2003)
	PI152225	*L*^{3} (tobacco mosaic virus)	Boukema (1980)
		pvr1 (TEV-C, TEV-F, PepMoV, PVY)	Kyle and Palloix (1997)
		Tsw (tomato spotted wilt tospovirus)	Black et al. (1991), Boiteux (1995), Jahn et al. (2000)
	PI159236	*L*^{3} (tobacco mosaic virus)	Boukema (1980)
		pvr1 (TEV-C, TEV-F, PepMoV, PVY)	Kyle and Palloix (1997)
		Pvr7 (PepMoV)	Grube et al. (2000)
		Tsw (tomato spotted wilt tospovirus)	Black et al. (1991), Boiteux (1995), Jahn et al. (2000)
	PI315008	*L*^{3} (tobacco mosaic virus)	Boukema (1980)
	PI315023	*L*^{3} (tobacco mosaic virus)	Boukema (1980)
	PI315024	*L*^{3} (tobacco mosaic virus)	Boukema (1980)
	PBC932	*co1*, *co2*, *co3* (*Colletotrichum capsici*)	Pakdeevaraporn et al. (2005), Mahasuk et al. (2009a)
C. pubescens	PI235047	Bacterial spot (*Xanthomonas campestris* pv. *vesicatoria*)	Sahin and Miller (1998)

accessions (Bosland 2000; Hernández-Verdugo et al. 2001; Suzuki et al. 2003) depending on the strains and types of viruses. What matters is the introgression of the identified resistance sources into cultivated or elite varieties to be of wide practical use. However, among those germplasm screened, *C. chinense* accessions appeared to be mostly used for viral resistance breeding, since the accessions contained resistances to several viruses including potyviruses (as already mentioned above), Tomato spotted wilt virus (TSWV; Black et al. 1991; Boiteux et al. 1993), and tobacco mosaic virus (Boukema 1980). For example, PI 159236 and PI 152225 were used to generate interspecific populations to identify resistance genes and mapping for the location of the identified genes.

4.5.3 Phytophthora capsici *Resistance*

P. capsici Leon. causes fruit and root rot, and shoot blight, and is considered to be a serious pseudo-fungal disease of chili worldwide. *P. capsici* is a soilborne pathogen and able to infect host plant at any developmental stage causing sudden wilt to the plant. According to Bosland (2008), the meaning in Greek of *P. capsici* is "plant destroyer of *Capsicum*." The first report of the resistance to *P. capsici* was by Kimble and Grogan (1960); however due to the specificity and complexity of the pathogen, successful chili breeding for *P. capsici* resistance has been hindered. *Capsicum* develops separate resistance systems to control the disease infecting different organs. In addition, races of *P. capsici* have been identified using differential chili host genotypes. Oelke et al. (2003) found nine races for Phytophthora root rot and four races for Phytophthora foliar blight.

Several sources of partial resistance to *P. capsici* have been identified (Kimble and Grogan 1960; Saini and Sharma 1978; Palloix et al. 1990; Gil Ortega et al. 1991, 1992; Lefebvre and Polloix 1996; Ares et al. 2005). Two wild accessions of *C. annuum,* CM334 originated from Mexico and AC2258 originated from Central America (derived from PI201234), contained high levels of resistance to *P. capsici* (Ares et al. 2005). The resistance in CM334 was reported to be controlled by two genes with dominant and recessive epistasis (Reifschneider et al. 1992). Smith et al. (1967) reported the resistance in AC2258 to be controlled by two dominant genes without additive effects. However, the number of genes involved in the resistances and the mode of gene actions were contradicted by other researchers due to the differences in the separation of susceptible and resistant phenotypes. Similarly, QTLs were detected in both resistant genotypes (Thabuis et al. 2001; Sugita et al. 2006; Minamiyama et al. 2007) using molecular mapping approach. Introgression of these resistances into elite varieties is being attempted.

More resistance sources of other important chili diseases are listed in Table 4.2.

References

Andrews J (1984) Peppers: the domesticated capsicums. University of Texas Press, Austin, TX, USA

Ares JLA, Martínez AR, Paz JF (2005) Resistance of pepper germplasm to *Phytophthora capsici* isolates collected in northwest Spain. Spn J Agric Res 3:429–436

AVRDC (1998) AVRDC Report 1997. AVRDC-The World Vegetable Center, Shanhua, Taiwan, pp 54–57

Barboza GE, Bianchetti LDB (2005) Three new species of *Capsicum* (Solanaceae) and a key to the wild species from Brazil. Syst Bot 30:863–871

Black LL, Hobbs HA, Gatti JM (1991) Tomato spotted wilt virus resistance in *Capsicum chinense* PI152225 and PI159236. Plant Dis 75:863

Boiteux LS (1995) Allelic relationships between genes for resistance to tomato spotted wilt tospovirus in *Capsicum chinense*. Theor Appl Genet 90:146–149

Boiteux LS, Nagata T, Dutra WP, Fonseca MEN (1993) Sources of resistance to tomato spotted wilt virus (TSWV) in cultivated and wild species of *Capsicum*. Euphytica 67:89–94

Bosland PW (2000) Sources of curly top virus resistance in *Capsicum*. HortScience 35:1321–1322

Bosland PW (2008) Think global, breed local: specificity and complexity of *Phytophthora capsici*. In: 19th International Pepper Conference, Atlantic City, NJ, USA

Boukema IW (1980) Allelism of genes controlling resistance to tobacco mosaic virus in *Capsicum chinense*. Euphytica 29:433–440

Cebolla-Cornejo J, Soler S, Gomar B, Soria MD, Nuez F (2003) Screening *Capsicum* germplasm for resistance to tomato spotted wilt virus (TSWV). Ann Appl Biol 143:143–152

Cecchini NM, Moscone EA (2002) Análisis cariotípico en dos variedades de *Capsicum baccatum* (Solanaceae) mediante bandeos cromosómicos AgNOR y de fluorescencia. Actas XXXI Congreso Argentino de Genética. J Basic Appl Genet, La Plata, Argentina, p 86

Cecchini NM, Scaldaferro MA, Barboza GE, Moscone EA (2003) Estudio de cromosomas mitóticos en dos ajíes silvestres de flores púrpuras (*Capsicum* – Solanaceae) mediante bandeo de fluorescencia. Actas IV Jornadas Argentino-

Chilenas de Genética. J Basic Appl Gen, Huerta Grande, Argentina, p 91
Cook AA, Guevara YG (1984) Hypersensitivity in *Capsicum chacoense* to race 1 of the bacterial spot pathogen of pepper. Plant Dis 68:329–330
DeWitt D, Bosland PW (1996) Peppers of the world: an identification guide. Ten Speed, Berkeley, CA, USA
Djian-Caporalino C, Pijarowski L, Januel A, Lefebvre V, Daubèze A, Palloix A, Dalmasso A, Abad P (1999) Spectrum of resistance to root-knot nematodes and inheritance of heat-stable resistance in pepper (*Capsicum annuum* L.). Theor Appl Genet 99:496–502
Djian-Caporalino C, Lefebvre V, Sage-Daubèze A-M, Palloix A (2007) Capsicum. In: Singh RJ (ed) Genetic resources, chromosome engineering and crop improvement: vegetable crops, vol 3. CRC, Boca Raton, FL, USA, pp 185–243
Dogimont C, Palloix A, Daubze A-M, Marchoux G, Selassie KG, Pochard E (1996) Genetic analysis of broad spectrum to potyviruses using doubled haploid lines of pepper (*Capsicum annuum* L.). Euphytica 88:231–239
Eshbaugh WH, Smith PG, Nickrent DL (1983) *Capsicum tovarii* (Solanaceae), a new species of pepper from Peru. Brittonia 35:55–60
FAOSTAT (2008a) Agricultural production. http://www.faostat.fao.org/
FAOSTAT (2008b) Agricultural trade. http://www.faostat.fao.org/
Gil Ortega R, Palazón Español C, Cuartero Zueco J (1991) Genetics of resistance to *Phytophthora capsici* in the Mexican pepper SCM-334. Plant Breed 107:50–55
Gil Ortega R, Palazón Español C, Cuartero Zueco J (1992) Genetic relationships among four pepper genotypes resistant to *Phytophthora capsici*. Plant Breed 108:118–125
Green SK, Kim JS (1994) Sources of resistance to viruses of pepper (*Capsicum* spp.): a catalog. Tech Bull No 20. AVRDC, Tainan, Taiwan
Greenleaf WH (1986) Pepper breeding. In: Bassett MJ (ed) Breeding vegetable crops. AVI, Westport, CO, USA, pp 67–134
Grube RC, Blauth JR, Arnedo AMS, Caranta C, Jahn MK (2000) Identification and comparative mapping of a dominant potyvirus resistance gene cluster in *Capsicum*. Theor Appl Genet 101:852–859
Heiser CB, Pickersgill B (1969) Names for the cultivated *Capsicum* species (Solanaceae). Taxon 18:277–283
Hernández-Verdugo S, Guevara-González RG, Rivera-Bustamante RF, Oyama K (2001) Screening wild plants of *Capsicum annuum* for resistance to pepper huasteco virus (PHV): presence of viral DNA and differentiation among populations. Euphytica 122:31–36
Hunziker AT (2001) Genera Solanacearum: the genera of Solanaceae illustrated, arranged according to a new system. Gantner, Ruggell, Leichtenstein, Germany
IBPGR (1983) Genetic resources of *Capsicum*: a global plan of action. IBPGR, Rome, Italy
IBPGR (1990) Directory of germplasm collections. In: Bettencourt E, Konopka J (eds) 4. Vegetables. IBPGR, Rome, Italy
Jahn M, Paran I, Hoffmann K, Radwanski ER, Livingstone KD, Grube RC, Aftergoot E, Lapidot M, Moyer J (2000) Genetic mapping of the *Tsw* locus for resistance to the tospovirus tomato spotted wilt virus in *Capsicum* spp. and its relationship to the same pathogen in tomato. Mol Plant Microbe Interact 13:673–682
Kim SH, Yoon JB, Do JW, Park HG (2007) Resistance to anthracnose caused by *Colletotrichum acutatum* in chili pepper (*Capsicum annuum* L.). J Crop Sci Biotechnol 10:277–280
Kim SH, Yoon JB, Do JW, Park HG (2008a) A major recessive gene associated with anthracnose resistance to *Colletotrichum capsici* in chili pepper (*Capsicum annuum* L.). Breed Sci 58:137–141
Kim SH, Yoon JB, Park HG (2008b) Inheritance of anthracnose resistance in a new genetic resource, *Capsicum baccatum* PI594137. J Crop Sci Biotechnol 11:13–16
Kimble KA, Grogan RG (1960) Resistance to *Phytophthora* root rot in pepper. Plant Dis Rep 44:872–873
Kyle MM, Palloix A (1997) Proposed revision of nomenclature for potyvirus resistance genes in *Capsicum*. Euphytica 97:183–188
Lafortune D, Béramis M, Daubèze A-M, Boissot N, Palloix A (2005) Partial resistance of pepper to bacterial wilt is oligogenic and stable under tropical conditions. Plant Dis 89:501–506
Lefebvre V, Polloix A (1996) Both epistatic and additive effects of QTLs are involved in polygenic induced resistance to disease: a case study, the interaction pepper – *Phytophthora capsici* Leonian. Theor Appl Genet 93:503–511
Lefebvre V, Palloix A, Caranta C, Pochard E (1995) Construction of an intraspecific integrated linkage map of pepper using molecular markers and double-haploid progenies. Genome 38:112–121
Lefebvre V, Pflieger S, Thabuis A, Caranta C, Blattes A, Chauvet JC, Daubeze AM, Palloix A (2002) Towards the saturation of the pepper linkage map by alignment of three intraspecific maps including known-function genes. Genome 45:839–854
Livingstone KD, Lackney VK, Blauth JR, Rv W, Jahn MK (1999) Genome mapping in Capsicum and the evolution of genome structure in the Solanaceae. Genetics 152:1183–1202
Mahasuk P, Khumpeng N, Wasee S, Taylor PWJ, Mongkolporn O (2009a) Inheritance of resistance to anthracnose at seedling, and green and red chili fruit stages. Plant Breed 128:701–706
Mahasuk P, Taylor PWJ, Mongkolporn O (2009b) Identification of two new genes conferring resistance to *Colletotrichum acutatum* in *Capsicum baccatum* L. Phytopathology 99:1100–1104
Minamiyama Y, Tsuro M, Kubo T, Hirai M (2007) QTL analysis for resistance to *Phytophthora capsici* in pepper using a high density SSR-based map. Breed Sci 57:129–134
Mongkolporn O, Montri P, Supakaew T, Taylor PWJ (2010) Differential reactions on mature green and ripe chili fruit infected by three *Colletotrichum* species. Plant Dis 94:306–310
Montri P, Taylor PWJ, Mongkolporn O (2009) Pathotypes of *Colletotrichum capsici*, the causal agent of chili anthracnose, in Thailand. Plant Dis 93:17–20
Moscone EA (1989) Estudios citotaxonómicos en las tribus Solaneae y Nicotianeae (Solanaceae) de América del Sur. University of Córdoba, Cordoba, Argentina

Moscone EA (1990) Chromosome studies on *Capsicum* (Solanaceae) I. Karyotype analysis in *C. chacoense*. Brittonia 42:147–154

Moscone EA (1992) Estudios de cromosomas meióticos en Solanaceae de Argentina. Darwiniana 31:261–297

Moscone EA (1993) Estudios cromosómicos en *Capsicum* (Solanaceae) II.Análisis cariotípico de *C. parvifolium* y *C. annuum* var. *annuum*. Kurtziana 229:9–18

Moscone EA (1999) Análisis cariotípico en *Capsicum baccatum var. umbilicatum* (Solanaceae) mediante bandeos AgNOR y de fluorescencia. Kurtziana 27:225–232

Moscone EA, Lambrou M, Hunziker AT, Ehrendorfer F (1993) Giemsa C-banded karyotype in *Capsicum* (Solanaceae). Plant Syst Evol 186:213–229

Moscone EA, Loidl J, Ehrendorfer F, Hunziker AT (1995) Analysis of active nucleolus organizing regions in *Capsicum* (Solanaceae) by silver staining. Am J Bot 82:276–287

Moscone EA, Lambrou M, Ehrendorfer F (1996) Fluorescent chromosome banding in the cultivated species of *Capsicum* (Solanaceae). Plant Syst Evol 202:37–63

Moscone EA, Baranyi M, Ebert I, Greilhuber J, Ehrendorfer F, Hunziker AT (2003) Analysis of nuclear DNA content in *Capsicum* (Solanaceae) by flow cytometry and Feulgen densitometry. Ann Bot 92:21–29

Moscone EA, Scaldaferro MA, Grabiele M, Cecchini NM, Sánchez García Y, Jarret R, Daviña JR, Ducasse DA, Barboza GE, Ehrendorfer F (2007) The evolution of chili peppers (*Capsicum* – Solanaceae): a cytogenetic perspective. Acta Hortic 745:137–170

Moury B, Palloix A, Selassie KG, Marchoux G (1997) Hypersensitive resistance to tomato spotted wilt virus in three *Capsicum chinense* accessions is controlled by a single gene and is overcome by virulent strains. Euphytica 94:45–52

Nee M, Bohs L, Knapp S (2006) New species of *Solanum* and *Capsicum* (Solanaceae) from Bolivia, with clarification of nomenclature in some Bolivian *Solanum*. Brittonia 58:322–356

Oelke L, Bosland PW, Steiner R (2003) Differentiation of race specific resistance to phytophthora root rot and foliar blight in *Capsicum annuum*. J Am Soc Hortic Sci 128:213–218

Pakdeevaraporn P, Wasee S, Taylor PWJ, Mongkolporn O (2005) Inheritance of resistance to anthracnose caused by *Colletotrichum capsici* in *Capsicum*. Plant Breed 124:206–208

Palloix A, Daubèze AM, Phaly T, Pochard E (1990) Breeding transgressive lines of pepper for resistance to *Phytophthora capsici* in a recurrent selection system. Euphytica 51:141–150

Paran I, van der Voort JR, Lefebvre V, Jahn M, Landry L, van Schriek M, Tanyolac B, Caranta C, Ben Chaim A, Livingstone K, Palloix A, Peleman J (2004) An integrated genetic linkage map of pepper (*Capsicum* spp.). Mol Breed 13:251–261

Park HK, Kim BS, Lee WS (1990a) Inheritance of resistance to anthracnose (*Colletotrichum* spp.) in pepper (*Capsicum annuum* L.) I. Genetic analysis of anthracnose resistance by diallel crosses. J Kor Soc Hortic Sci 31:91–105

Park HK, Kim BS, Lee WS (1990b) Inheritance of resistance to anthracnose (*Colletotrichum* spp.) in pepper (*Capsicum annuum* L.). II. Genetic analysis of resistance to *Colletotrichum dematium*. J Kor Soc Hortic Sci 31:207–212

Pernezny K, Roberts PD, Murphy JF, Goldberg NP (2009) Compendium of pepper diseases. APS, St. Paul, MN, USA

Pickersgill B (1971) Relationships between weedy and cultivated forms in some species of chili peppers (genus *Capsicum*). Evolution 25:683–691

Pickersgill B (1977) Chromosomes and evolution in *Capsicum*. In: Pochard E (ed) *Capsicum* 77, Comptes Rendus 2ème Congrès Eucarpia Piment. Institut National de la Recherche Agronomique, Avignon-Montfavet, France, pp 27–37

Pickersgill B (1991) Cytogenetics and evolution of *Capsicum* L. In: Tsuchiya T, Gupta PK (eds) Chromosome engineering in plants: genetics, breeding, evolution, Part B. Elsevier, Amsterdam, Netherlands, pp 139–160

Pozzobon MT, Schifino-Wittmann MT, Bianchetti LB (2006) Chromosome numbers in wild and semidomesticated Brazillian *Capsicum* L. (Solanaceae) species: do x = 12 and x = 13 represent two evolutionary lines? Bot J Linn Soc 151:259–269

Prince JP, Pochard E, Tanksley SD (1993) Construction of a molecular linkage map of pepper and a comparison of synteny with tomato. Genome 36:404–417

Rao GU, Ben Chaim A, Borovsky Y, Paran I (2003) Mapping of yield-related QTLs in pepper in an interspecific cross of *Capsicum annuum* and *C. frutescens*. Theor Appl Genet 106:1457–1466

Reifschneider FJB, Boiteux LS, Vecchia PTD, Poulos JM, Kuroda N (1992) Inheritance of adult-plant resistance to *Phytophthora capsici* in pepper. Euphytica 62:45–49

Rosellol S, Diezl MJ, Jorda C, Nuezl F (1996) Screening of *Capsicum chacoense* accessions for TSWV resistance by mechanical inoculation. Capsicum Eggplant Newsl 16:68–78

Sahin F, Miller SA (1998) Resistance in *Capsicum pubescens* to *Xanthomonas campestris* pv. *vesicatoria* pepper race 6. Plant Dis 82:794–799

Saini SS, Sharma PP (1978) Inheritance of resistance to fruit rot (*Phytophthora capsici* Leon.) and induction of resistance in bell pepper (*Capsicum annuum* L.). Euphytica 27:721–723

Smith PG, Kimble KA, Grogan RG, Millett AH (1967) Inheritance of resistance in peppers to *Phytophthora* root rot. Phytopathology 57:377–379

Sugita T, Yamaguchi K, Kinoshita T, Yuji K, Sugimura Y, Nagata R, Kawasaki S, Todoroki A (2006) QTL analysis for resistance to Phytophthora blight (*Phytophthora capsici* Leon.) using an intraspecific doubled-haploid population of *Capsicum annuum*. Breed Sci 56:137–145

Suzuki K, Kuroda T, Miura Y, Murai J (2003) Screening and field trials of virus resistant sources in *Capsicum* spp. Plant Dis 87:779–783

Tanksley SD (1984) Linkage relationships and chromosomal locations of enzyme-coding genes in pepper, *Capsicum annuum*. Chromosoma 89:352–360

Tanksley SD, Bernatzky R, Lapitan NL, Prince JP (1988) Conservation of gene repertoire but not gene order in pepper and tomato. Proc Natl Acad Sci USA 85:6419–6423

Thabuis A, Lefebvre V, Daubèze AM, Signoret P, Phaly T, Nemouchi G, Blattes A, Polloix A (2001) Introgression of partial resistance to *Phytophthora capsici* Leon. into a pepper elite line by marker assisted backcrosses. Acta Hortic 546:645–650

Than PP, Jeewon R, Hyde KD, Pongsupasamit S, Mongkolporn O, Taylor PWJ (2008) Characterization and pathogenicity of *Colletotrichum* species associated with anthracnose on chilli (*Capsicum* spp.) in Thailand. Plant Pathol 57:562–572

Tong N, Bosland PW (1997) Meiotic chromosome study of *Capsicum lanceolatum*, another 13 chromosome species. Capsicum Eggplant Newsl 16:42–43

Tong N, Bosland PW (1999) *Capsicum tovarii*, a new member of the *Capsicum baccatum* complex. Euphytica 109:71–77

Yoon JB, Park HG (2005) Trispecies bridge crosses, (*Capsicum annuum* x *C. chinense*) x *C. baccatum*, as an alternative for introgression of anthracnose resistance from *C. baccatum* into *C. annuum*. J Kor Soc Hortic Sci 46:5–9

Yoon JB, Yang DC, Do JW, Park HG (2006) Overcoming two post-fertilization genetic barriers in interspecific hybridization between *Capsicum annuum* and *C. baccatum* for introgression of anthracnose resistance. Breed Sci 56:31–38

Chapter 5
Citrullus

P. Nimmakayala, N. Islam-Faridi, Y.R. Tomason, F. Lutz, A. Levi, and U.K. Reddy

5.1 Introduction

Watermelon is an important crop in the United States, whose farm value is estimated at $340 million (http://www.watermelon.org). Economic and nutraceutical importance of this crop is rapidly increasing throughout the world. During the last century, the importance of watermelon is steadily increased and accounts for 2% of the world area devoted to vegetable production (FAO 1995; Levi et al. 2001a, b). Enhancing disease and pest resistance of watermelon cultivars and improving their response to environmental stress can be accomplished by widening genetic diversity through hybridization with wild *Citrullus* accessions (Levi and Thomas 2005).

The family Cucurbitaceae consists of two well-defined subfamilies, eight tribes, 118 genera, and about 825 species (Robinson and Decker-Walters 1997; Jarret and Newman 2000). Cultivated watermelon and its wild relatives belong to the genus *Citrullus* of the subfamily Cucurbitoideae, tribe Benincaseae Ser., Subtribe Benincasinae (Ser.) C. Jeffrey (Robinson and Decker-Walters 1997). The name of *Citrullus* was first coined by Forskal in the year 1775 but H. Schrader was the first who classified the genus systematically, which was adopted by the Eighth International Botanical Congress, 1954 to be included in the *Nomina Conservanda* (Fursa 1972). Some of the morphological traits of taxonomic importance in various species of *Citrullus* genus are pollen structure, anatomy of fruits, seed structure, presence or absence of nectary flowers, characteristics of embryos, and variations in chromosome karyotypes. Genus *Citrullus* includes *C. lanatus* (var. *lanatus* (Thunb.) Matsum and Nakai., var. *citrides* (Bailey) Mansf.), *C. ecirrhosus* Cogn., *C. rehmii* De Winter., *C. colocynthis* (L.) Schrad, and *Acanthosicycos naudinianus* (Sond.) C. Jeffrey (Robinson and Decker-Walters 1997; Jarret and Newman 2000). Genus *Citrullus* in a wild state is distributed mostly in xerophytic habitats of the northern (*C. colocynthis*) and southern Africa (*C. lanatus*, *C. ecirrhosus*, *C. rehmii*, and *A. naudinianus*). *C. rehmii* and *C. lanatus* are monoecious annuals, whereas *C. colocynthis*, *C. ecirrhosus*, and *A. naudinianus* are perennials (Jarret and Newman 2000).

According to Meeuse (1962) and Pitrat et al. (1999), the species *Citrullus lanatus* ($n = 11$) originated in the Kalahari region of Namibia and Botswana (Bates and Robinson 1995; Robinson and Decker-Walters 1997; Ellul et al. 2007). Cultivated watermelon includes three subspecies: *C. lanatus* subsp. *lanatus* (Shrad. Ex Eckl. et Zeyh.), *C. lanatus* subsp. *vulgaris* (Shrad. Ex Eckl. et Zeyh.) Fursa, and *C. lanatus* subsp. *mucosospermus* Fursa (Levi et al. 2001a, b). On the other hand, now all these three species are under the var. group *lanatus* (Jeffrey 2001). Currently, the species *C. lanatus* (Thunb. Matsum and Nakai) includes two botanical varieties, namely var. *lanatus* (Bailey) and var. *citroides* (Mansf). Cultivated watermelons belong to var. *lanatus* and have endocarps in wide ranging colors. The var. *citroides* is cultivated in southern Africa, and also called "Tsamma" or "citron" melon, whose rind is used as preservative in pickles (Whitaker and Davis 1962; Fursa 1972; Whitaker and Bemis 1976; Burkill 1985; Jarret et al. 1997; Jeffrey 2001). The citron fruits have green- or white-colored flesh and their taste

U.K. Reddy (✉)
Department of Biology and Gus R. Douglass Institute, West Virginia State University, Institute, WV 25112–1000, USA
e-mail: ureddy@wvstateu.edu

C. Kole (ed.), *Wild Crop Relatives: Genomic and Breeding Resources, Vegetables*,
DOI 10.1007/978-3-642-20450-0_5, © Springer-Verlag Berlin Heidelberg 2011

varies from bland to bitter. Seed production fields should be isolated from weedy citron types since these two botanical varieties cross readily (Wehner 2008). The species *Citrullus colocynthis* (Schrad) is a perennial herb known as bitter apple and is a desert species with a rich history as a medicinal plant (Dane et al. 2007). T.W. Whitaker considered *C. colocynthis* to be a likely ancestor of watermelon as it is morphologically similar to *C. lanatus*, and is freely intercrossable and produces fertile hybrids (Wehner 2007). Dane et al. (2007) reported divergent lineages of *C. colocynthis* that are from tropical Asia and Africa, now widely distributed in the Saharo-Arabian phylogeographic region of Africa and in the Mediterranean region. In earlier reports, isozyme and random amplified polymorphic DNA (RAPD) markers were used extensively in molecular diversity and phylogenetic analyses in *Citrullus* spp. (Zamir et al. 1984; Navot and Zamir 1986; Biles et al. 1989; Levi et al. 2001a, b).

5.2 Phylogenetic Relationships

Phenetic relationships among the main *Citrullus* species and subspecies were examined by using isozymes (Zamir et al. 1984; Navot and Zamir 1987) and nuclear DNA markers (Jarret et al. 1997; Levi et al. 2000). Dane et al. (2004) summarized domestication of cultivated watermelon to be ancient and its cultivation dates back to pre-historic times. It was grown by the ancient Egyptians (Robinson and Decker-Walters 1997). Dane and Lang (2004) further reviewed about introduction of watermelons to Europe by the Moors during their invasion of Spain and to the Americas in the seventeenth century on slave ships and have been cultivated ever since in the western hemisphere. The domesticated watermelon is classified as *C. lanatus* var. *lanatus*, whereas wild citron, which is common in central Africa, is classified as var. *citroides* (Bailey) Mansf. Citron is a preserving melon as its rind is used to make pickles (Dane et al. 2004). Wehner (2008) reported that the fruits of *citroides* are used for food for livestock in Africa. In West Africa, especially Nigeria, Egusi-type watermelons with bitter fruit (*citroides*) are cultivated for their seeds as they have high edible oil content. Dane et al. (2004) reported cpDNA variation after studying 55 *C. lanatus* types and 15 *C. colocynthis* that are from diverse geographical areas. This study revealed insertion–deletion sites (Indels) at *ndh*A, *trn*S–*trn*fM, and *trn*C–*trn*D regions of cpDNA along with single nucleotide polymorphisms (SNPs) to separate *lanatus* and *colocynthis* species. Dane and Lang (2004) also reported diagnostic SNPs at *ndh*F and *trn*C–*trn*D of chloroplast to distinguish between the var. *lanatus* and var. *citroides*, respectively. In this study, several indels at *ndh*A, *trn*S–*trnf*M and *trn*C–*trn*D regions, and several substitutions at restriction sites were characterized between *colocynthis* and *lanatus*. Dane and Liu (2007) critically studied var. *lanatus* and var. *citroides* using PCR-based restriction fragment length polymorphism (RFLP) such as cleaved amplified polymorphic sequence (CAPS) of chloroplast regions to conclude that they had a common ancestor and resolved subspecies-specific haplotype fixation. This study identified that the ancient *citroides*-type haplotype originated in Swaziland and South Africa and followed colonization routes from these areas to all over the world. Levi and Thomas (2005) used 20 cpDNA and 10 mitochondrial DNA probes for RFLP analysis for phylogenetic analysis. A combined analysis of large data sets (3,089 AFLPs and 127 SSR alleles) by Nimmakayala et al. (2010) provided strong evidence of phylogenetic signal that clearly resolved a tree with three clusters of *lanatus*, *citroides*, and *colocynthis* supported by significant bootstrap values. In this study, tree topologies inferred by Neighbor-Joining analysis have resolved the phylogenic relationships among the species with special reference to established taxonomic classification. Further, boundaries of various taxa belonging to *citroids*, *lanatus*, and *colocynthis* could be drawn. Clustering pattern of principal coordinate analysis (PCA) with the shared polymorphisms using the subsets of data between any two taxa combinations helped to elucidate the introgression and interrelationships among the species. This research resolved two major groups of *lanatus* taxa, one of which has undergone wide introgressions with the taxa of *citroids* and *colocynthis*.

Dane et al. (2007) characterized phylogeography of the species *C. colocynthis*, a non-hardy species, which is predominantly drought-resistant perennial herbaceous vine, now widely distributed in the Sahar-Arabian region in Africa and also in the Mediterranean region. This species was characterized by angular stems, lobed leaves, solitary pale yellow flowers, and can produce up to 15–30 fruits (Dane et al. 2007). Seeds are small, smooth, and brownish in color (Robinson and Decker-Walters 1997). Dane et al. (2007) further summarized that these species were known since Biblical times as bitter apple and were used to extract deadly poison. Fruits are widely used medicinally as laxative because of colocynthin content. The seeds are edible and used to make bread as well as extract oil (17–19% oil with 80–85% unsaturated fatty acids) (Dane et al. 2007). The oil is edible and also useful for candle light, medicinal, and industrial purposes (Zohary and Hopf 2000). *C. colocynthis* primarily accumulates citrulline under drought conditions, which contributes to oxidative stress tolerance (Yokota et al. 2002). Dane et al. (2007) characterized several polymorphic intergenic cpDNA and a relatively large intron (0.6 kb) of G3*pdh* to resolve geographical structure among the world collections of *C. colocynthis*. The study revealed the migration of the species from Africa into Middle East and Far East. This study also revealed divergent haplotypes in *C. colocynthis* population based on differential patterns of adaptation.

Jarret and Newman (2000) amplified internal transcribed spacer regions (ITS1 and ITS2) of the 18S–25S nuclear ribosomal DNA on: *C. lanatus* (var. *lanatus* (Thunb.) Matsum and Nakai., var. *citrides* (Bailey) Mansf.), *C. ecirrhosus* Cogn., *C. rehmii* De Winter., *C. colocynthis* (L.) Schrad, and *A. naudinianus* (Sond.). Cladistic and phenetic analysis in this study resulted in robust tree placing the species *C. rehmii* closure to the clade of *C. lanatus*. This study confirmed the species status to *C. rehmii* and indicated its closeness to the cultivated watermelon than to the species *C. colocynthis*. Phenetic analysis in this study resolved the branch separating *C. rehmii* and *C. lanatus* from the species *C. ecirrhosus* with high bootstrap values. The terminal placement of annual species *C. rehmii* and *C. lanatus*, relative to the xerophytic perennials *C. ecirrhosus* and *C. colocynthis* in this investigation, supported the argument of Jobst et al. (1998) concerning the derivation of annual species from perennial forms. Leaves of *C. rehmii* more closely resembled to those of *C. lanatus* rather than to those of *C. ecirrhosus*, in which the leaves were distinctly rigid with strongly recurved margins (Meeuse 1962). The fruits of *C. ecirrhosus* are hard and bitter with ellipsoid shape and maturity duration of 60 days after anthesis (Meeuse 1962). The fruits of *C. rehmii* resembled more to *C. lanatus*, in which the fruits are not hard and bitter but globose in shape and mature within 30 days post-anthesis (Jarret and Newman 2000). Martyn and Netzer (1991) reported that the species *C. ecirrhosus* harbors several important genes for disease resistance.

5.3 Genetic and Genomic Resources

5.3.1 Molecular Markers, ESTs, and Unigenes

A large number of fruit-related expressed sequence tags (ESTs) were developed by Ok et al. (2000) and Levi et al. (2009). Several genomic and EST-specific simple sequence repeat (SSR) markers were developed in *Citrullus* species. Very interestingly, the SSRs developed for watermelon will amplify as well as show polymorphism indicating their transportability across the other species/genera of cucurbits including melon, cucumber, squash, and pumpkin (Figs. 5.1 and 5.2). The gel picture presented in Fig. 5.1 represents amplification pattern of a fruit-specific EST SSR across various cucurbit genera. Figure 5.2 is from the summary of transportability of 124 SSRs, showing amplifications and polymorphic levels across the genera.

Nimmakayala et al. (2010) amplified 127 alleles using a set of 42 SSRs in 31 watermelon accessions. A range of 2–15 alleles were amplified per SSR. The number of specific alleles within the group was 20, 13, and 7 specific to var. *lanatus*, var. *citroides*, and

Fig. 5.1 Gel showing resolution of a microsatellite across various genera Cucurbitaceae family. 1. *Citrullus colocynthis* (PI386016), 2. *Citrullus lanatus* var. *citroides* (PI482252), 3. *Citrullus lanatus* var. *lanatus* (PI 248178), 4. *Lagenaria siceraria*, 5. *Momordica charantia*, 6. *Cucumis melo* var. *aestivalis*, 7. *Cucumis melo* var. *europius*, 8. *Cucurbita moschata*, 9. *Cucurbita pepo*, 10. *C. pepo* ssp. *Texanana*, 11. *C. pepo* ssp. 12. *Fraterna*, and 13. *C. pepo* ssp. *Ozarkana*

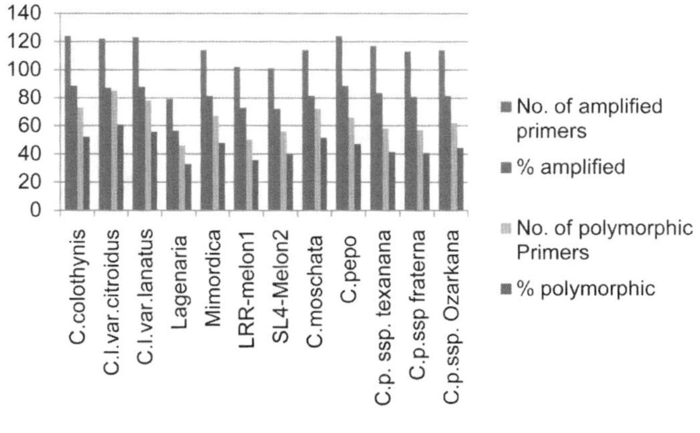

Fig. 5.2 Polymorphic levels of various microsatellites used across various *Citrullus* species and the other cucurbit genera

C. colocynthis, respectively. The shared alleles between *lanatus* and *citroides* were 38, *citroides* and *colocynthis* were 17 and *colocynthis* and *lanatus* were 25. SSRs are simple-to-use, multiallelic, and codominant marker systems that are sequence-based and produce highly repeatable amplifications. The SSRs in this study generated important diagnostic markers that are species-specific and can be of immense use for resolving species conflicts that are reported to exist between *lanatus* and *citroides*. A large set of fruit-specific ESTs and assembled unigene resources for various cucurbit crops are available at http://www.icugi.org. However as stated, the SSRs mined from these ESTs (about 315 can be accessed at http://www.icugi.org) are usable and when tested on reference accessions of *Citrullus* spp., all the SSRs amplified single products and a large number of them were polymorphic (U. Reddy unpublished data). This resource will help to develop syntenic maps across the cucurbit species and also will aid to identify heterologous locations for important horticultural and disease-resistant traits. In addition, Dane et al. (2004) identified diagnostic markers using cpDNA haplotypes for *lanatus* and *citroides* types and used them to track lineages with *C. rehmii* and *C. ecirrhosus*. Jarret et al. (1997) developed seven SSRs and used them to amplify 32 watermelon genotypes; they found that SSR-derived polymorphisms are very efficient in discriminating among various species. Guerra-Sanz (2002) identified 19 SSRs from cDNA sequence data.

We recently isolated SSR markers in large scale from a single run of watermelon genomic DNA using 454 Life Sciences sequencing technology. We have characterized a total of 2,143 contigs that contain a total of 2,727 SSRs from a pool of 13,176 contigs of 454 sequencing reads. We identified 1,025 SSR motifs that could be used as potential molecular markers based on their longer repeat lengths and quality of flanking sequence for primer design (U. Reddy unpublished data). Out of 2,727 total isolated microsatellite regions, 1,346 were dinucleotide repeats (DNRs), 980 trinucleotide repeats (TNRs), 287 tetranucleotide repeats (TTNRs), 83 pentanucleotide repeats (PNRs), and 24 SSRs with hexanucleotide repeats. Dinucleotide repeats constituted 49.36% of the total identified repeats. The most common motif type of DNRs was TA/AT (67.68% of DNRs) followed by AG/CT (24.45% of DNRs) and AC/GT (7.88% of DNRs) in watermelon genome.

5.3.2 Genetic Maps and QTLs

Densely-saturated genetic maps are very important in breeding programs of crop plants. They are useful for locating genes or quantitative trait loci (QTL) of various traits (Lee 1995). Extensive linkage maps have been constructed for such cucurbits as melon (Baudracco-Arnas and Pitrat 1996; Wang et al. 1997; Brotman et al. 2000; Oliver et al. 2000, 2001; Perin et al. 2002) and cucumber (Park et al. 2000; Staub and Serquen 2000). However in watermelon, only a few linkage maps with minimal coverage have been reported (Navot and Zamir 1986; Navot et al. 1990; Hashizume et al. 1996; Xu et al. 2000; Hawkins et al. 2001).

The first genetic map of watermelon was constructed by Navot and Zamir (1986) in a segregating population of *C. lanatus* × *C. colocynthis*, and later extended by Navot et al. (1990) into seven linkage groups covering a length of 354 cM. Hashizume et al. (1996) constructed a linkage map of 11 linkage groups spanning only 524 cM with 58 random amplified polymorphic DNA (RAPD), one isozyme, one restriction fragment length polymorphism (RFLP), and two morphological markers. In 2003, Hashizume et al. constructed another linkage map using an F_2 population with 477 RAPD, 53 RFLP, 23 intersimple sequence repeat (ISSR), and one isozyme markers that covered 2,384 cM. Levi et al. (2001a, b) constructed a linkage map of 17 linkage groups using 155 RAPD markers and a sequenced characterized amplified region (SCAR) marker covering 1,295 cM in a backcross population [PI 296341 (*C. lanatus* var. *citroides*) × New Hampshire Midget (NHM, *C. lanatus* var. *lanatus*)] × NHM. Another linkage map was constructed by Levi et al. (2002) using a testcross population [Griffin 14113 (*C. lanatus* var. *citroides*) × NHM] × PI 386015 (*C. colocynthis*) with 141 RAPD, a SCAR, and 27 ISSR markers segregating in 25 linkage groups covering a total distance of 1,166 cM. Lately, they added another 114 amplified fragment length polymorphism (AFLP) markers to this map (Levi et al. 2006). However, a significant part of the genome (watermelon genome size 425 Mb; Arumuganathan and Earle 1991) has not been saturated yet, and a good number of markers therefore are still needed for construction of a high-density map.

Many economically important traits of crop plants are inherited as quantitative traits. The phenotypes appear to be conditioned by several loci with strong environmental effects. Quantitative traits were usually analyzed using biometrical models before the discovery of mapping techniques, and the biometrical approach cannot completely explain the effects of individual QTL controlling a trait. In recent years, the availability of DNA markers coupled with biometric methods has helped to make considerable progress in QTL mapping. The joint analysis of marker segregation and phenotypic values of individuals or lines in QTL mapping enables the scientists to detect and locate the loci governing quantitative traits (Asins 2002). QTL analysis not only provides DNA markers for efficient marker-assisted selection (MAS) in plant breeding, but also resolves the interacting environmental effects on important yield-related traits.

So far the detection and mapping of QTL for interesting agronomic traits are hindered by the scarcity of molecular markers (Danin-Poleg et al. 2002). Unfortunately, there are very few markers identified so far in watermelon which are linked to important agronomic traits. A molecular marker linked to resistance to *Fusarium* (Xu et al. 2000) and two other QTLs controlling fruit traits, viz. rind hardness and Brix of flesh juice (Hashizume et al. 2003), have been reported. Special efforts are currently underway to identify QTLs for important traits as well as to develop recombinant inbred lines that will facilitate extensive phenotypic evaluation at multiple locations (U. Reddy unpublished data).

5.4 Fluorescent In Situ Hybridization and Chromosome Organization in var. *lanatus* and var. *citroides*

Hereunder is presented the first attempt of fluorescent in situ hybridization (FISH) in var. *lanatus* (PI 270306) and its wild counterpart var. *citroides* (PI 244018), using 18S–28S rDNA and 5S rDNA probes. Well-separated somatic chromosomes were prepared from root meristems, using enzyme digestion technique for hybridization following the standard techniques (Islam-Faridi et al. 2007). Chromosome spreads that are presented below present very interesting chromosome organization between cultivated watermelon (Figs. 5.3 and 5.4) and its wild counterpart *C. lanatus*

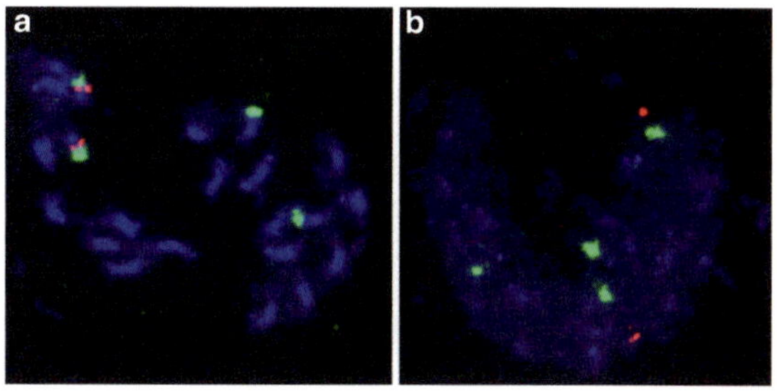

Fig. 5.3 FISH with 18S–28S rDNA (*green signals*) and 5S rDNA (*red signals*) on watermelon (var. *lanatus*, accession # PI 270306) chromosome spread. (**a**) Late prophase chromosome spread cell, and (**b**) interphase cell of var. *lanatus*

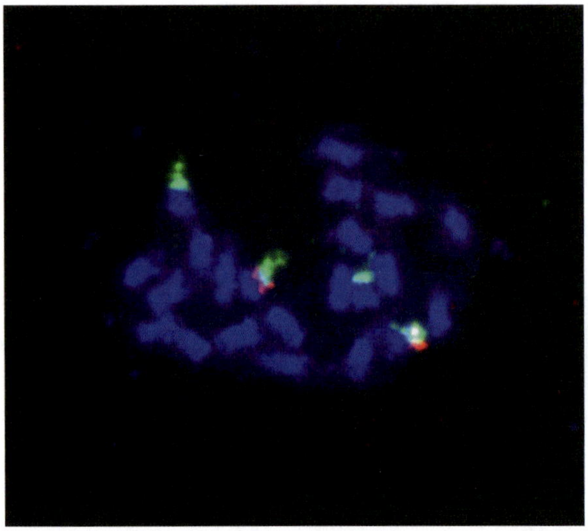

Fig. 5.4 A prometaphase stage in var. *lanatus*

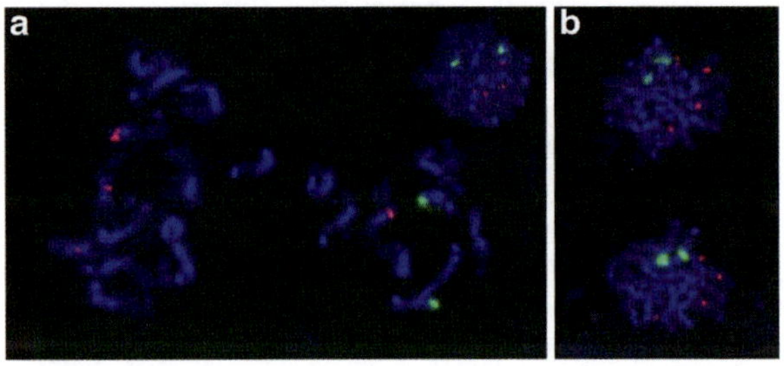

Fig. 5.5 FISH with 18S–28S rDNA (*green signals*) and 5S rDNA (*red signals*) probes on *citroides* chromosome spreads. (**a**) Mid prophase chromosome spread and showing two major 18S–28S rDNA signals, four 5S rDNA signals, and an interphase nucleus also showing two green signals from 18S–28S rDNA probe, (**b**) A late anaphase chromosome spread with the same kind of hybridization pattern

var. *citroides*. In *lanatus* spread, we noticed that there are two 18S–28S rDNA sites and one 5S rDNA site (Figs. 5.3a, b and 5.4). The 5S rDNA site is located interstitially and appears to be syntenic to one of the 18S–28S rDNA sites. As revealed by the interphase FISH (Fig. 5.3b), the sites of 18S–28S rDNA and 5S rDNA site are not linked. In contrary, there were three different sites of rDNA in *Citroides* accession (PI 244018) (Fig. 5.5a, b), one was for 18S–28S rDNA and two were for 5S rDNA, and all were on three different chromosomes. These results clearly indicate that there are major structural differences between these subspecies.

5.5 Conclusions

Enhancing disease and pest resistance in watermelon cultivars to improve their response to drought and other biotic resistance is possible through hybridization with wild *Citrullus* accessions as all the cultivars are freely crossable and produce fertile hybrids. Since the EST, unigene, and SSRs are highly conserved across the *Citrullus* genera, it is possible to develop dense maps with the positions of map locations of important traits. Generating extensive molecular cytogenetics resources, such as FISH, will allow integrating the mapping information, BACs/physical maps, and other probes that allow precise karyotype analysis of various species. Mapping endeavors such as high-throughput genotyping, developing reciprocal recombinant inbred populations, multiple environment evaluation, and QTL localization will speed up the introgression process as well as launch watermelon breeding in the new era that would impact the productivity and quality of watermelon cultivars.

Acknowledgments Funding support is provided by USDA-CSREES Research (2007-38814-18472).

References

Arumuganathan K, Earle E (1991) Estimation of nuclear DNA content of plants by flow cytometry. Plant Mol Biol Rep 9:221–231

Asins MJ (2002) Present and future of quantitative trait locus analysis in plant breeding. Plant Breed 121:281–291

Bates DM, Robinson RW (1995) Cucumbers melon and watermelons. In: Smart J, Simmonds NW (eds) Evolution of crop plants, 2nd edn. Longman, London, UK, pp 89–96

Baudracco-Arnas S, Pitrat M (1996) A genetic map of melon (*Cucumis melo L.*) with RFLP, RAPD, isozyme, disease resistance and morphological markers. Theor Appl Genet 93:57–64

Biles CL, Martyn RD, Wilson HD (1989) Isozymes and general proteins from various watermelon cultivars and tissue types. HortScience 24:810–812

Brotman Y, Silberstein L, Kovalski I, Klingler J, Thompson G et al (2000) Linkage groups of *Cucumis melo*, including resistance gene homologues and known genes. Acta Hortic 510:441–448

Burkill HM (1985) The useful plants of west tropical Africa, vol 1, 2nd edn. Royal Botanic Gardens, Kew, UK

Dane F, Lang P (2004) Sequence variation at chloroplast DNA regions of watermelon and related wild species. Implications

for the evolution of Citrullus haplotypes. Am J Bot 91: 1922–1929
Dane F, Liu J (2007) Diversity and origin of cultivate and citron type watermelon (*Citrullus lanatus*). Genet Resour Crop Evol 54:1255–1265
Dane F, Lang P, Bakhtiyarova R (2004) Comparative analysis of chloroplast DNA variability in wild and cultivated *Citrullus* species. Theor Appl Genet 108:958–966
Dane F, Liu J, Zhang C (2007) Phylogeography of the bitter apple, *Citrullus colocynthis*. Genet Resour Crop Evol 54: 327–336
Danin-Poleg Y, Tadmor Y, Tzuri G, Reis N, Hirschberg J, Katzie N (2002) Construction of a genetic map of melon with molecular markers and horticulture traits, and localization of genes associated with ZYMV resistance. Euphytica 125:373–384
Ellul P, Lelivelt C, Naval MM, et al (2007) Watermelon. Biotechnology in agriculture and forestry, vol 60: Transgenic crops. Springer, Berlin, pp 129–165
FAO (1995) Production year book for 1994. No 48. Food and Agriculture Organization of the United Nations, Rome, Italy
Faridi M, Dana Nelson C, Thomas LK (2007) Reference karyotype and cytomolecular map for loblolly pine. Genome 2: 241–251
Fursa TB (1972) K sistematike roda *citrullus* Schrad. Bot Z 57 (1):31–41
Guerra-Sanz JM (2002) *Citrullus* simple sequence repeats markers from sequence databases. Mol Ecol Notes 2:223–225
Hashizume T, Shimamoto I, Harushima Y, Yui M, Sato T (1996) Construction of a linkage map for watermelon (*Citrullus lanatus* (Thunb.) Matsum. & Nakai) using random amplified polymorphic DNA (RAPD). Euphytica 90: 256–273
Hashizume T, Shimamoto I, Hirai M (2003) Construction of a linkage map and QTL analysis of horticultural trails for watermelon (*Citrullus lanatus*) using RAPD, RFLP and ISSR markers. Theor Appl Genet 106:779–785
Hawkins LK, Dane F, Kubisiak TL, Rhodes BB, Jarret RL (2001) Linkage mapping in watermelon population segregating for Fusarium wilt resistance. J Am Soc Hortic Sci 126: 344–350
Jarret RL, Newman M (2000) Phylogenetic relationships among species of *Citrullus* and the placement of *C. rehmii* De Winter as determined by internal transcribed spacer (ITS) sequence heterogeneity. Genet Resour Crop Evol 47:215–222
Jarret RL, Merrick LC, Holms T, Evans J, Aradhya MK (1997) Simple sequence repeats in watermelon (*Citrullus lanatus* (Thunb.) Matsum & Nakai). Genome 40:433–441
Jeffrey C (2001) Cucurbitaceae (*citrullus*). In: Hanelt P and IPGCPR (eds) Mansfeld's encyclopedia of agricultural and horticultural crops. Springer, New York, NY, USA, pp 1533–1537
Jobst J, King K, Hemleben V (1998) Molecular evolution of the internal transcribed spacers (ITS1 and ITS2) and phylogenetic relationship among the species of the family *Cucurbitaceae*. Mol Phylogenet Evol 9:204–219
Lee M (1995) DNA markers and plant breeding programs. Adv Agron 55:265–344
Levi A, Thomas CE (2005) Polymorphisms among chloroplast and mitochondrial genomes of *Citrullus* species and subspecies. Genet Resour Crop Evol 52:609–617
Levi A, Thomas CE, Keinath AP, Wehner TC (2000) Estimation of genetic diversity among *Citrullus* accessions using RAPD markers. Acta Hortic 510:385–390
Levi A, Thomas CE, Keinath AP, Wehner TC (2001a) Genetic diversity among watermelon (*Citrullus lanatus* and *Citrullus colocynthis*) accessions. Genet Resour Crop Evol 48:559–566
Levi A, Thomas CE, Wehner TC, Zhang X (2001b) Low genetic diversity indicates the need to broaden the genetic base of cultivated watermelon. HortScience 36:1096–1101
Levi A, Thomas CE, Joobeur T, Zhang X, Davis A (2002) A genetic linkage map for watermelon derived from a test cross population: (*Citrullus lanatus var. citroids* x C. *lanatus var. lantus*) x *Citrullus colocynthis*. Theor Appl Genet 105:555–563
Levi A, Thomas CE, Trebitsh T, Salman A, King J, Karalius J, Newman M, Reddy OUK, Xu Y, Zhang X (2006) An extended linkage map for watermelon based on SRAP, AFLP, SSR, ISSR, and RAPD markers. J Am Soc Hortic Sci 131:393–402
Levi A, Wechter P, Davis A (2009) EST-PCR markers representing watermelon fruit genes are polymorphic among watermelon heirloom cultivars sharing a narrow genetic base. Plant Genet Resour 7(01):16–32
Martyn RD, Netzer D (1991) Resistance to races 0, 1, and 2 of Fusarium wilt of watermelon in *Citrullus* sp. PI-296341-FR. HortScience 26:429–432
Meeuse ADJ (1962) The Cucurbitaceae of southern Africa. Bothalia 8:111
Navot N, Zamir D (1986) Linkage relationships of 19 protein-coding genes in watermelon. Theor Appl Genet 72:274–278
Navot N, Zamir D (1987) Isozyme and seed protein phylogeny of the genus *Citrullus* (Cucurbitaceae). Plant Syst Evol 156:61–68
Navot N, Sarfatti M, Zamir D (1990) Linkage relationships of genes affecting bitterness and flesh color in watermelon. J Hered 81:162–165
Nimmakayala P, Tomason YR, Jeong J, Vajja G, Levi A, Reddy UK (2010) Genetic reticulation and interrelationships among Citrullus species as revealed by joint analysis of shared AFLPs and species specific SSR alleles. Plant Genet Resour 8:16–25
Ok S, Chung YS, Um BY, Park MS, Bae JM, Lee SJ, Shin JS (2000) Identification of expressed sequence tags of watermelon (Citrullus lanatus) leaf at the vegetative stage. Plant cell Rep 19:932–937
Oliver M, Garcia-mas J, Morales M, Dolcet-Sanjuan R, Carmen de Vincente M, Gomez H, van Leewen H, Monfort A, Puigdomenech P, Arus P (2000) The Spanish melon genome project: construction of a saturated genetic map. Acta Hortic 510:375–378
Oliver M, Garcia-mas J, Cardus M, Pueyo N, Lopez-Sese A, Arroyo M, Gomez-Paniahue H et al (2001) Construction of a reference linkage map for melon. Genome 44:836–845
Park YH, Sensoy S, Wye C, Antonise R, Peleman J, Havey MJ (2000) A genetic map of cucumber composed of RAPD, RFLPs, AFLPs, and loci conditioning resistance to papaya ring spot and zucchini yellow mosaic viruses. Genome 43: 1003–1010
Perin C, Hagen LS, De Conto V, Katzir N, Danin-Poleg Y, Portnoy V et al (2002) A reference map of *Cucumis melo* based on two recombinant inbred line population. Theor Appl Genet 104:1017–1034

Pitrat M, Chauvet M, Foury C (1999) Diversity, history and production of cultivated cucurbits. Acta Hortic 492:21–28

Robinson RW, Decker-Walters DS (1997) Cucurbits. CABI, Wallingford, Oxfordshire

Staub JE, Serquen FC (2000) Towards an integrated linkage map of cucumber: map merging. Acta Hortic 510:357–366

Wang YH, Thomas CE, Dean RA (1997) A genetic map of melon (*Cucumis melo* L.) based on amplified fragment length polymorphism (AFLP) markers. Theor Appl Genet 95:791–798

Wehner TC (2007) Watermelon. North Carolina State University, Department of Horticulture Science, Raleigh, NC, USA, pp 381–418

Wehner TC (2008) Watermelon. In: Prohens J, Nuez F (eds) Handbook of plant breeding; Vegetables I: Asteraceae, Brassicaceae, Chenopodiaceae, and Cucurbitaceae. Springer, New York, NY, USA, pp 381–418

Whitaker TW, Bemis WB (1976) Cucurbits. In: Simmonds NW (ed) Evolution of crop plants. Longman, London, UK, pp 64–69

Whitaker TW, Davis GN (1962) The Cucurbits: botany, cultivation and utilization. Interscience, New York, NY, USA

Xu Y, Zhang HY, Kang GB, Wang YJ, Chen H (2000) Studies of molecular marker-assisted selection for resistance to fusarium wilt in watermelon (*Citrullus lanatus*) breeding. Acta Genet Sin 27:151–157

Yokota A, Kawasaki S, Iwano M, Nakamura C, Miyake C, Akashi K (2002) Citrulline and DRIP-1 protein in drought tolerance of wild watermelon. Ann Bot 89:825–832

Zamir D, Navot N, Rudich J (1984) Enzyme polymorphism in *Citrullus lanatus* and *C. colocynthis* in Israel and Sinai. Plant Syst Evol 146:163–167

Zohary D, Hopf M (2000) Domestication of plants in the old world. Oxford University Press, Oxford, UK

Chapter 6
Cucumis

Jin-Feng Chen and Xiao-Hui Zhou

6.1 Introduction

The genus *Cucumis* L. is one of the important genera of flowering plants. It includes cucumber (*C. sativus* L.) and melon (*C. melo* L.), two of the most economically important and widely cultivated vegetable crops in the world (Pitrat et al. 1999). However, cucumber and melon suffer from a range of devastating fungal, bacterial, viral, and insect diseases (Whitaker and Davis 1962). The wild *Cucumis* species are of economic interest because they are a reservoir of potentially useful genes such as biotic stress resistances. A comprehensive knowledge of these wild species is of great importance for conservation and utilization of genetic resources that can be employed for cucumber and melon improvement.

6.2 Basic Botany of the Genus *Cucumis*

6.2.1 *Taxonomy*

Cucumis belongs to the family Cucurbitaceae, subfamily Cucurbitoideae, and is currently placed in the tribe Benincaseae (Jeffrey 2005). According to Kirkbride (1993), the genus *Cucumis* is represented by 32 species, among which two species of *Cucumis*, *C. sativus* L. (Cucumber) and *C. melo* L. (melon), are of great commercial importance. Besides cucumber and melon, the species *C. anguria* (West Indian gherkin) and *C. metuliferus* (African horned cucumber) are commercially cultivated in several areas as well (Garcia-Mas et al. 2004). Other wild species originating mostly from arid and/or semi-arid regions of Africa are cultivated as ornamental plants (e.g., *C. dipsaceus* – "hedgehog gourd" and *C. myriocarpus* – "gooseberry gourd") (Rubatzky and Yamaguchi 1997).

Taxonomy of *Cucumis* was first described by Linnaeus in 1753 (Ghebretinsae et al. 2007). According to this classification, the genus *Cucumis* contains seven species, all of which were cultivated or economically useful. There have been a number of taxonomic placements of *Cucumis* since the work of Linnaeus (Pangalo 1950; Jeffrey 1962, 1967, 1980, 1990; Kirkbride 1993; Schaefer 2007).

The most comprehensive placement of *Cucumis* was proposed by Kirkbride (1993). On the basis of his investigations, the genus *Cucumis* was divided into two subgenera with different geographical origin and basic chromosome numbers. Subgenus *Melo* (30 spp., $n = 12$) was originated in Africa and was partitioned into two sections (*Melo* and *Aculeatosi*), whereas subg. *Cucumis* (two spp., $n = 7$) was originated in Asia. A detailed taxonomic depiction of the genus *Cucumis* elaborated by Kirkbride (1993) is given in Table 6.1. However, this taxonomic treatment was challenged by the rediscovery of *C. hystrix*, a wild *Cucumis* species of Asian origin possessing 24 chromosomes. *C. hystrix* ($2n = 24$) was successfully crossed with cucumber (*C. sativus*, $2n = 14$) (Chen et al. 1997). A new species *C.* × *hytivus* Chen and Kirkbride was proposed in 2000 followed by chromosome doubling of the F_1 hybrid (Chen and Kirkbride 2000).

Recently, on the basis of molecular phylogenetic studies, 19 species of five genera including *Cucumella*,

J.-F. Chen (✉)
State Key Laboratory of Crop Genetics and Germplasm Enhancement, Nanjing Agricultural University, No. 6 Tongwei Road, Nanjing 210095, China
e-mail: jfchen@njau.edu.cn

C. Kole (ed.), *Wild Crop Relatives: Genomic and Breeding Resources, Vegetables*,
DOI 10.1007/978-3-642-20450-0_6, © Springer-Verlag Berlin Heidelberg 2011

Table 6.1 Taxonomy of the genus *Cucumis* (Kirkbride 1993)

Cucumis spp.	Chromosome number (*n*)
Subgenus *Melo*	
Section *Aculeatosi*	
Serie *Myriocarpus*	
C. myriocarpus	
subsp. *myriocarpus*	12
subsp. *leptodermis*	12
C. africanus	12
C. quintanilhae	–
C. heptadactylus	24
C. calahariensis	–
Serie *Angurioidei*	
C. anguria	
var. *anguria*	12
var. *longaculeatus*	12
C. sacleuxii	12
C. carolinus	–
C. dipsaceus	12
C. prophetarum	
subsp. *prophetarum*	12
subsp. *dissectus*	12
C. pubituberculatus	–
C. zeyheri	12 (24)
C. prolatior	–
C. insignis	12
C. globosus	12
C. thulinianus	–
C. ficifolius	12 (24)
C. aculeatus	24
C. pustulatus	12, 48, 72
C. meeusei	24
C. jeffreyanus	–
C. hastatus	–
C. rigidus	–
C. baladensis	–
Serie *Metuliferi*	
C. metuliferus	12
C. rostratus	–
Section *Melo*	
Serie *Hirsuti*	
C. hirsutus	12
Serie *Humifructosi*	
C. humifructosi	12
Serie *Melo*	
C. melo	
subsp. *melo*	12
subsp. *agrestis*	12
C. sagittatus	12
Subgenus *Cucumis*	
C. sativus	7
C. hystrix	12

Dicoelospermum, *Mukia*, *Myrmecosicyos*, and *Oreosyce* have been transferred to *Cucumis*, resulting in 14 new combinations, two changes in status, and three new names (*Cucumis indicus*, *C. kirkbrideana*, and *C. oreosyce*). A complete morphological key to all these species now included in *Cucumis* was provided in Schaefer (2007).

6.2.2 Morphology

The genus *Cucumis* includes annual and perennial taxa, and the fruit morphology is the most important character within the genus (Fig. 6.1). Following morphological descriptions are given by combining the data from Kirkbride (1993), Rubatzky and Yamaguchi (1997), and Kristkova et al. (2003).

6.2.2.1 Plants

Herbs, exceptionally semi-shrubs, usually having a trailing or climbing growth habit, are monoecious, or rarely dioecious or andromonoecious; root systems are rarely woody (*C. trigonus*) and extensive, but usually shallow and rarely tuberous (*C. kalahariensis*); stems are angled, sulcate, not aculeate or rarely aculeate (*C. aculeatuc* and *C. ficifolius*), and variously pubescent or rarely glabrous, with non-breakaway hairs or rarely breakaway hairs (*C. sacleuxii*); nodes are geniculate or not geniculate. Each node has a single leaf and a simple tendril (sometimes curling), except that of *C. humifructus,* which has a fasicicle of five to eight tendrils, and that of *C. rigidus,* which lacks them; tendrils of *C. insignis* are either simple or bifid. Tendrils are variously pubescent, rarely glabrous, or rarely aculeate.

6.2.2.2 Leaves

Simple and petiolate. Petioles vary in length (with regard to the length of a leaf blade). They are not aculeate or rarely aculeate and variously pubescent or rarely glabrous, with non-breakaway hairs or rarely breakaway hairs. The majority of species have a uniform type of pubescence on the petioles. *C. sagittatus* and *C. thulinianus* have two pubescence types uniformly mixed over the entire petiole, and *C. myriocarpus* has three types separated into distinct zones

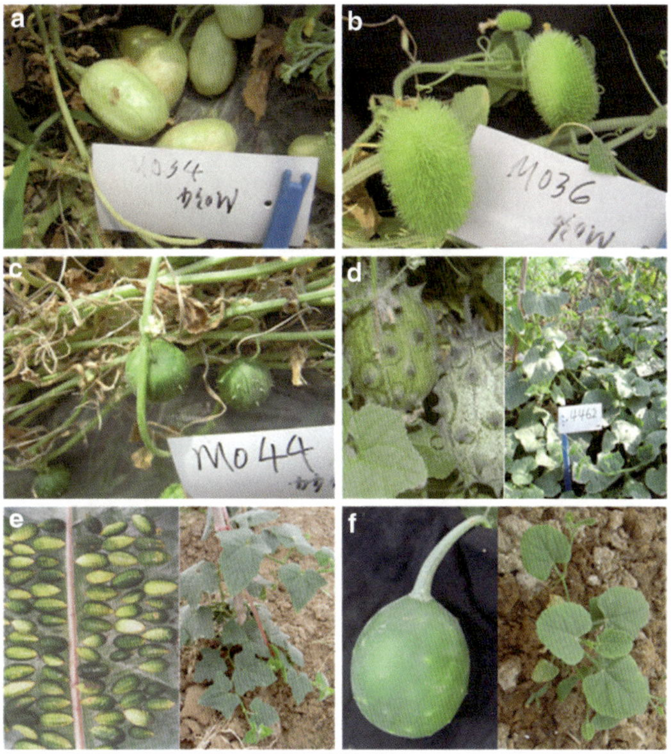

Fig. 6.1 Fruit of some species of the genus *Cucumis*. (**a**) *C. anguria*. (**b**) *C. Dipsaceus*. (**c**) *C. myriocarpus*. (**d**) *C. metuliferus*. (**e**) *C. hystrix*. (**f**) *C. Figarei*

on the petioles, retrose–strigose on the base, hirsute in the middle, and antrorse–strigose at apex; leaf blades are 3- or 5-palmately lobed, trilobite, pentalobate, heptalobate, or entire; Central leaf lobe is symmetrical, entire, or sometimes pinnatifid; lateral leaf lobes are asymmetrical, or sometimes symmetrical, entire, or sometimes pinnatifid.

6.2.2.3 Inflorescences and Flowers

The inflorescence is unisexual and most species are monoecious. *C. humifructus* has only androgynous inflorescences (i.e., inflorescence with both male and female flowers and the female flower below the male ones), and *C. metuliferus* has mainly unisexual inflorescence and a few gynecandrous ones (i.e., inflorescence with both female and male flowers and the female flower above the male ones).

Male inflorescence consists of solitary flowers, or fasciculate, racemose, paniculate, or rarely modified compound dichasial from 1 to 18 flowered, sessile, or rarely pedunculate. Male inflorescences are often multiflowered and rarely branched. When the inflorescences are branched, the male flowers are always pedicellate. Male flowers are 5-merous; pedicel is terete or rarely sulcate in cross section, variously pubescent or rarely glabrous, and without bracteoles or rarely subtended by a bracteole (*C. heptadactylus*). Calyx consists of five or rarely four lobes, linear to oblong, or narrowly to broadly triangular in outline, acute to narrow in the apex, and variously pubescent or rarely glabrous. Corolla is yellow, infundibular, or rarely campanulate and is variously pubescent or rarely glabrous. Corolla is fused into a basal tube. Corolla leaves are elliptic to broad, ovate to shallow, obovate to narrowly, or rarely oblong or broadly triangular in outline, narrowly to broadly acute or obtuse, and sometimes also mucronate at the apex. Three stamens are free, with separation from the free portion of the hypanthium above the ovary. Two of them are 2-thecate and one is 1-thecate. Filaments are terete or radially compressed in cross section and are glabrous or with basal puberculence and glabrous apically. Anther theceae is sigmoid and glabrous with the edges shortly pubescent. Anther connective is extended, obovate, oblong to narrow, transversely broadly oblong, or ovate, unilobate or rarely bilobate, obtuse or rarely acute in the apex, minutely papillate, sometimes smooth, or rarely glabrous, fimbriate, or crenulate at the apex. Disk is cylindrical or rarely consisting of three papillae and is glabrous.

Female flowers are solitary or rarely in fasciclate inflorescences; sessile flowers arise from leaf axils, very often from secondary branches. They are pedicellate and 5-merous. Pedicel is terete or sulcate in cross section and is variously pubescent, with non-breakaway hairs or rarely with breakaway hairs. Hypanthium is hourglass-shaped. The constricted portion and the lower bulge fused to the ovary. The upper bulge of hypanthium is free from the ovary. Free portion of hypanthium is campanulate. Ovary has three to five placentas with numerous horizontal ovules. Calyx has five, occasionally four or six lobes of the same shape as male flowers. Corolla is yellow and infundibular, with the same shape and types of pubescence as male flower. Corolla tube is present or absent. Three staminodes are present or rarely absent, separating from the free portion of hypanthium above the ovary. Style is terete in cross section, glabrous, subtented by a circular disk, or rarely lacking one. Stigma is copular, lobate, or sometimes entire or sublobate, with one to six or rarely nine finger-like projections on the margin.

6.2.2.4 Fruits and Seeds

Fruits are pendulous; fruit is spherical, oval, oblong, elongated, or blocky in shape and variable in size; fruit surface varies in the number and size of scattered spiny tubercles (warts), or sharp soft hairs. It can be smooth and glabrous, sometimes deeply ridged or covered with a corky (reticulate) netting (e.g., for *C. melo*); skin color varies from pale to very dark green, sometimes with longitudinal indentations or stripes. In maturity, the skin color is white cream to orange brown. Inferior flesh color can be white, green, pink, or orange. The fruit stalk is referred to as a pedicel. The pedicel is sulcate or sometimes terete in cross section and is variously pubescent or rarely glabrous.

Mature seeds have white, cream to yellow color. They are smooth, compressed, ovoid to elliptic, immarginate, with an acute edge, and unwinged or rarely apically winged. *C. humifructus* develops its fruits below ground.

6.2.3 Cytology

Cytologically, the genus *Cucumis*, like all other Cucurbits, is a less studied genus (Ramachandran and Narayan 1985). Most *Cucumis* species are diploid with 12 pairs of chromosomes ($2n = 24$): *C. africanus*, *C. anguria*, *C. dipsaceus*, *C. ficifolius*, *C. hirsutus*, *C. humifructus*, *C. metiluferus*, *C. myriocarpus*, *C. melo*, *C. prophetarum*, *C. pustulatus*, *C. sagittatus*, *C. sacleuxii*, *C. zeyheri*, and *C. hystrix*. Among these species, three have also been reported to be polyploid: *C. ficifolius*, $2n = 48$ (Dane and Tsuchiya 1979; den Nijs and Visser 1985), *C. pustulatus*, $2n = 48$ (Ramachandran 1984; den Nijs and Visser 1985; Ramachandran and Narayan 1985) or $2n = 72$ (Dane and Tsuchiya 1979), and *C. zeyheri*, $2n = 48$ (Dane and Tsuchiya 1976, 1979; Varekamp et al. 1982; Ramachandran 1984; den Nijs and Visser 1985; Ramachandran and Narayan 1985). *C. sativus* is the only species of *Cucumis* reported to have a chromosome count of $2n = 14$ (Fig. 6.2).

There are two base chromosome numbers in *Cucumis*: $x = 7$ and $x = 12$. Two different hypotheses have been put forward to explain the relationship between the two basic chromosome numbers. The fragmentation hypothesis suggests that $x = 12$ has derived from $x = 7$ by fragmentation of particular chromosomes followed by de novo regeneration of centromeres (Bhaduri and Bose 1947; Ayyangar 1967). The fusion hypothesis, on the other hand, says that the basic number $x = 7$ might have arisen from $x = 12$ possibly by unequal translocation or fusion of non-homologous chromosomes (Trivedi and Roy 1970). Comparative genomics between *C. melo* and *C. sativus* may clarify

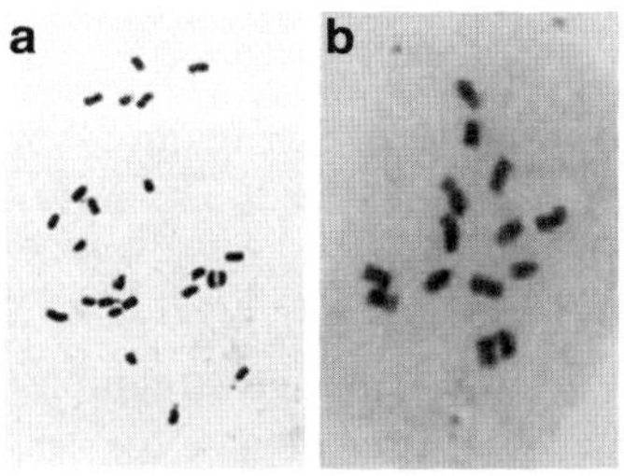

Fig. 6.2 (a) Chromosome numbers of *C. hystrix*. (b) Chromosome numbers of *C. sativus*

the phylogeny of these species (Danin-Poleg et al. 2001).

As for the karyotype of *Cucumis*, most studies have focused on the cultivated species: cucumber and melon (Figs. 6.3 and 6.4). However, discriminatory information from karyotype analysis for detailing relationships in *Cucumis* has been difficult to access due to the small chromosome size and poor stainability. Ramachandran and Seshadri (1986) used C-banding and pachytene analysis to compare the genomes of cucumber and muskmelon (*C. melo* L.), but their study did not differentiate chromosomes by measurement, and their description of the chromosome morphology and C-banding figures are equivocal.

Chromosomal DNA amounts varied in different species of *Cucumis* (Ramachandran and Narayan 1985). The DNA amounts varied from 1.373 to 2.483 pg in diploids and from 2.846 to 3.886 pg in tetraploids. DNA amount was not correlated with chromosome number and periodicity. Tetraploids were found to have double the quantity of nuclear DNA of diploids.

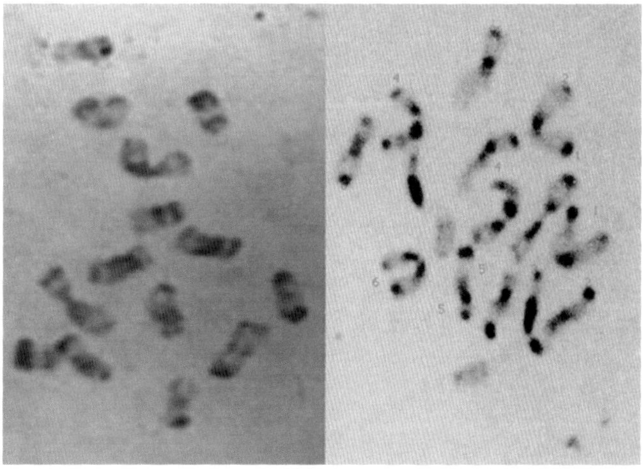

Fig. 6.3 Chromosome C-banding of *Cucumis sativus* L

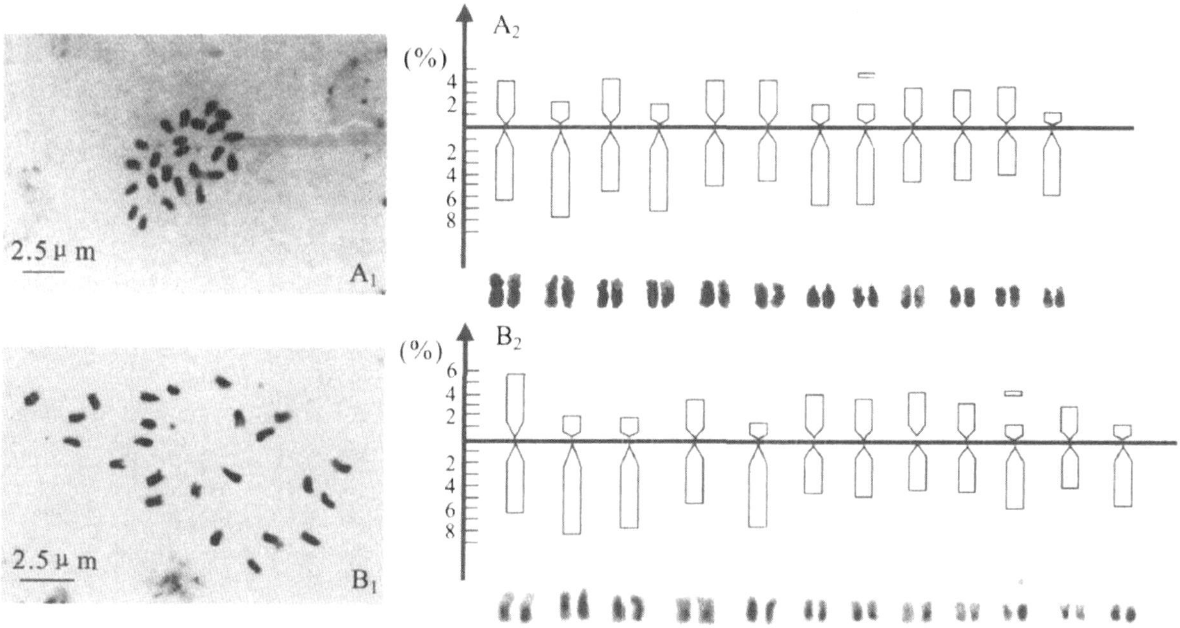

Fig. 6.4 Karyotypes (A1, A2) and their ideograms (B1, B2) of Jiashi and Huangjin melon according to Zhang et al. (2005)

6.2.4 Origin and Distribution

The center of origin for *Cucumis* species is likely Africa for the most wild species with chromosome $2n = 24$, while the Middle East and southern Asia has been considered an important center of diversification for melon and cucumber, respectively (Dane et al. 1980; McCreight et al. 1993; Staub et al. 1999).

Cucumis species occurred in a large scale from 38°N to 37°S of the Old World, with more species in the southern hemisphere than in the northern hemisphere. With latitude rising, *Cucumis* species sharply decreased in the northern hemisphere. There were abundant species near the equator. *Cucumis* species occurred in about 75 countries, but 90.6% of them were from Ethiopia, Kenya, Somalia, South Africa, and Tanzania. Somalia had more rare species in absolute and relative. Thirty-three of 75 countries had only one species (Liu 2007). Table 6.2 presents the details of the distribution of *Cucumis* species.

6.3 Germplasm Conservation

There are some centers, which are engaged in the conservation of *Cucumis* germplasm worldwide. In the United States, plant germplasm is maintained and evaluated by the US National Plant Germplasm System (NPGS). In Europe, the International Plant Genetic Resources Institute (IPGRI) coordinates institutional germplasm holdings. In China, the Crop Germplasm Resources Institute of Chinese Academy of Agricultural Sciences (CAAS) is responsible for the germplasm conservation.

Germplasm information can be found in some good germplasm resources information network web servers, such as Germplasm Resources Information network of United States (http://www.ars-grin.gov), Chinese Crop Germplasm Resources Information System (http://icgr.caas.net.cn), N.I. Vavilov Research Institute of Plant Industry of Russia (http://www.vir.nw.ru), and the European Central Cucurbits Database (http://www.comav.upv.es).

According to the American germplasm resources information network, the regional plant introduction (PI) station of NPGS at Ames, Iowa, houses about 1,486 *C. sativus* accessions of worldwide origin and currently lists 3,074 accessions in its melon inventory. The collection of wild species of *Cucumis* includes 48 *C. africanus* L. f. accessions, 50 *C. anguria* L. accessions, 10 *C. anguria* var. *anguria* accessions, 17 *C. anguria* var. *longaculeatus* J. H. Kirkbr. accessions, one *C. asper* Cogn. accession, one *C. canoxyi* Thulin & Al-Gifri accession, six *C. dipsaceus* Ehrenb. ex Spach accessions, seven *C. ficifolius* A. Rich. accessions, one *C. heptadactylus* Naudin accession, three *C. hirsutus* Sond. accessions, one *C. meeusei* C. Jeffrey accession, 42 *C. metulifer* E. Mey. ex Naudin accessions, 21 *C. myriocarpus* Naudin accessions, two *C. myriocarpus* subsp. *leptodermis* (Schweick.) C. Jeffrey & P. Halliday accessions, three *C. myriocarpus* subsp. *myriocarpus* accessions, three *C. prophetarum* L. accessions, seven *C. pustulatus* Hook. f. accessions, four *C. sagittatus* Peyr. accessions, one *C. subsericeus* Hook. f. accession, eight *C. zambianus* Widrlechner et al. accessions, and nine *C. zeyheri* Sond. accessions. The collection also has 88 accessions labeled *Cucumis* sp., which may include some *C. melo* or *C. metuliferous* accessions (http://www.ars-grin.gov/cgi-bin/npgs/html/genform.pl; Table 6.3).

According to the *AD HOC* meeting on Cucurbit genetic resources held in Turkey in 2002, the number of accessions of *Cucumis* species, including landraces, breeding material, and wild relatives, maintained in European collections was 14,333. Among them, there are 33 accessions of *C. anguria*, 31 accessions of *C. dipsaceus*, 11 accessions of *C. ficifolius*, 7,553 accessions of *C. melo*, 11 accessions of *C. metuliferus*, 12 accessions of *C. myriocarpus*, 5,896 accessions of *C. sativus*, 10 accessions of *C. zeyheri*, and 776 accessions of *C.* spp. (Table 6.4).

In China, it is reported that there are 1,506 accessions of *C. sativus* (Sheng et al. 2006) and 1,003 accessions of *C. melo* (Ma et al. 2003); however, the collection of wild species is not clear.

Although *Cucumis* accessions are held by numerous collections around the World, the most wild species are less formal collections for research purposes and through personal exchanges among scientists throughout the world. Those wild species are often not documented or represented in the NPGS base collection or IPGRI collection, so samples acquired through personal contact could be important.

Table 6.2 Distribution of the genus *Cucumis*

Cucumis spp.	Distribution
C. mriopcarpus	Lesotho, Mozambique, South Africa, Zambia, Botswana, Zimbabwe
C. africanus	Namibia, South Africa (Cap Province, Natal Transvaal), Angola, Lesotho, Zimbabwe, Botswana
C. quintanilhae	Botswana (southern most), South Africa (northern most Trasvaal)
C. heptadactylus	South Africa (Cap Province, Transvaal, Orange Free State)
C. kalahariensis	Botswana (central and northwestern), Namibia (northeastern)
C. anguria	Angola, Botswana, Cape Verde Islands, Malawi, Mozambique, Namibia, South Africa, Sierra Leone, Swaziland, Tanzania, Zaire, Zambia, Zimbabwe
C. sacleuxii	Kenya, Madagasar, Tanzania, Uganda, Zaire
C. carolinus	Ethiopia (southeast), Kenya (northeast)
C. dipsaceus	Ethiopia, Kenya, Somalia, Tanzania, Uganda, possibly native to Sudan and southern Egypt
C. prophetarum	Egypt, Mali, Mauritania, Senegal, Somalia, Sudan, Iran, Iraq, Nigeria (northern), Israel, Oman, Saudi Arabia, Socotra, Syria, Unite Arab Emirates, Jordan, India (northeastern), Pakistan. Chad, Ethiopia, Niger, Kenya, Rwanda, Tanzania, Uganda, South Yemen, Yemen
C. pubituberculatus	Central, coastal Somalia
C. zeyheri	Lesotho, Mozambique, South Africa (Cap Province, Natal, and Transvaal), Swaziland, Zambia, Zimbabwe
C. prolatior	Central and south central Kenya
C. insignis	Ethiopia
C. globosus	Tanzania
C. thulinianu	Somalia, only near Erigavo
C. ficifolius	Ethiopia Kenya, Rwanda, Tanzania, Uganda, Zaire (eastern most areas near the Rift Valley)
C. jeffreyanus	Ethiopia, Kenya, Somalia
C. aculeatus	Rarely from southern Ethiopia, Kenya, Rwanda, Tanzania, Uganda, Zaire (eastern)
C. pustulatus	Chad, Ethiopia, Kenya, Nigeria, Sudan, Tanzania, and Uganda. Southwest Asia (Saudi Arabia and Yemen)
C. meeusei	Botswana (northern), South Africa (northern Cape Province), Namibia (northern)
C. hastatus	Somalia (southern)
C. baladensis	Somalia
C. rigidus	Namibia, southern most area along Orange River, South Africa, northwestern Cap Province along the Orange River
C. metuliferu	Botswana, Ethiopia, Mozambique, Kenya, Malawi, Namibia, Senegal, South Africa, Sudan, Swaziland, Tanzania, Uganda, Zaire, Zimbabwe, Cameral African Republic, Liberia, Burkina, South Yemen, Yemen, Angola
C. rostratu	Ivory Coast, Nigeria
C. hirsutus	Botswana, Burundi, Congo, Kenya, Malawi, South Africa, Sudan, Swaziland, Tanzania, Zaire, Zambia, Zimbabwe, Mozambique, Angola
C. humifrucutus	Angola, Ethiopia (very rarely), Kenya, Namibia, South Africa (Transvaal), Zaire, Zambia, Zimbabwe
C. melo	Angola, Benin, Burkina, Cameroon, Cape Verd Islands, Central African Republic, Chad, Egypt, Ethiopia, Gambia, Ghana, Guinea-Bissau, Ivory Coast, Kenya, Madagascar, Malawi, Maldive Islands, Mail, Mauritania, Mozambique, Niger, Nigeria, Senegal, Seychelles, Sierra Leone, Somalia, South Africa, Sudan, Tanzania, Uganda, Zaire, Zambia, Zimbabwe, Iran, Iraq, Oman, Saudi Arabia, South Yemen, Yemen, Afghanistan, Bangladesh, Burma, China, India, Japan, Korea, Nepal, Pakistan, Sri Lanka, Thailand, Malaysia, Indonesia, New Guinea, Philippines, Australia, Fiji Islands, Guam, New Britain, Papua New Guinea, Samoa, Solomon Islands, Tonga Islands
C. sagittatus	Namibia, and South Africa (only northwestern Cape province), Angola (southwestern corner next to Namibia)
C. sativus	Burma, China (Yunnan province, GuangXi, GuiZhou), India, Sri Lanka, Thailand
C. hystrix	Burma, China (Yun Nan province), India (Assam), Thailand, Myanmar

Table 6.3 Number of accessions of *Cucumis* species stored in the regional plant introduction station of NPGS at Ames according to the American germplasm resources information net work (http://www.ars-grin.gov/cgi-bin/npgs/html/genform.pl)

Species	Number
C. africanus L.	48
C. anguria L.	50
C. asper Cogn.	1
C. canoxyi Thulin & Al-Gifri	1
C. dipsaceus Ehrenb. ex Spach	6
C. ficifolius A. Rich.	7
C. heptadactylus Naudin	1
C. hirsutus Sond.	3
C. meeusei C. Jeffrey	1
C. metulifer E. Mey. ex Naudin	42
C. myriocarpus Naudin	21
C. myriocarpus subsp. *leptodermis*	2
C. myriocarpus subsp. myriocarpus	3
C. prophetarum L.	3
C. pustulatus Hook. f.	7
C. sagittatus Peyr.	4
C. subsericeus Hook. f.	1
C. zambianus Widrlechner et al.	8
C. zeyheri Sond.	9
Cucumis sp.	88

Table 6.4 Number of accessions of *Cucumis* species stored in the main European genebanks and breeders' collections (from Ad-hoc meeting on Cucurbit genetic resources in Europe held in Turkey, 2002)

Collection curator (Country)	*Cucumis*
Genebanks	
T. Piskunova (Russian Federation)	4,931
A. Börner (Gatersleben, Germany)	975
F. Nuez (Valencia, Spain)	798
E. Křístková (Czech Republic)	967
L. Krasteva (Bulgaria)	1,247
A. Küçük (Turkey)	632
L. Horváth (Hungary)	383
M. Carravedo (BGHZ, Spain)	777
W. Dooijeweert (The Netherlands)	790
T. Kotlińska (Poland)	390
N. Polignano (Italy)	143
R. Farias (BPGV, Portugal)	119
J. Berenji (Yugoslavia)	–
Breeders' collections	
M. Pitrat (France)	605
M. L. Gómez-Guillamón (CSIC, Spain)	561
S. Strajeru (Romania)	280
K. Abak (Turkey)	301
Total	13,899

6.4 Evolution and Phylogenetic Relationships

Knowing the closest relatives and natural composition of the genus *Cucumis* L. is important simply because of the ongoing efforts by plant breeders worldwide to improve melon and cucumber with traits from wild relatives (Renner and Schaefer 2008). Quite a few studies, using morphological, cytology, and molecular characters such as isozymes, random amplified polymorphic DNA (RAPD), chloroplast simple sequence repeat (cpSSR), and internal transcribed spacer (ITS), have been carried out to determine the *Cucumis* phylogeny.

Early studies on karyomorphological investigations of 13 species in the genus *Cucumis* L. indicated that South African annual species are the primitive and identified five distinct groups in taxa with $2n = 24$. They are (1) *C. leptodermis* and *C. africanus*; (2) *C. ficifolius*, *C. hookeri*, and *C. dipsaceus*; (3) *C. myriocarpus*, *C. zeyheri*, *C. prophetarum*, *Cucumis* species CUCU44/74, and *C. anguria*; (4) *C. metuliferus*; and (5) *C. sagittatus* and *C. melo* (Singh and Yadava 1984).

The crossability between species, chromosome pairing, and pollen fertility in F_1 hybrids were also investigated for assessing species relationships and *Cucumis* phylogeny. Deakin et al. (1971) were the first to produce a comprehensive and monographic account on relative cross-compatibility between *Cucumis* species and pollen fertility of their F_1 hybrids. Their studies involved 14 species including cultivated *C. melo* L. On the basis of these data obtained, they grouped *Cucumis* species into four major groups. Singh and Yadava (1984) had investigations on interspecific crossability in eight *Cucumis* species ($2n = 24$, *C. melo, C. dipsaceus, C. anguria* var. *anguria, C. anguria* var. *longipus*, *C. myriocarpus, C. zeyheri*, *C. prophetarum*, and *C. species*). Information on chromosome pairing and pollen fertility of the hybrids from 15 combinations had been utilized for tracing the phylogenetic relationships among these taxa.

Esquinas-Alcazar (1977) studied the alloenzyme variation and relationships in the genus *Cucumis* and divided the genus *Cucumis* into four groups. (1) Ser. *angurioidei*: *C. aculeatus*, *C. africanus*, *C. anguria*, *C. dipsaceus*, *C. ficifolius*, *C. heptactylus*, *C. myriocarpus*,

C. pustulatus, and *C. zeyheri*; (2) Ser. *metuliferi*: *C. aculeatus*, *C. metuliferus*, and *C. sagittatus* (some accessions); (3) Ser *melo*: *C. melo* and *C. sagittatus* (some accessions); and (4) Ser. *sativus: C. sativus*.

From the evolutionary and systematic point of view, Perl-Treves and Galun (1985) compared the phylogenies of *Cucumis* based on cpDNA and nuclear-coded isozymes. The comparison was carried out for 21 *Cucumis* species and the two phylogenies were found to share the main dendrogram features, which also agreed well with most taxonomic data available on *Cucumis*. Accordingly, most of the African *Cucumis* species form a close group ("*Anguria* group," *C. africanus*, *C. anguria*, *C. dipsaceus*, *C. ficifolius*, *C. heptactylus*, *C. meesusei*, *C. myriocarpus*, *C. prophetarum*, *C. pustulatus*, and *C. zeyheri*), which was distant from the melon (*C. melo*), and from a few other distant species (*C. humifructus*, *C. metuliferus*, and *C. sagittatus*), all of which were far apart from each other. The cucumber (*C. sativus*) was the most distant species within the genus.

In 1989, *C. hystrix* Chakr., a wild *Cucumis* species, was rediscovered in Yunnan, China, by Jinfeng Chen (Chen et al. 1994). Subsequent research revealed that *C. hystrix* has $2n = 24$ instead of $2n = 14$ as in the Asian members. This finding challenges the basic chromosome number theory that African *Cucumis* have $n = 12$ and that Asian *Cucumis* have $n = 7$. Isozyme patterns (Chen et al. 1995) suggested that *C. hystrix* has closer genetic affinities with *C. sativus* than with *C. melo*, even though *C. hystrix* and *C. melo* possess the same number of chromosomes.

Chung et al. (2006) used nine chloroplast SSR (cpSSR) markers to investigate the phylogenetic relationships among African *Cucumis* species ($x = 12$) accessions, *C. melo* accessions, *C. sativus* accessions, and *C. hystrix* accessions. Sequence variation analysis identified a group of African *Cucumis* species and a group composed of *C. melo*, *C. sativus*, and *C. hystrix* species leading to the conclusion that *C. hystrix* is the progenitor species of *C. sativus*, or that they at least share a common ancestral lineage.

Zhuang et al. (2006) investigated the phylogenetic relationships in *Cucumis* species using RAPD. Their focus was mainly on the analysis of genetic relationship among *C. hystrix*, *C. sativus,* and *C. melon* and *C.* × *hytivus,* a new synthetic species. On the basis of results, a modified taxonomic system was proposed that *C. hystrix* should remain in subgen. *Cucumis*, although it had a chromosome number different from that of *C. sativus*. With the interspecific hybrids *C.* × *hytivus* as the third species, subgen. *Cucumis* was thus made up by three species. Although the basic chromosome number and geographic location theorized were challenged by the proposed system, the use of it will likely assist in the exploitation of the wild *Cucumis* species in Asia.

Using sequences of the internal transcribed spacer (ITS) 1 and 2 regions of the nuclear ribosomal RNA genes, Jobst et al. (1998) evaluated the phylogenetic relationships among different members of the family Cucurbitaceae. Six *Cucumis* species along with *C. melo* and *C. sativus* were analyzed in the study and the results obtained by ITS sequence data were highly congruent with isoenzyme data of Puchalski and Robinson (1990). Garcia-Mas et al. (2004) defined phylogenetic relationships among *Cucumis* species also using the nuclear ribosomal DNA ITS region and microsatellite markers. In their study, the genus *Cucumis* was splited into five groups: cucumbers, melons, *C. metuliferus*, a group containing 12 wild species of *Cucumis*, and Oreosyce Africana, and a fifth group comprising *C. sagittatus* and *C. globosus*.

Kocyan et al. (2007) presented a multilocus chloroplast phylogeny for the Cucurbitaceae that included all putative close relatives of *Cucumis*. Their results support a paraphyletic *Cucumis*, with *Cucumela*, *Dicaelospermum*, *Mukia*, *Myrmecosicyos*, and *Oreosyce* nested among species of this genus. Although evolutionary relationships are not completely resolved, following the discovery by Koeyan and coworkers that *Cucumis* as traditionally circumscribed (Kirkbride 1993) was highly unnatural, two molecular phylogenetic studies reinvestigated species relationships in a much more broadly circumscribed *Cucumis* (Ghebretinsae et al. 2007; Renner et al. 2007).

Ghebretinsae et al. (2007) presented a comprehensive molecular phylogeny of *Cucumis* and the traditionally related genera based on sequences from both nuclear and chloroplast genomes. Their study used a much more complete sampling of species within *Cucumis* than did previous studies and includes representatives of *Cucumella*, *Dicaelospermum*, *Mukia*, *Myrmecosicyos*, and *Oreosyce*. Their combined phylogenetic analyses did not support Kirkbride's (1993) subdivision of *Cucumis* into two subgenera based principally upon chromosome number and geographical distribution. *Cucumis* sensu Kirkbride (1993) is paraphyletic and the Asian *Cucumis* (subg. *Cucumis* of Kirkbride) forms a well-supported clade nested within African *Cucumis* s.s. They identified six clades within the *Cucumis* complex, designated as Clades I, II, III, IV, V, and VI. Clade I comprised all non-domesticated African *Cucumis* s.s. after the exclusion of *C. hirsutus*, *C. humifructus*, *C. metuliferus*, *C. rostratus*, and *C. sagittatus*; *C. hystrix* and *C. sativus* were included in Clade II. Clade III comprised *C. melo* and *C. sagittatus*. Clade IV consisted of *C. metuliferus* and *C. rostratus*. Clade V comprised four species of *Cucumella* and *Oreosyce*. Clade VI consisted of *C. hirsutus* and *C. humifructus*.

Based on these two recent molecular phylogenetic studies of *Cucumis*, Renner and Schaefer (2008) summarize what is now known about phylogenetic relationships in *Cucumis*. The phylogeny of *Cucumis* resulting from combined nuclear and chloroplast data (Fig. 6.5) implied that the deepest divergence lies between the common ancestor of *C. hirsutus*/*C. humifructus* and the stem lineage of the remainder of the genus. The area of origin of *Cucumis* cannot be inferred because its sister genus, *Muellerargia*, has one species in *Madagascar* and the other one in tropical Australia and Indonesia. The next closest relatives are in African/Asian clades including the genera *Coccinia*, *Zehneria*, *Neoachmandra*, and *Peponium*, but their exact position is still unresolved. The earliest divergence events in *Cucumis* likely took place in Africa. However, contrary to the traditional classification (Kirkbride 1993), which grouped *C. melo* with the African *C. hirsutus*, *C. humifructus*, and *C. sagittatus*, melon instead is closest to an Australian/Asian clade.

6.5 Interspecific Hybridization Among Cucumis Species

6.5.1 Progress of Interspecific Hybridization

Interspecific hybridization is used to improve crops by transferring specific traits, such as pest and stress resistance, from their wild relatives (Bowley and Taylor 1987). Interspecific hybrids in the Cucurbitaceae have been produced in several genera, including *Cucumis* (Deakin et al. 1971), *Citrullus* (Valvilov 1925), *Luffa* (Singh 1991), and *Cucurbita* (Weeden and Robinson 1986). In the genus *Cucumis*, an amphidiploid was reported from the cross of *C. anguria* L. and *C. dipsaceus* E. ex S. (Yadava et al. 1986). However, in the Cucurbitaceae only in *Cucurbita* interspecific hybridization has been successfully utilized for crop improvement (Robinson and Decker-Walters 1997).

Cucumis contains two species of economic importance: melon (*C. melo* L., $2n = 24$) and cucumber (*C. sativus* L., $2n = 14$). The importance of wild *Cucumis* species has long been recognized because they possess resistance to pathogens, such as powdery mildew, downy mildew, anthracnose, and Fusarium wilt (Leppick 1966; Lower and Edwards 1986; Kirkbride 1993). Genetic variation is relatively limited in cucumber; thus, efforts to create interspecific hybrids have become more critical and meaningful.

The first recorded attempt to make crosses between cucurbits by removing the male flowers and transferring pollen by hand was made by Naudin. He was unsuccessful in obtaining a cross between *C. melo* and *C. myriocarpus* (Naudin 1859). Many more studies have also been attempted to make interspecific crosses in *Cucumis* (Betra 1953; Andrus and Fassuliotis 1965; Deakin et al. 1971; Chelliah and Sambandam 1972; Fassuliotis 1977; Dane et al. 1980; Kho et al. 1980; Visser and den Nijs 1983; Singh and Yadava 1984; den Nijs and Visser 1985; den Nijs and Custers 1990; Chatterjee and More 1991, etc.).

The first comprehensive crossability analysis of the genus was published by Deakin et al. (1971), who observed that crosses among wild species were frequently possible, but that all attempts to cross any of these with the two cultivated species, *C. sativus* and *C. melo*, failed. Raamsdonk et al. (1989) summarized the data in two crossing polygons. The species *C. heptadactylus*, *C. humifructus*, *C. melo*, and *C. sativus* have never been successfully crossed with any other species of *Cucumis* to produce a fertile F_1 generation. The following species can be crossed to a limited extent among themselves: *C. africanus*, *C. anguria*, *C. dipsaceus*, *C. ficifolius*, *C. metuliferus*, *C. myriocarpus*, *C. prophetarum*, *C. pustulatus*, and *C. zeyheri* (Kirkbride 1993). Some wide-cross attempts between cultivated and wild *Cucumis* species are presented in Table 6.5.

A successful interspecific hybridization between *C. hystrix* and *C. sativus* was reported (Chen et al. 1997). It was the first reproducible cross between a

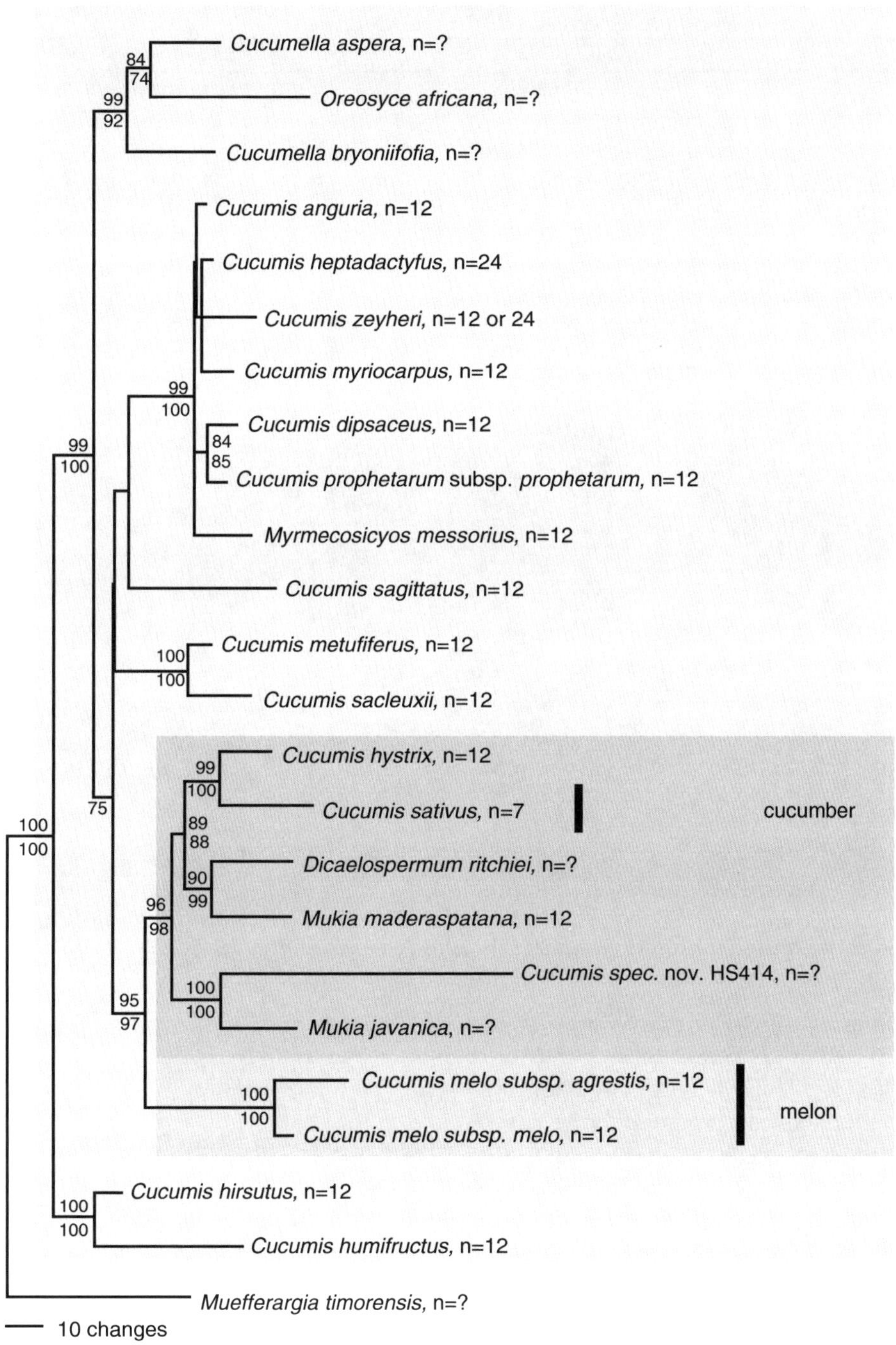

Fig. 6.5 Parsimony tree for *Cucumis* based on combined chloroplast and nuclear DNA sequences according to Renner et al. (2007)

cultivated *Cucumis* species and a wild relative, and it represented a breakthough in interspecific hybridization in *Cucumis*. The success of this cross was even more surprising because the parental species have different chromosome numbers. The original F_1 hybrid ($2n = 19$), obtained by embryo rescue following pollination of *C. sativus* by *C. hystrix* (Fig. 6.6), has 7 chromosomes from *C. sativus* and 12 from *C. hystrix* and was both male- and female-sterile. To restore fertility, reciprocal crosses were made and the

Table 6.5 Wide-cross attempts between cultivated and wild *Cucumis* species

Cross	Result	Source
C. sagittatus × *C. melo*	Embryos only	Deakin et al. (1971)
C. metuliferus × *C. melo*	Embryos only	Fassuliotis (1977)
C. sativus × *C. melo*	Globular stage embryos only	Niemirowicz-Szczytt and Kubicki (1979)
C. metuliferus × *C. melo*	Fertile F_1	Norton and Granberry (1980)
C. prophetarum × C. melo	Fruit with non-viable seeds	Singh and Yadava (1984)
C. zeyheri × *C. sativus*	Fruit with non-viable seeds	Custers and Den Nijs (1986)
C. sativus × *C. metuliferus*	Embryos only	Franken et al. (1988)
C. melo × *C. metuliferus*	Embryos only	Soria et al. (1990)
C. sativus × *C. hystrix*	Sterile plant ($2n$ and $4n$)	Chen et al. (1997)
C. hystrix × *C. sativus*	Fertile plants ($4n$)	Chen et al. (1998)

Fig. 6.6 Production and chromosome counting of interspecific hybrids F_1 between *C. hystrix* and *C. sativus*. (**a**) Embryo of hybrid. (**b**) Regeneration from the young embryo. (**c**) Acclimatized plants. (**d**) Metaphase chromosomes of interspecific hybrid F_1 ($2n = 19$)

chromosome numbers of the progeny were successfully doubled (Chen et al. 1998). Pollen grains were produced by these progeny when *C. hystrix* was used as the seed parent; the plants produced fertile flowers and set fruit with viable seeds, indicating that fertility was restored. This restoration of fertility marked the creation of a new synthetic species, which has close phylogenetic relationships with its parental species, but is distinctively different from each parent. The new species (*Cucumis* × *hytivus* Chen and Kirkbride; Fig. 6.7) has genome HHCC, where H represents the genome of *C. hystrix* and C represents the genome of *C. sativus* and chromosome number $2n = 4x = 38$ (Chen and Kirkbride 2000). This synthetic species might be useful as a new *Cucumis* crop. In addition, as a *C. hystrix* × *C. sativus* hybrid, it might be useful as a bridging species for transfer of useful traits to cucumber. Figure 6.8 shows the polygon of crossability in *Cucumis* species.

6.5.2 Major Problems in Interspecific Hybridization

6.5.2.1 Hybridization Barriers

Many experiments have indicated the presence of a strong barrier to interspecific hybridization in *Cucumis*. The nature of cross-incompatibility between cultivated *Cucumis* species and their wild relatives is not well understood. Incompatibility is characterized by delayed growth of pollen, or arrested pollen tube growth to reach the ovules (Kishi and Fujishita 1969), as well as lack of cell division of the zygote, and abortion of the endosperm (Kishi and Fujishita 1970).

Several traditional approaches in interspecific hybridization have been used to overcome the hybridization barriers in *Cucumis*. These include growth

Fig. 6.7 (**a**) *Cucumis* × *hystivus* plant in the field. (**b**) Chromosome numbers of *Cucumis* × *hytivus*. (**c**) Fruit of *Cucumis* × *hystivus*

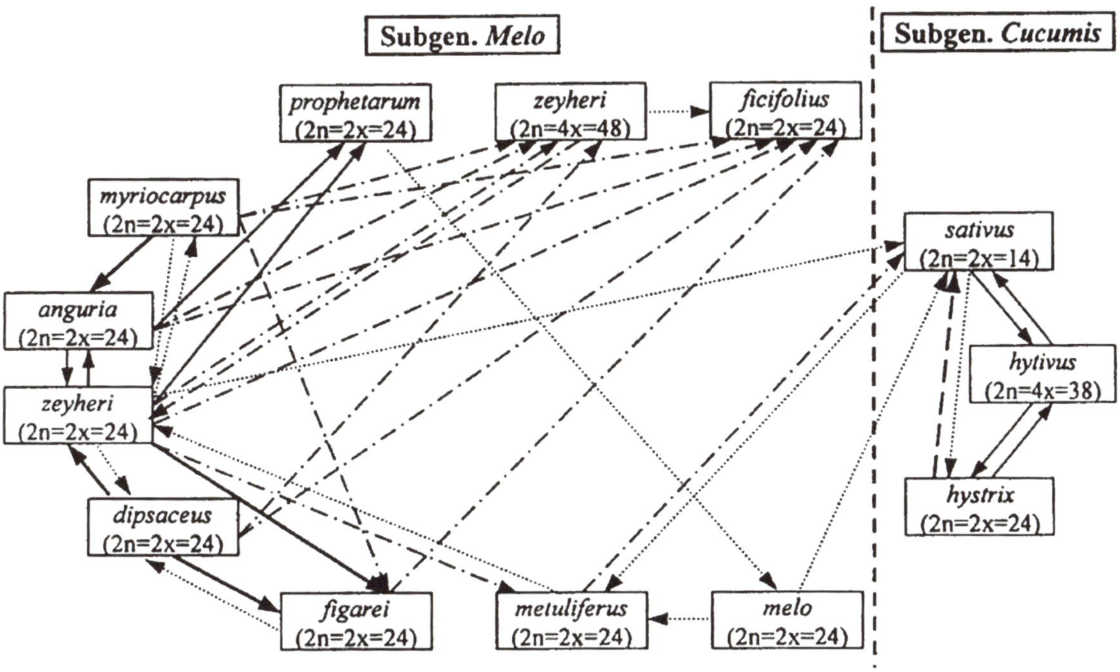

Fig. 6.8 Polygon of crossability in *Cucumis* species according to Zhuang (2003). *Arrows* point to the female parent; moderately to strongly self-fertile and cross-fertile hybrids (*thick solid line*); sparingly self-fertile and moderately cross-fertile hybrids (*thin solid line*); self-fertile, usually not cross-fertile hybrids (*dashed* and *dotted line*); inviable seeds or seedlings (*dashed line*); self-sterile and cross-sterile hybrids (*thick dashed line*); self-sterile and cross-fertile hybrids (*long dashed line*); *absence of a line* indicates that seed fruits were not obtained

regulator application (Custers and Den Nijs 1986), pollen irradiation (Beharav and Cohen 1994), use of mentor pollen (Kho et al. 1980), and bud pollination (Chatterjee and More 1991). Biotechnological techniques, such as somatic hybridization, have also been applied as possible tools for overcoming these barriers in *Cucumis* (Tang and Punja 1989; Chatterjee and More 1991). Likewise, fusion of *C. sativus* and *C. melo* protoplasts has been attempted, but the results indicated that successful hybridization is still unpredictable (Fellner et al. 1996).

The interspecific hybrid between *C. sativus* and *C. hystrix* represents an important step in interspecific hybridization in *Cucumis*. If *C. hystrix* and *C. melo* are cross-compatible and if the F_1 derived from either interspecific hybridization can be made fertile through crossing and/or chromosome doubling, then *C. hystrix* could act as bridge species between *C. melo* and *C. sativus*.

6.5.2.2 Post-fertilization Abortion and Embryo Rescue

In higher plants, post-zygotic failure of hybrid embryos is not often due to incompatibility between the parental chromosomes, but incompatibility problems in the endosperm. In such cases, embryos from interspecific hybridization have to be rescued; otherwise, they will fail due to embryo abortion and/or endosperm degeneration. Successful embryo rescue in tissue culture allows further advances in interspecific hybridization.

Embryos can sometimes be rescued, even if they are immature or lack endosperm (Laibach 1925). In *Cucumis*, fruits with non-viable seeds were obtained in the cross between *C. prophetarum* L. and *C. melo* (Singh and Yadava 1984). The barriers between these two species may be post-zygotic. If the embryo rescue technique had been employed, the experiment might have been successful.

6.5.2.3 Sterility in F_1 Hybrids

A common problem on utilizing germplasm of wild species for crop improvement was sterility in F_1 hybrids. In many cases, this sterility was associated with meiotic abnormalities and was a large obstacle that followed hybridization and hindered utilization.

The ability to cross *C. sativus* and *C. hystrix* offered the promise of introgressing desirable characters from *C. hystrix* to *C. sativus*. However, self-pollination and backcrossing of the F_1 plants to either parent was unsuccessful because the original hybrid was both male- and female-sterile, probably because of the non-functional gametes containing odd chromosome numbers. When chromosomes were doubled, each chromosome had a homologous partner for pairing during meiosis; if there were no cytoplasmic incompatibility, the chromosome-doubled F_1 hybrid might have produced viable gametes, and fertility restoration was anticipated.

External application of chemical agents is the usual way to double chromosome number. Among various agents, colchicine was one of the antimitotic substances most frequently used for this purpose (Chen and Staub 1997). Colchicine at an aqueous solution of 0.05–0.5% (w/v) is believed to be most effective for many plant species. Since colchicine is poisonous to plants, germinating seeds or young seedlings are often preferred for treatment because they grow rapidly and recover more readily than more mature plants do.

When the experimental material does not respond well to chemical treatment, in vitro chromosome doubling (spontaneous polyploidy as a consequence of tissue culture) could be an alternative (D'Amato 1977). When and how the polyoloidization happened in tissue culture was not entirely clear, but it occurred at a low rate during plant formation from axillary buds (Adelberg et al. 1994), callus (Osifo et al. 1989) and culture of protoplasts (Tabei et al. 1992). Polyploidization can be generalized as a universal phenomenon in melon tissue culture (Ezura et al. 1992), although genotype is an important factor in determining the rate of chromosome doubling (Adelberg and Chen 1998).

6.6 Utilization of Wild Species

Although there are many wild *Cucumis* species, not all of them have been fully and properly characterized and documented. Still a few species have been actually used in breeding programs, despite the fact that several of them have been reported to have one or more excellent features to be used as donors (Fassuliotis 1967; Norton and Granberry 1980). Intergroup incompatibility has been assigned as the main reason for this (Deakin et al. 1971; Fassuliotis and Nelson 1988). However, the research work on transfer of useful genes from wild relatives was strengthened with the successful interspecific hybridization between *C. hystrix* and *C. sativus*.

6.6.1 Resistance in C. hystrix

C. hystrix is a wild species of *Cucumis* subgen. *Cucumis*, which originated in Asia (Kirkbride 1993). It has a sour taste that is not really like cucumber. As it has been described before that although it bears morphological similarity with cucumber, its diploid chromosome number is 24, the same number as in melon (Chen et al. 1997). Since the first interspecific hybridization successfully made in 1997 and development of the synthetic allotetraploid species (*C.* × *hytivus* Chen and Kirkbride, $2n = 4x = 38$) in 2000, several disease screens were undertaken to characterize the response of *C. hystrix* and its progenies derived from the interspecific hybridization to common cucurbit diseases (Chen and Lewis 2000; Chen et al. 2004b).

Fig. 6.9 Gummy stem blight resistance shown in *C. hystrix* (**a**, **b**) in the field. *C. hystrix* showing susceptibility to WMV-2 (**c**) and resistance to PRV, CMV-C, and ZYMV (**d**)

6.6.1.1 Gummy Stem Blight

Resistance evaluations were made in a field at Cornell University using the highly virulent *Didymella bryoniae* isolate NY1. Resistance was found in *C. hystrix* plants in the field (Fig. 6.9a, b). Few symptoms were found on stem and slightly more on the leaves. There was no segregation observed among the plants.

6.6.1.2 Downy Mildew

Resistance tests were conducted in Nanjing Agricultural University, China. The disease index in *C. hystrix* was 5.3, indicating that it is highly resistant to downy mildew. This resistance was partially transmitted to the *C. hytivus*, and the progenies from backcross. Compared to the susceptible cucumber cultivar "Jinlu," all the materials derived from this interspecific hybridization possess at least moderate resistance.

6.6.1.3 Viruses

C. hystrix was evaluated for resistance to four viruses: CMV-C, WMV-2, PRV, and ZYMV in the field at Cornell University. *C. hystrix* plants generally suffered from the WMV-2 inoculation (Fig. 6.9c), but showed stronger resistance to PRV and moderate resistance to CMV-C and ZYMV (Fig. 6.9d).

6.6.1.4 Nematodes

Nematode resistance tests were carried out. The three groups (*C. hystrix*, *C. sativus*, and reciprocal interspecific hybrids) varied greatly in their response to *M. incognita*. *C. hystrix* had a high level of resistance to *M. incognita* with mean gall index of 1.8 (Fig. 6.10). In contrast, cucumbers were confirmed as being highly susceptible possessing a mean gall index of 4.8–5.0. The interspecific F_1 hybrid was intermediate in resistance to the two parents, with a mean gall index 3.4. The transmission of resistance was observed in backcross progeny of the chromosome-doubled F_1 to cucumber.

6.6.2 Synthesis, Characterization, and Utilization of Novel Germplasm

6.6.2.1 Allotetraploid

The first repeatable interspecific hybridization between cucumber (*C. sativus* L., $2n = 14$) and *C. hystrix* ($2n = 24$) was successfully made through embryo rescue (Chen et al. 1997). Hybrid plants ($2n = 19$; 12 from *C. hystrix* and 7 from cucumber) were sterile, but morphologically uniform. The multiple-branching habit, densely brown hairs (on corolla and pistil), orange–yellow corolla, and ovate fruit of

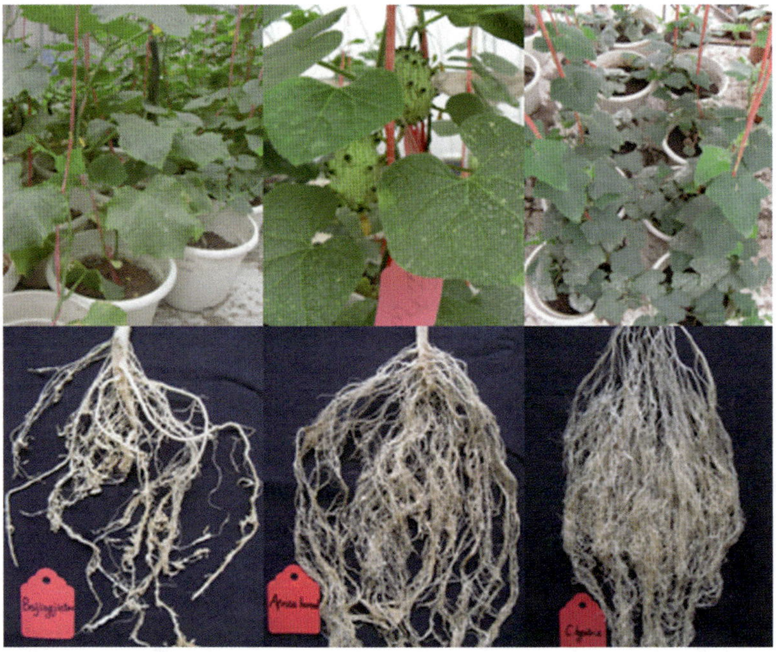

Fig. 6.10 *C. hystrix* showing resistance to root-knot nematode *M. incognita* (*right*), susceptible control "Beijingjietou" (*left*), and resistant control *C. metuliferus* (*middle*)

F_1 hybrid plants were similar to that of the *C. hystrix* parent, and the appearance of the first pistillate flower was more similar to that of *C. sativus* parent. The diameter and internode length of the stem and the shape and size of leaves and flowers were intermediate when compared to the parents.

To restore fertility, chromosome doubling of the F_1 hybrid plants was carried out (Chen and Staub 1997). Sixty-two chromosome-doubled plants were obtained. The chromosome-doubled F_1 plants were morphologically distinct from the parents and other progeny in traits such as a curve on leaf margins and shorter and stronger internodes. The fruits at two ploidy levels vary in morphology (Fig. 6.11). While diploid fruit ($2n = 19$, seedless) was longer and spindle-like in shape, the tetraploid fruit was shorter and column-shaped. After fertility selection, two primary allotetraploid produced fertile flowers and set fruit with viable seeds (Chen et al. 1998). The restoration of fertility in the chromosome-doubled F_1 hybrid marks the creation of a new combination of genomes and a new synthetic species that did not exist previously.

Several photosynthetic characters of the hybrid species *Cucumis* × *hytivus* Chen and Kirkbride under weak light condition were studied (Qian et al. 2002). The light compensation point of allotetraploid was 11.25 $\mu E\ m^{-2}\ s^{-1}$. After treatment with low intensity light for 2 weeks, the leaf contents of chlorophyll a and b increased, while the value of chlorophyll a/b decreased, indicating that the allotetraploid has good tolerance to low irradiance.

Zhuang et al. (2002) investigated the responses of seedlings of the new species *Cucumis* × *hytivus* and its progenies from backcross with cucumber to chilling injury. The abnormal metabolism was observed in *C. hytivus* as it was subjected to low temperature treatment; however, the progenies from backcrossing the new species to cucumber showed high tolerance to chilling injury.

Chen et al. (2007) studied the genomic events in the early generations of the synthesized allotetraploid. Extensive genomic changes were detected by amplified fragment length polymorphism (AFLP) analysis. The changes mainly involved loss of parental restriction fragments and gain of novel fragments. The total detectable changes were from 11.1 to 32.1%, and the frequency of losing parental fragments was much higher than that of gaining novel fragments. Although no significant differences were detected in the reciprocal crosses, the data showed that the frequency of sequence loss in *C. sativus* was two times higher than that in *C. hystrix*. The results demonstrated that the sequence elimination was the major event of genomic changes, and it might provide the physical basis

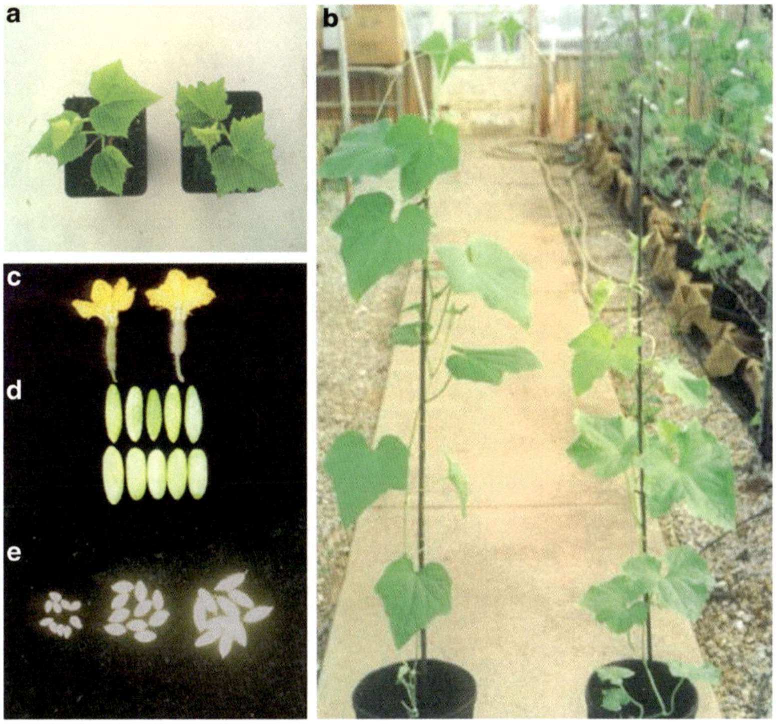

Fig. 6.11 Morphological comparison between the interspecific hybrid and synthetic allotetroploid *C.* × *hytivus* (**a** and **b**). The interspecific hybrid F_1 diploid, sterile, hybrid plant from embryo rescue (*left*) and its chromosome-doubled tetraploid, fertile plant (*right*). (**c**) Female flowers of F_1 hybrid (*left*) and allotetraploid (*right*). (**d**) Fruit of F_1 hybrid (*left*) and allotetraploid (*right*). (**e**) Seeds harvested from the allotetraploid and its diploid progenitors

for the diploid-like meiotic behavior in the diploidization of the newly formed allopolyploids.

In order to explore the molecular involvement of epigenetic phenomena, cytosine methylation was investigated in *C.* × *hytivus* by using methylation-sensitive amplified polymorphism (MSAP) (Chen and Chen 2008). Twofold difference in the level of cytosine methylation in the reciprocal F_1 hybrids and in the allotetraploid was observed. Pattern analysis found that 2.0–6.4% of total sites changed in both the F_1 hybrids and the allotetraploid compared to their corresponding parents. 68.2–80.0% of the changed sites showed an increase in cytosine methylation, and most of the methylated sites were from the maternal parent. The extent of cytosine methylation pattern changes was greatly decreased during selfing process, suggesting stability in advanced generations.

Changes of gene expression played an important role in the evolution of plant allopolyploids. Characters of the changes of gene expression between the allotetraploid *C.* × *hytivus* and its diploid parental species were analyzed by using cDNA-AFLP technique. The results indicated that most genes from parents could be stably expressed in the allotetraploid, while some genes expressed differentially. A total of 36 (3.37%) differentially expressed transcripts were detected and classified into three types: no expression of genes from parents, expression of genes from one parent, and novel expression of new genes. The majority was the expression of genes from one parent. The data also showed that the genes from female parent were easier to be changed. Those results indicated a rapid change in gene expression in early generations of *C.* × *hytivus*.

In Cucurbitaceae, amphidiploidy was reported from *C. maxima* × *C. moschata* (Pearson et al. 1951) and *C. anguria* × *C. dipsaceus* (Yadava et al. 1986). There were no successful efforts on the two most commercially important *Cucumis* spp.: cucumber and melon. *C.* × *hytivus* as a synthesized allopolyploid can be a useful model system to study polyploidization and may also serve as a genetic bridge in *Cucumis* and thus is a source for broadening the genetic base of *C. sativus*.

6.6.2.2 Allotriploid

A primary allotriploid cucumber ($2n = 3x = 26$; HCC) was obtained through backcrossing the synthetic allotetraploid (HHCC, maternal parent) to cultivated cucumber (CC, paternal parent) (Chen et al. 2003). The allotriploid showed many novel characters, such as strong heterosis, sequential fruit set, tolerance to low temperature and low light, and high nutritional value of the fruit. In morphology, fruit weight, number of branches, ovary length, fruit length, and spine and mature fruit of progeny more closely resembled their maternal *C.* × *hytivus* parent (Fig. 6.12); developmental vigor (i.e., relative growth rate) and leaf color (i.e., dark green vs. reseda of *C.* × *hytivus*) were more similar to and characteristic of the paternal, *C. sativus*, parent. Progeny were more parthenocarpic than either parent bearing abundant seedless fruit (i.e., 84.8% developed fruit under greenhouse conditions). While their parents, *C.* × *hytivus* and *C. sativus,* were lower (26.7 and 45.2%, respectively) (Table 6.6).

Fig. 6.12 (**a**) Fruit of *Cucumis* allotriploid plants derived from a cross between *C.* × *hytivus* and *C. sativus*. (**b**) Chromosome number of allotriploid plants

Recently, 19 cucumber cultivars were used to make crosses with *C.* × *hytivus* to identify crosses producing allotriploid fruits with viable seeds. Only three crosses had fruits with viable seeds. However, this marked significant progress toward commercial use of this seedless novel cucumber. The putative allotriploid plants were confirmed by molecular and cytological analysis.

6.6.2.3 Monosomic Alien Addition Lines

Two monosomic alien addition lines (MAALs) (14 CC + 1 H, $2n = 15$) were recovered among 252 regenerated plants, when the allotriploid was treated with colchicine to induce polyploidy (Chen et al. 2004a). Both the putative MAALs, plant numbers 87 and 517, grew more slowly and were easily differentiated morphologically from the allotriploids, *C.* × *hytivus*, and *C. sativus*. The leaf shape of these plants was palmate and hastate, respectively, in contrast to the pentagon shape observed in allotriploid and the wavy leaf edges observed in the allotetraploid *C.* × *hytivus*. Fruit length of plant numbers 87 and 517 was 23.6 ± 0.9 and 25.4 ± 1.5 cm, respectively, much longer than the allotriploids (12.1 ± 2.1 cm) (Fig. 6.13). Both plants showed white spines, characteristic of *C. sativus*, in contrast to black spines observed in progenies of the interspecific hybrids from the female parent, *C. hystrix*, indicating that the gene for black spines is not located on the alien chromosome in either plant number 87 or 517; both

Table 6.6 Morphological characteristics of *Cucumis hytivus*, *C. sativus* L. cv. Beijing jietou, and allotriploid plants derived from a *C. hytivus* × *C. sativus* mating

Traits[a]	*Cucumis hytivus*	Allotriploid plantlets	*Cucumis sativus*
Average number of branches	4.6 ± 1.9[b]	4.9 ± 1.2	1.3 ± 0.6
Ovary length (cm)	1.3 ± 0.2	1.9 ± 0.2	5.4 ± 0.7
Average fruit length (cm per fruit)	7.5 ± 1.4	12.1 ± 2.0	51.7 ± 3.5
Average fruit weight (gm per fruit)	35.4 ± 4.3	123.2 ± 5.8	322.4 ± 14.4
Mean pollen grain stainability (%)	21.9 ± 2.6	9.8 ± 2.5	94.4 ± 4.28
Fruit length: diameter ratio	2.2 ± 0.3	3.4 ± 0.2	12.5 ± 0.9
Parthenocarpic rate (%)	26.7	84.8	45.2
Fruit spine color	Black	Black	White
Leaf color	Reseda	Dark green	Dark green
Chromosome number	38 ($4n$)	26 ($3n$)	14 ($2n$)

[a]Primary branch number was recorded before season-end vegetative decline; ovary length and pollen grain stainability were recored at anthesis; five mature fruits were used for the determination of fruit weight, length and diameter fruit index, and spin color and leaf color was observed during the vegetative growth period.
[b]Data are shown with mean ±SE.

MAALs also showed a multibranching character reminiscent of *C. hystrix*.

Alien chromosome addition lines harboring one single chromosome from the wild species *C. hystrix* might be used as a bridge to transfer genes of interest originating in *C. hystrix* to individual chromosomes of *C. sativus* via recombination or translocation events.

6.6.2.4 Introgression Lines

C. sativus–hystrix introgression lines were obtained among progenies of interspecific hybrids after the allotetraploid was backcrossed to cultivated cucumber.

Fig. 6.13 (**a**) Fruit of the monosomic alien addition lines (MAAL). (**b**) Chromosome number of MAAL ($2n = 15$)

These introgression lines were genetically stable and differed from each other as well as their two parents in many morphological traits (Fig. 6.14). Substantial genomic changes were detected in introgression lines, including *C. hystrix*-specific fragments, deletion of fragments originally observed in *C. sativus*, and novel bands not from both parents (Zhou et al. 2009).

Zhou et al. (2009) used SSR markers to detect the introgression from *C. hystrix* to *C. sativus*, and one locus at 210 bp was revealed and assigned to introgressive fragment of *C. hystrix* genome in introgression line 56. This line was characterized to have small-sized leaf, short fruit, and multibranching habit, which were closer to the wild parent (*C. hystrix*), and had fertility as high as that of cultivated cucumber. Moreover, this line was found to have a desirable response to downy mildew and Fusarium wilt. Several molecular markers linked to these introgressed traits are now being developed. The introgression lines could be used as vehicle for transferring desirable characters from *C. hystrix* and valuable for the improvement of cucumber cultivars.

6.6.2.5 New Pickling Cucumber F_1 Hybrid

A new pickling cucumber line 7012A was developed through subsequent backcrossing of the hybrid to cucumber and selection from the selfed progenies

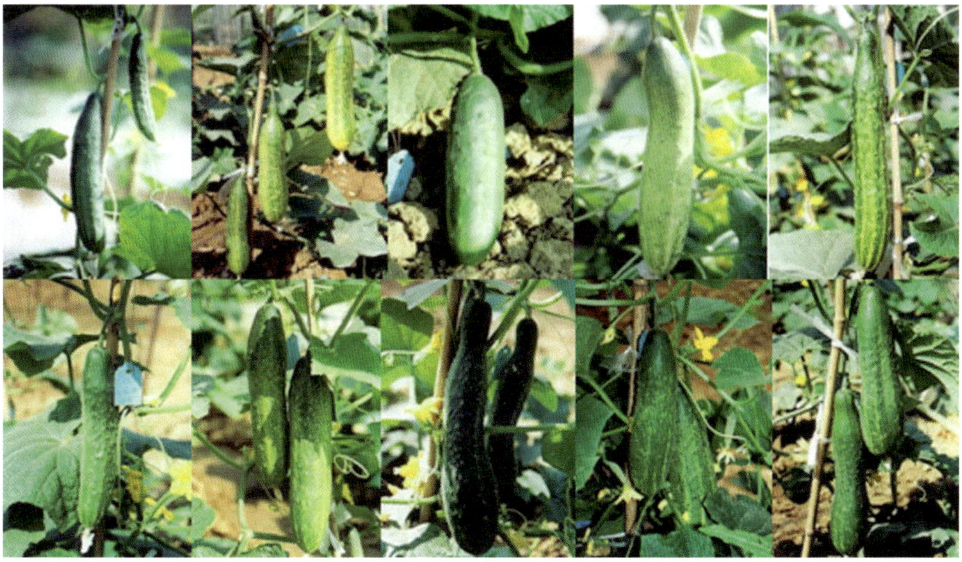

Fig. 6.14 Variation in fruit shape of introgression lines

Fig. 6.15 Fruit of Ningjia #1

(Chen et al. 2005). This line was used to cross with an elite American pickling cucumber line (7011A) from the University of Wisconsin to produce F_1 (Fig. 6.15). The results indicated that the F_1 has significant heterosis over its parents in yield and growth vigor. The plants set uniform fruits with good quality. Fruits could be set on both the main and lateral branches. It has highest late yield and its total yield was higher than all the other cultigens tested and this F_1 was subsequently designated as Ningjia #1.

6.7 Genomic Resources

The development of genomic tools in *Cucumis* species has been very limited in spite of the importance of these species. In recent years, there has been an effort toward the development of genomic resources mainly in melon and cucumber, the cultivated species of *Cucumis*.

The Cucurbit genomics database of the International Cucurbit Genomics Initiative offers comprehensive information on express sequence tags (ESTs) and maps developed in the cucurbits. The version 2 of melon EST collection contains 34,451 ESTs and published genes, representing 16,128 unigenes. In cucumber, 4,331 ESTs are also available from leaf, fruit, and flower tissues representing around 3,000 unigenes. Several genetic maps are available for melon and cucumber. They have been constructed using different types of molecular markers and populations; however, these maps are still far from saturation (Garcia-Mas 2008).

Several bacterial artificial chromosome (BAC) libraries have been constructed in melon (Luo et al. 2001; van Leeuwen et al. 2003) and cucumber (Nam et al. 2005). A fosmid library of cucumber was also constructed (Havey et al. 2008). These libraries can be accessed from the website of the Clemson University Genomics Institute (http://www.cugi.org). Construction of a BAC library of *C. hystrix* is now undergoing in China.

The Cucumber Genome Initiative has reported the progress of the sequencing of the cucumber genome. The sequencing of the cucumber genome has been completed and about 30,000 genes were annotated. A genetic map was constructed, consisting of 1,200 diversity array technology (Dart) markers and 1,000 SSR (microsatellite) markers, distributed over seven linkage groups. By fluorescent in situ hybridization (FISH) mapping of 60 SSR-anchored fosimids, an integrated genetic and cytogenetic map was constructed and chromosomal rearrangements between subspecies of cucumber were discovered (Huang et al., pers comm). The cucumber genome sequence will provide a rich resource for investigating the pathways of gene and genome evolution.

6.8 Recommendations for Future Actions and Future Prospects

Wild species are an important reservoir of useful genes and offer great potential to incorporate such genes into commercial cultivars for resistance to major diseases, insects, and tolerance to various abiotic stresses. Moreover, many of the useful alien genes are different from those of the cultivated species and are thus useful in broadening the sources of resistance/tolerance to various stresses. However, despite the importance of wild species of *Cucumis*, very little information is available on the studies of wild species. Morever, use of exotic germplasm for the development of lines and populations with unique traits has been limited. The application of genetic markers to germplasm management or for marker-assisted selection has not been clearly defined in *Cucumis*. Therefore, we hope that future attention should be paid to research on the wild species.

As a first step, germplasm acquisition of all the *Cucumis* species available from every country should be initiated. Our objective should not be limited to the maintenance of germplasm. Instead, what is required is a well-organized research program on germplasm characterization, utilization, and enhancement.

An important long-term objective for *Cucumis* breeders is the introduction of genes from wild relatives. Some wild relatives, such as *C. metuliferus* E. Meyer ex Naudin (nematode resistance) and *C. figarei* Naudin (virus resistance), have long been attractive to scientists. However, the limits to utilization of wild relatives depend on the breeders' ability to produce interspecific hybrids, but hybrids may be sterile. Future research should focus on producing fertile interspecific hybrids through biotechnological or other approaches.

The rapid advancement in the molecular genetics has now opened a new era of technologies and the application of MAS in crop breeding has become more and more important. There is thus a need to develop molecular markers tightly linked to the useful traits of wild species. Generation and utilization of novel genetic stocks such as introgression lines, substitution lines, and deletion lines will further facilitate the genetic analysis of complex agronomic traits and the introgression of desirable genes into commercial cultivars. Construction of the bacterial artificial chromosome (BAC) library of wild species of *Cucumis*, along with the tightly linked markers of desirable traits, may greatly promote the identification and isolation of useful genes for cucumber and melon improvement.

Wild relatives are useful sources of characters desired to improve cultivated *Cucumis* species. Although many of these wild species are resistant to pests and diseases or adapted to adverse environments, the utilization of wild relatives is limited due to the cross-incompatiability problems. However, the successful development of the synthetic allotetraploid species *C.* × *hytivus* Chen and Kirkbride via chromosome doubling of an interspecific hybrid between cucumber and *C. hystrix* Chakr. provides a step toward transferring desirable genes from the wild species *C. hystrix*. Future works emphasized on map-based cloning of the downy mildew and root-knot nematode resistance genes from *C. hystrix* will be of great value for the improvement of cucumber and melon.

References

Adelberg JW, Chen JF (1998) Genetic control of regeneration was altered during one-week ripening of immature melon cotyledons on liquid/membrane system. Presented at IAPTC World Congress on Cell Culture, Jerusalem, Israel

Adelberg JW, Rhodes BB, Skorupska HT, Bridges WC (1994) Explant origin affects the frequency of tetraploid plants from tissue cultures of melon. HortScience 29:689–692

Andrus CF, Fassuliotis G (1965) Crosses among *Cucumis* species. Veg Improv Newsl 7:3

Ayyangar KR (1967) Taxomomy of Cucurbitaceae. Bull Natl Inst Sci India 34:380–396

Beharav A, Cohen Y (1994) Effect of gamma radiation on vitality and fertilization ability of *Cucumis melon* and *C. metuliferus* pollen. Cucurbit Genet Coop Rep 17:94–96

Betra S (1953) Interspecific hybridization in the genus *Cucumis*. Sci Cult 18(9):445–446

Bhaduri PN, Bose PC (1947) Cyto-genetical investigations in some common cucurbits, with special reference to fragmentation of chromosomes as a physical basis of speciation. J Genet 48:237–256

Bowley SR, Taylor NL (1987) Introgressive hybridization. In: Christie BR (ed) CRC handbook of plant science in agriculture, vol 1. CRC, Boca Raton, FL, USA, pp 23–59

Chatterjee M, More TA (1991) Interspecific hybridization in *Cucumis* spp. Cucurbit Genet Coop 14:69–70

Chelliah S, Sambandam CN (1972) Inheritance of resistance to the fruit fly, Dacus cucurbitae C., in the interspecific cross between *Cucumis callosus* (Rottl.) Cogn. and *Cucumis melo* L. Auara 4(5):169–171

Chen LZ, Chen JF (2008) Changes of cytosine methylation induced by wide hybridization and allopolyploidy in *Cucumis*. Genome 51:789–799

Chen JF, Kirkbride J (2000) A new synthetic species *Cucumis* (Cucurbitaceae) from interspecific hybridization and chromosome doubling. Brittonia 52:315–319

Chen JF, Lewis S (2000) New source of nematode resistance was identified in *Cucumis*. Cucurbit Genet Coop Rep 23:32–35

Chen JF, Staub JE (1997) Attempts at colchicine doubling of an interspecific hybrid of *Cucumis sativus* L. × *C. hystrix* Chakr. Cucurbit Genet Coop Rep 20:24–26

Chen JF, Zhang S, Zhang X (1994) The xishuangbanna gourd, a traditional cultivated plant of the Hanai People, Xishuangbanna, Yunnan, China. Cucurbit Genet Coop Rep 17:18–20

Chen JF, Isskiki S, Tashiro Y, Miyazaki S (1995) Studies on a wild cucumber from China (*Cucumis hystrix* Chakr.) I. Genetic distances between *C. hystrix* and two cultivated *Cucumis* species (*C. sativus* L., and *C. melo* L.) based on isozyme analysis. J Jpn Soc Hortic Sci 64:264–265

Chen JF, Staub JE, Tashiro Y, Isshiki S, Miyazaki S (1997) Successful interspecific hybridization between Cucumis sativas L. and C. hystrix Chakr. Euphytica 96:413–419

Chen JF, Adelberg JW, Staub JE, Skorupska HT, Rhodes BB (1998) A new synthetic amphidiploid Cucumis from a *C. sativus* × *C. hystrix* F_1 interspecific hybrid. In: McCreight J (ed) Cucurbitaceae'98 – Evaluation and enhancement of cucurbit germplasm. ASHS, Alexandria, VA, USA, pp 336–339

Chen JF, Luo XD, Staub JE, Jahn MM, Qian ChT, Zhuang FY, Ren G (2003) An allotriploid derived from a amphidiploid × diploid mating in *Cucumis* I: production, micropropagation and verification. Euphytica 131:235–241

Chen JF, Luo XD, Qian ChT, Molly MJ, Staub JE, Zhuang FY, Lou QF, Ren G (2004a) *Cucumis* monosomic alien addition lines: morphological, cytological, and genotypic analyses. Theor Appl Genet 108:1343–1348

Chen JF, Moriarty G, Jahn M (2004b) Some disease resistance tests in *Cucumis hystrix* and its progenies from interspecific hybridization with cucumber. In: Lebeda A, Paris HS (eds) Proceedings of 8th EUCARPIA meet on cucurbit genetics and breeding, Palacký University, Olomouc, Czech Republic, pp 189–196

Chen LZ, Chen JF, Staub J, Qian CT (2005) A new pickling cucumer F1 hybrid bred from interspecific hybridization. China Veg 3:4–6

Chen LZ, Lou QF, Zhuang Y, Chen JF, Zhang XQ, Wolukau JN (2007) Cytological diploidization and rapid genome changes of the newly synthesized allotetraploids *Cucumis hytivus*. Planta 225:603–614

Chung SF, Staub JE, Chen JF (2006) Molecular phylogeny of *Cucumis* species as revealed by consensus chloroplast SSR marker length and sequence variation. Genome 49:219–229

Custers JBM, Den Nijs APM (1986) Effects of aminoethoxyvinylglycine (AVG), environment, and genotype in overcoming hybridization barriers between *Cucumis* species. Euphytica 35:639–647

D'Amato F (1977) Cytogenetics of differentiation in tissue and cell cultures. In: Reinert J, Bajaj YPS (eds) Plant cell, tissue and organ culture. Springer, Berlin, Germany, pp 34–393

Dane F, Tsuchiya T (1976) Chromosome studies in the genus *Cucumis*. Euphytica 25:367–374

Dane F, Tsuchiya T (1979) Meiotic chromosome and pollen studies of polyploidy *Cucumis* species. Euphytica 28:563–567

Dane F, Denna DW, Tsuchiya T (1980) Evolutionary studies of wild species in the genus *Cucumis*. Z Pflanzenzucht 85:89–109

Danin-Poleg Y, Reis N, Tzuri G, Katzir N (2001) Development and characterization of microsatellite markers in *Cucumis*. Theor Appl Genet 102:61–72

Deakin JR, Bohn GW, Whitaker TW (1971) Interspecific hybridization in *Cucumis*. Econ Bot 25:195–211

den Nijs APM, Custers JBM (1990) Introducing resistance into cucumbers by interspecific hybridization. In: Bates DM, Robinson RW, Jeffery C (eds) Biology and utilization of the Cucurbitaceae. Cornell University Press, Ithaca, NY, USA, pp 382–396

den Nijs APM, Visser DL (1985) Relationships between African species of the genus *Cucumis* L. estimated by the production, vigour and fertility of F1 hybrids. Euphytica 34:279–290

Esquinas-Alcazar JT (1977) Alloenzyme variation and relationships in the genus *Cucumis*. PhD Dissertation, University of California, Davis, CA, USA

Ezura H, Amagai H, Yoshioka K, Oosawa K (1992) Highly frequent appearance of tetraploidy in regenerated plants, a universal phenomenon in tissue culture of melon (*Cucumis melo* L.). Plant Sci 85:209–213

Fassuliotis G (1967) Species of *Cucumis* resistant to the root-knot nematode, *Meloidogyne incognita acrita*. Plant Dis Rep 51:720–723

Fassuliotis G (1977) Self-fertilization of *Cucumis metuliferus* Naud. and its cross-compatilibity with *C. melo* L. J Am Hortic Soc 102(3):336–339

Fassuliotis G, Nelson BV (1988) Interspecific hybrids of *Cucumis metuliferus* × *C. anguria* obtained through embryo culture and somatic embryogenesis. Euphytica 37:53–60

Fellner M, Binarova P, Lebeda A (1996) Isolation and fusion of *Cucumis sativas* and *Cucumis melo* protoplasts. In: Gomez-Guillamon ML, Soria C, Cuartero J, Tores JA, Fernandez-Munoz R (eds) Cucurbits towards 2000, Proceedings of 6th EUCARPIA meeting on cucurbit genetics and breeding, Malaga, Spain, pp 202–209

Franken J, Custers JBM, Bino RJ (1988) Effects of temperature on pollen tube growth and fruit set in reciprocal crosses between *Cucumis* sativus and *C. metuliferus*. Plant Breeding 100:150–153

Garcia-Mas J (2008) Current genomic resources in melon and other cucubits. In: Pitrat M (ed) Proceedings of 9th EUCARPIA Cucurbitaceae Symposium, Avignon, France, pp 201–205

Garcia-Mas J, Monforte AJ, Arus P (2004) Phylogentic relationships among *Cucumis* species based on the ribosomal internal transcribed spacer sequence and microsatellite markers. Plant Syst Evol 248:191–203

Ghebretinsae AG, Thulin M, Barber JC (2007) Relationships of cucumbers and melons unraveled: molecular phylogenetics of *Cucumis* and related genera (Benincaseae, Cucurbitaceae). Am J Bot 94:1256–1266

Havey MJ, Meyer JDF, Deleu W, Garcia-Mas J (2008) Construction of a fosmid library of cucumber (*Cucumis sativus*) and comparative analyses of the eIF4E and eIF(iso)4E regions from cucumber and melon (*Cucumis melo*). Mol Genet Genom 279(5):473–480

Jeffrey C (1962) Notes on Cucurbitaceae including a proposed new classification for the family. Kew Bull 15:337–371

Jeffrey C (1967) On the classification of Cucurbitaceae. Kew Bull 20:417–426

Jeffrey C (1980) A review of the Cucurbitaceae. Bot J Linn Soc 81:233–247

Jeffrey C (1990) Systematics of Cucurbitaceae: an overview. In: Bates DM, Robinson RW, Jeffrey C (eds) Biology and utilization of the Cucurbitaceae. Cornell University Press, Ithaca, NY, USA, pp 3–9

Jeffrey C (2005) A new system of Cucurbitaceae. Bot Zhurn 90:332–335

Jobst J, King K, Hemleben V (1998) Molecular evolution of the internal transcribed spacers (ITS1 and ITS2) and phylogenetic relationships among species of the family Cucurbitaceae. Mol Phylogenet Evol 9:204–219

Kho YO, den Nijs APM, Franken J (1980) Interspecific hybridization in *Cucumis* L. II. The crossability of species, an investigation of in vivo pollen tube growth and seed set. Euphytica 29:661–672

Kirkbride JH (1993) Biosystematic monograph of the genus *Cucumis* (Cucurbitaceae). Parkway, Boone, NC, USA

Kishi Y, Fujishita N (1969) Studies on interspecific hybridization in the genus *Cucumis*. I. Pollen germination and pollen tube growth in selfings and incompatible crossings. J Jpn Soc Hortic Sci 38:329–334

Kishi Y, Fujishita N (1970) Studies on interspecific hybridization in the genus *Cucumis* II Pollen tube growth, fertilization and embryogenesis of post-fertilization stage in incompatible crossings. J Jpn Soc Hortic Sci 39:51–57

Kocyan A, Zhang LB, Schaefer H, Renner SS (2007) A multilocus chloroplast phylogeny for the Cucurbitaceae and its implications for character evolution and classification. Mol Phylogenet Evol 44:553–577

Kristkova E, Lebeda A, Vinter V, Blahousek O (2003) Genetic resources of the genus *Cucumis* and their morphological description (English-Czech version). HortScience (Prague) 30(1):14–42

Laibach F (1925) Das Taubwerden von Bastardsamen und die kunstliche Aufzucht fruh absterbender Bastardembryonen. Z Bot 17:417–459

Leppick EE (1966) Searching gene centers of the genus *Cucumis*. Euphytica 15:323–328

Liu Q (2007) Distribution and molecular phylogenentics of *Cucumis* species. MA Dissertation, Nanjing Agricultural University, PR China

Lower RL, Edwards MD (1986) Cucumber breeding. In: Basset MJ (ed) Breeding vegetable crops. AVI, Westport, CN, USA, pp 173–207

Luo M, Wang YH, Frisch D, Joobeur T, Wing RA, Dean RA (2001) Melon bacterial chromosome (BAC) library construction using improved methods and identification of clones linked to the locus conferring resistance to melon Fusarium wilt (*Fom-2*). Genome 44:154–162

Ma SW, Wang JM, Qiu JT (2003) Present situation of collection and presevation of watermelon and melon germplasm in China. China Watermelon Muskmelon 5:17–19

McCreight JD, Nerson H, Grumet R (1993) Melon, *Cucumis melo* L. In: Kalloo G, Bergh BO (eds) Genetic improvement of vegetable crops. Pergamon, New York, USA, pp 267–294

Nam YW, Lee JR, Song KH, Lee MK, Robbins MD, Chung SM, Staub JE, Zhang HB (2005) Construction of two BAC libraries from cucumber (*Cucumis sativus* L.) and identification of clones linked to yield component quantitative trait loci. Theor Appl Genet 111(1):150–161

Naudin C (1859) Revue des Cucurbitaceae cultiveesau museum en 1859. Ann Sci Nat Ser 4 Bot 12:79–164

Norton JD, Granberry DM (1980) Characteristics of progeny from an interspecific cross of *Cucumis melo* with *C. metuliferus*. J Am Soc Hortic Sci 105:174–180

Niemirowicz-Szczytt K, Kubiki B (1979) Cross fertilization between cultivated species of genera *Cucumis* L. and *Cucurbit L*. Genetica Polonica 20:117–125

Osifo E, Webb JK, Henshaw GG (1989) Variation amongst callus derived potato plants *Solanum bredidens*. J Plant Physiol 134:1–4

Pangalo KJ (1950) Melons as the independent genus *Melo* Adans. Bot Z (Moscow and Leningrad) 35:571–580

Pearson OH, Hopp R, Bohn GW (1951) Notes on species crosses in *Cucurbita*. Proc Am Soc Hortic Sci 57:310

Perl-Treves R, Galun E (1985) The *Cucumis* plastome: physical map, intrageneric variation and phylogenetic relationships. Theor Appl Genet 71:417–429

Pitrat M, Chauvet M, Foury C (1999) Diversity, history, and production of cultivated cucurbits. Acta Hortic 492:21–28

Puchalski T, Robinson RW (1990) Electrophoretic analysis of isozymes in *Cucurbita* and *Cucumis* and its application for phylogenetic studies. In: Bates DM, Robinson RW, Jeffrey C (eds) Biology and utilization of the Cucurbitaceae. Cornell University Press, Ithaca, NY, USA, pp 60–76

Qian CT, Chen JF, Zhuang FY, Xu YB, Li SJ (2002) Several photosynthetic characters of the synthetic species *Cucumis hytivus* Chen and Kirkbride under weak light condition. Plant Physiol Commun 38:336–338

Raamsdonk LWD, den Nijs APM, Jongerius MC (1989) Meiotic analyses of *Cucumis* hybrids and an evolutionary evaluation of the genus *Cucumis* (Cucurbitaceae). Plant Syst Evol 163:133–146

Ramachandran C (1984) Nuclear DNA variation in *Cucumis* species. Cucurbit Genet Coop 7:97–98

Ramachandran C, Narayan RKJ (1985) Chromosomal DNA variation in *Cucumis*. Theor Appl Genet 69:497–502

Ramachandran C, Seshadri VS (1986) Cytological analysis of the genome of cucumber (*Cucumis sativus* L.) and muskmelon (*Cucumis melo* L.). Z Pflanzenzücht 96:25–38

Renner SS, Schaefer H (2008) Phylogenetics of *Cucumis* (Cucurbitaceae) as understood in 2008. Cucurbitaceae 2008. In: Pitrat M (ed) Proceedings of 9th EUCARPIA meeting on genetics and breeding of Cucurbitaceae. INRA, Avignon, France, 21–24 May 2008, pp 53–58

Renner SS, Schaefer H, Kocyan A (2007) Phylogenetics of *Cucumis* (Cucurbitaceae): Cucumber (*C. sativus*) belongs in an Asian/Australian clade far from melon (*C. melo*). BMC Evol Biol 7:58

Robinson RW, Decker-Walters DS (1997) Interspecific hybridization. In: Robinson R, Decker-Walters DS (eds) Cucurbits. CABI, Wallingford, Oxon, UK, pp 51–55

Rubatzky VE, Yamaguchi M (1997) World vegetables: principles, production, and nutritive values, 2nd edn. Chapman and Hall, New York, NY, USA

Schaefer H (2007) *Cucumis* (Cucurbitaceae) must include *Cucumella*, *Dicoelospermum*, *Mukia*, *Myrmecosicyos*, and *Oreosyce:* a recircumscription based on nuclear and plastid DNA data. Blumea 52:165–177

Sheng D, Li XX, Wang HP, Song JP (2006) Research progress and prospects on cucumber germplasm. China Veg Suppl 150:77–81

Singh BP (1991) Interspecific hybridization in between new and old-world species of *Luffa* and phylogenetic implication. Cytologia 56:35–365

Singh AK, Yadava KS (1984) An analysis of interspecific hybrids and phylogenetic implications in *Cucumis* (Cucurbitaceae). Plant Syst Evol 147:237–252

Soria C, Gomez-Guillamon ML, Esteva J, Nuez F (1990) Ten interspecific crosses in the genus Cucumis: a preparatory study to seek crosses resistant to melon yellowing diseases. Cucurbit Genet Coop Rpt 13:31–33

Staub JE, Serquen FC, Horejsi T, Chen JF (1999) Genetic diversity in cucumber (*Cucumis sativus* L.): IV. An evaluation of Chinese germplasm. Genet Resour Crop Evol 46:297–310

Tabei Y, Nishio T, Kanno T (1992) Shoo regeneration from cotyledonary protoplasts on melon (*Cucumis melo* L.cv. Charentais). J Jpn Soc Hortic Sci 61:317–322

Tang FA, Punja ZK (1989) Isolation and culture of protoplasts of *Cucumis sativas* and *Cucumis metuliferus* and methods for their fusion. Cucurbit Genet Coop Rep 12:29–32

Trivedi RN, Roy RP (1970) Cytological studies in *Cucumis* and *Citrullus*. Cytologia 35:561–569

Valvilov N (1925) Inter-genetic hybrids of melons, watermelons and squashes. Plant Breed 14:3–35

van Leeuwen H, Monfort A, Zhang HB, Puigdomenech P (2003) Identification and characterisation of a melon genomic region containing a resistance gene cluster from a constructed BAC library. Microcolinearity between Cucumis melo and Arabidopsis thaliana. Plant Mol Biol 51:703–718

Varekamp HQ, Visser DL, den Nijs APM (1982) Rectification of the names of certain accessions of the IVT- Cucumis collection. Cucurbit Genet Coop 5:59–60

Visser DL, den Nijs APM (1983) Variation for interspecific crossability of *Cucumis anguria* L. and *C. zeyer* Sond. Cucurbit Genet Coop 6:100–101

Weeden NF, Robinson RW (1986) Allozyme segregation ratios in the interspecific cross *Cucurbita maxima* × *C. ecuadorensis* suggest that hybrid breakdown is not causes by minor alterations in chromosome structure. Genetics 114:593–609

Whitaker TW, Davis GN (1962) Cucurbits: botany, cultivation and utilization. Interscience, New York, USA

Yadava KS, Singh AK, Roy RP, Jha UC (1986) Cytogenetics in *Cucumis* L. VI Synthetic amphidiploids. Nucleus 29:58–62

Zhang YB, Chen JF, YI HP, Feng JX, Wu MZ (2005) Staining and slide-preparing technique of mitotic chromosomes and its use in karyotype determination of *Cucumis melo* L. Acta Bot Boreal-Occident Sin 25(9):1735–1739

Zhou XH, Qian CT, Lou QF, Chen JF (2009) Molecular analysis of introgression lines from *Cucumis hystrix* Chakr. to *C. sativus* L. Sci Hortic 119:232–235

Zhuang FY, Chen JF, Qian ChT, Li ShJ, Ren G, Wang ZhJ (2002) Responses of seedlings of *Cucumis* × *hytivus* and progenies to low temperature. J Nanjing Agric Univ 25:27–30

Zhuang FY, Chen JF, Staub JE, Qian CT (2006) Taxonomic relationships of a rare *Cucumis* species (*C. hystrix* Chakr.) and its interspecific hybrid with cucumber. HortScience 41(3):571–574

Chapter 7
Daucus

Dariusz Grzebelus, Rafal Baranski, Krzysztof Spalik, Charlotte Allender, and Philipp W. Simon

7.1 Introduction

The genus *Daucus* is a member of the Apiaceae (Umbelliferae) family, which is one of the most important families of angiosperms from an economical point of view. Apart from the cultivated carrot (*Daucus carota* subsp. *sativus* Hoffm.), this family comprises such important crop species as celery and celeriac (*Apium graveolens* L.) and parsley [*Petroselinum crispum* (Mill.) Nyman ex A. W. Hill], as well as many locally used vegetables and condiments: parsnip (*Pastinaca sativa* L.), fennel (*Foeniculum vulgare* Mill.), dill (*Anethum graveolens* L.), coriander (*Coriandrum sativum* L.), chervil (*Anthriscus cerefolium* (L.) Hoffm.), lovage (*Levisticum officinale* W.D.J. Koch), aniseed (*Pimpinella anisum* L.), cumin (*Cuminum cyminum* L.), caraway (*Carum carvi* L.), and many others. Several species are used in traditional medicine, including the members of *Ferula* and *Dorema*, the sources of aromatic resins: asafoetida, sagapenum, galbanum, sumbul, and ammoniacum (Korovin 1959). The usage of umbellifers, particularly as flavorings, spices, and in traditional medicine, is connected with aromatic compounds that may occur in all parts of the plant.

D. Grzebelus (✉)
Department of Genetics, University of Agriculture in Krakow, Al. 29 Listopada 54, 31-425 Kraków, Poland
e-mail: dgrzebel@ogr.ar.krakow.pl

7.2 *Daucus* Botany, Taxonomy, and Genetics

7.2.1 *Botany*

During the last half century, the genus *Daucus* was described and revised several times including a regional revision for the *Flora Europaea* (Heywood 1968), a checklist of species for the revision of tribe Caucalideae (Heywood 1982), and worldwide generic revision (Sáenz Laín 1981). Moreover, *Daucus* was subject to several molecular and morphological phylogenetic studies (Lee and Downie 1999, 2000; Lee et al. 2001; Spalik and Downie 2007). It includes 21–24 species (Table 7.1) that are widespread in the northern hemisphere, with few species in South America and Australia. These are usually herbaceous biennials, rarely annuals, growing from slender to very stout taproots. As in most members of umbellifer subfamily Apioideae, their leaves are pinnatisect, the inflorescence is a compound umbel, and the fruit is a schizocarp splitting into two one-seeded mericarps. The fruit is oblong to ovoid, dorsally compressed, with prominent longitudinal projections: three primary ribs on each mericarp, situated above vascular bundles, and two secondary ribs, situated between the primary ones (in valleculae), above secretory canals (vittae). Primary ribs are covered with 2–4 rows of unbranched, semi-erect or spreading hairs. On each secondary rib there is a row of spines that may be glochidiate or simple at the apex; these spines serve exozoochory.

C. Kole (ed.), *Wild Crop Relatives: Genomic and Breeding Resources, Vegetables*,
DOI 10.1007/978-3-642-20450-0_7,

Table 7.1 Comparisons of taxonomic treatments of *Daucus* by Heywood (1968, 1982) and Sáenz Laín (1981) with results of molecular phylogenetic studies concerning *Daucus* sensu lato clade (Lee and Downie 1999, 2000; Lee et al. 2001; Spalik and Downie 2007)

Heywood (1968, 1982)	Sáenz Laín (1981)	Molecular studies
sect. ***Daucus***	sect. ***Daucus***	***Daucus*** **I** subclade
D. carota L.	*D. carota*	*D. carota*[a]
subsp. *carota*	subsp. *carota*	subsp. *azoricus* Franco
subsp. *maritimus* (Lam.) Batt.	subsp. *maritimus*	subsp. *carota*
subsp. *major* (Vis.) Arcang.	subsp. *gummifer*	subsp. *drepanensis*
subsp. *sativus* (Hoffm.) Arcang.	subsp. *hispanicus*	subsp. *gadecaei*
subsp. *gummifer* Hook. f.	subsp. *maximus* (Desf.) Heywood (=*D. maximus*)	subsp. *gummifer*
subsp. *hispanicus* (Gouan) Thell.	*D. capillifolius*	subsp. *maritimus*
subsp. *hispidus* (Arcang.) Heywood	*D. crinitus*	subsp. *hispanicus*
subsp. *gadecaei* (Rouy & Camus) Heywood	*D. gracilis*	subsp. *maximus*
subsp. *drepanensis* (Arcang.) Heywood	*D. guttatus*	*D. aureus*
subsp. *rupestris* (Guss.) Heywood	*D. involucratus*	*D. biseriatus* Murb.[b]
D. capillifolius Gilli	*D. jordanicus*	*D. capillifolius*
D. gracilis Steinh.	*D. montevidensis*	*D. crinitus*
D. guttatus Sibth. & Sm.	*D. pusillus*	*D. gracilis*
D. involucratus Sibth. & Sm.	*D. sahariensis*	*D. muricatus*
D. jordanicus Post	*D. syrticus*	*D. sahariensis*
D. littoralis Sibth. & Sm.	*D. tenuisectus*	*D. syrticus*
D. maximus Desf.		*D. tenuisectus*
D. sahariensis Murb.	sect. ***Meoides***	*Athamanta della-cellae* Asch. & Barbey ex E. A. Durand & Barratte
D. syrticus Murb.	*D. setifolius*	*Pachyctenium mirabile* Maire & Pamp.
D. tenuisectus Coss. ex Batt.		*Pseudorlaya minuscula* (Pau ex Font Quer) Laínz
	sect. ***Anisactis***	*Pseudorlaya pumila* (L.) Grande
sect. ***Meoides*** Lange	*D. durieua* (= *D. hochstetteri*)	*Tornabenea annua* Bég. ex A. Chev.
D. crinitus Desf.	*D. glochidiatus*	*Tornabenea tenuissima* (A. Chev.) A. Hansen & Sunding
D. setifolius Desf.	*D. montanus*	
		Macaronesian Endemics group
sect. ***Anisactis*** DC.	sect. ***Chrysodaucus***	*Cryptotaenia elegans* Webb
D. durieua Lange	*D. aureus*	*Monizia edulis* Lowe
D. glochidiatus (Labill.) Fisher & C.A. Mey.		*Melanoselinum decipiens* Hoffm.
D. hochstetteri A. Braun ex Engl.	sect. ***Platyspermum***	
D. montanus Humb. & Bonpl. ex Spreng.	*D. bicolor* (= *D. broteri* Ten.)	***Daucus*** **II** subclade
	D. littoralis	*D. arcanus* Garcia-Martin & Silvestre
sect. ***Leptodaucus*** Thell.	*D. muricatus*	*D. bicolor*
D. pusillus Michx. (=*D. montevidensis* Link ex Spreng.)		*D. broteri*
	Incertae sedis	*D. conchitae*
sect. ***Chrysodaucus*** Thell.	*D. conchitae* Greuter	*D. durieua*
D. aureus Desf.		*D. glochidiatus*
		D. guttatus
sect. ***Platyspermum*** (Hoffm.) DC.		*D. involucratus*
D. muricatus (L.) L.		*D. montanus*
		D. pusillus

(*continued*)

Table 7.1 (continued)

Heywood (1968, 1982)	Sáenz Laín (1981)	Molecular studies
sect. ***Pseudoplatyspermum*** Thell.		
D. bicolor Sibth. & Sm.		***Agrocharis*** subclade
		A. incognita (C. Norman) Heywood & Jury
		A. melanantha Hochst.
		A. pedunculata (Baker f.) Heywood & Jury
		Hitherto unexamined
		D. jordanicus
		D. littoralis
		D. montevidensis
		D. setifolius
		Tornabenea spp.

[a]The subspecies of *D. carota* are listed to show the extent of sampling rather than to uphold their taxonomic position. Molecular studies confirmed that these taxa are very closely related and should be included within *D. carota*; however, most of these names should be reduced to synonymy
[b]Previously recognized in *Pseudorlaya*

7.2.2 Taxonomy

Daucus and its presumed relatives were classified by Heywood (1982) in the tribe Caucalideae along with other umbellifers characterized by the presence of secondary ribs and spines. Following the classification of Okeke (1978), Heywood (1982) recognized 21 species distributed among seven sections. The section *Daucus* includes the highly polymorphic *D. carota*, the nomenclatural type of the genus (Heywood 1968). Sáenz Laín (1981) reduced some of Heywood's subspecies of *D. carota* to synonymy and recognized only five sections. However, subsequent molecular and morphological studies demonstrated that both classification systems are incongruent with phylogeny. In these analyses, Caucalideae sensu Heywood (1982) were placed into two separate clades and the *Daucus* clade allied with members of tribe Laserpitieae. The fruits of the latter are also characterized by the presence of secondary ribs but they have wings rather than spines (Lee and Downie 1999, 2000; Lee et al. 2001). The clade comprising *Daucus* and its relatives as well as former members of the Laserpitieae was therefore recognized as subtribe Daucineae in tribe Scandiceae (Downie et al. 2000; Lee et al. 2001). These studies also revealed that the genus is not monophyletic as its inclusive branch of Daucinae, hereafter named *Daucus* sensu lato (s.l.) clade, includes four subclades with only two of them comprising species traditionally placed in the genus; these two branches are designated as the *Daucus* I and *Daucus* II subclades (Spalik and Downie 2007). The *Daucus* I subclade includes the wild ancestor of the cultivated carrot with all of its subspecies, several Mediterranean members of the genus, and some species that were traditionally placed in other genera: *Athamanta della-cellae*, *Pachyctenium mirabile*, and two members of *Pseudorlaya* (Table 7.1). Surprisingly, two representatives of *Tornabenea*, a genus encompassing 5–6 species endemic to Cape Verde (Brochmann et al. 1997; Schmidt and Lobin 1999), were placed among subspecies of *D. carota*. Contrary to the members of *Daucus*, the fruits of *Tornabenea* have wings and are wind dispersed; therefore, this genus was traditionally placed in tribe Laserpitieae (Pimenov and Leonov 1993). The *Daucus* I subclade is sister to a group of three Macaronesian endemics: *Monizia edulis* and *Melanoselinum decipiens* from Madeira and *Cryptotaenia elegans* from the Canary Islands. The *Daucus* II subclade comprises the remaining members of *Daucus* including its American and Australian representatives. It is sister to *Agrocharis*, an African member of Daucinae. As reconstructed with dispersal-vicariance analysis (Ronquist 1997) using nuclear DNA (nrDNA) internal transcribed spacer (ITS) phylogeny, the ancestral area of the *Daucus* s.l. clade includes the Mediterranean region. The colonization of Macaronesia by its members occurred at least twice from the western Mediterranean, as well as a single long-distance dispersal to North America, with subsequent dispersals to South America and to Australia (Spalik and Downie 2007).

Because of the aforementioned polyphyly, the current classification system of *Daucus* is untenable. Two solutions need to be considered (1) to adopt a broad definition of *Daucus* that encompasses all species placed in the *Daucus* s.l. clade, or (2) to split *Daucus* into smaller genera, at least to those equivalent to the *Daucus* I and *Daucus* II subclades. However, both solutions cause serious problems. With morphologically distant species included, particularly *M. edulis* and *M. decipiens*, it is difficult to define the *Daucus* s.l. clade based on morphological characters that may be used to construct a key. On the other hand, a detailed study of fruit morphology and anatomy failed to find any good characters delimiting the *Daucus* I and *Daucus* II subclades (A. Wojewódzka and K. Spalik unpublished data). Additionally, the most detailed phylogeny of the *Daucus* s. l. clade was based on analyses of a single marker, nuclear ribosomal DNA ITS (Spalik and Downie 2007), whereas the analyses including cpDNA restriction sites and *rps16* intron sequences were performed for a much smaller selection of taxa (Lee and Downie 2000). Several potential members of the clade have not been examined for molecular data yet (Table 7.1). Therefore, additional data from both molecular and morphological markers are necessary before any workable classification system of *Daucus* is proposed.

7.2.3 Nuclear Genome

The nrDNA content in *D. carota* is 0.5 pg per 1C nucleus and the genome is estimated at 473 Mbp (Arumuganathan and Earle 1991; Bennett and Leitch 1995). Bennett and Smith (1976) estimated the 1C DNA content of *D. carota*, including several subspecies, to range from 1.0 to 2.0 pg. Estimates for 1C content of other *Daucus* are *D. aureus* 1.1 pg, *D. blanchei* 4.7 pg, *D. crinitus* 3.9 pg, *D. littoralis* 3.0 pg, *D. montanus* 5.5 pg, *D. muricatus* 2.5 pg, *D. subsessilis* (*D. durieua*) 1.9 pg, and *D. syrticus* 1.2 pg. Ca. 40% of the *D. carota* genome is presumed to be highly repetitive. The chromosome number in *Daucus* species comprise $2n = 18$ (*D. carota*, *D. capillifolius*, *D. sahariensis*, *D. syrticus*), $2n = 20$ (*D. broteri*, *D. guttatus*, *D. littoralis*, *D. muricatus*), $2n = 22$ (*D. aureus*, *D. involucratus*, *D. tenuisectus*, *D. crinitus*, *D. montevidensis*, *D. pusillus*), $2n = 44$ (*D. glochidiatus*), or $2n = 66$ (*D. montanus*) (Rubatzky et al. 1999; Iovene et al. 2008). It is assumed that $x = 11$ is the basic chromosome number in the Apiaceae, and $x = 10$ and 9 are its derivatives (Pimenov et al. 2003). Fluorescent in situ hybridization (FISH) experiments revealed that the ten *Daucus* species investigated, including a putatively disomic tetraploid *D. glochidiatus*, invariably produced two signals for 5S and two signals for 18S–25S rDNA, thus indicating the presence of one locus for each rDNA type. Tentative karotypes were produced for eight species, i.e., *D. broteri*, *D. capillifolius*, *D. carota*, *D. crinitus*, *D. guttatus*, *D. littoralis*, *D. muricatus*, and *D. pusillus* (Iovene et al. 2008).

7.2.4 Biology and Genetics of D. carota

Daucus carota (wild carrot, Queen Anne's Lace, bird's nest weed), an outcrossing diploid species ($2n = 2x = 18$), is the most widespread species of the genus *Daucus*, presently appearing in temperate regions all over the world. Central Asia is usually indicated as the place of origin of domesticated carrot (Small 1978; Simon and Goldman 2007), while most subspecies occur rather in the Mediterranean and therefore this region is the endemism center of the species. *Daucus carota* usually develops seedlings in the autumn, which are then vernalized during winter to induce flowering next summer. Hence, *D. carota* can usually be described as a "winter annual" (Simon and Goldman 2007), even though some accessions can be annual, biennial, or triennial (Lacey 1986). The species shows great morphological plasticity, resulting in the presence of a range of distinct phenotypes (Small 1978), including cultivated carrot, which is believed to be directly derived from wild carrot (Simon 2000). It also shows a high level of molecular diversity, as revealed by investigations employing isozyme markers (St. Pierre et al. 1990) and DNA polymorphisms (Nakajima et al. 1998; Bradeen et al. 2002). Wild and cultivated carrots intercross freely, which has significant implications both for the historical development of the modern cultivated carrot and for the future of carrot breeding.

7.3 Conservation and Utilization of Wild *Daucus* Species

The diversity of plant life across the globe is under threat from a combination of human-mediated habitat alteration and a changing climate. The need to increase crop productivity to feed an evergrowing human population is a real challenge, made even harder by a changing environment and consequent alterations in the distribution and severity of plant pests and diseases. For carrots, like other crop plants, crop breeding and improvement programs can develop new varieties adapted to less favorable conditions but novel sources of genetic variation must be available to breeders. The wild species of *Daucus* potentially represent such a resource but their conservation and accessibility to researchers and breeders must be assured for future use. *Daucus* species, like other crop wild relatives, can be conserved in both an in situ and ex situ manner. A summary of current knowledge and conservation efforts is presented below.

7.3.1 In situ Conservation

In situ conservation depends mainly on the geographical distribution of a given species overlapping with a protected area. A global assessment of the degree to which wild *Daucus* species are protected in situ has yet to be published. Such an assessment would be a major undertaking involving the collation of distributional datasets and comparative analysis of protected areas as described by Maxted et al. (2008a). Similar investigations have been undertaken for endemic species (Riemann and Ezcurra 2005) and other crop wild relatives (for example, *Aegilops* spp; Maxted et al. 2008b). Such studies aid evidence-based conservation strategies, emphasizing the need for protected areas in regions containing high levels of diversity. An assessment based on globally comprehensive data for *Daucus* species is required to determine if current in situ (and ex situ) conservation strategies are adequate. The required datasets are currently collated in a piecemeal manner. National and regional examples of well-cataloged in situ conservation are in various stages of development. One example is The National Biodiversity Network Gateway (http://www.searchnbn.net), which describes species distribution and protected areas in the UK. It is then possible to determine whether a species record falls within or overlaps with a protected area. The NBN Gateway shows that four *Daucus* taxa are recorded in the UK at the time of writing; *D. carota* subsp. *carota*, *D. carota* subsp. *gummifer*, *D. carota* subsp. *sativus* (presumably feral populations of cultivated carrot) and *D. glochidiatus* (introduced via sheep's wool from Australia; Stace 1997). All of these species have a recorded spatial distribution that overlaps with one or more protected areas in the UK.

7.3.2 Ex Situ Conservation of Daucus Species

There are several major collections of *Daucus* germplasm in gene banks across the world (see Simon et al. 2008 for a summary). These collections are mainly based around the cultivated carrot, as this material has been of the most interest to plant breeders and researchers. Wild carrot (*D. carota* subsp. *carota*) is generally well represented as it is the closest relative of cultivated carrot. Most *Daucus* species tested have seed, which is described as in "orthodox" meaning that it can be successfully maintained under long-term standard conditions of low seed moisture and temperature (Hong et al. 1998).

The most significant collections of wild *Daucus* species are based in Europe and the USA. The European collection is decentralized; accessions are held at various gene banks in ten different countries. Material in the European collections originates from 46 countries in total. These collections are described and cataloged by the EURISCO database (http://eurisco.ecpgr.org), which provides a simplified method of searching for accessions of interest. The *Daucus* germplasm collection in the USA is managed by the Agricultural Research Service of the USDA and is also cataloged as an internet searchable database (see http://www.ars-grin.gov/npgs/index.html). A summary of the wild *Daucus* accessions in both the US and European collections can be seen in Table 7.2. Other smaller collections of wild *Daucus* material exist, including the *Daucus* working collection of the Julius-Kühn-Institute, held at the Institute of Horticultural Crops, Quedlinburg, Germany and the

Table 7.2 A summary of the wild *Daucus* accessions in the EURISCO and ARS-GRIN databases

Genus	Species	Subtaxon	Accessions at EURISCO	Accessions at ARS-GRIN	TOTAL
Daucus	*aureus*		4	2	6
Daucus	*bicolor*		22	–	22
Daucus	*broteri*		3	15	18
Daucus	*capillifolius*		–	–	–
Daucus	*conchitae*		–	–	–
Daucus	*crinitus*		1	3	4
Daucus	*durieua*		–	–	–
Daucus	*glochidiatus*		–	–	–
Daucus	*gracilis*		–	–	–
Daucus	*guttatus*		4	22	26
Daucus	*involucratus*		5	3	8
Daucus	*jordanicus*		–	–	–
Daucus	*littoralis*		3	–	3
Daucus	*montanus*		–	–	–
Daucus	*muricatus*		23	2	25
Daucus	*pusillus*		1	2	3
Daucus	*sahariensis*		–	3	3
Daucus	*setifolius*		–	–	–
Daucus	*syrticus*		1	6	7
Daucus	*tenuisectus*		–	–	–
Daucus	*carota*	Unknown/other wild	231	120	351
Daucus	*carota*	*carota*	111	21	132
Daucus	*carota*	*commutatus*	6	3	9
Daucus	*carota*	*drepanensis*	–	–	–
Daucus	*carota*	*gadecaei*	3	–	3
Daucus	*carota*	*gummifer*	9	1	10
Daucus	*carota*	*hispanicus*	2	–	2
Daucus	*carota*	*hispidus*	2	–	2
Daucus	*carota*	*major*	4	3	7
Daucus	*carota*	*maritimus*	7	2	9
Daucus	*carota*	*maximus*	18	4	22
Daucus	*carota*	*rupestris*	–	–	–
Daucus	Unknown/other		152	24	176
		Total wild Daucus Accessions			*848*

Databases were queried on 22 June 2010 and material with a sample status of "wild or natural" (EURISCO) and "wild" (ARS-GRIN) were selected. The taxonomy of samples was unified as far as possible following Heywood (1968, 1982) and Sáenz Laín (1981), with addition information from GRIN taxonomy for plants (http://www.ars-grin.gov/cgi-bin/npgs/html/index.pl). This was done to ensure synonyms were accounted for, and not to uphold any particular taxonomic system for the group

Millennium Seed Bank in the UK. A slightly different approach is used in the French "Carrot and other *Daucus* species" network. Here, accessions are conserved and characterized in a coordinated effort between public and commercial institutions (see Briard et al. 2007). Currently, ten wild *Daucus* taxa are conserved within the network (E. Geoffriau personal communication).

Representation of wild *Daucus* species (with the exception of *D. carota* subsp. *carota*) in the collections is patchy (see Table 7.2). Some species listed in the taxonomic treatments of Heywood (1968, 1982) or Sáenz Laín (1981), such as *D. gracillis*, *D. jordanicus*, *D. conchitae*, and *D. setifolius* are not represented at all. Some species have a reasonable number of accessions conserved, such as *D. bicolor*, *D. muricatus*, *D. guttatus*, and *D. broteri*. The differences in the taxonomic systems used make direct comparisons between the European and US collections difficult. For the purposes of this chapter, an attempt has been made to standardize the taxonomy of samples based on Heywood (1968, 1982) and Sáenz Laín (1981).

Synonyms were accounted for using the GRIN Taxonomy for Plants resource (http://www.ars-grin.gov/cgi-bin/npgs/html/index.pl). Samples with a taxonomic designation not listed in Table 7.2 or without an accepted synonym were included in either *Daucus* "unknown/other" or *D. carota* "unknown/other." This standardization was essential to begin to understand the depth of sampling of wild *Daucus* taxa, and is not intended in any way to be a formal taxonomic treatment of the group. It is likely that a small number of duplicate accessions exist between the two collections, as samples can arrive in gene banks via different donors, but originate from a single sample collected from a wild population. It is apparent that further sampling from across the geographic ranges of most species is desirable. This will ensure that the samples of wild *Daucus* maintained in these ex situ collections are representative of the natural genetic diversity of each species so that this diversity can be conserved and made available for future use.

As has been discussed earlier in this chapter, the taxonomy of this group of species is challenging and open to debate. Ex situ collections often rely on the taxonomic designation assigned to an accession by its original donor. As can be seen from Table 7.2, sometimes material is donated that has not been properly identified to species or subspecies level; a total of 147 of the 176 Daucus "unknown/other" accessions are listed simply as *Daucus* spp. There is a huge potential value in this material in terms of genetic diversity, but effort is required to correctly classify it so that it can be more easily used for research and breeding. Not all gene banks have the resources and expertise to validate the identity of accessions routinely. Additionally, as the taxonomy of a genus such as *Daucus* is revised, it is inevitable that material will occasionally be donated using out-of-date classifications, and some gene banks use different taxonomic standards. More work is needed to ensure that samples in ex situ collections are identified correctly using standard classifications. Table 7.1 at the beginning of this chapter lists non-*Daucus* species, which are suggested to lie within the *Daucus* sensu lato clade based on the results of molecular studies. None of the non-*Daucus* species listed are present in either the EURISCO or GRIN catalog, however, the genera *Athamanta* and *Cryptotaenia* are represented by a very small number of accessions within the European collections.

7.3.3 Threats to Daucus Species

Daucus species vary in their distribution patterns. Some occupy a relatively limited geographical area while others are more widespread. Examples of *Daucus* species with a limited geographical distribution are given in Table 7.3. Species with highly restricted distributions may be more at risk of genetic erosion and extinction in the wild. The closest relative to cultivated carrot (*D. carota* subsp. *carota*; wild carrot or Queen Anne's Lace) is very widely distributed across temperate regions of the world and as such is unlikely to require special conservation efforts at the moment.

The most obvious threat to wild *Daucus* species growing outside of protected areas is loss of habitat, either through adoption of land into agriculture, changes in agricultural practice, or urban development. The effect of climate change on the geographical range of *Daucus* species may mean that current in situ conservation efforts may not be adequate. Jarvis et al. (2008) modeled the effect of climate change on wild relatives of three crop species and suggested that where species are predicted to undergo significant reductions in range size, the collection and conservation of adequate samples in ex situ collections as well as in situ is vital.

Locally maintained cultivated selections of carrot landraces are being replaced by cultivars adapted for large-scale production, as is typical for most vegetables worldwide. As this trend continues, the ability to establish ex situ populations of wild *D. carota* will be enhanced, albeit to the detriment of future carrot breeders.

7.3.4 Conservation Summary

Wild *Daucus* species are currently conserved in both in situ and ex situ. The extent and adequacy of in situ

Table 7.3 *Daucus* species with a potentially limited geographic distribution

Species	Distribution
D. capillifolius	Morocco, Libya
D. gracilis	Algeria
D. jordanicus	Israel (Palestine)
D. tenuisectus	Morocco

conservation is unclear. Some wild species such as *D. carota* subsp. *carota* are distributed across temperate regions of the world and are unlikely to be under threat. However, this species is the closest relative of cultivated carrot and is the most likely to be genetically altered through introgression. Ex situ collections of wild *Daucus* exist, but several species are not present, and further work is required to correctly identify and classify existing accessions. A combined assessment of in situ and ex situ collections will identify species which are not adequately conserved or represented and this will allow future collecting exercises for ex situ conservation to be targeted efficiently.

7.4 Role in Elucidation of Origin and Evolution of Allied Crop Plants

7.4.1 Genetic Diversity of D. carota

Small (1978) investigated more than 400 *D. carota* populations and 100 herbarium specimens using a set of morphological traits. The *D. carota* complex was divided into two partially overlapping groups named subspecies aggregate *gingidium* and subspecies aggregate *carota*. The latter comprised a sharply divergent group of cultivated carrots and two subgroups of the wild *D. carota* (Fig. 7.1).

That study formed a basis for subsequent research aiming at characterization of genetic diversity using molecular tools. Several molecular marker techniques have been developed and widely utilized for investigation of genetic diversity in *D. carota*. St. Pierre et al. (1990) used eight isozyme systems, i.e., phosphoglucose isomerase (PGI), triose phosphate isomerase (TPI), leucine amino peptidase (LAP), alcohol dehydrogenase (ADH), glutamate dehydrogenase (GDH), malate dehydrogenase (MDH), phosphoglucomutase (PGM), isocitrate dehydrogenase (IDH), and 6-phosphogluconate dehydrogenase (6-PGD), to carry out an extensive study on the diversity of the wild and the cultivated *D. carota*, comprising 168 accessions from 32 countries. In total, they identified 16 putative loci coding for 34 alleles. A greater proportion of the total variability was found within than between populations. Subspecies *carota* was the most diverse, while subsp. *major* (the authors followed Heywood's classification) was the most uniform. St. Pierre et al. (1990) found no support for defining clearly separated groups, as proposed by Small (1978), neither in relation to the taxonomic classification nor to the geographic localization of the populations under investigation. Moreover, no sharp division between the wild and the cultivated *D. carota* could be inferred from the isozyme analysis. These conclusions were further supported by the results of a study utilizing amplified fragment length polymorphism (AFLP), inter-simple sequence repeat (ISSR), and other DNA markers (Bradeen et al. 2002) on a collection of 124 accessions, showing that the genetic diversity of *D. carota* was high and non-structured. However, unlike St. Pierre et al. (1990), Bradeen et al. (2002) reported that the cultivated and the wild accessions formed separate, only marginally overlapping clusters. The difference between the wild and the cultivated materials was even more pronounced in the study of Shim and Jørgensen (2000), who used AFLP markers to compare five Danish wild populations and five carrot cultivars. They reported that the wild *D. carota* were clearly different from the cultivated carrots. Moreover, the genetic distance between the wild populations reflected the geographic distance. Analysis of restriction fragment patterns of mitochondrial DNA was another approach used to study genetic diversity of *D. carota*. Ichikawa et al. (1989) used four restriction endonucleases, i.e., *Pst*I, *Sal*I, *Xba*I, and *Xho*I, and showed that wild carrot and *D. capillifolius*, a close relative to *D. carota*, had mtDNAs of different types than those characteristic to domesticated carrots. Mitochondrial genomes of 80 individuals representing wild *D. carota* populations of different origin were studied by Ronfort et al. (1995). Restriction fragment length polymorphism (RFLP) analysis of mtDNA digested with *Eco*RV and *Hind*III and hybridized to probes *cox*I, *atp*6, and *cox*II revealed the presence of 25 mitotypes, indicating a high level of intraspecific polymorphism of the *D. carota* mitochondrial genome. Interestingly, a larger portion of that variability could be attributed to intrapopulation diversity, with 4.4 mitotypes present on average in a single population. On the other hand, all mitotypes were specific to a single geographic region, i.e., France, Greece, and Crete. A summary of research programs using molecular markers for investigating genetic diversity in *Daucus* is given in Table 7.4.

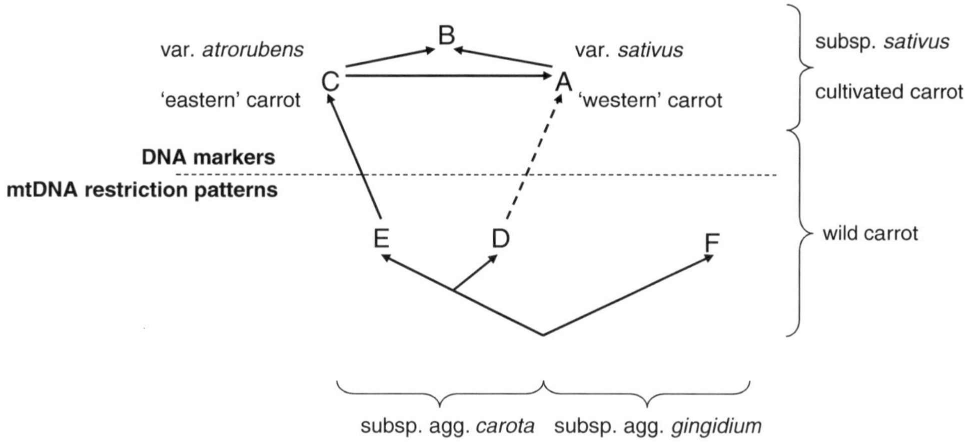

Fig. 7.1 Postulated relationships in the *Daucus carota* complex. The *graph* representing groups differentiated on the basis of morphological observations was adopted from Small (1978). *Letters* represent morphologically different groups within the *D. carota* complex: A – cultivated, orange-rooted, B – cultivated, intermediate between A and C, C – cultivated, non-orange (purple or yellow), D – wild, non-Asian, with unpubescent, highly dissected leaves, E – wild, Asian, with pubescent, poorly dissected leaves, F – wild, with relatively undissected, glossy leaves. The *dashed line* shows division between the wild and the domesticated *D. carota* revealed by DNA markers Shim and Jørgensen (2000), Bradeen et al. (2002), and mtDNA restriction patterns (Ichikawa et al. 1989)

7.4.2 Genetic Diversity in the Genus Daucus

A few studies utilizing molecular markers to investigate relationships between *Daucus* species have been reported. Nakajima et al. (1998) performed a small-scale investigation on genetic diversity within the genus *Daucus*, involving 26 accessions, including 6, 13, and 7 accessions representing *Daucus* species, wild *D. carota*, and cultivated *D. carota*, respectively. They used random amplified polymorphic DNA (RAPD) and AFLP systems, both on the nuclear and the mitochondrial genomes. Clusters resulting from the molecular marker experiments were generally in good agreement with the taxonomic classification; however, the comparison of the nuclear and the mitochondrial genomes suggested that some accessions might have had different evolutionary histories. They reported that results of RAPD and AFLP analysis was in good agreement, while in the above-mentioned study of Bradeen et al. (2002), poor correlation between AFLP and ISSR was observed. Vivek et al. (1999) analyzed cpDNA and mtDNA of nine *Daucus* species using RFLP polymorphisms identified with 14 heterologous probes of *Petunia* cpDNA and four probes of maize mtDNA, after restriction digestion with 10 nucleases. Thirty-seven nucleotide substitutions and one insertion were revealed on cpDNA, which allowed grouping of accessions concordant to their taxonomic classification. Analysis of mtDNAs was more efficient, allowing identification of 209 polymorphisms. Variation of the mitochondrial genome was higher than that of the chloroplast genome, enabling differentiation of accessions within the section *Daucus*. Notably, mitochondrial RFLP profiles differed for fertile and male-sterile cytoplasms.

7.4.3 Chemical Composition and Chemotaxonomy

Several attempts were commenced to broaden our knowledge on the Apiaceae family through investigation of chemotaxonomic relationships. The occurrence of characteristic phenolics, terpenes, saponins, fatty acids, polysaccharides, and proteins was proposed to delimit Apiaceae tribes and genera (Crowden et al. 1969; Harborne 1971; Hegnauer 1971). However, the relationship inferred from sequences of nuclear ribosomal DNA internal transcribed spacer regions (rDNA

Table 7.4 Summary of research projects investigating genetic diversity in the genus *Daucus* using molecular markers

Species examined	Genome investigated	Marker system used	Reference
D. carota *D. capillifolius* *D. pusillus*	mt, cp	Restriction fragment patterns	DeBonte et al. (1984)
D. carota *D. capillifolius* *D. pusillus*	Nuclear, cp	Isozyme Restriction fragment patterns	Matthews and Widholm (1985)
D. carota *D. capillifolius*	mt	Restriction fragment patterns	Ichikawa et al. (1989)
D. carota	Nuclear	Isozyme	St. Pierre et al. (1990)
D. carota	mt	RFLP	Ronfort et al. (1995)
D. carota	mt	RFLP	Bowes and Wolyn (1998)
D. carota *D. aureus* *D. guttatus* *D. muricatus* *D. pussilus*	Nuclear, mt	RAPD, AFLP	Nakajima et al. (1998)
D. carota *D. aureus* *D. broteri* *D. capillifolius* *D. glochidiatus* *D. guttatus* *D. littoralis* *D. muricatus* *D. pusillus*	mt, cp	RFLP	Vivek et al. (1999)
D. carota	Nuclear	AFLP	Shim and Jørgensen (2000)
D. carota	Nuclear, mt	AFLP, ISSR	Bradeen et al. (2002)

mt mitochondrial genome; *cp* chloroplast genome

ITS) did not support phylogenetic assumptions based on the distribution of those secondary metabolites (Katz-Downie et al. 1999). Chemotaxonomical studies at a species level were much less advanced as it is difficult to find key components of sufficient variation. In this context, studies of *Daucus* spp. are fragmentary and concern only selected constituents. The most complex data is available for fruits. This is related to the taxonomical relationship between *Daucus* and other members of the Apiaceae, among which there are several species rich in characteristic volatiles. Much less attention was given to root and leaf composition.

Polyacetylenes are unstable compounds and occur ubiquitously in Apiaceae including several *Daucus* species. Nine *Daucus* species have been evaluated and the presence of polyacetylenes has been positively verified (Table 7.5). The main polyacetylene of *Daucus* is falcarinol, previously known also as carotatoxin, possessing two –C ≡ C– bonds. It is usually accompanied by a derivative falcarindiol with the additional hydroxyl group. In roots of cultivated carrot falcarindiol-3-acetate was also identified, but it is not known whether this polyacetylene occurs in wild carrots. The distribution of polyacetylenes in root organ is not uniform. In cultivated carrot, they appear mainly in oil droplets localized in oil ducts running along the root length in periderm/pericycle parenchyma (Garrod and Lewis 1979; Czepa and Hofmann 2003, 2004). These observations were confirmed using the non-destructive Raman spectroscopic approach (Baranska and Schulz 2005). The same technique, applied to wild carrot, revealed a different polyacetylene distribution. In the root of *D. c.* subsp. *maritimus,* falcarindiol dominates and it is much more uniformly distributed across root phloem with higher amounts occurring additionally in the periderm. A similar pattern of distribution was also found in the subspecies *D. c.* subsp. *gummifer*, *D. c.* subsp. *commutatus,* and *D. c.* subsp. *halophilus* (Baranska et al. 2005). It is speculated that falcarindiol increases resistance of carrot cultivars to

Mycocentrospora acerina and *D. c.* subsp. *hispanicus* to root knot nematodes (Frese 1983).

Biennial and perennial Apiaceae develop storage roots that function as a sink of assimilates utilized for the development of stem and inflorescences in the second vegetation year. They usually contain sucrose accompanied by hexose, glucose, and fructose, as in the edible carrot (Hegnauer 1971). Starch and pectins can also function as storage carbohydrates. They are present in considerable amounts, for example in roots of *D. c.* subsp. *maritimus,* as revealed by spectrometric analyses. The presence of starch was also confirmed by histochemical iodine staining of a root section. In the analyzed specimen, starch accumulated in the pith (a parenchymatic center of the root), phloem, and peripheral parenchyma while xylem tissue was devoid of starch. The presence of starch in the pith was not observed in edible carrot, which may be explained by its much smaller core than in wild carrot. In contrast, pectins were uniformly distributed throughout root section with a noticeable decrease along the secondary cambium only. The intensity of pectin characteristic signals in the Raman spectrum of *D. c.* subsp. *maritimus* suggests that this linear polysaccharide has a low degree of esterification, e.g., it contains only a few esterified galacturonic acid groups (Baranska et al. 2005). Other saccharides are not reported in *Daucus* except of umbelliferose. This characteristic to Apiaceae family trisaccharide (α-D-galactopyranosyl-2′-α-D-glucopyranosyl-2-β-D-fructofuranoside) is a raffinose isomer and was found in three *D. carota* subspecies including *D. c.* subsp. *carota,* although in small amount only (Crowden et al. 1969; Goris 1983).

Daucus fruits are dry with a hard pericarp being highly lignified (Drude 1898; Baranski et al. 2006). The fruit wall covers the seed composed mainly of endosperm and small embryo. Thus, endosperm occupies the majority of the seed volume and serves as a storage tissue. *Daucus* endosperm is rich in proteins that account for 11–29% of the tissue (Kleiman and Spencer 1982) (Table 7.5). As in other Apiaceae, two classes of protein bodies are present in *D. c.* subsp. *carota* seed: globoid bodies with nitrogen-rich storage proteins of 2–10 μm in diameter and a second type

Table 7.5 Selected constituents of *Daucus* root, canopy, and seed

Species	Root polyacetylenes[a]	Canopy flavonoids	Seed							
			Flavonoids	Proteins [%]	Oil [%]	Fatty acids				
						PA [%]	LA [%]	C18:2 [%]	Others [%]	Iodine value
D. aureus	+	Lu	Ap Chr Lu	19	23	76	7	11	6	91
D. broteri		Lu		19	19	75	5	12	8	100
D. carota subsp. *carota*	+	Lu-7G	Ap Lu Qu	22	27	71	12	12	5	102
D. c. maritimus	+			17	18	70	12	11	7	188
D. c. maximus				19	19	73	10	11	6	150
D. crinitus	+	Lu	Lu Ka Qu	21	12	74	10	11	5	91
D. durieua		Lu Ka Qu	Lu							
D. glochidiatus		Qu	Qu							
D. guttatus	+			21	23	65	16	12	7	105
D. involucratus		Qu-3G								
D. littoralis			Lu Qu	11	10	80	4	10	6	95
D. montanus			Ap Lu Qu							
D. muricatus		Lu	Lu	18	11	75	7	11	7	92
D. pusillus	+		Ap, Lu Ka Qu	20	22	75	7	12	6	117
D. setifolius		Qu	Ap Lu Qu	29	16	51	13	25	11	106
D. syrticus			Ap Lu Qu							
Reference	Crowden et al. (1969)		Harborne (1971)	Kleiman and Spencer (1982)						

Flavonoids: *Ap* apigenin; *Chr* chrysoeriol; *Ka* kaempferol; *Lu* luteolin; *Qu* quercetin; *3G* 3-glucoside; *7G* 7-glucoside
Fatty acids: *PA* petroselinic acid ($C18:1^{\Delta}6$); *LA* linoleic acid ($C18:1^{\Delta}9$)
Blanks not studied
[a]Polyacetylenes were also detected in *D. c.* subsp. *gummifer*, *D. c.* subsp. *commutatus* and *D. c.* subsp. *halophilus* (Baranska et al. 2005)

with crystal inclusions. Both types of protein bodies never occur together in one cell. Crystals often aggregate forming large globoid druses that disappear within 5 days after seed germination. Additionally, calcium-rich inclusions are also present and they persist in germinating mericarps. Druses in wild carrot may be composed of up to 100 individual globoid crystals or calcium-rich crystals and have 6–10 μm and are much bigger than those in endosperm of caraway, dill, and anise (Spitzer and Lott 1982a). Globoid crystals are probably phytate molecules as they contain mainly P, K, and Mg elements (Spitzer and Lott 1982b). In contrast, calcium-rich crystals are identified as whewellite (calcium oxalate) (Spitzer and Lott 1982c).

Analysis of storage proteins revealed differences in protein composition between species. For this purpose, protein extracts can be separated using gel electrophoresis and stained with amido black. Nine to 12 anodic bands of different intensity were distinguished in extracts from *D. aureus*, *D. c.* subsp. *carota,* and *D. c.* subsp. *gadecaei.* The species also differed in enzymatic activity of esterases and peroxidases, both enzymes occurring in several izoforms (Crowden et al. 1969).

Other important components of endosperm and embryo are lipid vesicles (Spitzer and Lott 1982a). Fatty acids may constitute up to almost 30% of *Daucus* seed. The composition of fatty acids varies between species but the main compound found in all studied *Daucus* is petroselinic acid ($C18{:}1^{\Delta}6$), a unique and characteristic fatty acid in the Apiaceae family. *Daucus* species contain it in the amount of 50–80% of all fatty acids identified and belong to the taxa in the family with the highest proportion of petroselinic acid (Kleiman and Spencer 1982) (Table 7.5). The second important fatty acid is $C18{:}1^{\Delta}9$ linoleic acid that together with C18:2 lipids constitutes 14–38% of total fatty acids. Thus, C18:1 and C18:2 make up 90–95% of the fatty acid profile. Despite their high proportion, the iodine value, informing on the degree of lipid unsaturation, varies considerably between species from 90 for *D. aureus, D. crinitus, D. muricatus,* and *D. littoralis* to around 150 for *D. c.* subsp. *maximus* and almost 200 for *D. c.* subsp. *maritimus* thus indicating a varied composition of the remaining 5–10% of fatty acids.

Phenolic compounds are important plant metabolites often appearing in distinct taxa. Both flavones and flavonoles are present in *Daucus* species making them rather unusual in the Apiaceae family. Flavones are considered more evolutionarily "advanced" compounds and they progressively replace flavonols (Harborne 1971). The distribution of flavonoids varies depending on the species and plant organ (Table 7.5). The flavone luteolin is present in the canopy of most studied *Daucus* species. It can also exist as 7-glucoside. A second common compound is quercetin (flavonol), which was found in four species (Crowden et al. 1969). The presence of flavonoles is more conserved in seeds. Quercetin is found in almost all species and additionally, kaempferol occurs in *D. crinitus* and *D. pusillus.* The seeds usually contain flavone luteolin accompanied often by apigenin. Chrysoeriol was found only in *D. aureus.* These compounds occur in seed usually as flavone-7-glucosides and flavonol-3-glucosides. However, other glycosides may also be observed like 5-glucoside, 4′-glucoside, 7-rhamnosylglucoside, 7-diglucoside, or 7-glucosylglucuronide. Quercetin may also occur as 3-di- or 3-triglucoside in *D. littoralis* and 3-rutinoside in *D. crinitus* (Harborne 1971). Additionally, a unique phenylpropanoid myristicin was identified in *D. glochidiatus, D. montanus,* and *D. carota* (Harborne 1971). Furanocoumarins are also ubiquitous in Apiaceae, however, *Daucus* are rather poor in these compounds. A survey of nine species demonstrated umbelliferone (7-hydroxycoumarin) only in *D. c.* subsp. *carota* seed (Crowden et al. 1969).

Apiaceaeous plants are rich in volatiles, which are constituents of essential oils present in leaves, umbels, and seed. In most studied species, monoterpene hydrocarbons dominate and make up 25–75% of all volatile components. They are usually followed by sesquiterpene hydrocarbons while oxygenated terpenes do not usually exceed 10%. Another important fraction may be phenylpropanoids; *D. c.* subsp. *carota* essential oil can be composed of 45% of these compounds (Rossi et al. 2007). Almost 100 fractions were identified in *Daucus* essential oils making this genera one of the richest reservoirs of volatiles in Apiaceae. The composition of these oils is highly species-specific and according to Harborne (1971) can be of significance in chemotaxonomy, e.g., geranyl acetate and carotol are characteristic for *Daucus.* Recent chromatography-mass spectrometry studies of various *D. carota* subspecies revealed high variation of the essential oil composition at species level. The most common components are β-bisabolene, elemicin, (E)-methyl-isoeugenol, myrcene, α-pinene, and sabinene but their amounts vary depending on species and plant organ, and additionally on developmental stage (Table 7.6). A range of composition is reported in different studies, which can be related to the use of different plant material, grown in different conditions, and to varying in physiological/developmental state. Also some analyses were performed from leaf extracts while other studies utilized a mixture of leaves and umbels and this may contribute to the observed variation. It is therefore not surprising that values obtained by different authors and listed in Table 7.6 are inconsistent to some extent. In general, however, seeds contain more essential oils than stems and leaves and their amount increases with ripening. Thus among the most important components, β-bisabolene, elemicin, and (E)-methyl-isoeugenol content increases two- to fourfold in comparison to other components in ripe umbel of *D. c.* subsp. *carota*. For *D. c.* subsp. *halophilus,* only the content of elemicin increases significantly (over fourfold). On the other hand, limonene, myrcene, α-pinene, and terpine-4-ol are more pronounced in the umbel at flowering stage.

Spectroscopic measurements can provide information on chemical composition of complex biological samples. Using a wide wavelength range, the spectra carry information of all chemical compounds present in the specimen. Data collected from individual seeds using Fourier-transform Raman spectroscopy produced convincing results allowing designation of seeds into distinct groups based on the similarity of the obtained spectra. Comparison of these spectra showed that *D. pusillus* and *D. capilifolius* were the most homogenous species among those used in the study. It was, therefore, possible to discriminate these two species from each other and from the seeds of *D. broteri*. In contrast, *D. carota* was highly heterogeneous and delimitation at intrageneric level was highly problematic, which was in accordance to previous chemotaxonomical inferences (Baranski et al. 2006).

7.5 Role in Classical and Molecular Genetic Studies

The market value of the edible carrot is smaller than that of rice, wheat, potato. or tomato. However, there are some advantages of using *Daucus* in genetic investigations, e.g., a wide spectrum of phenotypic diversity among wild and cultivated *D. carota*; availability of interspecific *Daucus* hybrids; presence of a well-defined system of cytoplasmic male sterility (CMS); and ease of cell, tissue cultures, and plant regeneration. Classical genetic studies involving wild *Daucus* have not been frequent, owing to their reproductive biology, small and difficult to manipulate flowers, and significant inbreeding depression. Therefore, to date no reports are available on the development of *Daucus* cytogenetic stocks, such as substitution or addition lines, employing wild materials. Technology for production of *D. carota* doubled haploid lines is being developed (Adamus et al. 2004; Górecka et al. 2009), but it has not been used for wild *Daucus*.

7.5.1 Use as Parents in Wide Crosses

Hauser and Bjorn (2001) and Hauser and Shim (2007) noted the existence of hybrids between cultivated carrot and *D. carota* subsp. *carota* in fields in Denmark, and found that hybrids were likely to survive and produce offspring in natural grassland environments. Wijnheijmer et al. (1989) also described evidence of introgression from cultivated carrot into *D. carota* subsp. *carota* in the Netherlands. It is possible to create hybrids between cultivated carrot and other *Daucus* species such as *D. capillifolius* (Ellis et al. 1993). Also, manual pollination of a range of *Daucus* accessions with pollen of cultivated carrot resulted in seed set and subsequent development of fertile plants (Nothnagel et al. 2000). They used subspecies of *D. carota* and accessions labeled as *D. halophilus, D. hispidifolius,* and *D. maximus*. Steinborn et al. (1995) also reported on the production of intersubspecific hybrids between *D. carota* subsp. *sativus* and other *D. carota* subspecies. They analyzed the inheritance of plastid and mitochondrial genomes in progeny obtained from those crosses. RFLP analysis of cpDNA and mtDNA proved that the cytoplasm was inherited exclusively from the maternal plant and Vivek et al. (1999) demonstrated strictly maternal inheritance of a plastid RFLP in carrot. Recently, Camadro et al. (2008) reported the successful creation of hybrid embryos but not mature seed in crosses between cultivated carrot and *D. pusillus* under laboratory conditions, indicating incomplete reproductive barriers between the two species. The authors postulated that the hybrids could provide carrot breeders

Table 7.6 Percentage of the main components of essential oils extracted from dried aerial parts of *Daucus carota* subspecies

Compound	*D. c.* subsp. *carota*										*D. c.* subsp. *halophilus* (mean from two locations)		*D. c.* subsp. *hispanicus*[a]	
	Canopy			Umbel						Seed	Umbel		Canopy	Seed
				Flowering				Ripe			Flowering	Ripe		
α-Asarone	0.8	0.4		0.1				1.5					0.2	nd
β-Bisabolene	21.3	tr	0.1	2.4	0.5	0.6	0.8	5.5	0.1	0.1	5.3 ; 0.4[b]	1.8 ; 0.4[b]	0.3	nd
Carotol		nd	nd	0.4	nd			4.2	nd				0.1	nd
Elemicin	16.3	tr		4.0				10.9			6.1	27.5		
Germacrene B													4.1	nd
Germacrene D	0.5	1.7	0.7	2.0	0.1	2.5	0.1	0.7	0.1	0.1			6.9	0.2
11α H-Himachal-4-en-1β-ol	0.7	nd		0.3				3.2						
Limonene	1.7	7.0	5.3	4.7	3.7	5.3	3.6	2.1	2.3	3.6	11.4	6.0	5.7	2.1
(E)-Methyl-isoeugenol	21.8	0.1		9.4		tr		34.0			7.4 ; 0.7[b]	6.9 ; 0.5[b]		
Myrcene	2.0	8.6	5.2	6.6	3.1	7.0	6.6	3.0	2.5	4.0	4.0	2.3	2.3	3.0
(Z)-Ocimene	0.1	1.3	0.2	1.4	0.2	3.3	2.8	0.3	0.4	1.0	0.4	0.2	5.0	0.4
α-Pinene	15.9	34.8	30.0	27.7	42.0	40.7	42.8	16.0	17.2	21.1	14.3	11.2	10.8	12.2
Sabinene	2.7	18.7	30.1	7.3	19.5	18.3	37.7	7.7	40.5	45.3	31.1	28.3	26.8	60.6
Terpine-4-ol	0.2	2.7	6.1	1.5	2.9	4.8	1.1	0.7	4.9	4.5	4.5	2.1	4.8	5.4
Monoterpenes	25.3	39.5[c]	81.0		92.0	91.7	91.7		96.7	96.1	78.2	57.7	74.4	96.7
Sesquiterpenes	29.1	11.2[c]	4.0		7.4	7.4	7.6			2.2	8.3	5.1	18.7	
Phenylpropanoids	39.1	45.8[c]	[d]		[d]	[d]	[d]		[d]	[d]	10.3	32.3		
No. of identified compounds	31	69[c]	29		31	100	32		31	28	55	55	97	39
Yield (% fw)	0.6	0.1	0.1	0.2	0.7	0.3	0.3	0.6	1.1	0.7	0.4	0.8	0.1	1.2
Reference	Rossi et al. (2007)	Gonny et al. (2004)	Staniszewska et al. (2005)	Gonny et al. (2004)	Staniszewska et al. (2005)	Staniszewska and Kula (2001)	Góra et al. (2002)	Gonny et al. (2004)	Staniszewska et al. (2005)	Góra et al. (2002)	Tavares et al. (2008)	Tavares et al. (2008)	Flamini et al. (2007)	Flamini et al. (2007)

Blanks: Not identified by the authors; nd: Not detected in the sample; tr: Traces amounts; Underlined values: the predominant compounds

[a]The name *D. gingidium* L. subsp. *gingidium* is used by the authors

[b]Values from two locations are given as they differ considerably

[c]Values are obtained from a mixture of all aerial parts

[d]Phenylpropanoids are not distinguished from terpenes

with genes of interest, such as resistance or tolerance to abiotic and biotic stresses.

Apart from the gametic hybridization, efforts were carried out to produce somatic hybrids. Dudits et al. (1977) obtained a somatic hybrid between *D. carota* subsp. *sativus* and *D. capillifolius* with intermediate leaf morphology and white and fleshy roots. The hybrid plants usually contained 36 chromosomes, but other chromosome numbers, ranging from 34 to 54, were also observed, the latter likely originating from triple fusion of protoplasts. Matthews and Widholm (1985) analyzed chloroplast and mitochondrial composition of *D. carota* × *D. capillifolius* somatic hybrids by comparing cpDNA and mtDNA restriction patterns to both parental species. While plastid DNAs in the hybrids were identical to that of *D. carota* and different from *D. capillifolius*, mtDNA examination revealed the presence of numerous rearrangements in the hybrids, including the presence of novel, unique bands, absent in both parents. The hybrids contained two isoforms of homoserine dehydrogenase, of which one was *D. capillifolius* specific and the other one was unique to the hybrid (not *D. carota* specific). Analysis showed that expression levels of key enzymes involved in amino acid synthesis in the hybrids were slightly higher than in either parent. Asymmetric fusion between *D. carota* and *D. capillifolius* was performed by Ichikawa et al. (1987), where *D. carota* nuclei from iodoacetamide-treated protoplasts were fused with *D. capillifolius* cytoplasm from X-irradiated protoplasts. Putative cybrids were morphologically similar to *D. carota* and contained 17 ($2n - 1$) or 34 ($4n - 2$) chromosomes, as counted for four plants and one plant, respectively. Again, mtDNA restriction pattern analysis showed the presence of unique bands indicating possible rearrangements in the mitochondrial genome of the hybrids.

7.5.2 Molecular Genetic Linkage Maps

The most detailed presently available genetic linkage map of *D. carota* has been developed using a segregating F_2 population from an intersubspecific cross between the North American wild carrot QAL (*D. carota* subsp. *carota*) and the cultivated carrot inbred B493 (*D. carota* subsp. *sativus*). The map was originally developed using a set of 250 AFLP markers (Santos and Simon 2002), and then supplemented with gene-specific size and SNP (single nucleotide polymorphisms, Just et al. 2007), *DcMaster* and *Krak* transposon insertion sites (Grzebelus et al. 2007; Grzebelus and Simon 2009). Currently, the map comprises 381 markers and spans 1,242.0 cM and 1,474.7 cM, for the QAL and B493 genomes, respectively (Table 7.7).

The map was originally developed to investigate genetic factors determining carotenoid content in carrot roots. Santos and Simon (2002) identified quantitative trait loci (QTLs) affecting the concentration of different carotenoids, explaining from 15.8, 21.7, 26.4, 37.7, and 44.2% of the total phenotypic variance, for lycopene, ξ-carotene, β-carotene, α-carotene, and phytoene, respectively. Just et al. (2007) positioned on the map 22 genes involved in carotenoid biosynthesis and metabolism, providing gene-specific codominant polymorphisms for eight of the nine linkage groups. The map served as a tool for localization of transposon copies belonging to the *DcMaster* and *Krak* families of *PIF/Harbinger* elements. It was shown that insertion sites of both families were highly polymorphic between the wild and the cultivated *D. carota*. They were uniformly distributed in both genomes; however, differences in saturation between the QAL and the B493 genomes were observed for linkage groups 2 and 5, which were identified as those carrying the most important QTLs for carotenoid content. Hence, transposition events might in part be responsible for the emergence of carotenoid-colored roots.

Table 7.7 Summary information on the genetic linkage map of *D. carota* for the cross between the wild carrot QAL and the cultivated inbred line B493

Marker type	QAL	B493
AFLP dominant	100	97
AFLP codominant	4	4
Genic	24	24
SCAR	2	2
SSR	2	2
DcMaster Transposon Display	28	23
Krak Transposon Display	35	34
Total	195	186
Length of map (cM)	1,242.0	1,474.7
Average distance between markers (cM)	6.37	7.93

7.6 Role in Crop Improvement Through Traditional and Advanced Tools

7.6.1 *Utilization of Wide Crosses in Carrot Breeding*

The relatively poor valorization of *Daucus* genetic resources is a limiting factor for their wider utilization in carrot breeding programs. Few projects aiming at better characterization of genetic properties of existing germplasm collections have been undertaken. The major effort on evaluation of European *Daucus* collection, including domesticated carrot and its wild relatives, was conducted under the framework of the GENRES project in the years 2000–2003. Detailed characteristics of the project are available at http://www2.warwick.ac.uk/fac/sci/whri/research/gru/eucarrot/. The USDA has been evaluating the US *Daucus* collection for traits of potential interest to plant breeders and germplasm curators since 1986 and nine traits from diverse open-pollinated cultivars, wild carrot, and wild *Daucus* species that have found their way into modern cultivars have been summarized (Simon 2000). Two areas where wild germplasm greatly contributed in the improvement of modern cultivated carrot are discussed below.

7.6.1.1 Introduction of the Petaloid Type of Cytoplasmic Male Sterility into Cultivated Carrot

The petaloid cytoplasm, known as "Cornell petaloid" was first discovered by Munger in 1953. In the 1970s, another source of petaloid CMS, "Wisconsin petaloid," was found (Morelock et al. 1996), and more recently yet another source, "Guelph petaloid," was identified in Ontario, Canada (Wolyn and Chahal 1998). It is likely that, despite a large geographic distance, all of these sources are related, suggesting that a single evolutionary event was responsible for the origin of the "sterile" cytoplasm (Bowes and Wolyn 1998). In male sterile plants of this type, additional petals replace anthers. A specific combination of "sterile cytoplasm" and nuclear alleles is required to produce male sterile plants. The cytoplasmic factors determining CMS are located in the mitochondrial genome, but the exact mechanism governing male sterility in carrot is still not known. CMS was introduced into carrot breeding materials, providing a very efficient tool for a mass scale pollination control. The CMS system is nowadays widely used to assure cross-pollination in the course of F_1 cultivar seed production and molecular tools differentiating "sterile" and "normal" cytoplasms were developed (Szklarczyk et al. 2000; Bach et al. 2002).

Owing to the importance of CMS to the seed industry, efforts have been undertaken to obtain new sources of male sterility through intra- and interspecific crosses. Steinborn et al. (1995) examined several crosses of *D. carota* subsp. *sativus* with a range of wild *D. carota* subspecies. They confirmed the exclusively maternal inheritance of mitochondrial and chloroplast genomes and reported another source of CMS in *D. carota* subsp. *gummifer*. In addition to the "gummifer" CMS, Nothnagel et al. (2000) identified two more CMS sources – in *D. carota* subsp. *maritimus* and *D. carota* subsp. *gadecaei*. Feasibility of these new CMS systems for carrot breeding remains to be examined.

7.6.1.2 Introduction of Resistance to Carrot Fly from *D. capillifolius*

Carrot root fly (*Psila rosae*) is the most devastating pest of carrot and other umbellifers grown in temperate regions. Resistance sources from domesticated carrot can provide a maximum of 50% reduction in carrot fly damage, while much higher levels of resistance were observed in wild *Daucus* (Ellis 1999). *D. capillifolius* was found to be resistant to *P. rosae* due to its foliage and root characteristics. It was shown that crosses between *D. carota* subsp. *sativus* and *D. capillifolius* were readily obtained (McCollum 1975). In order to introduce the carrot fly resistance of *D. capillifolius* into cultivated carrot, crosses with partially resistant cv. 'Sytan' were produced (Ellis et al. 1993). As a result of the project, a set of F_5 and F_6 lines with significantly improved resistance to carrot fly were produced, however, none of those lines was as resistant as *D. capillifolius*. Finally, carrot cv. 'Flyaway' was released in 1993.

7.6.2 Genetic Transformation

There are no existing reports concerning genetic modification of *Daucus* wild relatives to date. However, cultivated carrot has been extensively engineered using modern biotechnological approaches. Carrot was one of the first species used in transgenic research probably mainly due to earlier successful establishments of in vitro techniques using carrot as a model plant. Nowadays, a large number of available reports indicate that carrot can be engineered using both vector and non-vector methods. Although there is no commercial release of any transgenic carrot, the research reports show a broad range of possible modifications that can be potentially fixed in future cultivars. The subject has been recently reviewed by Baranski (2008). Beyond carrot, several successful attempts of genetic transformation in various Apiaceae species are known. The published data indicate that optimized protocols can be developed for some species and then utilized for application purposes. A number of Apiaceae species are rich in phytochemicals of significant pharmaceutical, dietary, and culinary importance. Thus, it is not surprising that research towards high-throughput production of these compounds is evolving. Review of existing data shows that furanocoumarins, essential oils, saponins, and phenylpropanoids are secreted by hairy roots in bioreactors after successful *Agrobacterium rhizogenes*-mediated transformation. Although the amounts of these metabolites in hairy roots are often much lower than in organs of growing plants, an efficient hairy root biomass production may be a convenient source of phytochemicals (Baranski 2008).

Experience gained from carrot and related genera suggests that wild *Daucus* can be amenable to genetic transformation, too. Although these species have no commercial value they are known to be rich in metabolites of phytopharmaceutical importance. Hairy root culture in bioreactors thus may become an interesting alternative for novel utilization of these resources.

7.7 Scope for Domestication and Commercialization

Historically, oils extracted from *Daucus* species had a variety of medicinal uses like other members of the Apiaceae. Recent investigations have focused on the antimicrobial properties of oil components from different taxa such as *D. carota* subsp. *halophilus* (Tavares et al. 2008), *D. carota* subsp. *carota* (Ahmed et al. 2005; Kumarasamy et al. 2005; Rossi et al. 2007), and *D. reboudii* (Djarri et al. 2006).

Polyacetylenes containing unique carbon–carbon triple bonds, which are derivatives of a fatty acid – crepenynic acid, are the most characteristic compounds found in *Daucus* roots. Polyacetylenes are considered as bioactive compound of adverse, toxic effect to animals and human beings. However, recent studies suggest that in low doses some of them, e.g., falcarinol, can also act beneficially to health exhibiting anti-inflammatory activity and inhibiting proliferation of tumor cells both in vitro and in vivo (Christiansen and Brandt 2006).

7.8 Gene Flow Between Cultivated and Wild Carrot

As mentioned earlier, cultivated carrot can freely hybridize with other *D. carota* subspecies. The most common wild relative is *D. c.* subsp. *carota*, which invades open grasslands, meadows, road sides, and rural areas. Its geographical distribution is wide: Europe, North Africa with Canary Islands, North, South, and Central America, Siberia, Central and East Asia, and Australia. It commonly grows in the vicinity of production carrot fields and is a serious problem for seed producers, since it can pollinate cultivated carrot, resulting in impurity of commercial seed lots. Root producers often identify intercrossed individuals that possess characters of both cultivated and wild form (Wijnheijmer et al. 1989). Investigations indicated that white-rooted individuals bolting in the first year in commercial fields were hybrids between cultivated carrot (female) and wild (male) parents (Hauser and Bjorn 2001). The problem of contamination with wild pollen has been well recognized and seed production guidelines have adopted rules to minimize unintended pollination. Restrictions arise from the fact that carrot is cross-pollinated by a large number of insects (Lamborn and Ollerton 2000) and that pollen can be effectively transported for a long distance. The same applies to wild carrot since, for example, hoverflies (*Eristalis tenax* L.) fly to a distance of 60 m, but pollen of wild carrot is viable for 10 days so it can be effectively transported up to 600 m before pollination (Umehara et al. 2005). Thus for production of carrot foundation seed, to ensure 95% purity a minimum distance of at least 1,000 m is required from any other field with flowering plants and from habitats of wild relatives. Bees carry pollen in excess of 6 km in some instances so commercial seed production in the USA keep cultivated orange carrots with differing root shapes at least 3 km apart, and other root colors at least 5 km apart. Moreover, Magnussen and Hauser (2007) described the existence of hybrids in natural populations, implying that gene flow occurred from carrot fields to closely growing wild populations.

Gene flow to/from other *Daucus* species is rather unlikely. Although carrot can hybridize with *D. capillifolius*, successful seed set after crossing with few other *Daucus* species was achieved only after manual pollination or in laboratory conditions. Spontaneous hybridization with *D. capillifolius* might occur if both species are present in close proximity. Since *D. capillifiolius* grows on a very limited area in North Africa the appropriate measures of the species conservation may considerably lower the chance of its contamination.

Another route of gene flow can occur is via seed. Although cultivated plants usually do not flower since they are harvested in the first vegetation year, some plants can bolt as discussed earlier. Early bolting hybrids can set seed. Carrot seed can be dispersed by wind and animals, but anemophily is of little importance in terms of gene flow and invasiveness since the distances to which seed is transported by wind is short. Experiments using controlled air velocity showed that carrot seed is scattered no longer than 3 m from the source plant (Umehara et al. 2005). On the other hand, fruit morphology encourages dispersal by animals. The small seed size and large number of spines enable dry fruit to attach to animal fur as well as to human clothing. There is no data available to what extent endo- or epizoochory contributes to *Daucus* dispersal, but a recent experiment by Manzano and Malo (2006) showed that carrot seed adhered to sheep wool can be transported for 400 km by animals from the North to Southwest Spain. After 28 days of flock movement, 12% of seed adhering to sheep wool still remained attached and 7% was still attached after the next 5 months. This indicated that the impact of animal movement in some regions should be taken into consideration. Another possible mode of seed dispersal is carriage by car and farm equipment. Usually carrot seed is produced in one region and then transported to root production areas. Mud attached to car undersurfaces and wheels can also accommodate viable seeds and enables their transport over a very long distance (Schmidt 1989). Any of these seed dispersal mechanisms gives evidence that seed can be transported to, and potentially germinate in a new location. While the above examples suggest possible ways carrot seed can disperse, they do not indicate to what degree long distance dispersal occurs in nature and whether these examples are relevant to successful gene flow and introgression into wild habitats.

Wild carrot is a weedy plant that can enter new locations, but it still has not been determined how fast settlement can become established. Umehara et al. (2005) found new wild carrot plants about 10 m away from the original site of maternal plants after 1 year and about 100 m after 2 years, probably due to the cumulative effects of seed dispersal by animals and humans.

The persistence of seed is an important issue when considering gene flow. Carrot seed has some dormancy and does not always germinate under favorable conditions, thus it can persist in soil. The seed mortality of wild carrot, 1 year after sowing in June in the Oregon (USA) prairie, was high (45–65%), but not related to vertebrate predation, which can be informative for risk assessment to non-target organisms. Twenty-eight percent of the seed survived the first winter and most germinated the next year while about 40% of surviving seed was viable in the soil and germinated the second year. No viable and ungerminated seed was found after the second year (Clark and Wilson 2003). These data indicated that in some conditions carrot seeds can persist in soil for at least 2 years. This agrees with finding of Magnussen and Hauser (2007) who deduced the presence of F_2 or backcross hybrid plants in natural habitats using molecular approaches.

A successful hybridization event does not necessarily imply gene introgression into a wild population. The observations made by Hauser (2002) show that hybrids are less tolerant to frost than wild plants and that this susceptibility is inherited from the cultivated parent. Without an acclimatization period, hybrid plants died at -8°C as did cultivated carrot, and after acclimatization some hybrids survived but at a low rate (below 10%). Climatic conditions and genetic background certainly are of great importance, but the results clearly indicated that outbreeding depression resulting from gene escape from a cultivated to wild form of carrot can significantly limit gene flow so that hybrids produced have lower ability to survive in exposed habitats.

7.8.1 Potential Gene Flow from Transgenic Carrot to its Wild Relatives

Although wild *Daucus* have not been the subject of genetic modification, a large number of transgenic crops have been developed and are being introduced into environment. It is, therefore, important to discuss the potential consequences of gene flow from released transgenic carrot to wild *Daucus*. Cultivated carrot (*D. c.* subsp. *sativus*), the closest relative of the wild *D. carota*, has been transformed worldwide and can potentially be utilized in agriculture. Although, there are no commercial transgenic carrot cultivars released to date, 16 field trials were conducted in 1993–2002 by seed companies; three of which were registered in the Netherlands and 13 in various states of USA. Transgenic carrot was modified with tobacco chitinase, glucanase, and osmotin, and with petunia osmotin to search for increased *Alternaria dauci* resistance; and with cowpea cysteine proteinase inhibitor to search for root knot nematode resistance. Altered glyphosate tolerance and nutritional quality was also examined. For the latter two projects, information on the introduced genes was proprietary and consequently not available. In addition, all projects utilized the neomycine phosphotransferase II (*npt*II) gene conferring resistance to kanamycin derived antibiotics as a selectable marker (USDA Information System for Biotechnology). There has been no further official deliberate release of transgenic carrot into the environment since 2002, but this does not infer that no transgenic cultivars will be registered in future. As several field trials have already been completed, one may anticipate that transgenic carrot is being proposed for release into the environment.

The potential impact of transgenic cultivars on wild carrot includes consideration of several issues (1) Can cultivated carrot hybridize with the wild relatives? (2) Can hybrids grow and flower? (3) Can pollination occur between hybrids and wild relatives? (4) Can cultivated transgenic carrot escape agricultural system? (5) Can such refuges establish invasive populations? and (6) Can gene flow increase fitness? The first three questions relate to reproductive biology, seed dispersal, and persistence (see Sect. 7.8) while the remaining ones require information on what genes are introduced.

It is obvious that wild carrot can receive transgenes via gene flow directly from transgenic crop or via hybrids that can serve as bridge plants. How these events can affect wild habitats depends on the transferred gene construct and its impact on population fitness. Therefore, a comprehensive answer is possible only after a case-by-case assessment. As there is no proposed release of transgenic *Daucus* at present there is no opportunity to judge the potential impact of gene flow. Assuming that production of a novel transgenic carrot cultivar is allowed, the crop would certainly be cultivated for both root and seed production. As discussed earlier, root production is of little importance in gene flow as long as the plants are harvested before flowering. Appearance of bolting carrots can easily be controlled and appropriate measures can be applied to discard them from commercial fields before anthesis. Uncontrolled fields can be, however, a source of transgenic pollen, additionally to plants at seed production fields.

Evaluations of natural habitats of wild carrot growing in a close proximity to carrot production fields indicate that although some hybrids can be identified they are infrequent (Magnussen and Hauser 2007). Only 4% of wild plants collected from field edges and roadsides close to carrot fields were identified as hybrids using AFLP markers. The introgression of deleterious genes resulted in maladaptation of hybrids to the conditions in the wild and subsequent loss of genes from cultivated parent in favor of the wild parent (Hauser 2002). On the other hand, the release of disease resistant cultivars may have a different effect. Jensen et al. (2008) showed that both wild and cultivated carrots died after inoculation with *Sclerotinia sclerotiorum* and subsequent disease development, and that they had the same survival rate. This indicates that wild carrot is susceptible at a level similar to that of cultivated carrot. Wild carrot may therefore benefit from introgression of resistance genes introduced by transgenic carrot. Populations with transgenes may have increased adaptiveness and be quicker to colonize new habitats or increased survival in environment where *Sclerotinia* occurs. On the other hand, non-infested resistant wild carrot may not serve as a pathogen reservoir thus limiting its propagation and incidence of crop disease. Another pathogen, *Alternaria dauci*, is common causal agent of leaf blight recognized worldwide. It can also infest wild carrot and cause severe plant damage as shown by Schouten et al. (2002). However, wild populations examined in the Netherlands were hardly infested and showed a very high tolerance in field experiments. This is strong evidence that those populations may possess resistance mechanisms against *A. dauci*. Introgression of transgenes conferring resistance from an *A. dauci* resistant transgenic crop would then have negligible effects on the fitness of wild carrots, and would unlikely lead to any changes in the wild carrot population.

7.9 Recommendations for Future Actions

The existing collections of *Daucus* genetic resources have recently been significantly enriched in the germplasm originating from the Mediterranean, the Middle East, Central Asia, and northern Africa. The need for thorough evaluation and valorization of these resources is urgent, as they can provide a basis for developments in carrot breeding for the coming years. At present, several national and international programs are being realized, but there is a need for a formal platform integrating these activities. One of the primary targets for utilization of the genetic resources is the development of cultivars suitable for organic and sustainable production. Carrot is important in the rotation of organic vegetable production systems. Wild materials can provide sources of resistance to a range of biotic and abiotic stresses, providing additional quality to organic and conventional farming. Breeding carrots for higher productivity and quality in all farming systems, and greater adaptation of carrot to warmer climates with concentrated human populations will also benefit from diverse collections of well-characterized germplasm.

References

Adamus A, Kiełkowska A, Michalik B (2004) Carrot haploid production through induced parthenogenesis. In: Vollmann J, Grausgruber H, Ruckenbauer P (eds) Genetic variation for plant breeding. BOKU, Vienna, Austria, pp 451–454

Ahmed AA, Bishr MM, El-Shanawany MA, Attia EZ, Ross SA, Pare PW (2005) Rare trisubstituted sesquterpenes daucanes from the wild *Daucus carota*. Phytochemistry 66:1680–1684

Arumuganathan K, Earle ED (1991) Nuclear DNA content of some important plant species. Plant Mol Biol Rep 9:208–218

Bach IC, Olesen A, Simon PW (2002) PCR-based markers to differentiate the mitochondrial genomes of petaloid and male fertile carrot (*Daucus carota* L.). Euphytica 127: 353–365

Baranska M, Schulz H (2005) Spatial tissue distribution of polyacetylenes in carrot root. Analyst 130:855–859

Baranska M, Schulz H, Baranski R, Nothnagel T, Christensen LP (2005) In situ simultaneous analysis of polyacetylenes, carotenoids and polysaccharides in carrot roots. J Agric Food Chem 53:6565–6571

Baranski R (2008) Genetic transformation of carrot (*Daucus carota*) and other *Apiaceae* species. Transgen Plant J 2: 18–38

Baranski R, Baranska M, Schulz H, Simon PW, Nothnagel T (2006) Single seed Raman measurements allow taxonomical discrimination of *Apiaceae* accessions collected in gene banks. Biopolymers 81:497–505

Bennett MD, Leitch IJ (1995) Nuclear DNA amounts in Angiosperms. Ann Bot 76:113–176

Bennett MD, Smith JB (1976) Nuclear DNA amounts in Angiosperms. Philos Trans R Soc Lond B 274:227–274

Bowes CE, Wolyn DJ (1998) Phylogenetic relationships among fertile and petaloid male-sterile accessions of carrot, *Daucus carota* L. Theor Appl Genet 96:928–932

Bradeen JM, Bach IC, Briard M, le Clerc V, Grzebelus D, Senalik DA, Simon PW (2002) Molecular diversity analysis of cultivated carrot (*Daucus carota* L.) and wild *Daucus* populations reveals a genetically nonstructured composition. J Am Soc Hortic Sci 127:383–391

Briard M, Le Clerc V, Peltier D, Suel A, Granger J, Dubois C, Geoffriau E (2007) Genetic Resources Management with Networks: the example of the Network "Carrot and other *Daucus*". Acta Hortic 760:99–104

Brochmann C, Rustan OH, Lobin W, Kilian N, Schmidt KHA (1997) The endemic vascular plants of the Cape Verde. Tornabenea Parl Sommerfeltia 24:77–91

Camadro EL, Cauhépé MA, Simon PW (2008) Compatibility relations between the edible carrot *Daucus carota* and *D. pusillus*, a related wild species from the Argentinian Pampas. Euphytica 159:103–109

Christiansen LP, Brandt K (2006) Bioactive polyacetylenes if food plants of the *Apiaceae* family: occurrence, bioactivity and analysis. J Pharm Biomed Anal 41:683–693

Clark DL, Wilson MV (2003) Post-dispersal seed fates of four prairie species. Am J Bot 90:730–735

Crowden RK, Harborne JB, Heywood VH (1969) Chemosystematics of the *Umbelliferae* – general survey. Phytochemistry 8:1963–1984

Czepa A, Hofmann T (2003) Structural and sensory characterization of compounds contributing to the bitter off-taste of carrots (*Daucus carota* L.) and carrot puree. J Agric Food Chem 51:3865–3873

Czepa A, Hofmann T (2004) Quantitative studies and sensory analyses on the influence of cultivar, spatial tissue distribution, and industrial processing on the bitter off-taste of

carrots (*Daucus carota* L.) and carrot products. J Agric Food Chem 52:4508–4514
DeBonte LR, Matthews BF, Wilson KG (1984) Variation of plastid and mitochondrial DNAs in the genus Daucus. Am J Bot 71:932–940
Djarri L, Medjroubi K, Akkal S, Elomri A, Seguin E, Vérité P (2006) Composition of the essential oil of aerial parts of an endemic Apiaceae of Algeria: *Daucus reboudii* Coss. Flav Fragr J 21:647–649
Downie SR, Katz-Downie DS, Spalik K (2000) A phylogeny of Apiaceae tribe Scandiceae: evidence from nuclear ribosomal DNA internal transcribed spacer sequences. Am J Bot 87: 76–95
Drude O (1898) Umbelliferae. In: Engler A, Prantl K (eds) Die natürlichen Pflanzenfamilien, vol 3(8). Wilhelm Engemann, Leipzig, Germany, pp 63–250
Dudits D, Hadlaczky G, Lévi E, Fejér O, Haydu Z, Lázár G (1977) Somatic hybridisation of *Daucus carota* and *D. capillifolius* by protoplast fusion. Theor Appl Genet 51: 127–132
Ellis PR (1999) The identification and exploitation of resistance in carrots and wild Umbelliferae to the carrot fly, *Psila rosae* (F.). Integr Pest Manag Rev 4:259–268
Ellis PR, Hardman JA, Crowther TC, Saw PL (1993) Exploitation of the resistance to carrot fly in the wild carrot species *Daucus capillifolius*. Ann Appl Biol 122:79–91
Flamini G, Cioni PL, Maccioni S, Baldini R (2007) Composition of the essential oil of *Daucus gingidium* L. ssp. *gingidium*. Food Chem 103:1237–1240
Frese L (1983) Resistance of the wild carrot *Daucus carota* ssp. *hispanicus* to the root-knot nematode *Meloidogyne*-hapla. Gartenbauwissenschaft 48:259–265
Garrod B, Lewis BG (1979) Location of the antifungal compound falcarindiol in carrot root tissue. Trans Br Mycol Soc 72:515–517
Gonny M, Bradesi P, Casanova J (2004) Identification of the components of the essential oil from wild Corsican *Daucus carota* L. using C-NMR spectroscopy. Flav Fragr J 19: 424–433
Góra J, Lis A, Kula J, Staniszewska M, Wołoszyn A (2002) Chemical composition variability of essential oils in the ontogenesis of some plants. Flav Fragr J 17:445–451
Górecka K, Krzyżanowska D, Kiszczak W, Kowalska U, Górecki R (2009) Carrot doubled haploids. In: Tourayev A, Foster BP, Jain SM (eds) Advances in haploid production in higher plants. Springer, New York, USA, pp 231–239
Goris A (1983) Presence of umbelliferose in carrots roots: *Daucus carota* L. Qual Plant Foods Hum Nutr 33:87–89
Grzebelus D, Jagosz B, Simon PW (2007) The *DcMaster* Transposon Display maps polymorphic insertion sites in the carrot (*Daucus carota* L.) genome. Gene 390:67–74
Grzebelus D, Simon PW (2009) Diveristy of *DcMaster*-like elements of the *PIF/Harbinger* superfamily in the carrot genome. Genetica 135:347–353
Harborne JB (1971) Flavonoid and phenylpropanoid patterns in the Umbelliferae. In: Heywood VH (ed) The biology and chemistry of the Umbelliferae. Academic, London, UK, pp 293–314
Hauser TP (2002) Frost sensitivity of hybrids between wild and cultivated carrots. Conserv Genet 3:75–78
Hauser TP, Bjorn GK (2001) Hybrids between wild and cultivated carrots in Danish carrot fields. Genet Resour Crop Evol 48:499–506
Hauser TP, Shim SI (2007) Survival and flowering of hybrids between cultivated and wild carrots (*Daucus carota*) in Danish grasslands. Environ Biosaf Res 6:237–247
Hegnauer R (1971) Chemical patterns and relationships of *Umbelliferae*. In: Heywood VH (ed) The biology and chemistry of the Umbelliferae. Academic, London, UK, pp 267–277
Heywood VH (1968) *Daucus* L. In: Tutin TG, Heywood VH, Burges NA, Moore DM, Valentine DH, Walters M, Webb DA (eds) Flora europaea, vol 2. Cambridge University Press, Cambridge, UK, pp 373–375
Heywood VH (1982) Multivariate taxonomic synthesis of the tribe Caucalideae. Monogr Syst Bot 6:727–736
Hong TD, Linington S, Ellis RH (1998) Compendium of information on seed storage behaviour, vol 2. IPGRI, Rome, Italy
Ichikawa H, Tanno-Suenaga L, Immamura J (1987) Selection of *Daucus* cybrids based on metabolic complementation between X-irradiated *D. capillifolius* and iodoacetamide-treated *D. carota* by somatic cell fusion. Theor Appl Genet 74:746–752
Ichikawa H, Tanno-Suenaga L, Immamura J (1989) Mitochondrial genome diversity among cultivars of *Daucus carota* (ssp. *sativus*) and their wild relatives. Theor Appl Genet 77: 39–43
Iovene M, Grzebelus E, Carputo D, Jiang J, Simon PW (2008) Major cytogenetic landmarks and karyotype analysis in *Daucus carota* and other Apiaceae. Am J Bot 95:793–804
Jarvis A, Lane A, Hijmans RJ (2008) The effect of climate change on crop wild relatives. Agriculture, Ecosystems and Environment 126:13–23
Jensen BD, Finckh MR, Munk L, Hauser TP (2008) Susceptibility of wild carrot (*Daucus carota* ssp. *carota*) to *Sclerotinia sclerotiorum*. Eur J Plant Pathol 122:359–367
Just BJ, Santos CAF, Fonseca MEN, Boiteux LS, Oloiza BB, Simon PW (2007) Carotenoid biosynthesis structural genes in carrot (*Daucus carota*): isolation, sequence-characterization, single nucleotide polymorphism (SNP) markers and genome mapping. Theor Appl Genet 114:693–704
Katz-Downie DS, Valiejo-Roman CM, Terentieva EI, Troitsky AV, Pimenov MG, Lee B, Downie SR (1999) Towards a molecular phylogeny of *Apiaceae* subfamily *Apioideae*: additional information from nuclear ribosomal DNA ITS sequences. Plant Syst Evol 216:167–195
Kleiman R, Spencer GF (1982) Search for new industrial oils: XVI Umbelliflorae seed oils rich in petroselinic acid. J Am Oil Chem Soc 59:29–38
Korovin EP (1959) *Ferula* L. In: Vvedensky AL (ed) Flora Uzbekistana, vol 4. Izdatelstvo Akademii Nauk Uzbekskoi SSR, Tashkent, pp 399–438
Kumarasamy Y, Nahar L, Byres M, Delazar A, Sarker SD (2005) The assessment of biological activities associated with the major constituents of the methanol extract of ‘wild carrot’ (*Daucus carota* L.) seeds. J Herb Pharmacother 5: 61–72
Lacey EP (1986) The genetic and environmental control of reproductive timing in a short-lived monocarpic species *Daucus carota* (Umbeliferae). J Ecol 74:73–86

Lamborn E, Ollerton J (2000) Experimental assessment of the functional morphology of inflorescences of *Daucus carota (Apiaceae)*: testing the 'fly catcher effect'. Funct Ecol 14: 445–454

Lee BY, Downie SR (1999) A molecular phylogeny of Apiaceae tribe Caucalideae and related taxa: Inferences based on ITS sequence data. Syst Bot 24:461–479

Lee BY, Downie SR (2000) Phylogenetic analysis of cpDNA restriction sites and *rps16* intron sequences reveals relationships among Apiaceae tribes Caucalideae, Scandiceae and related taxa. Plant Syst Evol 221:35–60

Lee BY, Levin GA, Downie SR (2001) Relationships within the spiny-fruited umbellifers (Scandiceae subtribes Daucinae and Torilidinae) as assessed by phylogenetic analysis of morphological characters. Syst Bot 26:622–642

Magnussen LS, Hauser TP (2007) Hybrids between cultivated and wild carrots in natural populations in Denmark. Heredity 99:185–192

Manzano P, Malo JE (2006) Extreme long-distance seed dispersal via sheep. Front Ecol Environ 4:244–248

Matthews BF, Widholm JM (1985) Organelle DNA compositions and isoenzyme expression in an interspecific somatic hybrid of *Daucus*. Mol Gen Genet 198:371–376

Maxted N, Dulloo E, Ford-Lloyd BV, Iriondo JM, Jarvis J (2008a) Gap analysis: a tool for complementary genetic conservation assessment. Divers Distrib 14:1018–1030

Maxted N, White K, Valkoun J, Konopka J, Hargreaves S (2008b) Towards a conservation strategy for *Aegilops* species. Plant Genet Resour 6:126–141

McCollum GD (1975) Interspecific hybrid between *Daucus carota* x *D. capillifolius*. Bot Gaz 136:201–206

Morelock TEP, Simon PW, Peterson CE (1996) Wisconsin wild: another petaloid male-sterile cytoplasm for carrot. HortScience 31:887–888

Nakajima Y, Oeda K, Yamamoto T (1998) Characterization of genetic diversity of nuclear and mitochondrial genomes in *Daucus* varieties by RAPD and AFLP. Plant Cell Rep 17: 848–853

Nothnagel T, Straka P, Linke B (2000) Male sterility in populations of *Daucus* and the development of alloplasmic male-sterile lines of carrot. Plant Breed 199:145–152

Okeke SE (1978) Systematic studies in *Daucus* L. (Umbelliferae). PhD Thesis, Department of Botany, University of Reading, Berkshire, UK

Pimenov MG, Leonov MV (1993) The genera of the Umbelliferae. Royal Botanical Gardens, Kew, UK

Pimenov MG, Vasil'eva MG, Leonov MV, Dauschkevich JV (2003) Karyotaxonomical analysis in the Umbelliferae. Science, Enfield, NH, USA

Riemann H, Ezcurra E (2005) Plant endemism and natural protected areas in the peninsula of Baja California, Mexico. Biol Conserv 122:141–150

Ronfort J, Saumitou-Laprade P, Cuguen J, Couvet D (1995) Mitochondrial DNA diversity and male sterility in natural populations of *Daucus carota* ssp *carota*. Theor Appl Genet 91:150–159

Ronquist F (1997) Dispersal-vicariance analysis: a new approach to the quantification of historical biogeography. Syst Biol 46:195–203

Rossi PG, Bao L, Luciani A, Panighi J, Desjobert JM, Costa J, Cassanova J, Bolla JM, Berti L (2007) (E)-methylisoeugenol and elemicin: antibacterial components of *Daucus carota* L. essential oil against *Campylobacter jejuni*. J Agric Food Chem 55:7332–7336

Rubatzky VE, Quiros CF, Simon PW (1999) Carrots and related Umbelliferae. CABI Publishing. New York

Sáenz Laín C (1981) Research on *Daucus* L. (Umbelliferae). Anal Jard Bot Madrid 37:481–534

Santos CAF, Simon PW (2002) QTL analyses reveal clustered loci for accumulation of major provitamin A carotenes and lycopene in carrot roots. Mol Genet Genomics 268:122–129

Schmidt W (1989) Plant dispersal by motor cars. Vegetatio 80: 147–152

Schmidt KHA, Lobin W (1999) *Tornabenea ribeirensis* (Apiaceae): a new species from Sao Nicolau, Cape Verde Islands (West Africa). Feddes Rep 110:7–11

Schouten HJ, van Tongeren CAM, van den Bulk RW (2002) Fitness effects of *Alternaria dauci* on wild carrots in The Netherlands. Environ Biosaf Res 1:39–47

Shim SI, Jørgensen RB (2000) Genetic structure in cultivated and wild carrots (*Daucus carota* L.) revealed by AFLP analysis. Theor Appl Genet 101:227–233

Simon PW (2000) Domestication, historical development, and modern breeding of carrot. Plant Breed Rev 19:157–190

Simon PW, Goldman IL (2007) Carrot. In: Singh RJ (ed) Genetic resources, chromosome engineering, and crop improvement, vol 3, Vegetable crops. CRC, New York, USA, pp 497–517

Simon PW, Freeman RE, Vieira JV, Boiteux LS, Briard M, Nothnagel T, Michalik B, Kwon Y-S (2008) Carrot. In: Prohens J, Nuez F (eds) Handbook of plant breeding, Vegetables II: Fabaceae, Liliaceae Solanaceae and Umbelliferae. Springer, New York, USA, pp 327–357

Small E (1978) A numerical taxonomic analysis of the *Daucus carota* complex. Can J Bot 56:248–276

Spalik K, Downie SR (2007) Intercontinental disjunctions in *Cryptotaenia* (Apiaceae, Oenantheae): an appraisal using molecular data. J Biogeogr 34:2039–2054

Spitzer E, Lott JNA (1982a) Protein bodies in umbelliferous seeds. I. Structure. Can J Bot 60:1381–1391

Spitzer E, Lott JNA (1982b) Protein bodies in umbelliferous seeds II. Elemental composition. Can J Bot 60: 1392–1398

Spitzer E, Lott JNA (1982c) Protein bodies in umbelliferous seeds III. Characterization of calcium-rich crystals. Can J Bot 60:1399–1403

Stace CA (1997) New Flora of the British Isles - 2nd Edition. Cambridge University Press

St. Pierre MD, Bayer RJ, Weis IM (1990) An isozyme-based assessment of the genetic variability within the Daucus carota complex (Apiaceae, Caucalideae). Can J Bot 68: 2449–2457

Staniszewska M, Kula J (2001) Composition of the essential oil from wild carrot umbels (*Daucus carota* L. spp. *carota*) growing in Poland. J Essent Oil Res 13:439–441

Staniszewska M, Kula J, Wieczorkiewicz M, Kusewicz D (2005) Essential oils of wild and cultivated carrots – the chemical composition and antimicrobial activity. J Essent Oil Res 17:579–583

Steinborn R, Linke B, Nothnagel T, Börner T (1995) Inheritance of chloroplast and mitochondrial DNA in alloplasmic forms of the genus *Daucus*. Theor Appl Genet 91:632–638

Szklarczyk M, Oczkowski M, Augustyniak H, Börner T, Linke B, Michalik B (2000) Organization and expression of mitochondrial *atp9* genes from CMS and fertile carrots. Theor Appl Genet 100:263–270

Tavares AC, Goncalves MJ, Cavaleiro C, Cruz MT, Lopes MC, Canhoto J, Salgueiro LR (2008) Essential oil of *Daucus carota* subsp. *halophilus*: composition, antifungal activity and cytotoxicity. J Ethnopharmacol 119:129–134

Umehara M, Eguchi I, Keneko D, Ono M, Kamada H (2005) Evaluation of gene flow and its environmental effects in the field. Plant Biotechnol 22:497–504

Vivek BS, Ngo QA, Simon PW (1999) Evidence for maternal inheritance of the chloroplast genome in cultivated carrot (*Daucus carota* L. ssp. *sativus*). Theor Appl Genet 98: 669–672

Wijnheijmer EHM, Brandenburg WA, Ter Borg SJ (1989) Interactions between wild and cultivated carrots (*Daucus carota* L.). Euphytica 40:147–154

Wolyn DJ, Chahal A (1998) Nuclear and cytoplasmic interactions for petaloid male sterile accessions of wild carrot (*Daucus carota* L.). J Am Soc Hortic Sci 123:849–853

Chapter 8
Lactuca

Michael R. Davey and Paul Anthony

8.1 Introduction

Leafy vegetables, of which lettuce is one of the most important, form an essential component of a well-balanced diet. Globally, lettuce (*Lactuca sativa* L.) is grown as a salad crop and consumed mainly in the fresh state. Lettuce is a self-fertile, annual species of the family Asteraceae, with a chromosome complement of $2n = 2x = 18$. It exhibits considerable morphological and genetic variation, and is classified according to its leaf shape and size. Seven phenotypically distinct groups (morphotypes) are recognized (Kristkova et al. 2008), these being (1) Crisphead, Iceberg or Cabbage, (2) Butterhead, (3) Cos, (4) Leaf or Cutting, (5) Latin, (6) Stem or Asparagus and (7) the Oilseed Group (Ryder 1999). Crisphead cultivars (cvs.) have compact, large heads while Butterhead types have crumpled, soft-textured leaves. Cos types are characterized by long, oval, dark upright green leaves, forming oblong heads or hearts, while leaf types produce a rosette of loose leaves. Latin cvs. are intermediate between Butterhead and Cos types, forming loose heads with oval leaves. The young fleshy stems of stem-type cvs. are eaten after being cooked. Oil-seed type lettuces bolt rapidly and may be primitive forms of *L. sativa*.

Lettuce may have originated in the Mediterranean region, the Middle East and Southwest Asia. It was cultivated in Europe during the late 1400s, prior to introduction into America where Crisphead cvs. were selected (de Vries 1997). Lettuce ranks about 26th compared to other fruit and vegetables in terms of its contribution to the human diet, following tomato and orange with regard to bulk consumption in the US. At least 70% of US lettuce production occurs in the field in California, with a farm gate value for iceberg lettuce exceeding $9.8 million. Other regions of cultivation include Canada, northern Europe and Mexico, South America, South Africa, the Middle East, Japan, China and Southeast Australia. Nineteen million tons of lettuce was produced in 2002 (FAOSTAT website 2003), of which the US produced about 4 million tons and 2 billion lettuce heads. The consumption of lettuce is increasing annually because of its popularity as a vegetable, necessitating the generation of new varieties to meet consumer demands. Whilst sexual hybridization will continue to generate new varieties, molecular procedures and tissue culture-based techniques provide an important under-pin to classical breeding approaches.

8.2 Genetic Improvement of Lettuce by Traditional Breeding

8.2.1 Requirement for Germplasm Collections

In an extensive review, Ryder (2001) discussed some of the current and future issues relating to lettuce breeding. The breeding of cultivated lettuce is targeted to improve agronomically important traits, with increase in disease resistance being a major target. Other characteristics for modification include leaf shape and head formation, pigmentation, increasing

M.R. Davey (✉)
Plant and Crop Sciences Division, School of Biosciences, University of Nottingham, Sutton Bonington Campus, Loughborough LE12 5RD, UK
e-mail: mike.davey@nottingham.ac.uk

C. Kole (ed.), *Wild Crop Relatives: Genomic and Breeding Resources, Vegetables*,
DOI 10.1007/978-3-642-20450-0_8,

tolerance to biotic and abiotic stresses and the introduction of male sterility to facilitate sexual hybridization. Improved leaf quality and succulence, a decrease in latex and bitter taste, reduced nitrate accumulation and an extension of shelf-life following harvest, are also traits being addressed by breeders.

Well-administered plant collections are essential as a source of donor and recipient material for breeding. Many workers have collected plants from the wild in attempts to maximize the biodiversity of germplasm available for sexual hybridization. In order to investigate geographical distribution of species, spatial population structure, plant microevolution, ecology, domestication and genetic variability of host–parasite interactions, Lebeda et al. (2001) collected wild *Lactuca* species and the related genera, *Chondrilla* and *Mycelis*. Most species biodiversity was in France. *Lactuca serriola* f. serriola, *L. serriola* f. integrifolia, *L. saligna* and *L. viminea* subsp. chondrilliflora were common in Italy and France, but only *L. serriola* (prickly lettuce) was plentiful in Austria, the Czech Republic, Germany, the Netherlands and Switzerland. In subsequent studies to assess the geographical distribution of wild *Lactuca* species, Lebeda et al. (2004) reported that at least 98 wild *Lactuca* species have been described, with 17 species in Europe, 51 in Asia, 43 in Africa and 12 in the Americas (mainly in North America), although the documentation is poor that described taxonomic relationships, ecogeography and variability. Generally, more details are available relating to geographical distribution of *L. serriola*, *L. saligna*, *L. virosa*, *L. perennis*, *L. quercina* and *L. tatarica*, representing the primary, secondary and tertiary gene pools of *L. sativa*. Later, Lebeda et al. (2007) focused attention on *L. serriola* collected from the Czech Republic, Germany, the Netherlands and the United Kingdom. Seed was harvested in an east to west transect from 50 locations, giving a total of 800 seed samples. *L. serriola* f. serriola was dominant in the Czech Republic and Germany, both f. serriola and f. integrifolia were present in pure and mixed stands in the Netherlands, whereas *L. serriola* f. integrifolia was dominant in the United Kingdom. *L. serriola* was recorded at altitudes up to 410 m, in a range of habitats and in populations of different sizes from 20 to more than 1,000 plants in areas from 25 to 10,000 m^2 in size. Associated plant communities also differed. Some wild lettuce populations were noticeably infected with *Bremia lactucae* (downy mildew) and powdery mildew (*Golovinomyces cichoracearum*), although *B. lactucae* was recorded only in the Czech Republic; *G. cichoracearum* was more common in continental Europe. Other investigators (Beharav et al. 2006) screened 962 genotypes of *L. serriola*, 43 of *L. saligna* and 22 of *L. aculeata* from Israel, Jordan, East Turkey, Armenia, Kazakhstan and China for their resistance to lettuce downy mildew, the fungal strains used in these assessments being isolated from lettuce cultivated in Italy and Israel. Eighty-three genotypes of *L. serriola* and three genotypes of *L. saligna* were resistant to infection. Such genotypes may carry previously unknown disease resistance genes and could be suitable for incorporation into breeding programs. In particular, *L. serriola*, the probable progenitor of cultivated lettuce, is an important source of genetic material for improvement of cultivated lettuce.

Occasionally, germplasm maintained locally or in international collections may need to be reclassified at the taxonomic level. For example, Dolezalova et al. (2004) evaluated 49 accessions of 24 wild *Lactuca* species, a sexual hybrid (*L. serriola* × *L. sativa*) and the related species *Mycelis muralis*. Screening of chromosome number, relative DNA content and isozyme polymorphism, resulted in 17 accessions being reclassified, with *L. alpina*, *L. dentata*, *L. denticulata*, *L. homblei*, *L. livida*, *L. quercina* and *L. squarrosa* (syn. *L. indica*) being absent, although these species had been reported previously in the Czech National Germplasm Collection. Additionally, it was not possible to identify *L. homblei* unambiguously because of the lack of herbarium specimens. This species may have been *L. schweinfurthii* or *L. logenspicata*.

An accurate description of lettuce germplasm is essential to assist plant breeders. Thus, Kristkova et al. (2008) listed 55 descriptors, 15 of which were supported by figures. Such a list discriminates intraspecific variation found within *L. sativa*, verifies old cvs. and identifies duplication and omission in germplasm collections. Descriptors of wild *Lactuca* species are also relevant in studying morphological variability and relationships between *Lactuca* species. Markovic et al. (2007) stressed the importance of assessing wild species, including those of *Lactuca*, for their potential as gene donors, or as possible vegetables in their own right. These authors stated that only 30 or so vegetable species are cultivated compared to about 150 species, which could be exploited. Importantly, such wild species contain more vitamins and minerals than many existing cultivated crops. Markovic et al. (2007) indicated the diversity of germplasm from Serbia and Montenegro that could be developed in this way.

8.2.2 Hybridization of Cultivated Lettuce with Wild Lactuca Species

Any breeding program to improve specific characteristics, such as enhancing the nutritional value of vegetables, including lettuce, must commence with an assessment of variation within existing germplasm. In this respect, Mou (2005) evaluated 562 lettuce genotypes, including crisphead, leaf, romaine, butterhead, Latin and stem types, together with the wild species *L. saligna*, *L. serriola* and *L. virosa* for their carotenoids. Carotenoid antioxidants, such as β-carotene and lutein, have important health benefits. The author recorded a significant difference in pigment content of wild and cultivated lettuces, with more β-carotene and carotenoids in wild species compared to cultivated lettuce. These investigations are fundamental in identifying the most suitable germplasms for hybridization in attempts to increase the pigment content of cultivated lettuce.

Lettuce is an obligate self-pollinating species; pollen is shed from the anthers before stigma emergence, ensuring 100% self-pollination. Consequently, introgression of genes into *L. sativa* by sexual hybridization is not a trivial issue. Three approaches have been employed to prevent self-fertilization, these being the manual removal of anthers, misting with water designated flowers to remove self-pollen prior to cross-pollination and the exploitation of male-sterile lines as pollen recipients. All these techniques are tedious and time consuming. Male sterile lines must be generated, or identified if they occur naturally. Maisonneuve (2003) reported the results and some of the difficulties of crossing sexually *L. sativa* with the wild species *L. virosa*, the latter being a reservoir of genes for resistance to *B. lactucae*, lettuce mosaic virus (LMV), beet western yellows virus (BWYV) and aphids. This wild species has the same chromosome complement as *L. sativa*. This author had to employ embryo rescue (see also later) to generate interspecific F_1 hybrids. Although the latter were sterile, it was possible to effect backcrosses using lettuce as the male parent, resulting in limited seed production. A useful reference with supporting literature is that of Gibson et al. (2007), who described a procedure for male-sterile lines as pollen recipients, combined with *Megachile rotundata* bees to effect pollination. A synopsis of some of the earlier papers relating to lettuce breeding has been presented by Davey et al. (2007, 2008).

Other interspecific hybrids have been easier to generate. For example, Zdravkovic et al. (2003) hybridized *L. sativa* with the wild species, *L. saligna*, with the aim of producing cv(s) with the genes for resistance to virus diseases from the wild parent and acceptable morphological characteristics from *L. sativa*. Using the same parental combination, Jeuken and Lindhout (2004) developed backcross inbred lines (BILs) in which chromosome segments from the wild species, *L. saligna* were introgressed into *L. sativa*, through four to five backcrosses and one generation of selfing. Marker-assisted selection (see later) commenced in the fourth backcross generation. BIL association mapping enabled the location of 12 morphological traits and amplified fragment length polymorphism (AFLP) markers to be determined. Jeuken and Lindhout (2004) indicated that such BILs would be essential in future genetic studies in *Lactuca* which, indeed, has been the case. Thus, in subsequent experiments, the same research group compared quantitative trait loci (QTL) mapping approaches using BILs with F_2 populations (Jeuken et al. 2008). They detected and mapped quantitative resistances to downy mildew in a set of 29 BILs of *L. sativa* containing genome segments introgressed from *L. saligna*. Jeuken et al. (2008) identified six QTLs using BILs but only three in the F_2 approach, with two common QTLs. They highlighted the importance of exploiting more than one technique to dissect complex genetic characteristics.

Since downy mildew is a major disease of lettuce, considerable effort has been invested in evaluating the resistance of both cultivated and wild *Lactuca* species to this fungus. Lebeda et al. (2006) assessed the phenotypic and histological responses of *L. sativa* and its wild relatives *L. saligna* and *L. virosa* and sexual hybrids of *L. sativa* × *L. serriola*, to two races (CS2, CS9) of *B. lactucae*. With the exception of genotypes of cultivated lettuce, all accessions and hybrids exhibited incomplete or complete resistance to both races of the pathogen. Such studies that contribute to increasing understanding of host–pathogen interactions should facilitate progress in developing resistant cvs. of lettuce.

Big vein, an economically damaging disease of lettuce, is incited by Mirafiori lettuce big vein virus (MLBVV), the latter being transmitted by the soil-borne fungus *Olpidium brassicae*. Partial resistance to the disease is present in some cvs. of *L. sativa*, but complete resistance has been identified only in *L. virosa*. In an attempt to introgress complete viral resistance into *L. sativa*, Hayes and Ryder (2007) tested progenies of backcross hybrids between *L. virosa* and *L. sativa*. Complete resistance to MLBVV was not found, although these investigations suggested that *L. virosa* has alleles that, when introgressed into *L. sativa*, confer partial resistance and that such alleles are different to those already present in partially resistant cvs. of *L. sativa*.

Other investigators have attempted to increase aphid resistance in lettuce. For example, McCreight (2008) introgressed the *Nr* gene from *L. virosa* accession PIVT280 into European cvs. and US-adapted cvs. of *L. sativa*. Sources of resistance against new strains of lettuce aphid were also sought through glasshouse screening of 1,203 accessions of lettuce, including 1,047 of *L. sativa*, 7 of *L. perennis*, 18 of *L. saligna*, 125 of *L. serriola* and 6 of *L. virosa*. Two new sources of resistance were found in *L. serriola* PI491093 and *L. virosa* PI274378.

Germination of lettuce seed is inhibited by warm soil temperatures above 25°C, which affect seedling emergence and crop establishment. However, genetic variation for tolerance to high temperatures also exists amongst accessions of *L. sativa* and related wild species. For example, seeds of *L. serriola* UC96US23 exhibited 100% germination at 37°C, although seeds of *L. sativa* cv. Salinas were inhibited above 31°C. Argyris et al. (2008) developed a recombinant inbred line population from a cross between the cv. Salinas and UC96US23, followed by screening of the seedlings to germinate at high temperatures. A QTL for high-temperature germination (Htg6.1), together with other QTLs, were identified in this investigation and candidate genes related to seed dormancy mapped to check their colocation with the thermoinhibition QTL. Such studies should provide information for elucidating the physiology of thermoinhibition and the development of lettuce cvs. tolerant to establishment under warm conditions.

8.3 Application of Molecular Technologies to Lettuce Improvement

Several recent studies have focused on the application of molecular technologies to identify genes and markers to facilitate breeding programs. Earlier publications relating to traits for morphological characteristics and disease resistance by authors such as Ryder (1999), Ryder et al. (1999), Kesseli et al. (1994), Waycott et al. (1999) and Frijters et al. (1997) are discussed in Davey et al. (2007, 2008) and are not presented in detail again here. However, references not included in the earlier review, and more recent publications are presented in the present discussion.

8.3.1 Molecular Characterization of Lactuca Germplasms

Witsenboer et al. (1997) applied selectively amplified microsatellite polymorphic locus (SAMPL) analysis to 45 lettuce cvs. and five wild *Lactuca* species. These workers reported that SAMPL analysis is more applicable to intraspecific than to interspecific comparisons. Importantly, dendrograms generated by SAMPL analyses were similar to those based on restriction fragment length polymorphism (RFLP) and AFLP markers. In the following year, Koopman et al. (1998) investigated the phylogenetic relationship between *Lactuca* species and genera based on the internal transcribed spacer (ITS-1) sequences from 97 accessions representing 23 *Lactuca* species and related genera. A clade with *L. sativa*, *L. serriola*, *L. dregeana*, *L. altaica* and *L. aculeata* represented the primary gene pool. *L. virosa* and *L. saligna* were in the secondary pool with *L. virosa* possibly being of hybrid origin. Other species such as *L. quercina*, *L. viminea* and *L. sibirica* and *L. tatarica* represented the tertiary pool. Van de Wiel et al. (1999) isolated microsatellite loci to distinguish *L. sativa* cvs. and to screen the diversity of genetic resources. These microsatellite-amplifying primer sets from *L. sativa* yielded products in *L. serriola*, the nearest relative to cultivated lettuce, but were less applicable to the more distinct species, *L. saligna* and *L. virosa*.

The target region amplification polymorphism (TRAP) marker technique provided a powerful approach to fingerprint lettuce germplasm (Hu et al. 2005). These authors fingerprinted 53 *L. sativa* cvs. and three accessions from each of *L. saligna* and *L. serriola*. The markers used discriminated all cvs. and also revealed the evolutionary relationship amongst three lettuce species, indicating that *L. sativa* is more closely related to *L. serriola* than to *L. saligna*. Leaf and butterhead type lettuces had relatively high genetic variability, and iceberg type moderate variability, with romaine types the lowest. In a subsequent report to investigate polymorphism using 54 *L. sativa* cvs. representing five horticultural types important to North America, Simko and Hu (2008) used the same wild *Lactuca* germplasm as used by Hu et al. (2005) to identify three main subpopulations in *L. sativa* that were well separated from wild species. These authors emphasized the importance of combining data from molecular markers with statistical models in order to avoid spurious associations in populations under investigation. Other investigators have exploited AFLP markers. Jeuken et al. (2001) produced an interspecific AFLP marker map of *Lactuca* with 12 *Eco*RI/*Mse*I primer combinations and two independent F_2 populations of *L. sativa* × *L. saligna*. This map provided an opportunity for use in the mapping of QTL and marker-assisted selection. Similarly, Jansen et al. (2006) used AFLP marker data on 1,390 accessions representing six lettuce morphotypes, to determine genetic diversity within gene bank accessions of self-fertile species. In the same year, Syed et al. (2006) produced a detailed linkage map of lettuce using molecular makers based on a novel lettuce retrotransposon and the nucleotide binding site-leucine-rich repeat (NBS-LRR) family of disease resistance associated genes, combined with AFLP markers to generate a 458 locus genetic linkage map. Markers were mapped in an F_2 population from an interspecific cross between *L. sativa* and a wild isolate of *L. serriola* from northern Europe. This information of genetic diversity within accessions is relevant to the maintenance of accessions within collections. The data from these experiments correlated well with linkage maps generated previously. More recently, Truco et al. (2007) developed an integrated map of 2,744 markers from seven intra- and interspecific mapping populations. Such markers are useful for marker-assisted selection, candidate gene identification, studies of gene evolution and possible gene flow from cultivated into wild species.

Large plant populations are important in assessing disease resistance and the identification of germplasms most relevant for incorporation into breeding programs. In determining the frequency of disease resistance genes in natural plant populations, Kuang et al. (2006) analyzed the frequency and variation of the *Dm3* gene in 1,033 accessions of 49 natural populations of *L. serriola*. Inoculation with *B. lactucae* carrying the avirulence gene *Avr3* showed *Dm3* was present in only one of the wild accessions analyzed. The authors proposed that the low frequency of *Dm3* specificity in populations of wild species may be due to *Dm3* specificity having evolved recently, to deletions of the whole gene or to gene conversions. They concluded that the total number of resistance genes may be high in any species, indicating the scale of germplasm conservation required to ensure the availability of material suitable for sexual hybridization. In their more recent study, Kuang et al. (2008) screened 696 accessions from 41 populations of *L. serriola* using AFLP markers, and showed that eastern Turkish and Armenian populations were the most diverse. These populations may be located in the center of origin and diversity of *L. serriola*. Seven hundred and nine accessions screened by the microsatellite MSATE6 resulted in the detection of 366 haplotypes. Kuang et al. (2008) concluded that the extensive genome diversity in populations from these regions may have resulted from abiotic and biotic stresses in the regions of origin of *L. serriola*. *Lactuca* germplasm has also been evaluated at the cytological level. For example, Matoba et al. (2007) karyotyped *L. sativa* and its wild relatives *L. serriola*, *L. saligna* and *L. virosa*. Fluorescence in situ hybridization (FISH) of 5S and 18S rDNAs revealed that *L. sativa* was closer to *L. serriola* and *L. saligna* than *L. virosa*, the data correlating with the earlier molecular analyses of Witsenboer et al. (1997), van de Wiel et al. (1999) and Hu et al. (2005). Matoba et al. (2007) demonstrated that there was lack of telomere-mediated chromosomal rearrangements amongst the *Lactuca* chromosomes.

8.3.2 Characterization of Plastid Genomes

Other investigations have focused on plastid genomes. For example, Dhingra and Folta (2005) devised a universal high-throughput, rapid PCR-based technique to amplify, sequence and assemble plastid genome sequences from total plant DNA of diverse species, including DNA of lettuce. The technology facilitates assessments of variability across both cultivated and wild species, and is a useful tool in plastid structural genomics. Likewise, Kanamoto et al. (2004) reported the complete genome sequence of lettuce chloroplasts, while Timme et al. (2007) made a comparative analysis based on the sequence data of the plastid genomes of *Lactuca* and *Helianthus*, both members of the Asteraceae. The *Helianthus* chloroplast genome was 151,104 bp and the *Lactuca* genome 152,772 bp in size. These workers discussed new DNA sequences for use in species-level phylogenetics, including the trnY–rpoB, trnL–rpl32 and ndhc–trnV spacers. Such detailed analyses in the Asteraceae contribute to an improved understanding of plastid evolution in flowering plants. Additionally, studies of plastid genomes will be of relevance to vector construction and the targeting of genes in plastid transformation in lettuce and other vegetables.

8.3.3 Molecular Analysis for Product Quality

Investigations have also been carried out to predict the characteristics that should be integrated into future cvs. of lettuce to maximize product quality. Thus, Zhang et al. (2007) investigated shelf-life in cultivated lettuce using 60 F_9 recombinant inbred lines (RILs) from a cross between *L. sativa* cv. Salinas and the wild species *L. serriola* acc. UC96US23. QTL for shelf-life colocated most closely with those for leaf biophysical properties, the latter including plasticity, elasticity and break strength. These characteristics are targets for molecular breeding for improved shelf-life. The data generated by Zhang et al. (2007) indicated that the ideal lettuce cv. should have small cells with strong cell walls. Such an investigation confirms the way in which molecular finger printing technologies will facilitate marker-assisted breeding which, in the case of leafy vegetables, should result in an improvement in salad leaf quality.

8.3.4 Molecular Characterization of Lettuce Pathogens and Plant Resistance Genes

Since pathogens have the potential to devastate commercial production of lettuce, it is not unexpected that considerable effort has been invested in applying molecular approaches to evaluate the variability of these disease-inciting pathogens. Barak and Gilbertson (2003) studied genetic diversity of *Xanthomonas campestris* pv. vitians, the causal agent of bacterial leafspot of lettuce, using strains of the bacterium collected from different geographical locations. All strains were tested for pathogenicity on lettuce and their genetic diversity assessed by gas-chromatographic analysis of bacterial fatty acids, polymerase chain reaction (PCR) analysis of repetitive DNA sequences, DNA sequence analysis of the internal transcribed spacer region 1 (ITS1) of the ribosomal RNA and RFLP analysis of total genomic DNA. These authors concluded that although diversity was detected in bacterial strains from the same location, diversity was greatest amongst strains from different geographical regions. Targeting of virulent phenotypes of *B. lactucae* in natural populations of *L. serriola* from the Czech Republic, France and Germany, led Lebeda and Petrzelova (2004) to report variation both between spatially isolated populations and within populations of the pathogen. An extremely useful review of the genetic and molecular analyses conducted in the last 25 years on *B. lactucae* has been published by Michelmore and Wong (2008). As these authors emphasized, advancements in molecular technology now permit rapid progress in determining the molecular basis of host–pathogen specificity and fungicide insensitivity of this pathogen. In addition to molecular analyses, a recent report correlated metabolic changes in enzyme activity, including antioxidant enzymes, in genotypes of *L. sativa*, *L. serriola*, *L. saligna* and *L. virosa*, when challenged with *B. lactucae*. The most significant metabolic modifications in relation to reactive oxygen species was in the resistant *L. virosa* NVRS 10.001 602 that responded to pathogen attack by a rapid and extensive hypersensitive reaction (Sedlarova et al. 2007).

Other workers have targeted viruses in their molecular investigations. Lettuce big-vein associated virus (LBVaV) is frequently identified in plants infected with MLBVV. In attempts to gain more understanding of these viruses, Navarro et al. (2005a) sequenced the coat protein gene open reading frames of geographically distinct isolates of MLBVV and LBVaV, because big-vein disease is found globally on lettuce. There was extensive similarity between nucleotide and amino-acid sequences amongst LBVaV isolates. Interestingly, Spanish isolates clustered in a phylogenetic group, whereas English isolates were more similar to an isolate from the US. An Australian isolate was closely related to one from the Netherlands. The same research group also demonstrated that MLBVV and LBVaV can infect *Sonchus oleraceus* (common sowthistle), a weed frequently associated with Spanish lettuce crops (Navarro et al. 2005b). Molecular analyses confirmed the coat protein sequences of the viruses isolated from the weed to be closely related to the sequences of the viruses from the lettuce crop. These authors also showed that *S. oleraceus* can act as a source of *O. brassicae*. Subsequently, Hayes et al. (2006) used reverse-transcriptase polymerase chain reaction (RT-PCR) and nucleotide sequencing to determine infection by MLBVV and LBVaV in the *L. sativa* cvs. Great Lakes 65, Pavane, Margarita and the wild species *L. virosa* accession IVT280. Partial resistance to big-vein was found in Margarita and Pavane. Interestingly, the wild accession was immune to both viruses. In a more recent publication, Hayes et al. (2008) screened diverse accessions of *L. virosa* under glasshouse conditions for resistance to MLBVV, and the presence of the virus by RT-PCR and nucleic acid spot hybridization. The accessions of *L. virosa* that were identified with a low incidence of the virus and that did not show symptoms of the disease should be useful for breeding new big vein resistant cvs. of lettuce. Research has also been directed towards the characterization of phytoplasmas associated with phyllody in cultivated and wild lettuce, the phytoplasmas being transmitted by the leaf hopper, *Neoaliturus fenestratus* (Salehi et al. 2007). PCR and RFLP analyses showed no significant differences between phytoplasmas in *Lactuca* samples from Iran, but Iranian samples were different at the molecular level to material harvested from Lebanon.

8.3.5 Molecular Analyses of Disease Resistance Genes in Lettuce and Its Pathogens

Molecular technologies have also been used to evaluate disease resistance genes in lettuce and its pathogens. Ratnaparkhe et al. (1998) employed intersimple sequence repeat (ISSR) polymorphisms to detect markers associated with clusters of disease resistance genes. They concluded that ISSR markers provide information to design other primers for genome analysis. Disease resistance genes in plants are often in complex multigene families with the largest known cluster of disease resistance genes containing the resistance gene candidate 2 (RGC2) family of genes. Meyers et al. (1998) determined the structure of nine full-length genomic copies of genes within the RGC2 family, while Sicard et al. (1999) investigated the molecular diversity of disease resistance genes in both cultivated lettuce and the wild species *L. serriola*, *L. saligna* and *L. virosa* using molecular markers derived from resistance genes of the NBS-LRR type. Large numbers of haplotypes indicated the presence of numerous resistance genes in the wild species. The authors were able to discriminate between accessions that were resistant to *B. lactucae*, concluding that such markers are informative for marker-assisted selection and establishing relevant core germplasms for breeding.

Chin et al. (2001) analyzed recombination and spontaneous mutation at the major cluster of *Dm3* resistance genes in lettuce and reported that short-term evolution of this major cluster involves several genetic mechanisms, including unequal crossing over and gene conversion. In subsequent studies of NBS-LRR disease resistance genes, Plocik et al. (2004) compared the NBS domain sequences from members of the Asteraceae, including lettuce, chicory and sunflower. Amplification of NBS sequences from genomic DNA suggested that species in the Asteraceae share distinct families of resistance genes, composed of those related to both coiled-coil (CC) and toll-interleukin-receptor (TIR) homology domain-containing NBS-LRR genes. Their data suggested that NBS families in the Asteraceae are ancient and that, with time, gene duplication and loss have changed these gene subfamilies. Likewise, Kuang et al. (2004) analyzed RGC2 genes belonging to a large duplicated family of NBS-LRR-encoding disease resistance genes located at a single locus in *L. sativa*. They investigated the genetic events occurring during evolution of this locus and concluded that different forces impacted on different parts of the leucine rich repeat, resulting in heterologous rates of evolution within the major cluster of disease resistance genes.

8.4 Somatic Cell Technologies for Genetic Improvement of Lettuce

8.4.1 Tissue Culture and Somatic Hybridization

Tissue culture based technologies, including micropropagation, in vitro pollination, anther and ovule culture, embryo rescue, exposure of somaclonal variation, somatic hybridization to circumvent sexual incompatibility barriers and transformation are adjuncts to conventional plant breeding. Robust tissue culture-based procedures for the regeneration of fertile plants from cultured explants, cells and isolated protoplasts are fundamental to the exploitation of these technologies. Although there are species and genotype differences in vitro, cultivated lettuce and its wild relatives are amenable to culture in the laboratory, with shoot regeneration from cells, tissues and isolated protoplasts. Many of the earlier investigations involving plant regeneration from source tissues, as a basis for the above techniques, especially those to introgress genes from wild species into cultivated lettuce by somatic hybridization, have been reviewed (Davey et al. 2007, 2008). However, there have been no recent reports of somatic hybridization in lettuce to introgress genes from wild species into *L. sativa*. In summary, any new information is limited relating to tissue culture-based technologies, since the literature was last screened. As already mentioned, Maisonneuve (2003) exploited embryo rescue in the sexual hybridization of *L. sativa* with *L. virosa*, while Gibson et al. (2007) micropropagated plants from axillary buds of progeny resulting from sexual hybridization. These micropropagated plants were useful for assessing male-sterility under field conditions, whilst maintaining the same germplasm in vitro for subsequent incorporation into breeding programs. Mazier et al. (2007) reported a gene tagging method to induce mutants in *L. sativa*, using the retrotransposon Tnt1 from *Nicotiana tabacum*, transposition being stimulated by a simple in vitro procedure. This investigation showed that Tnt1 is a powerful tool for insertion mutagenesis in plants with large genomes, such as lettuce.

8.4.2 Transformation of Lettuce

In contrast to the lack of new information relating to gene transfer from wild species into *L. sativa* by somatic hybridization, there have been several recent reports of gene introduction into lettuce by transformation. Background information to the general procedures to transform *Lactuca* species and other leafy vegetables, together with examples of some of the pioneering studies to introduce and express genes for agronomically important traits, improved taste, nutritional value and anticancer activity, have been discussed by Davey et al. (2007, 2008). Extended shelf-life of harvested produce and engineering lettuce to synthesize high-value recombinant proteins were also presented. Consequently, the present discussion merely updates these previous reports. In extending the methodology to transform specific cvs. of lettuce, Deng et al. (2007) reported an *Agrobacterium tumefaciens*-based system for *L. sativa* var. *capatata*. Their protocol included the combination and concentrations of growth regulators used to regenerate transgenic plants, the selection system employed and molecular analysis of transgenic plants. *Agrobacterium* inoculation of leaf explants is a universal approach for introducing genes into lettuce.

8.4.2.1 Introduction into Lettuce of Genes Related to Human Nutrition

Since primates cannot synthesize ascorbic acid because of a mutation in the L-gulono-gamma-lactone oxidase (GLOase) gene, it is essential to acquire ascorbic acid from the diet. In experiments to increase ascorbic acid biosynthesis in lettuce, Kim et al. (2004) inoculated cotyledon-explants with *Agrobacterium* carrying a binary vector with a rat cDNA encoding GLOase gene. Transgenic plants accumulated up to four times more ascorbic acid than non-transformed plants, demonstrating that ascorbic acid in plants can be increased in plants by expressing a single gene from the relevant animal biosynthetic pathway. The iron storage protein, ferritin, was increased in lettuce. Lee et al. (2004) used human H- and L-chain ferritin genes, followed by a ferritin gene from rice (Lee et al. 2005) in their experiments. In attempts to induce lettuce to synthesize novel compounds of pharmaceutical interest, Cho et al. (2005) transformed lettuce with the amopha-4,11-diene synthase gene, a terpene synthase gene, from *Artemisia annua*. Unfortunately, transgenic plants did not synthesize artemisinin, the product encoded by this gene. Tocopherols, synthesized by photosynthetic organisms, are also essential components of the human diet. In order to evaluate the feasibility of increasing the tocopherol content of plants, particularly leafy vegetables, Lee et al. (2007) expressed genes from *Arabidopsis thaliana* encoding homogentisate phytyltransferase (HPT/VTE2) and tocopherol cyclase (TC/VTE1) in lettuce. Tocopherol was increased more than twofold through an increase in gamma-tocopherol, although chlorophyll content decreased by 20% from expression of HPT/VTE2, or increased up to 35% with TC/VTE1.

Other chimeric genes have been introduced into lettuce, as in the report of Kim et al. (2006) who expressed a synthetic cholera toxin B subunit (sCTB) in this leafy vegetable. The sCTB reached 0.24% of the total soluble protein in transgenic plants and, importantly, the sCTB conserved the antigen sites for binding and correct folding of the pentameric CTB structure. In other experiments by the same research group, Kim et al. (2007) studied the synthesis and assembly of *Escherichia coli* heat-labile enterotoxin B (LTB) that induces immune responses and that can be used as an adjuvant for coadministered antigens. As in previous work, plants synthesized and assembled the synthetic LTB into oligomeric structures of pentameric size. The binding of sLTB pentamers to intestinal epithelial cell membrane glycolipid receptors was confirmed by G(M1)-ganglioside-enzyme-linked immunosorbent assay (G(M1)-ELISA). The sLTB protein comprised 1.0–2.0% of the total soluble protein of lettuce leaf tissues. This work confirmed the feasibility of engineering edible plant-based vaccines to induce mucosal immunity. Deng et al. (2006) also concluded that plant-derived vaccines are becoming the focus of plant gene engineering because of their ease of production, low cost and aspects of biosafety associated with their synthesis in plants. These workers transformed lettuce with type O and A foot and mouth disease virus epitopes as their experimental system, using an established procedure for *Agrobacterium*-mediated gene transfer.

The extended period generally needed to generate transgenic plants in conventional protocols has stimulated workers to attempt to circumvent the requirement for tissue culture-based procedures. For example, Joh et al. (2005) infiltrated lettuce leaf explants with *A. tumefaciens* carrying the β-glucuronidase (*gus*) gene, the latter being used to monitor the success of the procedure. Similarly, Negrouk et al. (2005) vacuum infiltrated leaves of lettuce plants purchased from a retail outlet with *A. tumefaciens* carrying antibody genes. Leaves of plants infiltrated with agrobacteria synthesized 20–80 mg of antibody per kg fresh weight of leaf tissue over a 7-day period. The system was ideal for transient gene expression as it was inexpensive, efficient and amenable to rapid scale-up. Indeed, agroinfiltration could be a promising technology for synthesizing large quantities of high-value recombinant proteins in planta. Although the synthesis of economically important products has been focused, to date, in *L. sativa*, wild *Lactuca* species could be exploited in a similar way, especially when the products are extracted rather than being delivered by consumption of the raw or cooked plant material. Indeed, longer-term, wild *Lactuca* species may prove to be more effective in product synthesis than cultivated lettuce. Consequently, effort should also be directed to the transformation of wild germplasm.

8.4.2.2 Introduction into Lettuce of Agronomically Important Genes

Several workers have focused attention on the introduction of agronomically important characteristics into lettuce. Vanjildorj et al. (2005) over-expressed the *Arabidopsis ABF3* gene, encoding a transcription factor for the expression of abscisic acid response genes, in *L. sativa* and recorded enhanced drought and cold tolerances in transgenic plants. The *MEleaN4* gene, for a late embryogenesis abundant protein from *Brassica napus*, was inserted into lettuce (Park et al. 2005). Transgenic plants exhibited increased growth under salt stress and water deficit conditions and were delayed in wilting induced by water deficiency. Research has also targeted increased disease tolerance in lettuce. Rot in crops such as lettuce is incited by *Sclerotinia sclerotiorum* and is associated with the production of oxalic acid. In experiments to reduce oxalic acid, Dias et al. (2006) transformed lettuce with the *oxdc* decarboxylase gene from *Flammulina* species. Two lines of transgenic lettuce plants were symptomless following inoculation with *S. sclerotiorum* in a detached leaf assay; leaves from other plant lines showed delayed symptoms of the disease. Likewise, increased virus resistance has been a target for genetic manipulation. Thus, Kawazu et al. (2006) generated lettuce plants resistant to LBVaV by introducing the coat protein gene in sense or anti-sense orientation. Transgenic plants were inoculated with LBVaV using the vector, *O. brassicae*. When T_1 generation seedlings were tested for LBVaV resistance, one plant line with the coat protein gene in the antisense orientation was resistant to the virus. Importantly, this line was also resistant to MLBVV and big-vein expression, regardless of the presence or absence of LBVaV.

As already discussed, male sterile plants are of major advantage in sexual hybridization programs (Gibson et al. 2007). In this regard, Takada et al. (2007) introduced into lettuce the mutated melon ethylene receptor gene, *Cm-ERSI/H70A*, that contains a mis-sense mutation converting the His(70) residue to Ala in the melon receptor gene *Cm-ERS1*. Transgenic lettuce plants were male-sterile or had significantly reduced fertility, in line with earlier work on tobacco using the same gene construct. Such male-sterile plants will be valuable as recipients of donor pollen from wild species and other cvs. of *L. sativa*. Since the nitrogen status of lettuce is relevant to its commercial production not only in terms of nitrate accumulation in leaves but also to plant growth in general, it is not surprising that it has become a target for genetic manipulation. Giannino et al. (2008a) over-expressed asparagine synthase A of *E. coli* under the control of the pMAC promoter. Primary transformants had more, longer leaves than wild type plants with seed germination, bolting and flowering occurring earlier than in non-transformed plants. Plants exhibited an increase in asparagine, aspartic acid and glutamine but, importantly, a reduction in nitrate content.

8.4.3 Investigations of the Invasiveness of Sexual Hybrids and Transgene Containment in Lettuce

The possible invasiveness of *Lactuca* hybrids and the potential transfer of genes from crops to wild species and from transgenic plants to non-transformed crops and wild species are concerns that have attracted public attention. Hybridization between species may enhance the potential to become invasive. As Hooftman et al. (2005) emphasized, there have been few investigations that compare the performance of sexual hybrids with that of their wild relatives. Interestingly, lettuce provides a system to investigate plant performance. These authors generated second generation (S_1 and BC_1) hybrids between *L. sativa* and its cross-fertile wild relative, *L. serriola*. They reported that hybrids between crop and wild *Lactuca* species are phenotypically indistinguishable from the wild relative and will remain undetected when they occur in the field. Because germination of the seed produced by the hybrids and growth of the resulting seedlings are more vigorous than in *L. serriola*, such hybrids could become invasive and result in loss of genetic identity of native species. Consequently, the vigor of hybrids, together with other factors, should be taken into consideration in any risk assessments. In other investigations from the same laboratory, Hooftman et al. (2007a) studied the infection by *B. lactucae* of *L. serriola* containing genomic segments from *L. sativa* that encode resistance to the pathogen. They concluded that when the wild species gains resistance to *Bremia* from *L. sativa*, the fitness effects are low. Consequently, the implication that increased resistance to *Bremia* in *L. serriola* may be inherited from *L. sativa* is probably incorrect. Subsequently, Hooftman et al. (2007b, 2008) performed modeling experiments to access gene flow from lettuce to wild relatives, with hybrids of *L. sativa* and *L. serriola* again being used as one of the models. Over four hybrid generations, hybrid plants were phenotypically similar to *L. serriola*; survival rates were significantly greater in the first and second generations than in subsequent generations, this being interpreted as a breakdown of heterosis. Hooftman et al. (2007b) concluded that a rapid, full displacement of *L. serriola* at low rates of outcrossing is unlikely. They proposed that a more likely scenario is the formation of a population containing a range of genotypes. The data presented by the authors are relevant not only in predicting any long-term environmental effects of sexual hybrids between *L. sativa* and wild species, but also in making decisions and monitoring the possible spread of transgenes following the release into the glasshouse and/or field of genetically manipulated lettuce and other crops. Similarly, Giannino et al. (2008b) assessed the risks of transgene escape by investigating pollen-mediated transgene flow in Italian lettuce cvs. The transgenic plants used as gene donors carried the *A. thaliana KNAT1* gene, encoding alterations in leaf morphology, and the bacterial neomycin phosphotransferase gene (*npt*II) used to select the transgenic plants. In glasshouse trials, the intravarietal outcrossing rate was 0.41 ± 0.11% when transgenic and non-transformed plants were spaced at 15 cm. In field experiments, outcrossing was 0.49, 0.071 and 0.035% when recipient plants were placed 0.5, 11 and 22 m, respectively, from an island of transgenic Luxor donor plants. Such rates of outcrossing are similar to those reported in other crops, but these rates would undoubtedly be considerably less between transgenic *L. sativa* and some wild *Lactuca* species, because of the difficulty of such interspecies hybridization, even under ideal experimental conditions.

Although gene introduction into lettuce, and into crop plants in general, has focused on nuclear transformation, considerable progress has been made in plastid transformation, including lettuce. Lelivelt et al. (2005) used polyethylene glycol (PEG)-induced DNA uptake into isolated leaf protoplasts of the *L. sativa* cv. Flora to generate fertile, homoplasmic plants. Transgenes were targeted to the *trnA–trnI* intergenic region of the plastome, with the *aadA* gene for spectinomycin resistance as a selectable marker. The first seed generation transgenic (T_1) plants were spectinomycin resistant; hybridization of transgenic plants with male-sterile wild type plants confirmed that antibiotic resistance was not transmitted through pollen. Similarly, Kanamoto et al. (2006) generated transplastomic plants of the lettuce cv. Cisco, using the same *aadA* gene for selection, this being inserted between the *rbcL* and *accD* genes of the lettuce genome. Chloroplasts of transplastomic plants expressed transgene-encoded green fluorescent protein (GFP) to 30% of the total soluble protein.

The advantages have been stressed of chloroplast transformation to engineer valuable traits and the exploitation of plastids as bioreactors, especially for the synthesis of biopharmaceuticals, vaccines and industrial enzymes (Verma et al. 2008). These authors provided detailed protocols for the construction of chloroplast expression and integration vectors, with reference to lettuce as an excellent target for transformation. The advantages of plastid transformation include the potential for reproducible, high foreign gene expression and lack of gene silencing, because of the ability to integrate genes into specific, intergenic regions of the plastome. Verma et al. (2008) also stressed the relevance of chloroplast transformation compared to nuclear transformation in limiting gene flow within crops such as lettuce, and to related wild species, since plastids are inherited maternally in most cases.

8.5 Future Prospects for Lettuce Improvement

Because of the nutritional importance of leafy vegetables (Lucarini et al. 2006), the current and expected future demand for this produce will necessitate adequate supply of existing and new cultivars. Maximizing yield, quality and durability of these crops will be the focus of both conventional breeding and genetic manipulation approaches. Undoubtedly, molecular procedures will make a significant contribution to the selection of existing germplasm of *L. sativa* and wild *Lactuca* species for incorporation into sexual and somatic cell breeding programs. While conventional breeding will continue to play a key role in the introgression of genetic material from wild *Lactuca* species into cultivated lettuce, somatic hybridization by protoplast fusion will enable complex nuclear–cytoplasmic combinations to be achieved that are not possible through sexual hybridization. This is of special relevance in the introgression of genes from wild *Lactuca* species that are sexually incompatible with *L. sativa*. Furthermore, this approach will enable gene mobilization without the requirement for sophisticated recombinant DNA technology. Undoubtedly, the most immediate application of somatic cell technologies will be through transformation because of its reproducibility, rapidity and the ability to insert and express genes in existing germplasms. Alongside breeding programs is the requirement to continue to address issues such as the rapid deterioration of lettuce following harvest, and the need to minimize health risks from uncooked vegetables (Doyle and Erickson 2007) through a reduction in microbial contamination and good horticultural practice (Davey et al. 2008).

References

Argyris J, Dahal P, Truco MJ, Ochoa O, Still DW, Michelmore RW, Bradford KJ (2008) Genetic analysis of lettuce seed thermoinhibition. In: Proceedings of 4th international symposium on seed, transplant and stand establishment of horticultural crops – translating seed and seedling physiology into technology. Acta Hortic 782:23–33

Barak JD, Gilbertson RL (2003) Genetic diversity of *Xanthomonas campestris* pv. vitians, the causal agent of leafspot of lettuce. Phytopathology 93:596–603

Beharav A, Lewinsohn D, Lebeda A, Nevo E (2006) New wild *Lactuca* genetic resources with resistance against *Bremia lactucae*. Genet Resour Crop Evol 53:467–474

Chin DB, Arroyo-Garcia R, Ochoa OE, Kesseli RV, Lavelle DO, Michelmore RW (2001) Recombination and spontaneous mutation at the major cluster of resistance genes in lettuce (*Lactuca sativa*). Genetics 157:831–849

Cho DW, Park YD, Chung KH (2005) *Agrobacterium*-mediated transformation of lettuce with terpene synthase gene. J Kor Soc Hortic Sci 46:169–175

Davey MR, Anthony P, Van Hooff P, Power JB, Lowe KC (2007) Lettuce. In: Pua EC, Davey MR (eds) Biotechnology in agriculture and forestry, vol 59, Transgenic crops IV. Springer, Berlin, Germany, pp 221–249

Davey MR, Anthony P, Power JB, Lowe KC (2008) Leafy vegetables. In: Kole C, Hall TC (eds) Compendium of transgenic crop plants, vol 6, Transgenic vegetable crops. Wiley-Blackwell, Chichester, UK, pp 217–248

de Vries IM (1997) Origin and domestication of *Lactuca sativa* L. Genet Resour Crop Evol 171:233–248

Deng XL, Chang JL, He J, He GC (2006) Transformation of lettuce with FMDV epitopes fused gene mediated by *Agrobacterium*. Wuhan Zhiweuxue Yanjiu 24:476–479

Deng XL, Dhou Y, Chang JL (2007) Establishment of genetic transformation system and transgenic studies in lettuce (*Lactuca sativa* var. capatata). Acta Bot Yunnan 29:98–102

Dhingra A, Folta KM (2005) Amplification, sequencing and annotation of plastomes. Genomics 6:176

Dias BBA, Cunha WG, Morais LS, Vianna GR, Rech EL, de Capdeville G, Aragao FJL (2006) Expression of an oxalate decarboxylase gene from *Flammulina* sp. in transgenic lettuce (*Lactuca sativa*) plants and resistance to *Sclerotinia sclerotiorum*. Plant Pathol 55:187–193

Dolezalova I, Lebeda A, Tiefenbachova P, Kristkova E (2004) Taxonomic reconsideration of some *Lactuca* spp. germplasm maintained in world bank collections. In: 4th international symposium on taxonomy of cultivated plants, Toronto, Canada, 11–17 Aug 2002, pp 193–201

Doyle MP, Erickson MC (2007) The problems with fresh produce: an overview. J Appl Microbiol 105:317–330

FAOSTAT website (2003) FAO database. http://apps.fao.org/

Frijters ACJ, Zhang Z, Van Damme M, Wang GL, Ronald PC, Michelmore RW (1997) Construction of a bacterial artificial chromosome library containing large *Eco*RI and *Hin*dIII genomic fragments of lettuce. Theor Appl Genet 94:390–399

Giannino D, Nicolodi C, Testone G, Di Giacomo MA, Iannelli MA, Frugis G, Mariotti D (2008a) Pollen-mediated transgene flow in lettuce (*Lactuca sativa* L.). Plant Breed 127:308–314

Giannino D, Nicolodi C, Testone G, Frugis G, Pace E, Santamaria P, Guardasole M, Mariotti D (2008b) The expression of asparagines synthase A from *E. coli* affects the nitrogen status in leaves of lettuce (*Lactuca sativa* L.) and enhances vegetative growth. Euphytica 162:11–22

Gibson GD, Olivas N, Salm P (2007) Lettuce breeding method. International Patent Application No PCT/US2007/014063, dated 15 June 2007

Hayes RJ, Ryder EJ (2007) Introgression of novel alleles for partial resistance to big vein disease from *Lactuca virosa* into cultivated lettuce. HortScience 42:35–39

Hayes RJ, Wintermantel WM, Nicely PA, Ryder EJ (2006) Host resistance to Mirafiori lettuce big-vein virus and lettuce big-vein associated virus and virus sequence diversity and frequency in California. Plant Dis 90:233–239

Hayes RJ, Ryder EJ, Wintermantel WM (2008) Genetic variation for big-vein symptom expression and resistance to Mirafiori lettuce big-vein virus in *Lactuca virosa* L., a wild relative of cultivated lettuce. Euphytica 164:493–500

Hooftman DAP, Oostermeijer JGB, Jacobs MMJ, den Nijs HCM (2005) Demographic vital rates determine the performance advantage of crop-wild hybrids in lettuce. J Appl Ecol 42:1086–1095

Hooftman DAP, Nieuwenhuis BPS, Posthuma KI, Oostermeijer JGB, den NIjs HJCM (2007a) Introgression potential of downy mildew resistance from lettuce to *Lactuca serriola* and its relevance for plant fitness. Basic Appl Ecol 8:135–146

Hooftman DAP, Jong MJD, Oostermeijer JGB, den Nijs HJCM (2007b) Modelling the long-term consequences of crop-wild relative hybridization: a case study using four generations of hybrids. J Appl Ecol 44:1035–1045

Hooftman DAP, Gerard J, Oostermeijer B, Marquard E, den Nijs HJCM (2008) Modelling the consequences of crop-wild relative gene flow: a sensitivity analysis of the effects of outcrossing rates and hybrid vigour breakdown in *Lactuca*. J Appl Ecol 45:1094–1103

Hu JG, Ochoa OE, Truco MJ, Vick BA (2005) Application of the TRAP technique to lettuce (*Lactuca sativa* L.) genotyping. Euphytica 144:225–235

Jansen J, Verbakel H, Peleman J, van Hintum TJL (2006) A note on the measurement of genetic diversity within genebank accessions of lettuce (*Lactuca sativa* L.) using AFLP markers. Theor Appl Genet 112:554–561

Jeuken MJW, Lindhout P (2004) The development of lettuce backcross inbred lines (BILs) for exploitation of the *Lactuca saligna* (wild lettuce) germplasm. Theor Appl Genet 109:394–401

Jeuken M, van Wijk R, Peleman J, Lindhout P (2001) An integrated interspecific AFLP map of lettuce (*Lactuca*) based on two *L. sativa* × *L. saligna* F-2 populations. Theor Appl Genet 103:638–647

Jeuken MJW, Pelgrom K, Stam P, Lindhout P (2008) Efficient QTL detection for nonhost resistance in wild lettuce: backcross inbred lines versus F-2 population. Theor Appl Genet 116:845–857

Joh LD, Wroblewski T, Ewing NN, VanderGheynst JS (2005) High-level expression of recombinant protein in lettuce. Biotechnol Bioeng 91:861–871

Kanamoto H, Yamashita A, Okumura S, Hattori M, Tomizawa K (2004) The complete genome sequence of the *Lactuca sativa* (lettuce) chloroplast. Plant Cell Physiol 45:S39

Kanamoto H, Yamashita A, Asao H, Okumura S, Takase H, Hattori M, Yokota A, Tomizawa K (2006) Efficient and stable transformation of *Lactuca sativa* L. cv. Cisco (lettuce) plastids. Transgenic Res 15:205–217

Kawazu Y, Fujiyama R, Sugiyanta K, Sasaya T (2006) Transgenic lettuce line with resistance to both lettuce big-vein associated virus and Mirafiori lettuce virus. J Am Soc Hortic Sci 131:760–763

Kesseli RV, Paran I, Michelmore RW (1994) Analysis of a detailed genetic linkage map of *Lactuca sativa* (lettuce) constructed from RFLP and RAPD markers. Genetics 136:1435–1446

Kim BK, Park SY, Jeon BY, Hwang DY, Min BW (2004) Metabolic engineering increased vitamin C levels in lettuce by overexpression of a L-gulono-gamma-lactone oxidase. J Kor Soc Hortic Sci 45:16–20

Kim YS, Kim BG, Kim TG, Kang TJ, Yang MS (2006) Expression of cholera toxin subunit in transgenic lettuce (*Lactuca sativa* L.) using *Agrobacterium*-mediated transformation system. Plant Cell Tiss Org Cult 87:203–210

Kim TG, Kim MY, Kim BG, Kang TJ, Kim YS, Jang YS, Arntzen CJ, Yang MS (2007) Synthesis and assembly of *Escherichia coli* heat-labile enterotoxin B subunit in transgenic lettuce (*Lactuca sativa*). Protein Express Purif 51:22–27

Koopman WJM, Guetta E, van de Wiel CCM, Vosman B, van den Berg RG (1998) Phylogenetic relationships among *Lactuca* (asteraceae) species and related genera based on ITS-1 DNA sequences. Am J Bot 85:1517–1530

Kristkova E, Dolezalova I, Lebeda A, Vinter V, Novotna A (2008) Description of morphological characters of lettuce (*Lactuca sativa* L.) genetic resources. HortScience 35:113–129

Kuang H, Woo SS, Meyers BC, Nevo E, Michelmore RW (2004) Multiple genetic processes result in heterogeneous rates of evolution within the major gene cluster disease resistance genes in lettuce. Plant Cell 16:2870–2894

Kuang H, Ochoa OE, Nevo E, Michelmore RW (2006) The disease resistance gene Dm3 is infrequent in natural populations of *Lactuca serriola* due to deletions and frequent gene conversions at the RGC2 locus. Plant J 47:38–48

Kuang HH, van Eck HJ, Sicard D, Michelmore R, Nevo E (2008) Evolution and genetic population structure of prickly lettuce (*Lactuca serriola*) and its RGC2 resistance gene cluster. Genetics 178:1547–1558

Lebeda A, Petrzelova I (2004) Variation and distribution of virulence phenotypes of *Bremia lactucae* in natural populations of *Lactuca serriola*. Plant Pathol 53:316–324

Lebeda A, Dolezalova I, Kristkova E, Mieslerova B (2001) Biodiversity and ecogeography of wild *Lactuca* spp. in some European countries. Genet Resour Crop Evol 48:153–164

Lebeda A, Dolezalova I, Ferakova V, Astley D (2004) Geographical distribution of wild *Lactuca* species (Asteraceae, Lactuceae). Bot Rev 70:328–356

Lebeda A, Sedlarova M, Lynn J, Pink DAC (2006) Phenotypic and histological expression of different genetic backgrounds in interactions between lettuce, wild *Lactuca* spp., *L. sativa* × *L. serriola* hybrids and *Bremia lactucae*. Eur J Plant Pathol 115:431–441

Lebeda A, Dolezalova I, Kristkova E, Dehmer KJ, Astley D, van de Weil CCM, van Treuren R (2007) Acquisition and ecological characterization of *Lactuca serriola* L. germplasm collected in the Czech Republic, Germany, The Netherlands and United Kingdom. Genet Resour Crop Evol 54:555–562

Lee ZA, Kim HY, Chung KH, Park YD (2004) Introduction of two types of human ferritin gene into lettuce plants. J Kor Soc Hortic Sci 45:330–335

Lee ZA, Cho YG, Kim SH, Park YD (2005) Development of transgenic lettuce plants with rice ferritin gene. J Kor Soc Hortic Sci 46:351–355

Lee K, Lee SM, Park SR, Jung J, Moon JK, Cheong JJ, Kim M (2007) Overexpression of *Arabidopsis* homogentisate phytyltransferase or tocopherol cyclase elevates vitamin E content by increasing gamma-tocopherol level in lettuce (*Lactuca sativa* L.). Mol Cells 24:301–306

Lelivelt CLC, McCabe MS, Newell CA, deSnoo CB, van Dun KMP, Birch-Machin I, Gray JC, Mills KHG, Nugent JM (2005) Stable plastid transformation in lettuce (*Lactuca sativa* L.). Plant Mol Biol 58:763–774

Lucarini M, Lanzi S, d'Evoli L, Aguzzi A, Lombardi-Boccia G (2006) Intake of vitamin A and carotenoids from the Italian population – results of an Italian total diet study. Int J Vit Nutr Res 76:103–109

Maisonneuve B (2003) *Lactuca virosa*, a source of disease resistance genes for lettuce breeding: results and difficulties for gene introgression. In: van Hintum ThJL, Lebeda A, Pink D, Schut JW (eds) Eucarpia leafy vegetables conference, Noordwijkerhout, Netherlands, 19–21 Mar 2003, pp 61–67

Markovic Z, Zdravkovic J, Damjanovic M, Zdravkovic M, Djordjevic R, Zecevic B (2007) Diversity of vegetable crops in Serbia and Montenegro. In: Proceedings of 3rd Balkan symposium on vegetables and potatoes. Acta Hortic 729: 53–56

Matoba H, Mizutani T, Nagano K, Hoshi Y, Uchimiya H (2007) Chromosomal study of lettuce and its allied species (*Lactuca* spp., Asteraceae) by means of karyotype analysis and fluorescence in situ hybridization. Hereditas 144:235–243

Mazier M, Botton E, Flamain F, Blouchet JP, Courtial B, Chupeau MC, Chupeau Y, Maisonneuve B, Lucas H (2007) Successful gene tagging in lettuce with the Tnt1 retrotransposon from tobacco. Plant Physiol 144:18–31

McCreight JD (2008) Potential sources of genetic resistance in *Lactuca* spp. to the lettuce aphid *Nasanovia ribisnigri* (Mosely) (Homotera: Aphididae). HortScience 43:1355–1358

Meyers BC, Shen KA, Rohani P, Gaut BS, Michelmore RW (1998) Receptor-like genes in the major resistance locus of lettuce are subject to divergent evolution. Plant Cell 10:1833–1846

Michelmore R, Wong J (2008) Classical and molecular genetics of *Bremia lactucae*, cause of lettuce downy mildew. Eur J Plant Pathol 122:19–30

Mou BQ (2005) Genetic variation of beta-carotene and lutein contents in lettuce. J Am Soc Hortic Sci 130:870–876

Navarro JA, Torok VA, Vetten HJ, Pallas V (2005a) Genetic variability in the coat protein genes of lettuce big-vein associated virus and Mirafiori lettuce big-vein virus. Arch Virol 150:681–694

Navarro JA, Botella F, Marhuenda A, Sastre P, Sanchez-Pina M, Pallas V (2005b) Identification and partial characterisation of lettuce big-vein associated virus and Mirafiori lettuce big-vein virus in common weeds found amongst Spanish lettuce crops and their role in lettuce big-vein disease transmission. Eur J Plant Pathol 113:25–34

Negrouk V, Eisner G, Lee HI, Han KP, Taylor D, Wong HC (2005) Highly efficient transient expression of functional recombinant antibodies in lettuce. Plant Sci 169:433–438

Park BJ, Liu ZC, Kanno A, Kameya T (2005) Increased tolerance to salt- and water-deficit stress in transgenic lettuce (*Lactuca sativa* L.) by constitutive expression of LEA. Plant Growth Regul 45:165–171

Plocik A, Layden J, Kesseli R (2004) Comparative analysis of NBS domain sequences of NBS-LRR disease resistance genes from sunflower, lettuce and chicory. Mol Phylogenet 31:153–163

Ratnaparkhe MB, Tekeoglu M, Muehlbauer FJ (1998) Intersimple-sequence-repeat (ISSR) polymorphisms are useful for finding markers associated with disease resistance gene clusters. Theor Appl Genet 97:515–519

Ryder EJ (1999) Lettuce, endive and chicory – crop production, Science in Horticulture Series. CABI, Wallingford, UK

Ryder EJ (2001) Current and future issues in lettuce breeding. In: Janick J (ed) Plant breeding reviews, vol 20. Wiley, San Francisco, USA, pp 105–134

Ryder EJ, Kim ZH, Waycott W (1999) Inheritance and epistasis studies of chlorophyll deficiency in lettuce. J Am Soc Hortic Sci 124:636–640

Salehi M, Izadpanah K, Nejat N, Siampour M (2007) Partial characterization of phytoplasmas associated with lettuce and wild lettuce phyllodies in Iran. Plant Pathol 56:669–676

Sedlarova M, Luhova L, Petrivalsky M, Lebeda A (2007) Localisation and metabolism of reactive oxygen species during *Bremia lactucae* pathogenesis in *Lactuca sativa* and wild *Lactuca* spp. Plant Physiol Biochem 45:607–616

Sicard D, Woo SS, Arroya-Garcia R, Ochoa O, Nguyen D, Korol A, Nevo E, Michelmore R (1999) Molecular diversity at the major cluster of disease resistance genes in cultivated and wild *Lactuca* spp. Theor Appl Genet 99:405–418

Simko I, Hu JG (2008) Population structure in cultivated lettuce and its impact on association mapping. J Am Soc Hortic Sci 133:61–68

Syed NH, Sorensen AP, Antonise R, van de Wiel C, van der Linden CG, vañt Westende W, Hooftman DAP, den Nijs HCM, Flavells AJ (2006) A detailed linkage map of lettuce based on SSAP, AFLP and NBS markers. Theor Appl Genet 112:517–527

Takada K, Watanabe S, Sano T, Ma B, Kamada H, Ezura H (2007) Heterologous expression of the mutated melon ethylene receptor gene Cm-ERS1/H70A produces stable sterility in transgenic lettuce (*Lactuca sativa*). J Plant Physiol 164:514–520

Timme RE, Kuehl JV, Boore JL, Jansen RK (2007) A comparative analysis of the *Lactuca* and *Helianthus* (Asteraceae) plastid genomes: Identification of divergent regions and categorization of shared repeats. Am J Bot 94:302–312

Truco MJ, Antonise R, Lavelle D, Ochoa O, Kozik A, Witsenboer H, Fort SB, Jeuken MJW, Kesseli RV, Lindhout P, Michelmore RW, Peleman J (2007) A high-density, integrated genetic linkage map of lettuce (*Lactuca* spp.). Theor Appl Genet 115:735–746

Van de Wiel C, Arens P, Vosman B (1999) Microsatellite retrieval in lettuce (*Lactuca sativa* L.). Genome 42:139–149

Vanjildorj E, Bae TW, Riu KZ, Kim SY, Lee HY (2005) Overexpression of *Arabidopsis* ABF3 gene enhances tolerance to drought and cold in transgenic lettuce (*Lactuca sativa*). Plant Cell Tiss Org Cult 83:41–50

Verma D, Samson NP, Koya V, Daniell H (2008) A protocol for expression of foreign genes in chloroplasts. Nat Protoc 3:739–758

Waycott W, Fort SB, Ryder EJ, Michelmore RW (1999) Mapping morphological genes relative to molecular markers in lettuce (*Lactuca sativa* L.). Heredity 82:245–251

Witsenboer H, Vogel J, Michelmore RW (1997) Identification, genetic localization, and allelic diversity of selectively amplified microsatellite polymorphic loci in lettuce and wild relatives (*Lactuca* spp.). Genome 40:923–936

Zdravkovic J, Stankovic L, Stevanovic D, Duzaman E, Tuzel Y (2003) Possibilities of using wild lettuce forms originating from spontaneous Yugoslav flora in the selection for virus diseases of *Lactuca sativa* L. In: Proceedings of international symposium on sustainable use of plant biodiversity to promote new opportunities for horticultural production development, 6–9 Nov 2001, Antalya, Turkey, pp 243–245

Zhang FZ, Wagstaff C, Rae AM, Sihota AK, Keevil CW, Rothwell SD, Clarkson GJJ, Michelmore RW, Truco MJ, Dixon MS, Taylor G (2007) QTLs for shelf life in lettuce colocate with those for leaf biophysical properties but not with those for leaf developmental traits. J Exp Bot 58:1433–1449

Chapter 9
Solanum sect. *Lycopersicon*

Silvana Grandillo, Roger Chetelat, Sandra Knapp, David Spooner, Iris Peralta, Maria Cammareri, Olga Perez, Pasquale Termolino, Pasquale Tripodi, Maria Luisa Chiusano, Maria Raffaella Ercolano, Luigi Frusciante, Luigi Monti, and Domenico Pignone

9.1 Introduction

Tomatoes belong to the large and diverse family Solanaceae, which includes more than 3,000 species, occupying a wide variety of habitats (Knapp 2002). Recent taxonomic revision of the Solanaceae has reintegrated *Lycopersicon* into the genus *Solanum* with a revised new nomenclature (Peralta and Spooner 2001; Spooner et al. 2005; Peralta et al. 2008). The majority of taxonomists as well as most plant breeders and other users have accepted the reintegration of tomatoes to *Solanum* (e.g., Caicedo and Schaal 2004; Fridman et al. 2004; Schauer et al. 2005; Mueller et al. 2009; see also http://tgrc.ucdavis.edu/key.html).

Morphological characters, phylogenetic relationships, and geographical distribution have demonstrated that tomatoes (*Solanum* sect. *Lycopersicon* (Mill.) Wettst.) and their immediate outgroups in *Solanum* sect. *Lycopersicoides* (A. Child) Peralta and sect. *Juglandifolia* (Rydb.) A. Child form a sister clade to potatoes (sect. *Petota* Dumort.), with *Solanum* sect. *Etuberosum* (Bukasov & Kameraz) A. Child being sister to potatoes + tomatoes (Spooner et al. 1993). Analyses of multiple datasets from a variety of genes unambiguously establish tomatoes to be deeply nested in *Solanum* (Bohs and Olmstead 1997, 1999; Olmstead and Palmer 1997; Olmstead et al. 1999; Bohs 2005). However, tomatoes and their close relatives can be easily distinguished from any other group of *Solanum* species on the basis of shared features such as their bright yellow flowers and pinnatifid, non-prickly leaves.

'The plant group *Solanum* sect. *Lycopersicon* consists of 13 closely related species or subspecies: the cultivated tomato, *Solanum lycopersicum* (formerly *Lycopersicon esculentum*), which includes the domesticated tomato and wild or weedy forms of the cherry tomato (*S. lycopersicum* 'cerasiforme') (Peralta et al. 2008), and the wild species *Solanum arcanum*, *S. cheesmaniae*, *S. chilense*, *S. chmielewskii*, *S. corneliomulleri*, *S. galapagense*, *S. habrochaites*, *S. huaylasense*, *S. neorickii*, *S. pennellii*, *S. peruvianum*, *S. pimpinellifolium*) (Tables 9.1 and 9.2; Peralta et al. 2005; Spooner et al. 2005). Four species have been segregated from the green-fruited species *S. peruvianum* sensu lato (s.l.); two of them *S. arcanum* and *S. huaylasense* have been described as new species (Peralta et al. 2005) from Peru, while the other two *S. peruvianum* and *S. corneliomulleri* had already been named by Linnaeus (1753) and MacBride (1962), respectively. In addition, *S. galapagense*, another yellow-to orange-fruited species, was segregated from *S. cheesmaniae*; both species are endemic to the Galápagos Islands (Darwin et al. 2003; Knapp and Darwin 2007). All members of sect. *Lycopersicon* are closely related diploid species ($2n = 24$) (Peralta and Spooner 2001; Nesbitt and Tanksley 2002) and are characterized by a high degree of genomic synteny (Chetelat and Ji 2007; Stack et al. 2009) and are to some degree intercrossable (Taylor 1986). The group *Solanum* sect. *Juglandifolia* contains the two woody tomato-like nightshades *S. ochranthum* and *S. juglandifolium*. These two species are partially sympatric and they are morphologically similar, both being woody perennials with rampant, liana-like stems up to 30 m in length (Correll 1962; Rick 1988). Based on evidence from molecular sequence data sect. *Juglandifolia* is the

S. Grandillo (✉)
CNR-IGV, Institute of Plant Genetics, Division of Portici, National Research Council, Via Università 133, 80055 Portici, Naples, Italy
e-mail: grandill@unina.it

C. Kole (ed.), *Wild Crop Relatives: Genomic and Breeding Resources, Vegetables*,
DOI 10.1007/978-3-642-20450-0_9, © Springer-Verlag Berlin Heidelberg 2011

Table 9.1 Principal ecological, botanical, and reproductive features of the wild tomatoes (*Solanum* sect. *Lycopersicon*) and related *Solanum* species

Species	Geographic distribution	Habitat	Mating system[a]	Crossability to tomato[b]	Distinguishing morphological features[c]
S. lycopersicum "cerasiforme"	Adventive worldwide in the tropics and subtropics (near sea level – 2,400 m); perhaps native in Andean region	Usually mesic sites, often feral or weedy	SC-autogamous	BC	Plants semi-erect to sprawling; fruits red, 1.5–2.5 cm
S. cheesmaniae	Endemic to Galápagos Islands (sea level – 1,500 m)	Arid, rocky slopes, prefers shaded, cooler sites	SC-autogamous	BC	Plants semi-erect to sprawling, flowers very small, pale; fruit purple, greenish-yellow, or orange, 0.5–1.5 cm
S. galapagense	Endemic to Galápagos Islands (sea level – 650 m)	Arid, rocky outcrops and slopes, sometimes near shoreline	SC-autogamous	BC	Plants erect; leaves highly subdivided; internodes short; flowers small, pale, fruit orange (0.5–1 cm)
S. pimpinellifolium	Lowland Ecuador and coastal Peru (sea level – 500 m)	Arid, sandy places, often near sources of water or on the edges of farm fields	SC-facultative	BC	Plants semi-erect to sprawling, flower small-large; fruit red (0.5–1 cm)
S. chmielewskii	Inter-Andean valleys of central and southern Peru (1,600–3,100 m)	Rather moist, well-drained, rocky slopes	SC-facultative	UI	Plant sprawling or trailing; flowers small, pale; fruit green (1–1.5 cm)
S. neorickii	Inter-Andean valleys from Cusco to central Ecuador (1,500–2,500 m)	Rather moist, well-drained, rocky slopes	SC-autogamous	UI	Plants sprawling or trailing; flowers tiny, pale; fruit green; seeds tiny
S. arcanum	Northern Peru, coastal and inter-Andean valleys, middle watershed of Marañon (500–3,000 m)	Varied, but generally dry, rocky slopes	Mostly SI, rarely SC-facultative	UI, EL	Plants erect to prostrate, reduced leaflet no.; flowers mostly straight anther tubes and undivided inflorescences; fruit whitish-green with dark stripe
S. chilense	Southern Peru, northern Chile (50–3,500 m)	Very arid and sometimes saline, rocky slopes or washes	SI	UI, EL	Plants erect; leaves finely pubescent; anthers straight; inflorescences compound; peduncles long; fruit purplish-green
S. peruvianum	Mostly coastal central/southern Peru and northern Chile (sea level – 2,500 m)	Arid, sandy, or rocky dry washes, sometimes near agricultural fields	Mosty SI, rarely SC-facultative	UI, EL	Plants procumbent; anthers bent; inflorescence simple; fruit purplish-green
S. corneliomulleri	Western Andes of central/southern Peru (1,000–3,000 m)	Rocky or sandy slopes and dry washes	SI	UI, EL	Erect to decumbent; leaves glandular pubescent; fruit purplish-green
S. huaylasense	Limited to Callejon de Huaylas, and Río Fortaleza, Peru (1,000–2,900 m)	Rocky slopes and waste places	SI	UI, EL	Spreading, anthers straight, inflorescence compound; fruit purplish-green
S. habrochaites	Northwestern and western central Peru, western and southern Ecuador (40–3,300 m)	Varied, but generally mesic slopes or stream banks	Mostly SI, some SC-facultative	UI	Sprawling shrub or vine; densely pubescent; flowers large; anthers straight; fruit green with dark stripe, hairy
S. pennellii	Coastal valleys of central to southern Peru (near sea level to 1,920 m)	Very arid, sandy or rocky slopes, or dry washes	Mostly SI, some SC-facultative	UI	Spreading shrub; 2 leaves per sympodium[d]; leaflets broad, round; foliage sticky; anthers poricidal; pedicel usually articulated at base

S. juglandifolium	Temperate rainforests of Columbia and Ecuador (1,200–3,100 m)	Mesic slopes and stream banks	SI	UI (no hybrids)	Woody vine or rampant shrub; 8–10 leaves per sympodium[d]; leaves rough, rugose; anthers orange-yellow, poricidal; flowers scented; fruit green (to 2 cm); seeds winged
S. lycopersicoides and *S. ochranthum*	Restricted to narrow range in southern Peru and northern Chile (1,200–3,700 m)	Arid rocky slopes, usually south-facing	SI	UI, EL, HS	Woody, erect to sprawling shrub; anthers white, poricidal; style hooked; flowers scented; fruit green-black (to 1 cm)
	Montane forests of Peru, Ecuador and Colombia (1,200–3,200 m)	Well-watered sites such as riverbanks	SI	UI (no hybrids obtained)	Woody vine, to 15 m height; 6–12 leaves per sympodium[d]; anthers orange-yellow, poricidal; flowers scented; fruit yellowish-green (2–5 cm); seeds winged
S. sitiens	Minor ranges around Calama, northern Chile (2,500–3,500 m)	Hyperarid, rocky slopes, or ravines	SI	UI, EL, HS	Woody, erect shrub; anthers white, poricidal; flowers scented; fruit greenish-brown, dry, and brittle when ripe

[a]SC = self-compatible; SI = Self-incompatible; autogamous = self-pollinating; allogamous = outcrossing; facultative = may self-pollinate or outcross

[b]BC = bilaterally compatible (i.e., no barrier in either direction); UI = unilateral incompatibility (crosses succeed only when cultivated tomato is used as the female parent); EL = embryo lethality (can usually be overcome by embryo culture); HS = hybrid male-sterility; no hybrids = interspecific hybrids with tomato so far not obtained

[c]Except as noted, all spp. are indeterminate, herbaceous shrubs, with 3 leaves per sympodium; flowers have the standard "*Lycopersicon*" morphology – petals yellow; anthers yellow and fused, with a sterile anther appendage, and lateral pollen dehiscence – and lack floral scent

[d]Values based on Charkes Rick's notes at the time of collection, or observations made during regeneration by the TGRC

Table 9.2 Species recognized in *Solanum* section *Lycopersicon* (tomatoes) and allied species and their distribution

Species	Distribution	Previous name in *Lycopersicon*
Solanum arcanum Peralta	Northern Peru, inter-Andean valleys and coastal	*L. peruvianum* (L.) Mill., pro parte
Solanum cheesmaniae (L. Riley) Fosberg	Galápagos Islands	*L. cheesmaniae* L. Riley
Solanum chilense (Dunal) Reiche	Coastal Chile and southern Peru	*L. chilense* Dunal
Solanum chmielewskii (C.M. Rick, Kesicki, Fobes & M. Holle) D.M. Spooner, G.J. Anderson & R.K. Jansen	Southern Peru	*L. chmielewskii* C.M. Rick, Kesicki, Fobes & M. Holle
Solanum corneliomulleri J.F. Macbr.	Southern Peru (Lima southwards), western Andean slopes	*L. peruvianum* (L.) Mill., pro parte
Solanum galapagense S. Darwin & Peralta	Galápagos Islands	*L. cheesmaniae* L. Riley var. *minor* Hook.f.
Solanum habrochaites S. Knapp & D.M. Spooner	Montane Ecuador and Peru	*L. hirsutum* Dunal
Solanum huaylasense Peralta	Callejon de Huaylas, Peru	*L. peruvianum* (L.) Mill., pro parte
Solanum juglandifolium Dunal	Andean Colombia, Ecuador and Peru	–
Solanum lycopersicoides Dunal	Southern Peru and northern Chile	–
Solanum lycopersicum L.	Globally cultivated; native distribution unknown	*L. esculentum* Mill.
Solanum neorickii D.M. Spooner, G.J. Anderson & R.K. Jansen	Ecuador to Peru, inter-Andean valleys	*L. parviflorum* C.M. Rick, Kesicki, Fobes & M. Holle
Solanum ochranthum Dunal	Andean Ecuador and Peru	–
Solanum pennellii Correll	Peru to Chile, coastal and western Andean slopes	*L. pennellii* (Correll) D'Arcy
Solanum peruvianum L.	Coastal Peru to northern Chile	*L. peruvianum* (L.) Mill., pro parte
Solanum pimpinellifolium L.	Coastal Ecuador to Chile	*L. pimpinellifolium* (L.) Mill.
Solanum sitiens I.M. Johnst.	Southern Peru and northern Chile	–

For detailed distribution data, see maps and specimens cited in Peralta et al. (2008). Previous names in the genus *Lycopersicon* are given here for ease in cross-referencing the breeding literature

sister group of *Solanum* sect. *Lycopersicon* (see below). Sister to both groups is *Solanum* sect. *Lycopersicoides* (Child) Peralta, comprising the allopatric sister taxa *S. lycopersicoides* and *S. sitiens* (also previously called *S. rickii*). These four tomato-like nightshade species have in common several morphological features that make them intermediate between tomato and potato (Rick 1988; Stommel 2001; Smith and Peralta 2002). Tomato-like morphological characters that together differentiate them from most of other *Solanum* spp. include yellow corolla, pedicels articulated above the base, pinnately segmented non-prickly leaves, and lack of tubers (Correll 1962; Rick 1988). These four allied outgroup species are diploids ($2n = 24$), however strong reproductive barriers isolate them from the core tomato group (Correll 1962; Rick 1988; Child 1990; Stommel 2001; Smith and Peralta 2002). Overall, crosses between tomato and all but two (*S. ochranthum* and *S. juglandifolium*) of these wild species are possible, although with varying degrees of difficulty (Rick 1979; Rick and Chetelat 1995; Pertuzé et al. 2003).

Peralta et al. (2008) have treated the 13 species belonging to *Solanum* sect. *Lycopersicon*, along with the four closely related species (*S. juglandifolium*, *S. lycopersicoides*, *S. ochranthum*, *S. sitiens*) in the taxonomic series *Systematic Botany Monographs*.

Tomato is an economically important vegetable crop worldwide, which is consumed either fresh or in the form of various processed products (Robertson and Labate 2007). Depending on the type of use, different breeding objectives are pursued, which include improved yield, sensory and nutritional quality, as well as adaptation to biotic and abiotic stresses. As for any other crop, tomato improvement needs to rely on sufficient genetic diversity in order to be able to satisfy current and future breeding challenges. Cultivated tomato germplasm, however, relatively little genetic variation, resulting from its inbreeding mating system associated with severe genetic bottlenecks that are postulated to have occurred prior to, during, and after the domestication process (Rick and Fobes 1975; Rick 1987). In contrast, tomato wild species possess rich genetic variation and are potential sources for the improvement of many economically important traits (Rick 1987). In fact, despite its relative small size and its recent evolutionary age – the radiation of the tomato clade has been estimated as ca. 7 Mya (Nesbitt and Tanksley 2002) – members of *Solanum* sect. *Lycopersicon*, along with taxa in the related sects. *Juglandifolia* and *Lycopersicoides*, are adapted to a wide variety of environmental conditions, which correspond to a wide range of variation in terms of morphological, physiological, mating system, and biochemical characteristics.

The reduced genetic variation of cultivated tomato can in part explain the slow rate of tomato improvement that was achieved until about 1940, when the first use of wild species as a source of desired traits was reported (Bohn and Tucker 1940). Thereafter, the exploitation of the favorable attributes hidden in tomato wild species via interspecific crosses flourished, resulting in the increased yields observed in the following decades (Rick 1988).

However, despite the wealth of genetic variation and many agriculturally important traits that can be found in the found in the potentially useful tomato wild accessions stored in gene banks, breeders have so far been unable to fully exploit this rich reservoir (Tanksley and McCouch 1997). Most commonly, wild tomato species have been used as a source for major genes for disease and insect resistances, as shown by the numerous resistance genes derived from these wild relatives, which can be found in modern varieties (Plunknett et al. 1987; Robertson and Labate 2007). In contrast, their use for the improvement of complex traits important to agriculture, including yield, quality, and tolerance to biotic and abiotic stresses, has been more limited. Several problems are, in fact, associated with the utilization of wild species, which have in many cases deterred breeders from using them. These include pre- and post-mating barriers, the presence of several undesirable loci that might be transferred along with the traits of interest, a phenomenon known as "linkage drag," the complexity and the time necessary to recover the elite genetic background while selecting for the desired characters, and a generally inferior phenotype of the wild germplasm for many of the traits that breeders would like to improve.

Over the years, the application of various molecular genetic methodologies has provided the necessary tools to overcome some of the above-mentioned limitations to the use of wild species in tomato cultivar improvement, thus accelerating their utilization. The availability of DNA markers and of derived molecular linkage maps has allowed genetic dissection of the loci underlying quantitative traits, as well as gene tagging for monogenic traits. Once markers tightly linked to a target gene or quantitative trait loci (QTL) are identified, marker-assisted selection (MAS) can be used for a more efficient and precise transfer of the gene/QTL into any selected genetic background. The negative effects of linkage drag can also be reduced, since the use of molecular markers allows for more efficient identification of recombinant plants in which close linkages are broken (Tanksley 1993). Using molecular markers, gene banks can be more rationally and efficiently sampled by taking into consideration marker-based estimates of genetic variability within and between accessions. Finally, another important contribution of QTL mapping studies conducted in tomato using interspecific crosses, as well as in other crops, has been the clear demonstration that exotic (wild) germplasm is likely to be a source of agronomically favorable QTL alleles also for traits in which the wild relatives show an inferior phenotype (deVicente and Tanksley 1993; Eshed and Zamir 1995; Tanksley et al. 1996; Tanksley and McCouch 1997; Grandillo et al. 2008). These results suggest that in the wild relatives of our crops there are numerous favorable alleles that were "left behind" by the domestication and breeding processes and that these alleles can now be more efficiently "discovered" and transferred into elite germplasm, using innovative genomic-assisted breeding strategies (Tanksley and McCouch 1997; Zamir 2001; McCouch 2004; Grandillo et al. 2008). This implies that in order to be able to fully exploit the genetic potential of our crops' wild relatives we need to change our selection approaches from phenotype based to allele based (Tanksley and McCouch 1997). In this respect, tomato has once again proven to be a model system in terms of development and application of innovative concepts and breeding approaches that can allow a more efficient and wider utilization of related wild species, and thus lead to an enrichment of the genetic base of this crop and hence to an accelerated rate of genetic improvement.

Approaches based on molecular maps and the integrative power of QTL analysis, such as the "advanced backcross QTL (AB-QTL) mapping strategy" and "exotic libraries" or introgression line (IL) libraries, have allowed the identification of favorable QTL alleles for numerous traits of agronomical interest, and the development of pre-bred lines that could be used in MAS breeding programs (Tanksley and McCouch 1997; Zamir 2001; Grandillo et al. 2008). The IL concept has proven to be ideal for map-based cloning of QTL, as demonstrated by the first cloning of a QTL (Frary et al. 2000; Fridman et al. 2000), and to explore the genetic basis of heterosis for "real-world" applications, as shown by the development of a new leading hybrid of processing tomato (Lippman et al. 2007).

The numerous genetic and "omics" tools that are available for tomato and that are being developed within the International Solanaceae Genome Project (SOL), including the information derived from the tomato genome sequence (http://solgenomics.net/solanaceae-project/), are expected to further improve the efficiency with which wild tomato relatives will contribute to the improvement of this important crop.

Given the value of wild tomato germplasm as a source of favorable alleles necessary to satisfy present and future breeding challenges, there is the need to ensure the availability of this precious resource is preserved for future generations. Therefore, conservation initiatives have to be taken not only for the excellent ex situ collections available worldwide, but also to preserve populations in situ.

9.2 Basic Botany of the Tomato

9.2.1 Agricultural Status

The cultivated tomato (*S. lycopersicum*, previously *Lycopersicon esculentum*, see Table 9.2 for the equivalent names for tomatoes in *Solanum* and *Lycopersicon*) is a popular food and an important source of vitamins and antioxidants. Botanically a fruit but treated as a vegetable, tomatoes are rich in the carotenoids lycopene and β-carotene (provitamin A), which are reported to have anticancer properties. Tomatoes are also an important source of vitamin C – ca. 10% of total dietary intake of vitamin C in the USA (Gerrior and Bente 2002) – due to their use in a wide variety of food products.

While tomato is widely cultivated as an annual vegetable crop throughout the world, its wild relatives are of relatively minor agricultural significance. Fruits of the cherry tomato, *S. lycopersicum* "cerasiforme," are probably consumed more than any other species. These small-fruited tomatoes are common in the eastern foothills of the Peruvian Andes, where they not only apparently grow wild, but are also weedy or feral around cultivated fields and are commonly consumed (Rick and Holle 1990; see also Peralta et al. 2008). The wild "currant" tomato, *S. pimpinellifolium*, is popular with some home gardeners and seeds are available commercially. In the native region, fruits are occasionally picked from wild or weedy plants, but it is not a significant commercial crop. The other wild relatives are only marginally edible and are not consumed in significant quantities. However, there are reports by indigenous people in the Andean region of various medicinal uses of leaves or fruits from wild tomatoes. For example, *S. habrochaites* is reportedly used to treat skin ailments, altitude sickness, and "gas" problems, *S. chilense* for stomach ailments, and *S. ochranthum* as a purgative or as a soap substitute (C. M. Rick and R. T. Chetelat personal communication; http://tgrc.ucdavis.edu).

9.2.2 Geographic Distribution and Ecology

The wild tomatoes (*Solanum* sect. *Lycopersicon*) and allied *Solanum* spp. (sects. *Lycopersicioides* and *Juglandifolia*) are native primarily to the Andean region of South America, principally Peru, Chile, Ecuador (including the Galápagos Islands), and Colombia. Each species has a distinct geographic distribution, often overlapping with other tomato taxa, and reflecting their specific ecological adaptations and habitat preferences (Table 9.1). The western slopes of the Andes in Peru and Chile are extremely arid, and natural populations tend to be limited to the river drainages where there is adequate moisture. Starting at the lowest elevations, *S. pimpinellifolium* and *S. peruvianum* are usually encountered first. At mid elevations, *S. peruvianum* overlaps with or is replaced by *S. corneliomulleri* (formerly part of *L. peruvianum*, see Sect. 9.2), *S. habrochaites*, or *S. pennellii*. The valleys between the Andean cordilleras in the northern part of Peru are home to

S. arcanum and *S. huaylasense* (both formerly part of *L. peruvianum*), *S. chmielewskii, S. neorickii*, and *S. ochranthum.* A similar pattern is seen in Chile and parts of southern Peru, with *S. peruvianum* most common along the coast, and *S. chilense* found at some coastal sites, but mostly at mid to high elevations, where it overlaps with *S. lycopersicoides*, the latter extending to the highest altitudes.

The cherry tomato, *S. lycopersicum* "cerasiforme", is the most widely distributed, having spread out of its original region of distribution into Mesoamerica and beyond. It is now adventive in many subtropical or tropical regions of the world, where it is commonly weedy or feral. In mainland South America, "cerasiforme" is found mostly on the wetter, eastern side of the Andean cordillera. Populations on the western side are usually associated with cultivation. In the Galápagos Islands, "cerasiforme" and the closely related *S. pimpinellifolium* probably escaped from cultivation (Rick 1956) and have in some places become more common than the two native species, *S. cheesmaniae* and *S. galapagense* (Darwin et al. 2003; Nuez et al. 2004). *S. lycopersicum* "cerasiforme" has often been referred to as "var. *cerasiforme*" in the literature, but that name has never been validly published under the rules of botanical naming and thus should not be used (see Peralta et al. 2008). Cherry tomatoes have also been shown to be complex genetic admixtures of *S. lycopersicum* and *S. pimpinellifolium* (Ranc et al. 2008), thus their true native distribution is not known.

The wild currant tomato, *S. pimpinellifolium,* is found along the Pacific coast and at low to mid elevations on the western slopes of the Andes, from southern Peru (Dept. Tacna) to Ecuador (Prov. Esmeraldas). Most populations have been collected below 1,000 m, however many of these have disappeared in the wild due to intensive agriculture and urbanization (see below). A small number (but increasing, see Darwin et al. 2003) of populations are present on the Galápagos Islands, but probably represent recent introductions (note that this does not include the native populations C. M. Rick referred to in early publications as the "*pimpinellifolium* type" – these are now considered part of *S. cheesmaniae*, see Darwin et al. 2003). Often growing as a weed in and around farm fields, *S. pimpinellifolium* has been found in cultivated areas outside the native region. Unlike "cerasiforme," *S. pimpinellifolium* appears to be adapted to the relatively arid conditions of coastal Peru (Nakazato et al. 2008).

The Galápagos endemics *S. cheesmaniae* and *S. galapagense* are each found on several of the islands, although their numbers have been reduced in recent years by goats and other grazers. The more common of the two, *S. galapagense* is found on at least eight of the main islands: Bartolomé, Fernandina, Floreana (including Corona del Diablo and Gardner islets), Isabela, Pinta, Pinzón, Rabida, Santiago, and possibly Santa Cruz. It abounds in the arid, lower life zones, often on rocky outcrops of lava. Occasional populations grow near the shoreline and are tolerant of saline conditions (Rick 1973; Rush and Epstein 1981). Populations from the littoral zone are more common during El Niño years when rainfall is more abundant at lower elevations. For example, the Tomato Genetics Resource Center's (TGRC) sole accession of *S. galapagense* from the tiny Corona del Diablo islet was collected in 1972, an El Niño year – a repeat visit in 1986, a dry year, turned up nothing (R. Bowman personal communication). Most populations of *S. galapagense* are found below 200 m elevation, but on the larger islands may extend into the forested belt up to 650 m on the slopes of the volcanoes. The closely related *S. cheesmaniae* is known from seven islands: Baltra, Fernandina, Isabela, Pinzón, San Cristóbal, Santa Cruz, and Santa Fe. Populations can be found from approximately sea level to 1,500 m, including each of the main life zones, from the littoral to the summits of the volcanoes. Where the two species overlap, *S. cheesmaniae* tends to occupy the cooler, more shady sites, and *S. galapagense* the hotter, drier locations (Rick 1956).

The sister taxa *S. chmielewskii* and *S. neorickii* are concentrated in the inter-Andean valleys of Peru and Ecuador, and no populations of either species are known from the west slopes of the Andes or east of the main cordilleras (Rick et al. 1976). Less widespread, *S. chmielewskii* is found only in southern Peru (Depts. Apurimac, Ayacucho and Cusco) and the adjacent dry Sorata valley of northern Bolivia (Peralta et al. 2008). *S. chmielewskii* overlaps in Peru with *S. neorickii*, the latter extending into southern Ecuador (Provs. Azuay and Loja). Sympatric populations are known from a number of sites (Rick et al. 1976; http://tgrc.ucdavis.edu).

Populations of *S. arcanum* are also concentrated in the inter-Andean valleys – principally the watersheds

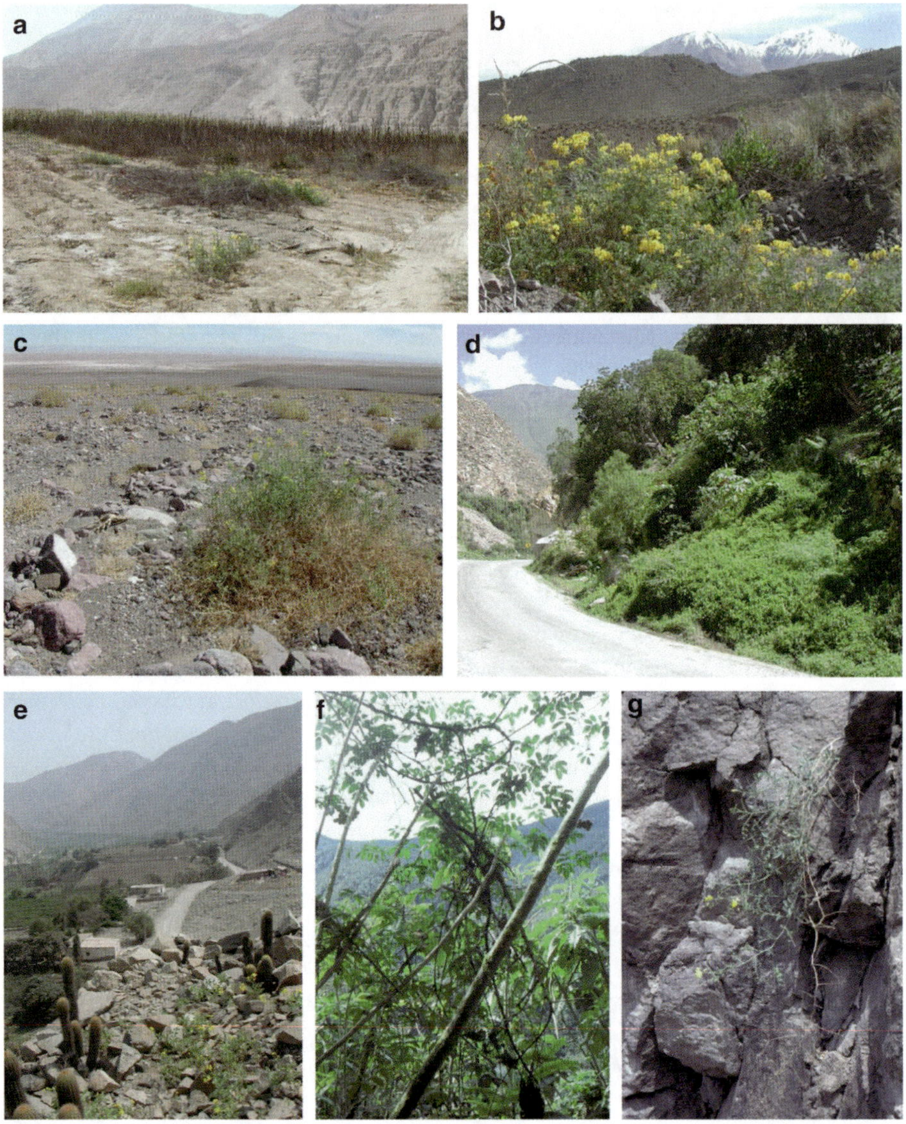

Fig. 9.1 Habitats of wild tomatoes and allied *Solanum* species growing in the native region. (**a**) *S. peruvianum* growing in an agricultural field (LA4318, Soro-Molinos, Arica and Parinacota, Chile); (**b**) *S. lycopersicoides* growing on exposed slopes at over 3,600 m (LA4323, Putre, Arica and Parinacota, Chile); (**c**) *S. chilense* growing in a dry wash (LA4334, Quebrada Sicipo, Antofagasta, Chile); (**d**) *S. habrochaites* growing in mesic site along road bank (LA2722, Puente Auco, Río Cañete, Lima, Perú); (**e**) *S. pennellii* on arid, rocky slope (LA1282, Sisacaya, Río Lurin, Lima, Perú); (**f**) *S. juglandifolium* growing in tropical forest (LA2134, Tinajillas, Zamora-Chinchipe, Ecuador); (**g**) *S. arcanum* plant scrambling down rock wall (LA2150, Puente Muyuno, Río Jequetepeque, Cajamarca, Perú). More information is available at http://tgrc.ucdavis.edu [Photos **a–c** by CM Jones, **d** and **e** by RT Chetelat, and **f** and **g** by CM Rick.]

of the Río Marañon, Río Chamaya, Río Chotano, and Río Moche – and coastal valleys, especially the Río Jequetepeque (Rick 1986c). In addition, populations of *S. arcanum* extend to the coast, at least in some years, as suggested by the many herbarium specimens collected in the "lomas" (Peralta et al. 2008). The altitudinal range for this species is thus quite broad, from below 500 m to nearly 3,000 m (Fig. 9.1g).

Populations of *S. peruvianum* are widespread in central and southern Peru, extending as far north as Dept. Cajamarca and south into the Regions of Arica/Parinacota and Tarapaca in Chile. Growing exclusively on the lower western slopes of the Andes and along the coast in lomas habitats, S. peruvianum has a narrow altitudinal range, from approximately sea level to 600 m (Peralta et al. 2008). It often grows in and around agricultural fields (Fig. 9.1a). The distribution of *S. corneliomulleri* is similar and overlapping, from central to southern Peru, but it occurs mostly at mid to high elevations on the western slopes of the Andes. The affiliated species *S. huaylasense* has a much more limited distribution, being found only in the watersheds of the Río Santa (Callejon de Huaylas region) and Río Fortaleza.

The geographic distribution of *S. chilense* extends from southern Peru (Dept. Arequipa) to northern Chile (Antofagasta Region), and from 80 to 3,600 m elevation. Its range overlaps with that of *S. peruvianum*, and the two are sympatric at several sites in Chile. In the drainages where both species are found, *S. chilense* tends to grow to higher elevations and in more arid situations, and generally avoids disturbed sites (Fig. 9.1c). A small number of marginal *S. chilense* populations have been collected as far north as Dept. Ica in Peru (Rick 1990) and are unusual in being polyploid (see below). At the other end of the distribution, the populations around Taltal, Chile, are the southernmost and are morphologically distinctive in several respects (Chetelat et al. 2009); leaves are exceptionally hairy and highly subdivided, and inflorescences are relatively short. Among the populations from coastal Chile, only the Taltal material grows to below 100 m elevation, a trend attributed to more abundant precipitation there than at sites to the north. The easternmost group of populations, located in the drainages to the east of the Salar de Atacama, is also recognizable morphologically from the rest of the species; leaves are glossy (nearly glabrous) green, with broad segments. The Atacama populations grow at higher elevations (up to 3,600 m) and at greater distance from the equator than of any other member of sect. *Lycopersicon* (exceeded only by *S. lycopersicoides*), and thus are a potential source of tolerance to low temperatures. Other abiotic stresses, to which *S. chilense* appears well adapted on the basis of its geographic distribution, include extreme aridity and soil salinity (Chetelat et al. 2009).

The geographic range of *S. habrochaites* extends from southern Ecuador (Prov. Manabi) to southern Peru (Dept. Ayacucho), and from 40 to 3,300 m elevation. In Peru, populations are found mostly at mid to high elevations in the river drainages, generally in less arid situations (Fig. 9.1d) and at higher elevations than *S. peruvianum,* with which it overlaps. In Ecuador, *S. habrochaites* is more broadly distributed (i.e., less restricted to river valleys), and some populations are morphologically distinctive (formerly recognized as *L. hirsutum* f. *glabratum*), with more slender stems, less upright growth, nearly glabrous leaves, and higher levels of anthocyanins compared to the more typical Peruvian material.

Populations of *S. pennellii* are found at relatively low elevations (10–1,940 m) along the coast, in Peru (Depts. Piura to Arequipa), and with a few collections known from northern Chile. This species is found on arid slopes and dry washes (Fig. 9.1e). The extreme drought tolerance of *S. pennellii* has been attributed to several factors: a tighter control of transpiration, increased water use efficiency (WUE), and tolerance of soil salinity (Yu 1972; Mittova et al. 2004; Xu et al. 2008). Populations from the northern margins (Bayovar and El Horador sites) are distinguishable from the rest of the species by their pedicel articulation, which is in the mid, instead of the basal, position. The populations from the vicinity of Nazca (Dept. Ica) differ from the rest of the species by their near absence of hairs on stems and leaves, relatively small leaflets with smooth (entire) margins, and more diminutive stature; on the basis of these traits they were recognized formerly as a subspecies (*L. pennellii* var. *puberulum*).

The sister taxa *S. juglandifolium* and *S. ochranthum* (comprising *Solanum* sect. *Juglandifolia*) are found at mid to high elevations in the valleys between the major cordilleras of the Andes. The natural range of *S. juglandifolium* is from northeastern Columbia (Dept. Santander) to southern Ecuador (Prov. Zamora-Chinchipe), and from ca. 1,200 to 3,100 m elevation (Rick 1988; Peralta et al. 2008). The large number of herbarium specimens collected for this species contrasts with the relatively few ex situ seed collections available – at the TGRC, eight accessions total, only one of which is from Colombia (http://tgrc.ucdavis.edu). Occupying a larger geographic range, *S. ochranthum* can be found from central Colombia to southern Peru (Dept. Apurimac). Its altitudinal range is relatively broad: 1,900–4,100 m, however most populations are in the 2,000–3,200 m range (Smith and Peralta 2002; Peralta et al. 2008). Where these two species occur in the same region, *S. ochranthum* is generally found at higher elevations than *S. juglandifolium* (Smith and Peralta 2002). The two species grow as rampant bushes or climbing lianas, with stems up to 30 m in length in the case of *S. ochranthum* (Rick 1988). Both prefer relatively mesic sites such as stream beds or tropical forest (Fig. 9.1f). The two have a similar morphology, but *S. juglandifolium* is generally more diminutive in the size of its plant parts, especially leaves, stems, and fruit; leaflets are also fewer in number, though broader in dimensions, and have a rough, scabrous surface texture compared to the softer feel of *S. ochranthum*.

The last two species to be considered herein, *S. lycopersicoides* and *S. sitiens*, form another pair of sister taxa (sect. *Lycopersicoides*). Both have narrow geographic ranges. Growing in no more than six river drainages, *S. lycopersicoides* is confined to deep canyons and slopes around the Chile/Peru frontier. While its altitudinal range is relatively broad (from 1,200 to 3,700 m), it tends to be more common at the higher elevations. This species overlaps with *S. chilense* and *S. peruvianum*, but grows higher – the highest of any tomato species, a likely indicator of low temperature adaptation – and more often on the cooler, less arid south-facing sides of the valleys (Fig. 9.1b). Endemic to Chile, *S. sitiens* grows only within a small part of the Atacama desert, on slopes to the northwest and south of Calama, and in a relatively narrow altitudinal belt of ca. 2,400–3,500 m. Often growing on exposed slopes, or in broad dry washes, *S. sitiens* occupies the most arid sites of any of the wild tomatoes (Chetelat et al. 2009). At many locations, it is the only perennial plant that can survive. Soil tests also point to an ability to tolerate high levels of salinity.

9.2.3 Geographic Distribution of Diversity

Genetic diversity within and between wild tomato populations is often structured in relation to their geographic distribution. Populations may be physically isolated from one another (fragmented) in their native distributions, with gene flow within a species restricted by distance and/or major geographic barriers such as deserts or mountain ranges. In addition, processes of adaptation to local conditions and genetic drift contribute to differentiation of populations. Thus, populations from one part of the geographic distribution – north to south, one river drainage to the next, low to high elevation, etc. – tend to be genetically differentiated from other populations.

The first detailed studies of natural variation in the wild tomatoes were those carried out by C. M. Rick and colleagues in the 1970s and 1980s. Using allozyme markers, they showed that diversity within populations of *S. pimpinellifolium* and *S. habrochaites* is highest in the geographic centers of their respective distributions, and that on the northern and southern margins, genetic variation tends to be depleted. For both species, the centers of highest diversity are in northern Peru. On the geographic margins, populations display changes in flower morphology and/or incompatibility systems that promote inbreeding over outcrossing. For example, the "central" populations of *S. pimpinellifolium* typically have relatively large flowers, with long anthers and exserted stigmas, all traits that in entomophilous flowers tend to increase the rate of outcrossing, and thus maintain diversity (Rick et al. 1977, 1978). The "southern" and "northern" populations on the other hand have relatively small flowers and stigmas that are only slightly exposed to visiting insects. A similar trend is seen in *S. habrochaites*; large flowers, exhibiting self-incompatibility (SI) – and thus strictly allogamous – in the center of the distribution, and smaller, self-compatible (SC) flowers on the margins of the distribution (Rick et al. 1979). Furthermore the northern and southern elements are morphologically distinctive from one another (see above) and show clear genetic differentiation. Crosses between the northern and southern SC races demonstrated that the loss of SI appears to have occurred via independent mutations in each group (Rick and Chetelat 1995).

Similar trends, though less pronounced, of north–south differentiation are seen in some of the other wild tomato species. For example, accessions of *S. pennellii* from the central region show the highest diversity and are strictly allogamous (SI). Self-compatibility (SC) occurs among accessions on the southern margin (Río Atico and Río Majes drainages), which tend to be highly inbred (Rick and Tanksley 1981). One of these, LA0716 from Puerto Atico, Peru, has been widely used for genetic studies in part because it is highly homozygous and polymorphic relative to the cultivated tomato, with which it can be easily hybridized. Accessions from the northern limits of the distribution, while retaining SI, are morphologically distinctive (see above). In the *S. peruvianum* complex, a highly diverse group recently subdivided to recognize three new species (Peralta et al. 2008), the vast majority of accessions are SI. Rare SC populations are found at or near the southern (LA4125, Río Camiña) and northern (LA2157, Río Chota) limits of the distribution (Rick 1986c; Graham et al. 2003). All populations of *S. chilense, S. lycopersicoides,* and *S. sitiens* are SI, yet in each case the marginal populations show evidence of genetic differentiation from populations in the center of their respective geographic distributions. Two accessions collected at/near the northern margins of *S. lycopersicoides* and *S. sitiens* (LA2387 and LA4114, respectively) are morphologically distinctive; they are the only accessions of either species that exhibit yellow anthers, white or cream colored anthers being the norm for both taxa (Chetelat et al. 2009). Studies of genetic relationships between populations also reveal a strong geographic structure, with northern, central, and southern elements identifiable in both species (Albrecht et al. 2010).

Within *S. chilense,* four geographic races can be readily distinguished morphologically: a northern ("Acari" race), central, southwestern ("Taltal" race), and southeastern ("Atacama") (Graham 2005). Each of the groups is geographically isolated from the others. Experimental hybridization between the northern, central, and southwestern assemblages results in reduced seed set, indicating partial reproductive barriers are developing in this species.

Genetic diversity within the cherry tomato, *S. lycopersicum* "cerasiforme," follows a different geographic pattern. Early studies with allozyme markers indicated that within the Andean region, the greatest diversity is found in the San Martín and Ayacucho areas (Rick and Holle 1990). Diversity decreased to the north and south. High levels of diversity within the Tarapoto (Dept. San Martín) region suggested the possibility of hybridization and introgression with cultivars. Subsequent studies with DNA-based markers provided further evidence for hybridization with introduced cultivars (Williams and St. Clair 1993), and supported the suggestion that the Andean region was the primary center of diversity for "cerasiforme" (Villand et al. 1998).

The preceding information related to geographic trends derives primarily from herbarium records and notes from plant collectors. These records provide a historical view of each species broadest natural range, and thus do not reflect recent changes, notably those caused by human influences. Wild tomatoes are threatened in their native area by a variety of anthropogenic factors, including loss of habitat, agricultural development, overgrazing, mining, and other aspects of urban expansion and economic development. In the coastal river valleys of Peru, modern agricultural practices appear to have contributed to the loss of many populations known from earlier collections. Wild tomatoes have largely disappeared from the lower stretches of river valleys and around cities. On the Galápagos Islands, the endemic tomato species have become rare – goats are a likely culprit – while non-native cherry and currant tomatoes are now common (Darwin et al. 2003; Nuez et al. 2004). Similar changes are occurring throughout the Andean region. Many populations known from herbarium specimens or genebank collections no longer exist in situ.

9.2.4 Morphology

Only a summary of morphological characters need be presented here, as a detailed description is available elsewhere (Peralta et al. 2008). The wild tomatoes and affiliated *Solanum* species have in common several basic morphological characteristics. Most grow as short-lived, herbaceous perennials in the native environment. It is common to find evidence of several years of growth. The base of plants often becomes woody, and some species appear to be capable of generating new shoots at or below the soil level. Most noteworthy in this regard is *S. sitiens*, plants of which are sometimes comprised mostly of dead branches, with only a few green shoots emerging from the crown (Chetelat et al. 2009).

Shoot growth is normally indeterminate, with each branch consisting of a repeating sequence of two or more leaves and an inflorescence, which together comprise a sympodium. At the base of each leaf, an axillary shoot is normally present. Growth of each sympodium terminates with the inflorescence, the next sympodium being produced by outgrowth of what would otherwise be an axillary meristem. The number of leaves between successive inflorescences – the sympodial index – is generally constant, once flowering begins in earnest. The sympodial index is

2–3 in species of sect. *Lycopersicon*; in the remaining species the alternation of leaves and flowers is less regular, and all tend to produce more leaves and fewer flowers. Plants of *S. ochranthum* and *S. juglandifolium* produce many leaves between inflorescences Peralta et al. (2008).

Plant habit also varies significantly among the species. A sprawling, decumbent growth habit is the most common (e.g., *S. pimpinellifolium, S. peruvianum,* and others). A more bushy, erect form of growth is seen in *S. lycopersicoides, S. sitiens, S. chilense,* and *S. galapagense*. A climbing vine-like growth habit is exhibited by *S. ochranthum* and *S. juglandifolium*; individual shoots of the former species can grow to 15 m or more, often clambering into or over trees and shrubs (Rick 1988).

Leaves are pinnately compound, with the number, size, shape, and relative dimensions of leaflets varying considerably between and within species. Leaflets may be further subdivided into secondary leaflets. Leaflets are connected, via petiolules, to the leaf rachis generally in pairs of primary lateral leaflets, with smaller interstitial leaflets in between. A petiole connects each leaf to the stem. Stipules or pseudostipular leaves are present at the base of the petiole in some species. Leaf surfaces are densely pubescent with several types of unbranched trichomes – unicellular, multicellular, and glandular – the density and types of hairs varying between and within species. Both *S. habrochaites* and *S. pennellii* are densely pubescent, yet each includes populations – previously recognized taxonomically as *L. hirsutum* f. *glabratum* and *L. pennellii* var. *puberulum* – that are much less hairy or nearly glabrous. Leaves of *S. juglandifolium* are rough textured and scabrous, with a prominent network of veins.

Flowers are born on cymose inflorescences, which may be simple (single cyme) or compound (more than one cyme), in the latter case with a variable number of dichotomous branch points. In some species, floral bracts are present at the base of the inflorescence and sometimes at each branch point within the inflorescence. Branched inflorescences are seen in *S. chilense, S. habrochaites, S. huaylasense, S. pennellii, S. peruvianum*, sect. *Juglandifolia,* and sect. *Lycopersicoides.* The other species more commonly produce unbranched inflorescences. Flowers are attached to the inflorescence by a pedicel that is articulated (i.e., position of the abscission zone) more or less midway between flower and inflorescence. Pedicel articulation in *S. pennellii* is strongly basal on the inflorescence, although some populations from the northern margin of its distribution are articulated in the middle. In the species of sect. *Lycopersicoides* and some *S. habrochaites* accessions, the pedicel joint is closer to the flower than the inflorescence rhachis.

The flower structure of the wild tomatoes resembles the typical *Solanum* flower in many respects. Flowers are composed of four whorls of organs: carpels, stamens, petals, and sepals. The innermost whorl normally consists of two carpels (the number may vary) fused together to form the pistil consisting of ovary, style, and stigma. The remaining whorls are generally five-parted, though this number also varies. The stamen whorl consists of stamens, which in *Solanum* sect. *Lycopersicon* are generally attached via interlocking hairs. Pollen is released through ovoid pores that quickly lengthen to longitudinal slits in the anthers. The tips of anthers are sterile (i.e., contain no pollen) except in *S. pennellii*. Anthers are various shades of yellow, and mostly straight or recurved downwards, as in *S. peruvianum* and *S. pennellii*. Style length and morphology vary considerably. In the outcrossing SI species, styles are longer that the anthers and stigmas are exserted several millimeter beyond the end of the anther cone. In the SC inbreeding species, stigmas are flush with the anther cone or slightly exserted. Styles are essentially straight in most species, but in *S. pennellii, S. lycopersicoides,* and *S. sitiens* styles are prominently bent or recurved where they protrude past the anthers. Petals and sepals are each fused to form a radially symmetrical (regular) corolla and calyx. No noticeable scent or nectar is produced.

A striking exception to this typical "Lycopersicon" flower structure is presented by *S. pennellii,* wherein anthers lack the sterile appendage and pollen grains are shed via terminal anther pores. Flowers are slightly irregular (zygomorphic), with the upper corolla segments being enlarged relative to the lower ones. Flowers of sect. *Lycopersicoides* and sect. *Juglandifolia* show additional structural differences. Pollen is shed via terminal pores which extend laterally. Anthers in sect. *Lycopersicoides* are white or cream colored, with occasional yellow variants in some populations. In sect. *Juglandifolia,* anthers are orange-yellow. Flowers of all four species are noticeably scented, the odor varying from species to species.

9.2.5 Cytology and Karyotype

The species considered herein are virtually all diploids, with $2n = 2x = 24$ chromosomes, like most other *Solanum* spp. The only reported exceptions are two cases of naturally occurring tetraploidy, both in *S. chilense* (Rick 1990). These appear to be marginal populations; one is from the northernmost locality for this species (LA1917, Llauta, Río Palpa, Dept. Ica, Peru), and is relatively infertile. Polyploidy is thus uncommon in the wild tomatoes.

Eleven of the 12 chromosome pairs are submetacentric. Chromosome 2 (the chromosomes are numbered 1–12 from longest to shortest at pachytene) is acrocentric, containing only a very short and heterochromatic short arm, which contains the nucleolus organizing region (NOR). At the pachytene stage of meiosis, each of the 12 chromosomes can be identified by the position of the centromere, the relative lengths of long and short arms, and the lengths of heterochromatic and euchromatic regions (Khush 1963; Sherman and Stack 1995).

The classical studies of chromosome morphology, based on light microscopy, revealed relatively little structural variation among the wild species. For example, hybrids between cultivated tomato and *S. pennellii*, two of the most distantly related species in sect. *Lycopersicon*, showed relatively few differences in chromosome structure at pachytene by light microscopy, and these were limited to the number and positions of heterochromatic knobs on certain chromosomes (Khush and Rick 1963). Other interspecific hybrids within the tomato clade gave a similar impression of overall colinearity in the early cytological work (reviewed by Chetelat and Ji 2007). However, this view is beginning to change, as new evidence of rearrangements and structural differences has emerged from higher resolution genetic and physical maps, and from improved cytological methods.

Comparative genetic linkage maps of the *S. lycopersicoides* and *S. sitiens* genomes show they differ from tomato by a paracentric inversion of the long arm of chromosome 10 (Pertuzé et al. 2002). This finding is consistent with the occasional inversion loops seen in *S. lycopersicum* × *S. lycopersicoides* hybrids (Menzel 1962), and the strongly suppressed recombination seen in this region (Chetelat et al. 2000). Surprisingly, this inversion was not seen with bacterial artificial chromosome-fluorescence in situ hybridization (BAC-FISH) (Szinay 2010). Assuming it is real, the 10L inversion must have occurred within the lineage leading to tomato, since the ancestral arrangement is found in all other Solanaceae examined to date (Livingstone et al. 1999; Doganlar et al. 2002a), but prior to divergence of the tomato species since their genomes appear to be collinear in this region. Interestingly, both *S. ochranthum* and *S. juglandifolium* have the inverted (i.e., tomato) orientation of chromosome 10L, suggesting they are more closely related to the tomatoes (Albrecht and Chetelat 2009) than are members of sect. *Lycopersicoides*. This interpretation is consistent with recent molecular phylogenies (Peralta et al. 2008, and see below), but contrasts with the evidence from crossing relationships, which point instead to sect. *Lycopersicoides* as being more tomato-like.

Short, proximal inversions were detected on chromosome 6S in *S. peruvianum* (Seah et al. 2004), chromosome 7S in *S. pennellii* (van der Knaap et al. 2004), and chromosome 12S in *S. chilense* (Szinay 2010), relative to cultivated tomato. A reciprocal whole arm translocation involving chromosomes 8 and 12 occurred in either *S. ochranthum* or *S. juglandifolium* (Albrecht and Chetelat 2009).

By studying the synaptonemal complexes of several interspecific tomato hybrids using electron microscopy, Stack et al. (2009) revealed a series of chromosome rearrangements, including inversions, translocations, length differences, and mismatched kinetochores. The number of structural rearrangements was generally consistent with phylogenetic expectations; *S. lycopersicum* × *S. pimpinellifolium* hybrids showed fewer structural changes than *S. lycopersicum* × *S. pennellii* hybrids, for instance. However, the *S. chmielewskii* hybrid revealed a greater than expected number of changes. Despite these examples of genome changes, overall gene order amongst the wild tomatoes and related *Solanum* is highly conserved, a fact that in large part explains their great practical usefulness.

9.2.6 Genome Size and Composition

Genome sizes have not been determined for all of the wild tomato species, but the data available are sufficient to indicate considerable variation. Estimates of the DNA content for the cultivated tomato, *S. lycopersicum,* vary from 1.88 to 2.07 pg/2C for a sample of six cultivars and 1.83 pg/2C for the closely related *S. cheesmaniae* (Arumuganathan and Earle 1991). The basal taxon in the core tomato clade, *S. pennellii*, has a larger genome size (2.47–2.77 pg/2C), while *S. peruvianum* is intermediate (2.27 pg/2C) (Arumuganathan and Earle 1991). Two other species, *S. habrochaites* and *S. pimpinellifolium*, have slightly smaller genomes (1.85 and 1.77 pg/2C, respectively) (Bennett and Smith 1976). The genome sizes of the sect. *Juglandifolia* species (1.75–1.96 pg/2C) are similar to the more compact tomato genomes, whereas those of the sect. *Lycopersicoides* group (2.43–2.69 pg/2C) are about 25% larger (Chetelat 2009).

In map units, the genome size of the tomatoes is approximately 1,200–1,400 centiMorgans (cM) (Tanksley et al. 1992; Frary et al. 2005). These values are based on recombination in F_1 interspecific *S. lycopersicum* × *S. pennellii* hybrids, and thus might be biased by sequence divergence or selection. Recombination rates in intraspecific maps appear to be similar, but a little lower (the lower marker polymorphism rate may be a contributing factor); a map for *S. peruvianum* contained 1,073 cM (van Ooijen et al. 1994), and one for *S. lycopersicum* only 965 cM (Saliba-Colombani et al. 2000).

The tomato genome is comprised of approximately 75% heterochromatin, most of which is located in the pericentromeric regions (Peterson et al. 1996). The remaining 25% of the genome is euchromatin and located in segments distal to the pericentromeric heterochromatin on each chromosome arm. The majority of expressed genes are thought to be located in the euchromatin fraction, an inference supported by several lines of evidence. Mapping of induced deletions to pachytene chromosomes showed that most mutant loci are in euchromatin (Khush and Rick 1968). Sequencing of BACs found a much higher gene density per unit DNA length in inserts from euchromatin than heterochromatin (van der Hoeven et al. 2002; Wang et al. 2006). Finally, recombination is generally higher in gene rich regions, whereas tomato heterochromatin is recombinationally inert. Mapping of recombination nodules on synaptonemal complexes showed that the pericentromeric heterochromatin portion of each chromosome is nearly devoid of crossovers (Sherman and Stack 1995), a result consistent with genetic evidence of crossover suppression around centromeres (Tanksley et al. 1992).

9.3 Evolutionary Relationships of *Solanum* Section *Lycopersicon* (Tomatoes) and Allied Species

9.3.1 The Generic Position of Tomatoes and Wild Relatives

Wild tomatoes (sensu stricto) traditionally were treated as members of the genus *Lycopersicon* Mill., mainly based on the anther morphology (D'Arcy 1972; Hunziker 2001). In the past decade, several molecular phylogenetic studies of the Solanaceae have unambiguously showed tomatoes to be deeply nested within *Solanum* (Spooner et al. 1993, 2005; Bohs and Olmstead 1997, 1999; Olmstead and Palmer 1997; Olmstead et al. 1999; Peralta and Spooner 2001; Bohs 2005). Data from chloroplast DNA (cpDNA) sequences strongly support a monophyletic *Solanum* (Bohs 2005; Weese and Bohs 2007) with the inclusion of all traditional segregate genera; *Cyphomandra* Mart. ex Sendtn. (Bohs 1995), *Lycopersicon* Mill. (Spooner et al. 1993), *Normania* Lowe, and *Triguera* Cav. (Bohs and Olmstead 2001). Some workers (e.g., Hunziker 2001) continue to maintain these taxa as distinct genera. The monophyletic *Solanum* is one of the ten most species-rich genera of angiosperms (Frodin 2004; see also Solanaceae Source, http://www.solanaceae-source.org), and contains several crops of economic importance such as the tomato (*S. lycopersicum*), the potato (*S. tuberosum* L.) and the aubergine or eggplant (*S. melongena* L.), as well as other minor crops (naranjilla, *S. quitoense* Lam.; tamarillo, *S. betaceum* Cav. and pepino, *S. muricatum* Aiton).

The tomatoes and their close relatives are easily distinguished from any other group of *Solanum* species by their bright yellow flowers and pinnatifid, non-spiny leaves; the only other species in the genus with yellow flowers is *S. rostratum* Dunal, a member of sect. *Androceras* (Nutt.) Whalen (1979). The tomatoes are most closely related to the potatoes and form a distinct clade (the Potato clade, sensu Bohs 2005; Weese and Bohs 2007) with relatively high (80%)

bootstrap support (Bohs 2005). Peralta et al. (2008) presented a phylogenetic classification of the group that simply states the hypothesis that tomatoes have more "predictivity" under *Solanum*; they also apply a Linnaean nomenclatural system (hierarchical) to provide the valid names of wild species under *Solanum*. Here we provide a short discussion on the history of generic classification of the tomatoes and their wild relatives in sects. *Lycopersicoides* and *Juglandifolia*, and discuss in detail both traditional taxonomic schemes for species-level relationships and modern statistically based studies of these relationships.

9.3.2 History of the Generic Classification of Tomatoes and Wild Relatives

In his first edition of *The Gardener's Dictionary* (Miller 1731) Philip Miller, the English botanist and curator of the Chelsea Physic Garden, used the generic name *Lycopersicon* and included a number of taxa with multilocular fruits ("roundish, soft, fleshy Fruit, which is divided into several Cells, wherein are contained many flat Seeds"), all color variants of the cultivated tomato (*S. lycopersicum*). In this same work, he also recognized *Solanum*, and included within it the eggplant as "*Solanum Americanum, spinosum, foliis Melongenae, fructu mammoro*" and the potato as "*Solanum tuberosum, esculentum*" (Miller 1731). His definition of *Lycopersicon* was confined to plants we would today recognize as cultivars of *S. lycopersicum*, the cultivated tomato.

In *Species Plantarum*, Linnaeus (1753) classified tomatoes in the genus *Solanum*, and described *S. lycopersicum* and *S. peruvianum*. Jussieu (1789), in his classification, also included tomatoes in *Solanum*. Miller (1754), however, continued to use both the generic name *Lycopersicon* and polynomial nomenclature in the abridged 4th edition of *The Gardener's Dictionary*. He expanded his definition of *Lycopersicon* by including "*Lycopersicon radice tuberose, esculentum*" (the potato) within it, using the following reasoning (Miller 1754): "This Plant was always ranged in the Genus of *Solanum*, or Nightshade, and is now brought under that Title by *Dr. Linnaeus*; but as *Lycopersicon* has now been established as a distinct Genus, on account of the Fruit being divided into several Cells, by intermediate Partitions, and as the Fruit of this Plant [the potato] exactly agrees with the Characters of the other species of this Genus, I have inserted it here." The editor of the posthumously published edition of *The Gardener's and Botanist's Dictionary* (Miller 1807), Thomas Martyn, merged *Lycopersicon* and *Solanum*, and recognized all Miller's species as members of *Solanum*. A number of classical and modern authors have recognized the genus *Lycopersicon* (e.g., Dunal 1813, 1852; Bentham and Hooker 1873; Müller 1940; Luckwill 1943; Correll 1958; D'Arcy 1972, 1987, 1991; Hunziker 1979, 2001; Rick 1979, 1988; Child 1990; Rick et al. 1990; Symon 1981, 1985; Taylor 1986; Warnock 1988; Hawkes 1990), but others continued to recognize the tomatoes as members of the genus *Solanum* (MacBride 1962; Seithe 1962; Heine 1976; Fosberg 1987).

9.3.3 Relationships of the Species of Tomatoes and Their Wild Relatives

The species of tomatoes have been treated quite differently by different authors, both in terms of species identity (current species recognized in the group and their distributions are presented in Table 9.2) and in terms of group membership and relationships. Figure 9.2 shows the chronology of the differing classifications through the twentieth century and compares them to the classification of Peralta et al. (2008) that is used here.

Müller (1940) and Luckwill (1943) produced the two most complete taxonomic treatments of wild tomatoes based on morphological concepts, and treated them under *Lycopersicon*. Müller (1940) divided *Lycopersicon* into two subgenera: subg. *Eulycopersicon* possessing glabrous, and red- to orange- to yellow-colored fruits, flat, obovate, and silky pubescent seeds, ebracteate inflorescences, and leaves without pseudostipules; subg. *Eríopersicon* with pubescent or hirsute green or greenish white to yellowish and purple-tinged fruits, frequently with a dark green, lavender, or purple stripe, thick, oblanceolate glabrous (pilose only at the apex) seeds, bracteate inflorescences, and leaves usually with pseudostipules. Luckwill (1943) hypothesized that the two subgenera might

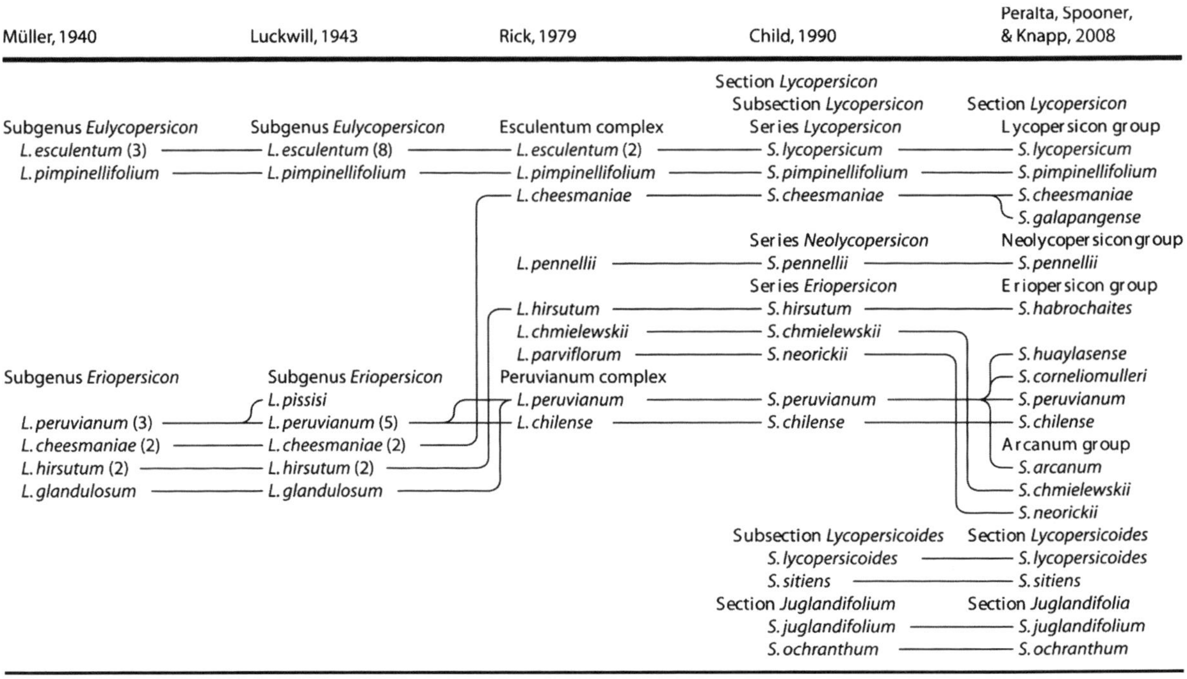

Fig. 9.2 Chronological flow chart of hypotheses of species boundaries and relationships of *Solanum* section *Lycopersicon*, section *Juglandifolia*, and section *Lycopersicoides* recognized by Müller (1940), Luckwill (1943), Child (1990), and Peralta et al. (2008). The *numbers* in *parentheses* represent the number of infraspecific taxa recognized by these authors. Modified and reproduced with permission from Syst Bot Monogr 84: 13, Fig 5 (2008)

have evolved from a simple ancestral form characterized by imparipinnate leaves with 5–7 entire leaflets, few interjected leaflets, probably no secondary leaflets, unbranched inflorescences, and undeveloped pseudostipules. He suggested that two lineages diverged from this ancestral form, one characterized by fruits with carotenoid pigments and the other by green fruits with anthocyanin pigments.

Rick (1979) recognized two "complexes" based on crossing relationships, the "Esculentum complex" and "Pervianum complex" (see Fig. 9.2). Rick (1986a) hypothesized that the races of his "*L. peruvianum*" found in the Río Marañón drainage in northern Peru were ancestral to all other wild tomatoes (*Solanum* sect. *Lycopersicon* as defined here), and that speciation and differentiation took place with migration to the south. Rick (1963) suggested that this distribution pattern pointed to a single origin of his broadly defined "*L. peruvianum*" with subsequent spread before or during the uplift of the central Andes.

Recent cladistic and phenetic studies of species boundaries and relationships within the tomatoes and wild relatives have used a combination of molecular and morphological data. Figure 9.3 shows abstracted summary trees based on cpDNA restriction sites (Palmer and Zamir 1982; Fig. 9.3a; Spooner et al. 1993; Fig. 9.3d), mitochondrial DNA (mtDNA) restriction sites (McClean and Hanson 1986; Fig. 9.3b), nuclear restriction fragment length polymorphisms (RFLPs) (Miller and Tanksley 1990; Fig. 9.3c), isozymes (Bretó et al. 1993; Fig. 9.3e), internal transcribed spacer (ITS) region of nuclear ribosomal DNA gene sequences (Marshall et al. 2001; Fig. 9.3f), nuclear DNA microsatellites (Alvarez et al. 2001; Fig. 9.3g), and morphology-based cladistics (Peralta and Spooner 2005; Fig. 9.3h). These phenetic and cladistic studies detailed below used a variety of statistical techniques and programs, the reader is referred to the primary literature for further details of specific algorithms used and parameters set.

The name *S. peruvianum* is used in three ways in the discussion of species relationships here. Firstly, *S. peruvianum* s.l. refers to the broadly circumscribed species complex prior to recognition of four species within it (Peralta et al. 2005). Second, *S. peruvianum* "north" and "south" refers to the geographic

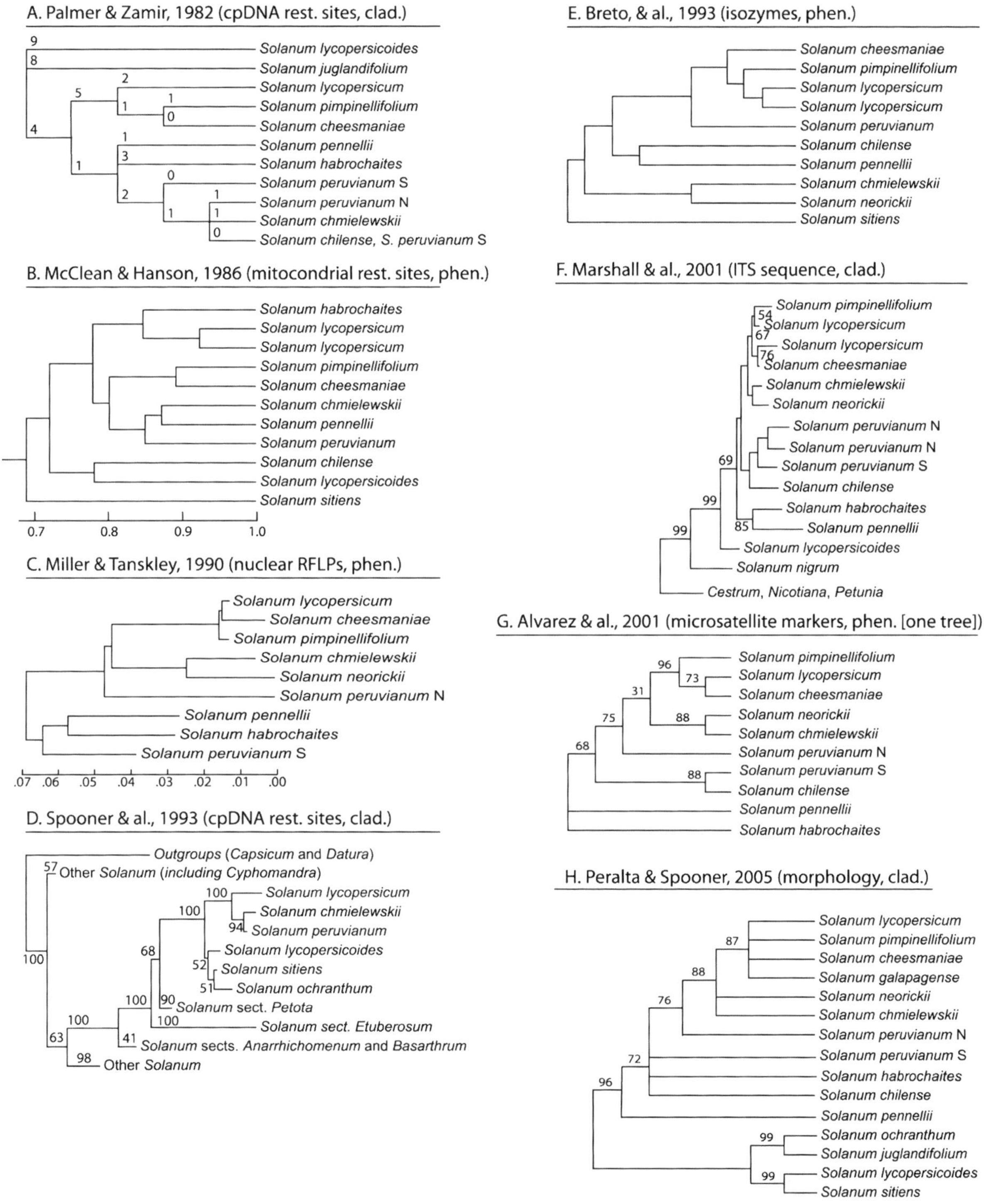

Fig. 9.3 An abstracted summary of cladistic (clad.) and phenetic (phen.) studies of tomatoes and outgroups using morphological, isozyme, and molecular data, including similarity coefficients (*lines below trees*, **b**, **c**) restriction sites supporting each branch (**a**), or bootstrap values over 50% (**d**, **f**, **g**, **h**); the study in **e** showed no statistics to support the *tree*. *Trees* are shortened when necessary to show summary results and use the *Solanum* equivalents of *Lycopersicon* names (see Table 9.2). The letters *N* and *S* following *S. peruvianum* indicate northern (N) and southern (S) accessions of that species corresponding to the companion GBSSI sequence study (Peralta and Spooner 2001), morphological study (Peralta and Spooner 2005) and AFLP study (Spooner et al. 2005) of tomatoes and outgroups (see text). Reproduced with permission from Taxon 54: 46, Fig. 2 (2005)

partitioning of *S. peruvianum* s.l. into two groups with the use of granule-bound starch synthase (GBSSI) (Peralta and Spooner 2001), morphological (Peralta and Spooner 2005), and amplified fragment length polymorphism (AFLP) data (Spooner et al. 2005). Third, in Peralta et al. (2008), based on the results of these three investigations and our examination of hundreds of additional herbarium specimens, *S. peruvianum* "north" was divided into *S. arcanum* and *S. huaylasense*, and *S. peruvianum* "south" into *S. corneliomulleri* and *S. peruvianum* s.str. (Peralta et al. 2005, 2008).

9.3.3.1 Chloroplast DNA Restriction Site Data

The cpDNA restriction site phylogenetic study of Palmer and Zamir (1982; Fig. 9.3a) was one of the first studies using this technique, and stimulated the use of chloroplast DNA in scores of other plant groups. The technique was soon refined to the use of heterologous probes, rather than total chloroplast banding patterns, to assess polymorphisms more accurately. Palmer and Zamir's (1982) study, using 25 restriction endonucleases, placed *S. lycopersicoides* (*Solanum* sect. *Lycopersicoides*) and *S. juglandifolium* (*Solanum* sect. *Juglandifolia*) as sister to tomatoes, and supported the monophyly of the red- to orange- to yellow-fruited species (*S. cheesmaniae*, *S. lycopersicum*, and *S. pimpinellifolium*). Palmer and Zamir's (1982) study was not able to place into separate clades the northern and southern populations of *S. peruvianum* or to solve the relationships of *S. chilense* and *S. chmielewskii*.

Spooner et al. (1993; Fig. 9.3d) examined cpDNA polymorphism of representatives of tomato, potato, other species of *Solanum*, and outgroups in *Capsicum* L. and *Datura* L. with a focus on examining outgroup relationships of tomato and potato. Their study showed tomatoes and their immediate outgroups in *Solanum* sect. *Lycopersicoides* and sect. *Juglandifolia* to form a sister clade to potatoes (sect. *Petota*), with *Solanum* sect. *Etuberosum* as the sister to all the above. These results stimulated the taxonomic recognition of all tomatoes in *Solanum*, which was also supported by other cpDNA restriction site and sequence data (Bohs and Olmstead 1997, 1999; Olmstead and Palmer 1997; Olmstead et al. 1999; Bohs 2005). These multiple datasets from a variety of genes unambiguously established tomatoes to be deeply nested in *Solanum*, and Spooner et al. (1993) made the necessary nomenclatural transfers. Treating tomatoes as members of *Solanum* is accepted by the majority of taxonomists as well as by most plant breeders and other users (e.g., Caicedo and Schaal 2004; Fridman et al. 2004; Schauer et al. 2005; Mueller et al. 2009; see also http://tgrc.ucdavis.edu/key.html).

9.3.3.2 GBSSI Sequence Data

Peralta and Spooner (2001) provided a GBSSI (granule-bound starch synthase, also often referred to as "waxy") gene sequence phylogeny of 79 accessions of tomatoes and outgroups, concentrating on the most geographically widespread and polymorphic species *S. peruvianum* s.l. These results (see Fig. 5 in Peralta and Spooner 2001) supported sect. *Juglandifolia* as sister to tomatoes; sect. *Lycopersicoides* as sister to tomatoes + sect. *Juglandifolia*; potatoes (sect. *Petota*) sister to tomatoes + sect. *Juglandifolia* + sect. *Lycopersicoides;* and sect. *Etuberosum* as sister to tomatoes + sect. *Juglandifolia* + sect. *Lycopersicoides* + sect. *Petota*. Within sect. *Lycopersicon*, there was a polytomy composed of *S. chilense*, *S. habrochaites*, and *S. pennellii*, and the central-southern Peruvian to northern Chilean populations of *S. peruvianum.* A sister clade contained the northern Peruvian populations of *S. chmielewskii*, *S. neorickii*, and *S. peruvianum*, and a monophyletic group composed of the SC and brightly colored (red- to orange- to yellow-fruited) species *S. cheesmaniae* (including accessions now recognized as *S. galapagense*), *S. lycopersicum*, and *S. pimpinellifolium*.

9.3.3.3 Internal Transcribed Spacer Region of Nuclear Ribosomal DNA Gene Sequences

Marshall et al. (2001) analyzed phylogenetic relationships of wild tomatoes with DNA sequences of the ITS region of nuclear ribosomal DNA (Fig. 9.3f). *Solanum lycopersicoides* was supported as sister to tomatoes (members of sect. *Juglandifolia* were not included in this study). *Solanum chilense* and S. *habrochaites* were supported as sister to all other tomatoes. *Solanum chilense* and northern and southern populations of

S. peruvianum formed a clade sister to *S. chilense* and *S. habrochaites*. *Solanum chmielewskii* and *S. neorickii* formed the next clade, followed by a clade of brightly colored-fruited species.

9.3.3.4 Morphological Phenetics and Cladistics

The phenetic morphological study of Peralta and Spooner (2005) used many of the same accessions as the GBSSI study described earlier. In total, 66 characters (50 quantitative and 16 qualitative) were measured for six individuals of 66 accessions, and averages of all six plants were taken as representative of the accession. Similarity matrices for the 61 characters found to be significantly different between at least two species were generated with various algorithms, and dendrograms were constructed with the unweighted pair group method (UPGMA) (see Figs. 6 and 7 in Peralta and Spooner 2005). The morphological distance phenogram had the best fit of the similarity matrix to the tree as determined by a cophenetic correlation coefficient (0.93), while the correlation matrix had a lower value (0.75). The distance phenogram defined four main groups. The outgroups, *S. lycopersicoides* and *S. sitiens*, cluster as the external branch (group D), followed by *S. galapagense*, and then a group of all three accessions of *S. pennellii* (group C). The SC, red- to orange- to yellow-fruited species (*S. lycopersicum*, *S. cheesmaniae*, and *S. pimpinellifolium*) form a third cluster (group A), but with the exclusion of the distinctive *S. galapagense*. The fourth group (B) includes the remaining species. Within group B, *S. neorickii* and two accessions of *S. chmielewskii* cluster together, to the exclusion of one accession of *S. chmielewskii* (LA1306) that grouped with all accessions of *S. arcanum*. All accessions of *S. chilense* formed a group that also contained one accession of *S. huaylasense* (LA1982). The three accessions of *S. habrochaites* formed a separate group. Two major groups were recognized within former *S. peruvianum*; the "northern" and the "southern." The "northern" *S. peruvianum* accessions are now recognized as the distinct species *S. arcanum* and *S. huaylasense*, and the "southern" ones as *S. peruvianum* s. str. and *S. corneliomulleri*.

The correlation UPGMA dendrogram had a lower cophenetic correlation (0.75; vs. distance, 0.93), but it placed *S. galapagense* with the other SC, red- to orange- to yellow-fruited species, and better grouped the former north and south populations of *S. peruvianum*. Unlike the distance phenogram, it placed the two outgroups, *S. lycopersicoides* and *S. sitiens*, as an internal branch with one of two main clusters (A). The three accessions of *S. habrochaites* formed a separate group, and also the three *S. pennellii* accessions clustered together. The other main branch (B) includes *S. arcanum*, *S. chilense*, *S. chmielewskii*, *S. corneliomulleri*, *S. huaylasense*, *S. neorickii*, and *S. peruvianum* s. str. This dendrogram, unlike the distance phenogram, shows better clustering of the former northern and southern *S. peruvianum* groups. Like the distance phenogram, *S. huaylasense* clustered with *S. chilense*, as part of a larger cluster that includes *S. corneliomulleri* and *S. peruvianum*. *S. arcanum*, *S. chmielewskii*, and *S. neorickii* cluster together.

Approximately one third of the morphological characters (24/66) could be scored as discrete for use in cladistic studies. A cladistic analysis of these characters in tomato and outgroups in sect. *Juglandifolia* and sect. *Lycopersicoides* supported *S. pennellii* as sister to all tomato species (see Fig. 8 in Peralta and Spooner 2005). The relationships among the self-incompatible (SI) species *Solanum chilense*, *S. habrochaites*, and *S. peruvianum* "southern" were not resolved. *Solanum peruvianum* "northern" appeared as sister to *S. chmielewskii* and *S. neorickii*. *Solanum chmielewskii* and *S. neorickii* always were sister to each other and these two sister to the monophyletic group formed by *S. cheesmaniae*, *S. galapagense*, *S. lycopersicum*, and *S. pimpinellifolium*.

9.3.3.5 AFLP Cladistics

Spooner et al. (2005) used four AFLP primer combinations to study the phylogenetic relationships of 65 accessions of tomato and outgroups, including most of the accessions corresponding to the GBSSI (Peralta and Spooner 2001) and morphological studies (Peralta and Spooner 2005) described earlier. A strict consensus tree of these 296 AFLP trees (see Fig. 7 in Spooner et al. 2005) support tomatoes (*Solanum* sect. *Lycopersicon*) and their immediate outgroup relatives in sect. *Juglandifolia* and sect. *Lycopersicoides* to form a sister clade to potatoes (sect. *Petota*) and further outgroups in sect. *Etuberosum*. *Solanum pennellii* and

S. habrochaites were part of a polytomy in sect. *Lycopersicon*. All red- or orange-fruited, SC species (*S. cheesmaniae*, *S. galapagense*, *S. lycopersicum*, *S. pimpinellifolium*) formed a well-supported clade. *Solanum chmielewskii*, *S. neorickii*, and four accessions of the SI *S. arcanum* from the Río Marañón drainage formed a clade. AFLP data, like the GBSSI and morphological data, show a clear separation of the northern and southern groups of *S. peruvianum* s. l., which includes *S. corneliomulleri* and *S. peruvianum* s. str. Only one accession from northern Peru (LA1984) grouped with the southern *S. peruvianum*. Interestingly, Rick (1986c) thought that this accession represented a "crossing bridge" between northern and southern populations of *S. peruvianum*. AFLP data, unlike morphological data, grouped *S. arcanum* with *S. huaylasense* instead of *S. chilense*.

9.3.3.6 Congruence Tests Among AFLP, cpDNA, GBSSI, ITS, and Morphological Studies

Spooner et al. (2005) tested congruence among AFLP, cpDNA (Palmer and Zamir 1982), GBSSI (Peralta and Spooner 2001), ITS (Marshall et al. 2001), and morphology (Peralta and Spooner 2005) datasets through three methods: (1) distance matrix-based comparisons (the Mantel test), (2) character-based comparisons (the incongruence length difference test (ILD), also called the partition homogeneity test of data partition congruence, of Farris et al. 1995), and (3) visual qualitative comparison of trees. Two comparative datasets were used: (1) A dataset containing 47 identical tomato accessions from AFLP and GBSSI studies and with one accession of *S. etuberosum* Lindl. as outgroup. (2) A smaller comparative dataset contained only 10 accessions that were common to all studies cited earlier (all tomato species were included except *S. neorickii* which was lacking from the cpDNA dataset; the northern and southern accessions of *S. peruvianum* were included as separate taxa; *S. lycopersicoides* was the common outgroup).

The distance-matrix test showed that all pairs of compared matrices were statistically correlated at $\alpha = 0.05$ except for GBSSI/ITS, GBSSI/morphology phenetics, and ITS/cpDNA. The matrix correlation coefficients of all comparisons varied greatly with AFLP/GBSSI the highest, and ITS/cpDNA the lowest. The character-based test showed the ITS/cpDNA, AFLP/GBSSI (both 10 and 48 taxon comparisons), the GBSSI/morphology, AFLP/ITS, GBSSI/ITS, AFLP/cpDNA, ITS/morphology, and AFLP/morphology datasets to be congruent. The other comparisons (cpDNA/morphology, cpDNA/GBSSI) proved to be incongruent.

9.3.3.7 Total Evidence Analysis of Chloroplast DNA, ITS, AFLP, and GBSSI

A combined AFLP and GBSSI Fitch tree (Spooner et al. 2005), consisting of 48 taxa and constructed with 1,652 characters, produced 34 most parsimonious 994-step trees with a consistency index of 0.35 and a retention index of 0.56. A strict consensus tree of these 34 trees (not shown) presented a topology very similar to that of the AFLP strict consensus tree (see Fig. 7 in Spooner et al. 2005; Peralta et al. 2008), including showing the relationship *S. chmielewskii*, *S. neorickii*, and four accessions of *S. arcanum*. A combined AFLP, GBSSI, cpDNA, ITS tree, and morphology analysis (10 taxa; 2,301 characters of which 148 were parsimony informative) produced two most-parsimonious 577-step trees with a consistency index of 0.816 and a retention index of 0.603. A strict consensus tree (Fig. 9.4) of these two trees showed (1) the brightly colored-fruited species as monophyletic, (2) *S. chmielewskii* and *S. arcanum* to be a sister clade to the above, (3) *S. chilense* and *S. peruvianum* s.s. and *S. corneliomulleri* to be a sister clade of the species above, (4) *S. habrochaites* and *S. pennellii* to be a well supported clade, but forming a polytomy. *Solanum lycopersicoides* was sister to tomatoes (sect. *Lycopersicon*). Members of sect. *Juglandifolia* were not included in this analysis.

9.3.4 Summary

The tomatoes and their wild relatives (sects. *Lycopersicoides*, *Juglandifolia* and *Lycopersicon*) are clearly monophyletic and sister to the potatoes (sect. *Petota*), with sect. *Etuberosum* clearly monophyletic and sister to potatoes + tomatoes s.l. Sect. *Lycopersicoides* (formerly recognized as a subsection of sect. *Lycopersicon*) is clearly monophyletic and sister to sect.

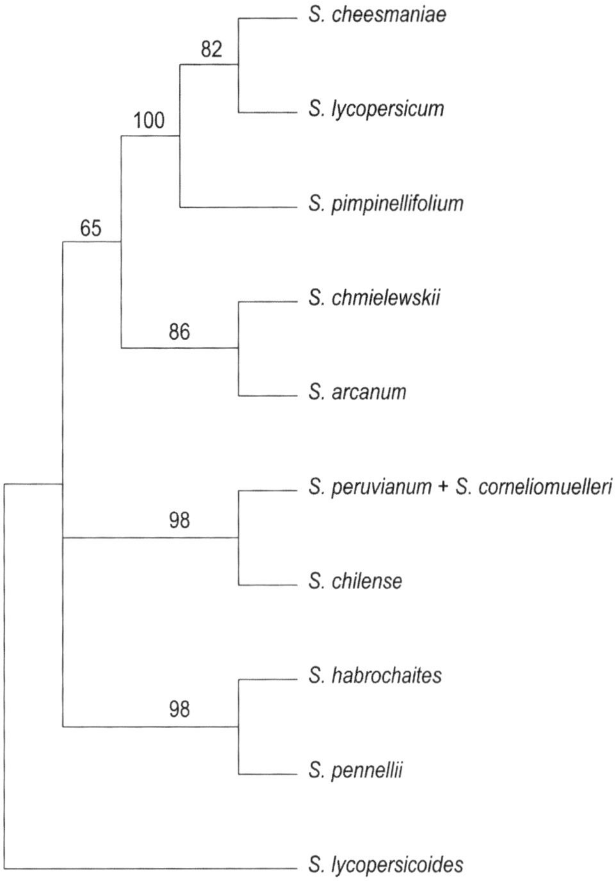

Fig. 9.4 The single combined AFLP, GBSSI, cpDNA, and ITS 530-step Fitch tree (10 taxa; 2,275 characters). The *numbers above* each *branch* represent bootstrap values over 50% (from Spooner et al. 2005). Modified and reproduced with permission from Syst Bot Monogr 84: 52, Fig. 18 (2008)

Juglandifolia + sect. *Lycopersicon*, and sect. *Juglandifolia* is clearly monophyletic and sister to sect. *Lycopersicon*.

Within sect. *Lycopersicon*, *S. pennellii* in most cases appears at the base of the trees as a polytomy with *S. habrochaites*, or sometimes forms a clade with this species. This relationship was considered unresolved by Peralta et al. (2008), although morphological data suggest that *S. pennellii* is sister to the rest of the tomatoes s.str. (sect. *Lycopersicon*); it is the only species in that group lacking the sterile anther appendage, the presence of which is a morphological synapomorphy of *S. habrochaites* and the rest of the core tomato clade. *S. pennellii* was placed by Peralta et al. (2008) in its own "group." Relationships within sect. *Lycopersicon* have been presented by Peralta et al. (2008) as informal species groups as given in Table 9.3. Such informal group systems of classification have been widely applied to *Solanum* by Whalen (1984), Knapp (1991, 2000, 2002), Bohs (1994, 2005), and Spooner et al. (2004). They are not intended to represent formal classification and are provisional names representing most highly supported ideas of relationships that are still unresolved.

Solanum huaylasense (a "northern" segregate of *S. peruvianum* s.l.) is grouped with *S. chilense*, *S. habrochaites*, *S. corneliomulleri* (a segregate of "southern" *S. peruvianum* s.l.), and *S. peruvianum* s. str. in the "Eriopersicon" species group (see Peralta et al. 2008). The SC green-fruited species *S. chmielewskii* and *S. neorickii* are related to *S. arcanum* (another northern segregate of *S. peruvianum* s.l.) as supported in almost all datasets and are recognized by Peralta et al. (2008) as the "Arcanum" species group. The four species with brightly colored fruits (*S. cheesmaniae*, *S. galapagense*, *S. lycopersicum*, *S. pimpinellifolium*) unambiguously form a closely related monophyletic group and are the closest relatives of the cultivated crop. These species with red to orange fruits could be recognized as a formal taxonomic group (as a series, for example), but this formal classification has not been taken up at present because of

Table 9.3 Classification of *Solanum* section *Lycopersicon* (tomatoes) and allied species (Peralta et al. 2008)

Section	Species group	Species
Section *Lycopersicoides*	–	*Solanum lycopersicoides*
–	–	*Solanum sitiens*
Section *Juglandifolia*	–	*Solanum juglandifolium*
–	–	*Solanum ochranthum*
Section *Lycopersicon*	"Neolycopersicon"	*Solanum pennellii*
–	"Eriopersicon"	*Solanum chilense*
–	–	*Solanum corneliomulleri*
–	–	*Solanum habrochaites*
–	–	*Solanum huaylasense*
–	–	*Solanum peruvianum*
–	"Arcanum"	*Solanum arcanum*
–	–	*Solanum chmielewskii*
–	–	*Solanum neorickii*
–	"Lycopersicon"	*Solanum cheesmaniae*
–	–	*Solanum galapagense*
–	–	*Solanum lycopersicum*
–	–	*Solanum pimpinellifolium*

Species within each group are in alphabetical order

ambiguity in the other species groups in sect. *Lycopersicon*.

9.4 Role in Development of Cytogenetic Stocks and Their Utility

The wild relatives of cultivated tomato have been used to develop several types of cytogenetic stocks. Of particular relevance here are the chromosome substitution and addition stocks. Other types of pre-breds, including ILs and backcross inbred lines (BILs), are not considered herein as they have been thoroughly covered in other recent reviews (Zamir and Eshed 1998a, b; Zamir 2001; Labate et al. 2007; Lippman et al. 2007; Grandillo et al. 2008).

Alien substitution lines and monosomic alien addition lines contain intact wild species' chromosomes in the genetic background of a standard tomato variety. The monosomic additions are trisomics ($2n + 1$), i.e., the foreign chromosome is added to a diploid tomato genome. In the substitution lines, the foreign chromosome replaces one or both of the corresponding tomato chromosomes (homeologs), and thus they are diploids. Monosomics ($2n - 1$) and other types of deficiency or deletion stocks are not commonly used in tomato because gametes carrying the deficient chromosomes generally fail to transmit through meiosis and gametogenesis (Khush and Rick 1966, 1968).

The first alien substitution lines in tomato contained chromosomes of *S. pennellii* in the background of *S. lycopersicum* (Rick 1969, 1971). They were obtained by backcrossing the wild parent to multiple marker stocks containing two or more morphological mutations, usually seedling expressed, located on a single chromosome. The *pennellii* chromosomes carried the wild type (dominant) alleles at each locus. Thus, selection for the non-mutant phenotype over several generations resulted in the replacement of one tomato chromosome by the homeologous chromosome of *S. pennellii*, as well as the progressive elimination of all other wild species chromosomes. After five or more backcross (BC) generations, homozygous substitutions were obtained by self-pollination. The method was rapid and inexpensive, but was limited by incomplete and uneven coverage of the chromosomes with convenient visual markers and dominance of the wild species alleles. For example, the chromosome 6 substitution was heterozygous; use of DNA-based markers (vastly more abundant) allowed isolation of the desired homozygous stock, and demonstrated a recombination event near the end of the chromosome that was not detected with the visual markers (Weide et al. 1993). A few alien substitution lines were also synthesized for *S. lycopersicoides* (Canady et al. 2005).

A complete set of monosomic alien addition lines in tomato was synthesized for *S. lycopersicoides*

(Chetelat et al. 1998) and a small number for *S. sitiens* (Pertuzé et al. 2003). These lines are relatively stable, because the extra wild species chromosomes tend to recombine at relatively low rates (Ji and Chetelat 2003). However, they are also relatively infertile and thus easily lost through poor seed set. The morphology of each monosomic addition is strikingly similar to the corresponding tomato trisomic. This observation is consistent with the observed colinearity of genetic maps for tomato and its wild relatives, including *S. lycopersicoides* (Pertuzé et al. 2002), suggesting a similar gene content of each chromosome.

Monosomic addition and substitution lines are potentially useful for a variety of genetic studies and breeding applications. In tomato, these stocks have been particularly useful for studies of genetic recombination between the alien chromosomes and their tomato homeologs. For example, in progeny of the *pennellii* substitution lines, recombination frequency was higher in early BC generations than in later ones, and higher in progeny of female than male meioses (Rick 1969, 1971). Similar trends were observed with the *lycopersicoides* substitution lines, which recombined at higher rates than either the monosomic additions or shorter, segmental introgressions, which recombined at only 0–10% of normal rates (Ji and Chetelat 2003; Canady et al. 2006). The low rates of genetic exchange between homeologous chromosomes may be due to competition by recombination within *homologous* chromosomes or chromosomal regions, a process that occurs in monosomic additions and segmental introgressions, but not heterozygous substitutions. Examination of chromosome pairing by genomic in situ hybridization (GISH) cytology indicated that the degree of pairing failure, as indicated by the formation of univalents, is correlated with the severity of recombination suppression. Pairing in the monosomic additions was more disrupted than the substitutions. Lines containing chromosome 10 of *S. lycopersicoides,* which carries a paracentric whole arm inversion relative to cultivated tomato, presented the most irregular pairing behavior. These results indicate that for future breeding purposes, substitution lines provide the best starting material for obtaining recombination events around a gene of interest.

9.5 Conservation Initiatives

9.5.1 Germplasm Collections

Tomato breeding and research can rely on a wide range of germplasm resources, which include extensive collections of wild forms and their derivatives (see recent reviews by Chetelat and Ji 2007; Ji and Scott 2007; Robertson and Labate 2007). The first collections of wild tomato species began in the eighteenth century in the region of their native distribution, which extends from northern Chile to southern Columbia and from the Pacific Ocean coast to the eastern foothills of the Andes, and the collection of this valuable material continues to this day.

Overall, there are more than 75,000 accessions of *Solanum* sect. *Lycopersicon* germplasm maintained in gene banks in more than 120 countries all around the world (for detail see review by Robertson and Labate 2007). The largest collections are hold at (1) the Asian Vegetable Research and Development Center (AVRDC), now referred to as The World Vegetable Center, located in Tainan, Taiwan; (2) the C. M. Rick Tomato Genetics Resource Center (TGRC), at the University of California-Davis; (3) the United States Department of Agriculture, Agricultural Research Service (USDA-ARS) Plant Genetic Resources Unit (PGRU) in Geneva, NY (Table 9.4).

The AVRDC was founded in 1971 with the mandate to increase vegetable production in the Asian tropics and is an international center affiliated with the Consultative Group of International Agricultural Resources (CGIAR). The first of the five research themes of the Center is "germplasm conservation, evaluation, and gene discovery". The AVRDC stores large amounts of germplasm including a vast collection of tomato, numbering ca. 7,500 accessions. Apart from the cultivated types (more than 6,000 accessions), the Center stores a collection of 725 accessions of wild tomato species (Table 9.4). *Solanum pimpinellifolium* and *S. peruvianum* are the most represented with 325 and 135 accessions, respectively. In addition, there are also almost 600 accessions of unidentified wild material, listed as *Lycopersicon* sp., and a few hundreds lines deriving from interspecific crosses (Ebert AW, pers. comm.). The Center has a very

Table 9.4 *Solanum* section *Lycopersicon* (tomatoes) and allied species genetic stocks maintained by The World Vegetable Center (AVRDC), the Tomato Genetic Resource Center (TGRC), and the USDA at Geneva, NY (USDA)

Species[a]	AVRDC[b]	TGRC	USDA
S. arcanum	3	44	4
S. cheesmaniae	17	44	7
S. chilense	47	112	1
S. chmielewskii	11	37	7
S. corneliomulleri	11	52	13
S. galapagense	17	29	5
S. habrochaites	82	120	63
S. huaylasense	0	14	0
S. neorickii	12	59	1
S. pennellii	65	65	10
S. peruvianum	135	78	122
S. pimpinellifolium	325	309	231
S. juglandifolium	0	8	0
S. lycopersicoides	0	23	0
S. ochranthum	0	9	0
S. sitiens	0	13	0
Subtotal	**725**	**1,016**	**464**
sp.	**595**	**0**	**0**
S. lycopersicum	6,067	2,349	5,884
S. lycopersicum "cerasiforme"	125	338	272
Subtotal	**6,192**	**2,687**	**6,156**
Total	**7,512**	**3,703**	**6,620**

[a]Previous names in the genus *Lycopersicon* are given in Table 9.2
[b]Erbert AW, pers. comm

useful web interface, with an information system (The AVRDC Vegetable Genetic Resources Information System or AVGRIS) for searching the data available for germplasm conserved at AVRDC's Genetic Resources and Seed Unit. A web version of AVGRIS is accessible at the URL http://203.64.245.173/avgris/ and provides all users a direct access to the stored germplasm data. Through this facility, it is possible to search for the accessions present in the gene bank and to have also access to a characterization data sheet per each accession.

Another excellent collection of wild tomato genetic resources is held by the C. M. Rick TGRC. The TGRC has been named in memory and honor of Dr. Charles M. Rick (1915–2002), Professor Emeritus of Vegetable Crops at the University of California, Davis, USA, who had originally built up much of the collection through his research and plant collecting activities (Rick 1979, 1986a, b). Dr. Rick had first recognized the potential value of wild germplasm as a useful reservoir of genes for the improvement of tomato. He undertook 15 expeditions to South America, between 1948 and 1995, in the Andean regions of Peru, Ecuador, and Chile and to the Galápagos Islands, establishing a first collection of some 700 samples of sect. *Lycopersicon* and related wild species of *Solanum*.

The TGRC is hosted by the Department of Plant Sciences of the University of California at Davis, and is integrated with the National Plant Germplasm System (NPGS), the latter 'storing backup seed samples of the TGRC collection and only very few samples that are not stored at the TGRC.

As regards the wild tomato germplasm, the TGRC maintains over 1,000 accessions of wild relatives that represent 13 species in *Solanum* sect. *Lycopersicon*, and the four related *Solanum* species *S. lycopersicoides*, *S. sitiens*, *S. juglandifolium*, and *S. ochranthum* (Chetelat 2006; Table 9.4). All the entries are reported with the *Lycopersicon* and the equivalent *Solanum* species name. This Center maintains a series of special purpose collections of selected wild and cultivated accessions with known or inferred tolerances to various environmental (abiotic and biotic) stresses that have been extensively utilized in tomato crop improvement (for detail see review by Robertson and Labate 2007; http://tgrc.ucdavis.edu/). A nice interface allows mapping of TGRC accession collection sites worldwide. The TGRC has a very useful website at the URL http://tgrc.ucdavis.edu/index.aspx, which is worth a visit.

In addition to wild tomato species the TGRC also stores over 1,000 monogenic mutants, including spontaneous and induced mutations affecting many aspects of plant development and morphology, disease resistance genes, and protein marker stocks (Labate et al. 2007). In addition, the collection contains hundreds of miscellaneous genetic and cytogenetic stocks such as trisomics, tetraploids, and translocations, as well as derivatives of wild species such as pre-bred stocks that are very valuable for mapping and breeding purposes. The pre-bred stocks include ILs, BILs, alien substitution lines, and alien addition lines. The IL populations originated from *S. pennellii* LA0716 (Eshed and Zamir 1995; Liu and Zamir 1999), *S. habrochaites* LA1777 (Monforte and Tanksley 2000a), and *S. lycopersicoides* LA2951 (Canady et al. 2005); the BILs were derived from the cross *S. lycopersicum* × *S. pimpinellifolium* LA1589

(Doganlar et al. 2002a). Moreover, the center stores a few alien substitution lines representing seven of the 12 *S. pennellii* LA0716 chromosomes (Rick 1969; Weide et al. 1993); four *S. lycopersicoides* (LA2951) chromosomes (Chetelat and Meglic 2000; Ji and Chetelat 2003); and ten alien addition lines, each containing one extra chromosome from *S. lycopersicoides* LA1964 added to the tomato genome (Chetelat et al. 1998).

The USDA-PGRU germplasm collection focuses on *S. lycopersicum* accessions, which constitute ca. 90% of the more than 6,600 accessions held by this center for tomato, including a large number of modern, vintage, and primitive cultivars along with breeding lines. The collection also contains 464 accessions of wild species, the majority of which are *S. peruvianum* s.l. (see Sect. 9.3) and *S. pimpinellifolium* (http://www.ars.usda.gov). Also the USDA-PGRU collection is duplicated in the National Center for Genetic Resources Preservation (NCGRP) located at Fort Collins, Colorado.

9.5.2 Modes of Preservation and Maintenance

Conservation of genetic resources requires several steps including germplasm collection, maintenance, distribution, characterization, and evaluation. In order to avoid loss of genetic diversity (or genetic erosion) within any given collection and to maintain genetic identity of accessions conserved therein, it is necessary to develop standard methodologies during all these steps, and large numbers of plants or seed are needed. Deployment of these methodologies mainly depends on the breeding system of the species, with cross-pollinated species requiring larger samples. The cultivated tomato is self-pollinated, while the other taxa can vary from self-pollinated to obligately cross-pollinated, showing different rates of outcrossing (Table 9.1). In most gene banks, *S. lycopersicum* is maintained by regenerants from relatively few (e.g 6–24) plants, with accessions usually planted in the field without pollination control. This allows the production of a sufficiently large amount of seed for storage, which can significantly reduce the chances of cross-pollination, or mix-ups, by increasing the time between regenerations. In contrast, for the cross-pollinated species such as most wild taxa, prevention of genetic drift and contamination requires the use of larger samples and controlled pollination. Generally, up to 50 plants are used for regeneration to obtain a representative sample by reducing the effects of genetic drift and selection during the regeneration process (Robertson and Labate 2007).

Seed production must be monitored in order to ensure sufficient production of quality, disease-free seed for maintenance and distribution. For long-term storage of species with orthodox seed, such as tomato, a temperature of -20°C and at a moisture content of $5 \pm 1\%$ is suggested (Robertson and Labate 2007). In some cases, the use of cryopreservation (conservation using liquid nitrogen) of seed for long-term genetic conservation has been suggested, although additional studies are necessary in order to determine whether there is any advantage to this for orthodox seed (Robertson and Labate 2007).

9.5.3 In Situ Conservation

The wild species of tomato are well preserved ex situ through national gene banks, but there is an urgent need to preserve threatened populations in situ as well. Throughout the native region, wild tomatoes are impacted in various ways by human activities. In the highlands, grazing by goats, sheep, and other herbivores is a constant threat. At low elevations, intensive agricultural development and urbanization have had a dramatic impact. A recent trip to Peru organized by one of the authors (RTC) provided clear evidence of genetic erosion via the loss or displacement of local populations.

During this expedition, conducted in April–May 2009, several river valleys (Pisco, Cañete, Lurin, Rimac, Chillon, Pativilca, and Jequetepeque) were explored. In the lower stretches of the valleys, intensive agriculture and urban development are common. With increasing elevation, the environment becomes more rural and agricultural systems more traditional and less intensive. As might be expected, the wild species growing predominantly at low elevations were more severely impacted than those growing at the higher, less disturbed sites.

Of particular concern, populations of *S. pimpinellifolium* have virtually disappeared from low elevation sites. Only two populations of this species were found below 1,000 m, whereas at least 25 populations were previously collected from the same river valleys. This represents a loss of up to 23 populations of this species in only seven valleys surveyed. The drainages around Lima (Ríos Rimac, Lurin, and Cañete) were most severely affected, due to urbanization and development. Similar trends are occurring around the other major cities in coastal Peru. North of Lima, intensive, modern agricultural practices, including sugarcane cultivation and the widespread use of herbicides, has resulted in the elimination of many local populations of wild tomatoes known from earlier collections. Wild tomatoes are also threatened by climate change. In three valleys (Cañete, Chillon, and Lurin), *S. pimpinellifolium* was found growing above 1,000 m elevation for the first time.

These examples of genetic erosion in *S. pimpinellifolium* are troubling for several reasons. First, this species is closely related to the cultivated tomato – its fruit are edible and sometimes consumed – and has served as a source of disease resistances and other desired traits used by plant breeders to develop improved varieties. The first disease resistance genes bred into the cultivated tomato, *Verticillium* and *Fusarium* wilt resistances, originated in *S. pimpinellifolium,* and it continues to be an important source of such genes. Secondly, the area in which the loss of *S. pimpinellifolium* populations seems most severe, the northern half of Peru, is/was the center of genetic diversity for this species. Accessions collected in the north have/had larger flowers with exserted stigmas, traits which tend to promote cross pollination and maintenance of genetic diversity (Rick et al. 1977).

Although excellent ex situ collections are available to support future breeding and research on tomato, they are subject to their own long-term risks, such as unstable funding, catastrophic loss, and genetic changes (inbreeding depression, artificial selection, etc.). For this reason, there will always be a need to preserve populations in situ. The appropriate authorities in national governments of the countries of origin – mainly Ecuador, Peru, and Chile – should be helped to take steps to protect their native tomatoes from further loss. International organizations, such as the CGIAR, are urged to get involved to initiate and/or support such conservation efforts. Without action, the wealth of wild germplasm in the tomato relatives may not be available to future generations.

9.6 Role in Classical and Molecular Genetic Studies

9.6.1 Genetic Variation

The cultivated tomato (*S. lycopersicum*) is highly autogamous and, despite its wide range of fruit shape, size, and color diversity, its genetic diversity is so reduced that it lacks many genes required for breeding purposes (Rick 1976). Genetic erosion of this crop has resulted from repeated genetic bottlenecks (due to a combination of natural self-pollination, reproduction in small populations, and natural and artificial selection), associated with the domestication process, the early history of improvement in Europe and North America, and modern breeding practices (Rick 1986a).

The level of genetic erosion of the primary tomato gene pool has been measured using different types of markers including allozymes (Rick and Fobes 1975) and RFLPs (Helentjaris et al. 1985; Miller and Tanksley 1990). The study conducted by Miller and Tanksley (1990), estimated that the level of genetic variation of cultivated varieties can be lower than 5% of that available in nature (Miller and Tanksley 1990). Due to this lack of genetic diversity, it is very difficult to identify polymorphisms within the cultivated tomato gene pool, even using sensitive molecular marker techniques (García-Martínez et al. 2006 and references therein).

In contrast, higher levels of variability exist in primitive cultivars of the native area and even more in the wild species, with particularly large genetic diversity observed within the SI species like *S. chilense* and *S. peruvianum* s.l. (Rick 1982, 1988). Interestingly, more genetic variation has been found within a single accession of the SI species than in all accessions of any of the SC species (Miller and Tanksley 1990; Städler et al. 2005). Given the low level of polymorphism among autogamous species, the study of their relationships necessitates the use of more variable molecular markers, such as micorsatellites and single nucleotide polymorphisms (SNP) (Alvarez et al. 2001; Yang et al. 2004).

Wild tomato species in sect. *Lycopersicon* occupy a wide variety of habitats ranging from sea level to above 3,000 m in altitude, and from temperate deserts to wet tropical rainforests (see Sect. 9.2.2). Accordingly, these wild species span a broad variation in terms of morphology, physiology, mating system, and biochemistry, which is of potential value for the improvement of cultivated tomato. In addition, in spite of the severe crossing barriers that separate the four tomato-like nightshade taxa in sects. *Juglandifolia* and *Lycopersicoides* from tomatoes (*Solanum* sect. *Lycopersicon*), these allied species are considered very promising to broaden the genetic variability available for tomato improvement (Rick 1988). In fact, even though they have not been thoroughly tested, the specificity of their habitats suggests that they might harbor novel traits that are lacking in the sect. *Lycopersicon* species (see Sect. 9.2.2). These include tolerance to extreme aridity, excessive moisture and freezing temperatures, as well as resistance to certain diseases and insects (Rick 1988; Rick and Yoder 1988).

9.6.2 Wide Hybridizations

The use of wild species as sources of traits of interest can be hindered by blocks to hybridization and hybrid sterility that might occur at the beginning of the breeding program. These limitations can vary enormously and generally are more severe as the phylogenetic distance between the parental species of the cross increases. Thus, while there are no problems for crosses between *S. lycopersicum* and the closely related wild species *S. cheesmaniae*, *S. galapagense,* and *S. pimpinellifolium*, at the other extreme, crosses with *S. chilense* or *S. peruvianum* s.l., are more difficult and require some type of embryo or ovule rescue; intermediate situations characterize the crosses with *S. chmielewskii*, *S. habrochaites*, *S. pennellii*, and others (Rick and Chetelat 1995). On the other hand, the exploitation of the genetic variability stored in the tomato-like nightshades *S. juglandifolium*, *S. lycopersicoides*, *S. ochranthum,* and *S. sitiens* has been more limited, as severe reproductive barriers isolate them from the core tomato group (Rick 1988; Child 1990; Stommel 2001; Smith and Peralta 2002). Within the group of the four tomato-like nightshades, the only successful cross is the one between *S. lycopersicoides* and *S. sitiens*. In this case, the easily synthesized F_1 hybrids display normal meiotic behavior and high fertility (Pertuzé et al. 2002; Ji et al. 2004). Of the four species, only *S. lycopersicoides* is cross-compatible with *S. lycopersicum* (Rick 1951, 1979; Pertuzé et al. 2002); F_1 hybrids are readily obtained using embryo culture, although they are generally infertile due to meiotic abnormalities (Menzel 1962). *Solanum lycopersicoides* has also been hybridized unilaterally with other taxa of sect. *Lycopersicon*, including *S. cheesmaniae*, *S. chilense,* and *S. pimpinellifolium*. In contrast, *S. sitiens* does not cross directly to tomato in either direction (Rick 1979, 1988), but it can be indirectly hybridized with cultivated tomato using polyploid and bridging line methods (e.g., by using *S. lycopersicum* × *S. lycopersicoides* derivatives as bridge) (DeVerna et al. 1990; Pertuzé et al. 2003). As a result, it has been possible to introgress to varying degrees chromosomes or chromosome segments from *S. lycopersicoides* and *S. sitiens* into the tomato genome (Chetelat and Meglic 2000; Ji and Chetelat 2003; Pertuzé et al. 2003; Canady et al. 2005). For *S. lycopersicoides* a complete set of monosomic alien addition lines in tomato was synthesized by Chetelat et al. (1998), and a set of ILs are now available (Chetelat and Meglic 2000; Canady et al. 2005); gene transfer from *S. sitiens* to *S. lycopersicum* has been obtained through chromosome addition, substitution, and recombination in the progeny of complex aneuploid hybrids (Pertuzé et al. 2003). In contrast, the other two tomato-like nightshades, *S. ochranthum* and *S. juglandifolium* appear to be sexually incompatible with cultivated tomato in all combinations tested (Rick 1988); although somatic hybrids between *S. lycopersicum* and *S. ochranthum* have been obtained by protoplast fusion, they are highly sterile and have not, so far, provided a means for gene transfer (Stommel 2001).

In spite of these difficulties, recent comparative genetic linkage maps based on an interspecific F_2 *S. ochranthum* × *S. juglandifolium* population obtained via embryo culture indicate that, consistent with the status of the sect. *Juglandifolia* as the nearest outgroup to the tomatoes, these two taxa are more closely related to cultivated tomato than predicted from crossing relationships (Peralta et al. 2008; Albrecht and Chetelat 2009). These results are encouraging from the standpoint of tomato breeding, as they suggest that with further attempts at hybridization there might be more opportunity for germplasm introgression with cultivated tomato than previously assumed (Albrecht and Chetelat 2009).

9.6.3 Development of Classical and Molecular Genetic Linkage Maps

High-density molecular linkage maps provide useful tools for genome studies, gene/QTL mapping and cloning, varietal development, and many other potential applications.

The analysis of linked genes in tomato began in the early 1900s, when Jones (1917) interpreted the results of Hedrick and Booth (1907) as linkage between *dwarfness* (*d*) and fruit shape. At the beginning, genetic linkage analysis of tomato was slow, but accelerated with the availability of seedling mutants, advanced mapping stocks, and a complete set of trisomics; these cytogenetic stocks have been extremely valuable in the assignment of genes to chromosomes and chromosome arms, or even to restricted regions in the arms (Stevens and Rick 1986; Rick and Yoder 1988). As a result, by mid-1970s, a total of 258 morphological and physiological markers had been assigned to tomato chromosomes (Linkage Committee 1973; Rick 1975). Subsequently, isozyme markers started to be used, and in 1980 Tanksley and Rick published an isozyme linkage map comprising 22 loci mapped on nine of the 12 tomato chromosomes. Isozyme mapping in tomato was accomplished using interspecific F_2 and BC populations along with the trisomic technique. In the late 1980s, the last comprehensive "classical" linkage map of tomato was published, which included ~400 morphological, physiological, isozyme, and disease resistance genes mapped onto the 12 tomato chromosomes (Stevens and Rick 1986; Mutschler et al. 1987). In mid-1980s, DNA-based RFLP markers were starting to be mapped also in tomato (Bernatzky and Tanksley 1986; Tanksley and Bernstzky 1987), and by the 1990s, this species had become one of the first plants for which RFLPs were used to generate a high-density linkage map (Tanksley et al. 1992). The map was based on a *S. lycopersicum* cv. "VF36-*Tm2a*" × *S. pennellii* (LA0716) F_2 population of 67 plants and comprised 1,030 RFLP markers. This map, referred to as the Tomato-EXPEN 1992, has been updated periodically and includes DNA markers, isozyme markers, and some morphological markers (Pillen et al. 1996b; http://solgenomics.net/). Although *S. pennellii* is not closely related to the cultivated tomato, the presence of the SC accession (LA0716) has favored its use as a parental line for many mapping studies (Tables 9.5–9.8).

Over the years numerous molecular linkage maps have been developed using different mapping populations, and, due to the limited genetic variation inherent in domesticated tomato, most of them derived from interspecific crosses between the cultivated tomato and most of the tomato wild species belonging to sect. *Lycopersicon,* with recent examples involving also the allied wild species *S. lycopersicoides* (Table 9.5). Other maps have been developed using crosses between species belonging to sect. *Lycopersicoides* (*S. sitiens* × *S. lycopersicoides*; Pertuzé et al. 2002) and sect. *Juglandifolia* (*S. ochranthum* × *S. juglandifolium*; Albrecht and Chetelat 2009).

For some interspecific crosses, particularly those between the cultivated tomato and the closely related wild species *S. pimpinellifolium*, *S. cheesmaniae,* and *S. galapagense*, identification of a sufficient number of polymorphic markers has been a serious limitation; however, albeit with more difficulties, genetic maps have been developed (Table 9.5). In addition, despite the low level of genetic variation found within *S. lycopersicum,* molecular linkage maps have been constructed also using intraspecific crosses (Table 9.5).

Several of these maps were developed with the objective of mapping genes/QTLs for traits of interest, and in many cases polymerase chain reaction (PCR)-based markers, including random amplified polymorphic DNAs (RAPDs), AFLPs, simple sequence repeats (SSRs), sequence characterized amplified regions (SCARs), and cleaved amplified polymorphic sequences (CAPSs), were developed and integrated with the RFLP maps (Table 9.5; see also reviews by Ji and Scott 2007; Labate et al. 2007). For some tomato chromosomes, the integration of the molecular map with classical maps has been accomplished using interspecific progenies that segregated for morphological and molecular markers (short arm of chromosome 1: Balint-Kurti et al. 1995; chromosome 3: van der Biezen et al. 1994; chromosome 6: Weide et al. 1993, Van-Wordragen et al. 1996; chromosome 7: Schumacher et al. 1995; chromosome 11: Wing et al. 1994).

Table 9.5 Summary of molecular linkage maps developed in *Solanum* sect. *Lycopersicon* and allied species

Initial cross	Mapping population[a]	Population size[b]	Type of markers[c]	No. of markers[d]	Online[e]	Reference[f]
S. lycopersicum × *S. arcanum*						
E6203 × LA1708	BC_3	241	RFLP, PCR	174 (171; 3)		Fulton et al. (1997)
S. lycopersicum × *S. chmielewskii*						
UC82B × LA1028	BC_1	237	RFLP, ISO, MO	70 (63; 5; 2)	NCBI	Paterson et al. (1988)
CH6047 × LE777	F_2	149	AFLP, CAPS/SCAR/ CG*FL*, SSR	255 (136; 81; 38)	SGN	Jiménez-Gómez et al. (2007)
S. lycopersicum × *S. galapagense*						
UC204B × LA0483	F_2	350	RFLP	71		Paterson et al. (1991)
UC204B × LA0483	F_7-RIL	97	RFLP, MO, ISO	135 (132; 2; 1)		Paran et al. (1995)
S. lycopersicum × *S. habrochaites*						
E6203 × LA1777	BC_1	149	RFLP	135	SGN, NCBI	Bernacchi and Tanksley (1997)
E6203 × LA1777	NIL, BIL	111	RFLP	95		Monforte and Tanksley (2000a)
T5 × LA1778	BC_1	196	RFLP	89		Truco et al. (2000)
Hunt 100 × LA0407	BC_2S_5- BIL	64	RFLP, PCR	63 (58; 5)		Kabelka et al. (2002)*
NC84173 × PI 126445	BC_1	145	RFLP, PCR/RGA[g]	171 (142; 29)		Zhang et al. (2002)
NC84173 × PI-126445	BC_1 (SGe)	76	RFLP, PCR/RGA[g]	179 (145; 34)		Zhang et al. (2003b)
Moneymaker × LYC4	F_2	174	AFLP, CAPS	269 (218; 51)		Finkers et al. (2007a)
Moneymaker × LYC4	IL	30	AFLP, CAPS	491 (457; 34)		Finkers et al. (2007b)
Ferum × PI 247087	BC_2S_1	130	AFLP, RFLP, SSR, CAPS, MO	217 (138; 36; 26; 15; 2)		Stevens et al. (2007)*
S. lycopersicum × *S. lycopersicoides*						
VF36 × LA2951	BC_1	84	RFLP, ISO, MO	93 (71; 20; 2)	NCBI	Chetelat et al. (2000)
VF36 × LA2951	BIL	311	RFLP, ISO, MO	139 (110; 22; 7)		Chetelat and Meglic (2000)
S. lycopersicum × *S. neorickii*						
E6203 × LA2133	BC_2	170	RFLP, PCR, MO	133 (131; 1; 1)		Fulton et al. (2000)
S. lycopersicum × *S. pennellii*						
LA1500 × LA0716	BC_1, F_2	46; 46[h]	RFLP, ISO	112 (76; 36)	NCBI	Bernatzky and Tanksley (1986)
VF36-*Tm2a* × LA0716 (high-density map of tomato; Tomato-EXPEN 1992)	F_2	67	RFLP, ISO, MO,	1,030; 1,050[i]	SGN, NCBI	Tanksley et al. (1992), Pillen et al. (1996b)
VF36-*Tm2a* × LA0716	F_2	67 (42)	SSR	11		Broun and Tanksley (1996)
VF36-*Tm2a* × LA0716	F_2	67 (42)	AFLP	909		Haanstra et al. (1999b)
VF36-*Tm2a* × LA0716	F_2	67	SSR	19; 20[i]		Areshchenkova and Ganal (1999, 2002)
Vendor *Tm2a* × LA0716	two rec. BC_1	78; 115[h]	RFLP	85		deVicente and Tanksley (1991)
Vendor *Tm2a* × LA0716	F_2	432	RFLP	98		deVicente and Tanksley (1993)
Allround × LA0716	F_2	84 (44)	RFLP, SSR	74 (51; 23)		Arens et al. (1995)
Allround × LA0716	F_2	84 (80)	RFLP, AFLP	707		Haanstra et al. (1999b)
M82 × LA0716	IL	50	RFLP	375	SGN	Eshed and Zamir (1995)

(*continued*)

Table 9.5 (continued)

Initial cross	Mapping population[a]	Population size[b]	Type of markers[c]	No. of markers[d]	Online[e]	Reference[f]
M82 × LA0716	IL	75	RFLP, t-NBS[g]	~665 (~590;75)		Pan et al. (2000)
M82 × LA0716	IL	72	AFLP, SSR, SNP	218 (159; 52; 7)		Suliman-Pollatschek et al. (2002)
M82 × LA0716	IL	52	T-DNA	140		Gidoni et al. (2003)
M82 × LA0716	IL	75	RFLP, CG*CB*	637 (614; 23)		Liu et al. (2003)
M82 × LA0716	IL	75	RFLP, CG*FSC*	~669 (~590; 79)		Causse et al. (2004)
M82 × LA0716	IL	50	SNP	20		Yang et al. (2004)
M82 × LA0716	IL	50	SSR, CAPS	122 (63; 59)		Frary et al. (2005)
M82 × LA0716	IL	75	CG*AA*	13		Stevens et al. (2007)
E6203 × LA1657	BC_2	175	RFLP	150		Frary et al. (2004a)
LA0925 × LA0716 (Tomato-EXPEN 2000)	F_2	83	RFLP, CAPS/COS/COSII, SSR, SNP	2,506	SGN, NCBI	Fulton et al. (2002b); http://solgenomics.net/
LA0925 × LA0716	F_2	83	SSR, CAPS	152 (76; 76)		Frary et al. (2005)
LA0925 × LA0716	F_2	83	RFLP, CAPS, SSR/TES/TGS, TEI	2,116	KAZUSA	Shirasawa et al. (2010)
S. lycopersicum × *S. pimpinellifolium*						
M82 × LA1589	BC_1	257	RFLP, RAPD, SSR, MO	120 (115; 53; 6; 2)	SGN	Grandillo and Tanksley (1996b)
NC84173 × LA0722	BC_1	119	RFLP	151		Chen and Foolad (1999)
Yellow Pear × LA1589	F_2	82	RFLP	82		Ku et al. (1999)
Giant Heirloom × LA1589	F_2	200 (114)	RFLP, CAPS	90		Lippman and Tanksley (2001)
Sun 1642 × LA1589	F_2	100	RFLP, SNP	108 (106; 2)	TMRD	van der Knaap and Tanksley (2001), Yang et al. (2004)
E6203 × LA1589	BC_2F_6-BIL	196	RFLP, MO	127 (126;1)	SGN	Doganlar et al. (2002b)
Long John × LA1589	F_2	85	RFLP	97		van der Knaap et al. (2002)**
Yellow Stuffer × LA1589	F_2	200	RFLP	93		van der Knaap and Tanksley (2003)
Banana Legs(BL), Howard German (HG) × LA1589	BLF_2, HGF_2, $HGBC_1$	99; 130; 100[h]	RFLP, PCR	111; 111; 108[h]		Brewer et al. (2007)***
Rio Grande × LA1589	F_2	94	SSR, RFLP, CAPS	181 (77; 68; 36)	SGN	Gonzalo and van der Knaap (2008)*
NCEBR-1 × LA2093	F_2	172	RFLP, CR-EST, RGA[g]	250 (115; 94; 41)		Sharma et al. (2008)
NCEBR-1 × LA2093	F_7-RIL	170	RFLP, CR-EST, CAPS, SSR[g, j]	294 (132; 132; 16; 14)		Ashrafi et al. (2009)
S. lycopersicum "cerasiforme" × *S. cheesmaniae*						
E9 × L3	F_6-RIL	115	SCAR, SSR	114		Villalta et al. (2005)
S. lycopersicum "cerasiforme" × *S. pimpinellifolium*						
E9 × L5	F_6-RIL	142	SCAR, SSR	132		Villalta et al. (2005)
S. ochranthum × *S. juglandifolium*						
LA3650 × LA2788	Pseudo-F_2	66	COS/COSII, RFLP, SSR	132 (96; 19; 17)		Albrecht and Chetelat (2009)

S. sitiens × *S. lycopersicoides*						
LA1974 × LA2951	F_2	82	RFLP	101		Pertuzé et al. (2002)
S. lycopersicum "cerasiforme" × *S. lycopersicum*						
Cervil × Levovil	F_7-RIL	153	AFLP, RFLP, RAPD, MO	377 (211; 132; 33; 1)		Saliba-Colombani et al. (2000)
S. arcanum × *S. arcanum*						
LA2157 × LA2172	BC_1	268	RFLP	94	NCBI	Van Ooijen et al. (1994)
S. pimpinellifolium × *S. pimpinellifolium*						
LA1237 × LA1581	F_2	147	RFLP	47		Georgiady et al. (2002)

Maps developed for specific chromosomes are not included (see also review by Labate et al. 2007)
[a]*IL* introgression line; *BIL* backcross inbred line; *RIL* recombinant inbred line; *SGe* selective genotyping
[b]In parenthesis: no. of genotyped plants
[c]*AFLP* amplified fragment length polymorphism; *CAPS* cleaved amplified polymorphic sequence; *COS* Conserved Ortholog Set; *COSII* Conserved Ortholog Set II; *ISO* isozyme; *MO* morphological; *PCR* PCR-based, not specified; *RAPD* random amplified polymorphic DNA; *RFLP* restriction fragment length polymorphism; *RGA* resistance gene analog; *SCAR* sequence characterized amplified region; *SNP* single nucleotide polymorphism; *TES* tomato EST-derived SSR; *TGS* tomato genome-derived SSR; *TEI* tomato EST-derived intronic polymorphism; *CGAA* candidate genes associated with ascorbic acid biosynthesis; *CGC* candidate carotenoid genes; *CGFL* candidate genes for flowering; *CGFSC* candidate genes for fruit size and composition; *CR-EST* candidate resistance/defense-response EST; *t-NBS* tomato-nucleotide binding site-leucine rich repeat (NBS-LRR) sequences
[d]In parenthesis: no. of markers per marker type
[e]*SGN* www.sgn.cornell.edu/cview; *NCBI* www.ncbi.nlm.nih.gov/mapview; *TMRD* (Tomato Mapping Resources Database) www.tomatomap.net; *KAZUSA* http://www.kazusa.or.jp/tomato/
[f]*Linkage map not shown; **Linkage maps shown only for four chromosomes; ***A composite map shown for the two F_2 populations
[g]The approximate locations of disease resistance genes (R genes) and QTL are shown
[h]Per population
[i]Per study
[j]The approximate locations of fruit quality-related genes are shown

Table 9.6 Summary of disease resistance genes and QTLs mapped in *Solanum* sect. *Lycopersicon* and allied species, using molecular markers

Source of resistance/tolerance	Disease	Pathogen	Gene/QTL[a]	Chrom. location[b]	Mapping population[c]	Marker type[d]	References
S. arcanum LA2157	Bacterial canker	*Clavibacter michiganensis* ssp. *michiganensis*	5 (Q)	1, 6–8, 10	(three) BC_1	RFLP	Sandbrink et al. (1995)
S. arcanum LA2157	Bacterial canker	*C. michiganensis* ssp. *michiganensis*	3 (Q)	5, 7, 9	F_2	RFLP, SCAR	Van Heusden et al. (1999)
S. arcanum LA2157	Early blight	*Alternaria solani*	6 (Q)	1, 2, 5–7, 9	F_2, F_3	AFLP, SSR, SNP	Chaerani et al. (2007)
S. arcanum LA2172	Powdery mildew	*Oidium neolycopersici*	*Ol-4*	6	pseudo-F_2, BC_2S_1	AFLP, PCR	Bai et al. (2004, 2005)
S. arcanum LA2157	Nematode (root-knot)	*Meloydogine* spp.	*Mi-9*	6S	F_2, F_3	RFLP, PCR	Veremis et al. (1999), Veremis and Roberts (2000), Ammiraju et al. (2003), Jablonska et al. (2007)
S. cheesmaniae LA0422	Blackmold	*Alternaria alternata*	5(Q)*: Bm-2a, 2c, 3, 9, 12*	2 (two), 3, 9, 12	BC_1S_1, BC_1S_2	RFLP, CAPS	Cassol and St. Clair (1994), Robert et al. 2001
S. chilense LA0458	Cucumber mosaic	Cucumber mosaic virus (CMV)	*Cmr*	12	BC_1-inbreds	ISO, RFLP	Stamova and Chetelat (2000)
S. chilense LA1969	Tomato yellow leaf curl	Tomato yellow leaf curl virus (TYLCV)	*Ty-1*	6S	BC_1S_1, NIL	RFLP, ISO	Zamir et al. (1994), Ji et al. (2007a)
S. chilense LA1932, LA2779, LA1938	Tomato yellow leaf curl, tomato mottle virus	TYLCV, tomato mottle virus (ToMoV)	3 (Q)	6	(three) F_2	RAPD	Griffiths and Scott (2001), Ji and Scott (2005), Agrama and Scott (2006), Ji et al. (2007a)
S. chilense LA2779, LA1932	Tomato yellow leaf curl, tomato mottle	TYLCV, ToMoV	*Ty-3*	6L	$F_{2,}$ ABLs	RAPD/SCAR, CAPS	Ji et al. (2007a), Jensen (2007)
S. chilense LA1932	Tomato yellow leaf curl, tomato mottle	TYLCV	*Ty-4*	3L	ABLs	PCR	Ji et al. (2009a)
S. chilense LA0458	Powdery mildew	*Leveillula taurica*	*Lv*	12C	(two)$F_{2,}$ BC_1	RAPD, RFLP	Chungwongse et al. (1994, 1997)
S. habrochaites PI 134417	Alfalfa mosaic	Alfalfa mosaic virus (AMV)	*Am*	6S	F_2, rec. BC_1	RFLP, AFLP	Parrella et al. (2004)
S. habrochaites PI 247087	Potyviruses	*Potato virus Y* (PVY) and *tobacco etch virus* (TEV)	*pot-1*	3S	F_2/F_3, BC_1	AFLP, RFLP	Parrella et al. (2002), Ruffel et al. (2005)
S. habrochaites PI 126445	Tobacco/tomato mosaic	Tobacco/tomato mosaic virus (TMV/ToMV)	*Tm-1*	2C	F_2 NILs	RAPD, SCAR	Holmes (1957), Levesque et al. (1990), Tanksley et al. (1992), Ohmori et al. (1996), Ishibashi et al. (2007)

S. habrochaites B6013	Tomato yellow leaf curl	TYLCV	*Ty-2*	11L	F_2, F_3 from H24 line	RFLP	Hanson et al. (2000, 2006), Kalloo and Banerjee (2000), Ji et al. (2007b, 2009b)
S. habrochaites LA0407	Bacterial canker	*Clavibacter michiganensis* ssp. *michiganensis*	2 (Q): *Rcm2.0, Rcm5.1*	2, 5	BC_2S_5-BILs F_2 from BIL	RFLP, PCR	Kabelka et al. (2002), Coaker and Francis (2004)
S. habrochaites PI 126445	Early blight	*Alternaria solani*	7-13 (Q): *EBR1.1-12.2*	1–6, 8–12	BC_1, BC_1S_1	RFLP, RGA	Foolad et al. (2002), Zhang et al. (2002, 2003b)
S. habrochaites LYC4	Gray mold	*Botrytis cinerea*	3, 10 (Q): *Rbcq1,2,4, 6,9,11,12*	1–3, 4 (two), 6, 9 (two), 11, 12	F_2, ILs	AFLP, CAPS, SCAR	Finkers et al. (2007a, b)
S. habrochaites LA1033	Late blight	*Phytophthora infestans*	4–9 (Q)	ns[f]	F_2, BC_1F_1	AFLP	Lough (2003)
S. habrochaites LA2099	Late blight	*Phytophthora infestans*	13, 18 (Q): *lb1a-lb12b*	All	rec. BC_1, NIL, sub-NILs	RFLP	Brouwer et al. (2004), Brouwer and St Clair (2004)
S. habrochaites PI 370085	Leaf mold	*Cladosporium fulvum* (syn. *Passalora fulva*)	*Cf-4*	1S – *MW* locus	F_2 from NILs	RFLP	Jones et al. (1993), Balint-Kurti et al. (1994), Parniske et al. (1997), Thomas et al. (1997)[e], Rivas and Thomas (2005)
S. habrochaites G1.1560	Powdery mildew	*Oidium neolycopersici*	*Ol-1*	6L	F_2, BC_1S_1, BC_1S_2	RFLP, RAPD, SCAR	Van der Beek et al. (1994), Huang et al. (2000a, b), Bai et al. (2005)
S. habrochaites G1.1290	Powdery mildew	*Oidium neolycopersici*	*Ol-3*	6L	ABL	RFLP, SCAR	Huang et al. (2000b), Bai et al. (2005)
S. habrochaites PI 247087	Powdery mildew	*Oidium neolycopersici*	*Ol-5*	6L	BC_2, BC_2S_1 from ABL	PCR	Bai et al. (2005)
S. lycopersicoides LA2951	Gray mold	*Botrytis cinerea*	7 (Q): 5 R + 2 Su	1–5, 9, 11	ILs	CAPS, RFLP	Davis et al. (2009)
S. lycopersicum cv. 'Hawaii 7998'	Bacterial spot	*Xanthomonas campestris* pv. *vesicatoria*	*rx-1, rx-2, rx-3*	1S, 1L, 5	BC_1	ISO, RFLP	Yu et al. (1995)
S. lycopersicum cv. MoneyMaker	Bacterial spot	*Xanthomonas campestris* pv. *vesicatoria* (avrBs4-expressing strain)	*Bs4*	5S	F_2	AFLP. RFLP, STS	Ballvora et al. (2001), Schornack et al. (2004)[e]

(*continued*)

Table 9.6 (continued)

Source of resistance/tolerance	Disease	Pathogen	Gene/QTL[a]	Chrom. location[b]	Mapping population[c]	Marker type[d]	References
S. lycopersicum cv. "Hawaii 7996"	Bacterial wilt	*Ralstonia solanacearum* race 1	several (Q)	3, 4 (two), 6 (two), 8, 10, 11	F_2, F_{2-3}, F_3	RFLP, RAPD	Thoquet et al (1996a, b), Mangin et al. (1999)
S. lycopersicum cv. "Hawaii 7996"	Bacterial wilt	*Ralstonia solanacearum* race 3	4 (Q): *Bwr-3, Bwr-4, Bwr-6, Bwr-8*	3, 4, 6, 8	F_2:3, F_8-RIL	RFLP	Carmeille et al. (2006)
S. lycopersicum	Antracnose ripe rot	*Colletotrichum coccodes*	6 (Q)	Various	F_2	RAPD, AFLP	Stommel and Zhang (1998, 2001)
S. lycopersicum	Leaf mold	*Cladosporium fulvum*	*Cf-1*	1S – *MW*	ns[f]	ns[f]	Langford (1937), Kerr and Bailey (1964), Jones et al. (1993), Rivas and Thomas (2005)
S. lycopersicum 'Peru Wild'	Verticillium wilt	*Verticillium dahliae* race 1	*Ve1, Ve2*	9S	$F_{2:3}$, RIL, IL	RAPD, RFLP	Kawchuk et al. (1998), Diwan et al. (1999), Kawchuk et al. (2001)[e]
S. lycopersicum "cerasiforme" L285	Bacterial wilt	*Ralstonia solanacearum*	3 (Q)	6, 7, 10	F_2, F_3	RFLP, RAPD	Danesh et al. (1994)
S. lycopersicum "cerasiforme": PI 187002	Leaf mold	*Cladosporium fulvum* (syn. *Passalora fulva*)	*Cf-5*	6S	F_2 from NILs	RAPD, RFLP	Jones et al. (1993), Dickinson et al. (1993), Dixon et al. (1998)[e], Rivas and Thomas (2005)
S. lycopersicum "cerasiforme" LA1230	Powdery mildew	*Oidium neolycopersici*	*ol-2*	4C	F_2	AFLP, RAPD, SCAR	De Giovanni et al. (2004), Bai et al. (2008)
S. neorickii G1.1601	Gray mold	*Botrytis cinerea*	3 (Q): *pQTL3, pQTL4, pQTL9*	3, 4, 9	F_3, BC_3S_1, BC_3S_2	AFLP, CAPS, SCAR	Finkers et al. (2008)
S. neorickii G1.1601	Powdery mildew	*Oidium neolycopersici*	3 (Q): *Ol-qtl1, -qtl2, -qtl3*	6L, 12 (two)	F_2, F_3	AFLP, CAPS, SCAR	Bai et al. (2003)
S. pennellii LA0716	Bacterial spot	*Xanthomonas campestris pv. vesicatoria* (race T3)	*Xv4*	3	F_2, IL	RFLP, CAPS	Astua-Monge et al. (2000)
S. pennellii LA0716	*Alternaria* stem canker	*Alternaria alternata* f. sp. *lycopersici*	*Asc*	3L	F_2, BC_1	RFLP	Van der Biezen et al. (1995), Mesbah et al. (1999), Brandwagt et al. (2000)
S. pennellii LA0716	Fusarium wilt	*Fusarium oxysporum f. sp. lycopersici* (race 1)	*I-1*	7	BC_1	RFLP	Sarfatti et al. (1991), Scott et al. (2004)

S. pennellii LA0716 and PI 414773	Fusarium wilt	*Fusarium oxysporum f. sp. lycopersici* (race 3)	*I-3*	7L	BC_1, IL	RFLP, RGA, AFLP, SCAR, CAPS	Bournival et al. (1989, 1990), Scott and Jones (1989), Sarfatti et al. (1991), Tanksley and Costello (1991), Sela-Buurlage et al. (2001), Hemming et al. (2004), Scott et al. (2004), Lim et al. (2008)
S. pennellii LA0716	Fusarium wilt	*Fusarium oxysporum f. sp. lycopersici* (race 2)	*I-5, I-6*	2, 10	IL	RFLP	Sela-Buurlage et al. (2001)
S. peruvianum PI 128650	Tobacco mosaic	Tobacco Mosaic Virus (TMV)	*Tm-2a* (*Tm-2*^{2}) and *Tm-2*	9C	F_2, NILs	RFLP	Young et al. (1988), Pillen et al. (1996a), Lanfermeijer et al. (2003, 2005)[e]
S. peruvianum s.l. (ns)[f,g]	Tomato spotted wilt	Tomato Spotted Wilt (TSWV)	*Sw-5*	9L	NIL, NIL-BC, F_2	RFLP, RAPD	Stevens et al. (1995), Brommonschenkel and Tanksley (1997), Brommonschenkel et al. (2000)[e]
S. peruvianum s.l. (ns)[g]	Tomato yellow leaf curl	TYLCV	5(Q): *Ty-5* (major) + 4 (minor)	1, 4 (major), 7, 9, 11	F_2	PCR	Anbinder et al. (2009)
S. peruvianum s.l. (ns)[g]	Corky root rot	*Pyrenochaeta lycopersici*	*py-1*	3S	NIL, F_2	RAPD, RFLP	Doganlar et al. (1998)
S. peruvianum s.l. (ns)[g]	Fusarium crown and root rot	*Fusarium oxysporum* f. sp. *radici-lycopersici*	*Frl*	9	F_2	NA	Vakalounakis et al. (1997)
S. peruvianum PI 128657	Nematode (root knot) and potato aphid	*Meloidogyne* spp. and *Macrosiphum euphorbiae*	*Mi-1* (*Meu-1*)	6S	NILs, (four) F_2	RFLP, ISO	Medina-Filho (1980), Klein-Lankhorst et al. (1991), Messeguer et al. (1991), Milligan et al. (1998)[e], Rossi et al. (1998), Vos et al. (1998)
S. peruvianum PI 126443-1MH	Nematode (root knot)	*Meloidogyne* spp	*Mi-3, Mi-5*	12S	BC, F_2	RAPD, RFLP	Yaghoobi et al. (1995), Veremis and Roberts (1996a, b)
S. pimpinellifolium "hirsute INRA"	Tomato yellow leaf curl	TYLCV	1 (Q)	6	F_4	RAPD	Chaguè et al. (1997)
S. pimpinellifolium (ns)	Bacterial speck	*Pseudomonas syringae* pv. 'tomato'	*Pto + Prf*	5S	NIL, F_2	RFLP, RAPD	Martin et al. (1991, 1993)[e], Salmeron et al. (1996)
S. pimpinellifolium PI 79532	Fusarium wilt	*Fusarium oxysporum* f. sp. *lycopersici* (race 1)	*I*	11S	BC_1	RFLP	Paddock (1950), Sela-Buurlage et al. (2001), Scott et al. (2004)

(continued)

Table 9.6 (continued)

Source of resistance/tolerance	Disease	Pathogen	Gene/QTL[a]	Chrom. location[b]	Mapping population[c]	Marker type[d]	References
S. pimpinellifolium PI 126915	Fusarium wilt	*Fusarium oxysporum* f. sp. *lycopersici* (race 2)	*I-2* complex locus (*I2C* gene family)	11L	NIL, F_2	MO, RFLP,	Laterrot (1976), Sarfatti et al. (1989), Segal et al. (1992), Ori et al. (1997), Simons et al. (1998)[e], Sela-Buurlage et al. (2001)
S. pimpinellifolium PI 79532	Gray leaf spot	*Stemphyllium spp.*	*Sm*	11	F_2	RFLP	Dennett (1950), Behare et al. (1991)
S. pimpinellifolium L3708	Late blight	*Phytophtora infestans*	2 (Q)	6, 8	F_2	RFLP	Frary et al. (1998)
S. pimpinellifolium WVa700	Late blight	*Phytophtora infestans*	*Ph-2*	10L	Several	AFLP, RFLP	Moreau et al. (1998)
S. pimpinellifolium L3708	Late blight	*Phytophtora infestans*	*Ph-3*	9L	F_2	AFLP, RFLP	Chungwongse et al. (2002)
S. pimpinellifolium (ns)[f]	Leaf mold	*Cladosporium fulvum* (syn. *Passalora fulva*)	*Cf-2*	6S	ns[f]	ns[f]	
S. pimpinellifolium PI 126915	Leaf mold	*Cladosporium fulvum* (syn. *Passalora fulva*)	*Cf-9*	1S – *MW*	F_2, NIL, BC_1	RFLP	van der Beek et al. (1992), Jones et al. (1993, 1994), Balint-Kurti et al. (1994), Parniske et al. (1997), Rivas and Thomas (2005)
S. pimpinellifolium PI 126947	Leaf mold	*Cladosporium fulvum* (syn. *Passalora fulva*)	*Cf-ECP2*, *Cf-ECP3*	1S – *OR*	F_2	CAPS	Haanstra et al. (1999a), Yuan et al. (2002)
S. pimpinellifolium CGN15529	Leaf mold	*Cladosporium fulvum* (syn. *Passalora fulva*)	*Cf-ECP5*	1S – *AU*	TC, F_3	CAPS	Haanstra et al. (2000)
S. pimpinellifolium LA1547, LA1683	Leaf mold	*Cladosporium fulvum* (syn. *Passalora fulva*)	*Cf-ECP1*, *Cf-ECP4*	1S – *MW*	NILs, F_2	PCR	Soumpourou et al. (2007)
S. pimpinellifolium LA0121	Potato cyst nematode	*Globodera restochiensis*	*Hero*	4S	NIL, F_2	RAPD, RFLP	Ganal et al. (1995), Ernst et al. (2002)[e]

[a]*(Q)* QTL; *R* resistant, *Su* susceptible; for the approximate locations of some of the listed disease resistance genes (R genes) and QTLs see Pan et al. (2000); Zhang et al. (2002, 2003a); Ashrafi et al. (2009)
[b]*L* long arm of chromosome; *S* short arm of chromosome; *C* centromeric region; *MW Milky Way* locus; *OR Orion* locus; *AU Aurora* locus
[c]*ABL* advanced breeding lines; *TC* testcross; *rec.* reciprocal
[d]*STS* sequence tagged site; for other markers abbreviations see legend to Table 9.5
[e]Studies reporting the cloning of the corresponding gene(s)
[f]*ns* accession number, or other data, not specified
[g]Not being available the accession number, these genotypes have been referred to as *S. peruvianum* s.l

Table 9.7 Summary of QTL mapping studies conducted in *Solanum* sect. *Lycopersicon* for morphological, yield-, fruit quality- and reproductive-related traits

Wild or donor parent	Main traits analyzed	No. traits evaluated[a]	No. QTL[b]	Mapping population (pop. size)[c]	Marker type[d]	No. Markers	References[e]
S. arcanum LA1708	Yield, fruit quality, horticultural	35 (29)	166	BC_3/BC_4 (200)	RFLP, PCR, MO	174	Fulton et al. (1997)
S. arcanum LA1708	Biochemical related to flavor	15	103	BC_3/BC_4 (200)	RFLP, PCR, MO	174	Fulton et al. (2002a)
S. chmielewskii LA1028	Fruit weight, brix, pH	3	15	BC_1 (237)	RFLP, ISO, MO	70	Paterson et al. (1988, 1990), Frary et al. (2003)
S. chmielewskii LA1028	Brix	1	ns[f]	LA1563 (BC_5S_5), derived F_2	RFLP	60	Osborn et al. (1987)
S. chmielewskii LA1028	Yield, brix, pH	6	ns	LA1500-LA1503, LA1563 (BC_5S_5), derived F_2/F_3	RFLP, ISO	132	Tanksley and Hewitt (1988)
S. chmielewskii LA1028	Yield, brix, fruit quality	13	ns	LA1500-LA1503, LA1563 (BC_5S_5), BILs	RFLP	9	Azanza et al. (1994)
S. chmielewskii CH6047	Flowering time	2	8	F_2 (149)	AFLP, CAPS/SCAR/CG, SSR	225	Jiménez-Gómez et al. (2007)
S. chmielewskii LA1840	Fruit weight and composition	16 (14)	103	ILs (20)	COSII, SSR	133	Prudent et al. (2009)
S. galapagense LA0483	Frui size, brix, pH	3	29	F_2/F_3 (350)	RFLP	71	Paterson et al. (1991)
S. galapagense LA0483	Fruit quality	3	73*	F_8 RILs (97)	RFLP, MO, ISO	135	Goldman et al. (1995)
S. galapagense LA0483	Morphological	7	41*	F_8 RILs (97)	RFLP, MO, ISO	135	Paran et al. (1997)
S. habrochaites LA1777	Sexual compatibility factors and floral morphology	9	23	BC_1 (149)	RFLP	135	Bernacchi and Tanksley (1997), Chen and Tanksley (2004), Chen et al. (2007)
S. habrochaites LA1777	Yield, fruit quality, horticultural	19	121	BC_2/BC_3 (315/200)	RFLP	122	Bernacchi et al. (1998a)
S. habrochaites LA1777 and *S. pimpinellifolium* LA1589	Yield, fruit quality, horticultural	12	22	NILs	RFLP	ns[f]	Bernacchi et al. (1998b), Monforte and Tanksley (2000b), Monforte et al. (2001), Yates et al. (2004)
S. habrochaites LA1777	Biochemical related to flavor	15	34	BC_2/BC_3 (315/200)	RFLP	122	Fulton et al. (2002a)
S. habrochaites LA1777	Aroma volatiles	40 (27)	30	ILs, BILs (89)	RFLP	95	Mathieu et al. (2009)
S. habrochaites LA1777	Hybrid incompatibility, floral morphology	25	22	ILs, BILs (71)	RFLP	95	Moyle and Graham (2005), Moyle (2007)
S. habrochaites LA0407	Stem vascular morphology	5	1	BILs (BC_2S_5), $F_{2:3}$ (64)	RFLP, PCR	67	Coaker et al. (2002)
S. habrochaites LA0407	Fruit color	3	13	BILs (BC_2S_5)/F_3, F_4 (64)	RFLP, PCR	63; 394	Kabelka et al. (2004)
S. habrochaites PI-247087	Ascorbic acid	2	5	BC_2/BC_2S_1 (130/79,68)	AFLP, RFLP, SSR, MO, CGAA	217	Stevens et al. (2007)

(*continued*)

Table 9.7 (continued)

Wild or donor parent	Main traits analyzed	No. traits evaluated[a]	No. QTL[b]	Mapping population (pop. size)[c]	Marker type[d]	No. Markers	References[e]
S. habrochaites LYC4 (IL5-1 and IL5-2 lines) and *S. habrochaites* (IVT-line 1)	Parthenocrapy, stigma exsertion	2	5	(two) BC_5S_1, F_2	CAPS, COS, SSR	34	Gorguet et al. (2008)
S. lycopersicum "cerasiforme" Cervil inbred line	Aroma volatiles (18), fruit quality	32 (26)	81	F_7-RILs (144)	RFLP, AFLP, RAPD, MO	103	Saliba-Colombani et al. (2001), Causse et al (2002, 2007), Lecomte et al. (2004a, b), Chaïb et al. (2006)
S. lycopersicum "cerasiforme" Cervil inbred line	Sensory attributes (12)	12	49	F_7-RILs (144)	RFLP, AFLP, RAPD, MO	103	Causse et al. (2001, 2002, 2007), Lecomte et al. (2004a, b), Chaïb et al. (2006)
S. lycopersicum "cerasiforme" Cervil inbred line	Ascorbic acid	2	6	F_7-RILs (144)	RFLP, AFLP, RAPD, MO	103	Stevens et al. (2007)
S. neorickii LA2133	Yield, fruit quality, horticultural	30	199	BC_2/BC_3 (170)	RFLP, PCR, MO	133	Fulton et al. (2000)
S. neorickii LA2133	Biochemical related to flavor	15	52	BC_2/BC_3 (170)	RFLP, PCR, MO	133	Fulton et al. (2002a)
S. pennellii LA0716	Fruit weight, seed weight, stigma exsertion, leaflet shape	4	21	BC_1 (400)	ISO	12	Tanksley et al. (1982)
S. pennellii LA0716	Morphological (plant, flower, leaf)	11	74	F_2 (432)	RFLP	98	deVicente and Tanksley (1993)
S. pennellii LA0716	Yield, fruit quality	6	104	ILs/HILs/ILs × Tester (49/50/50)	RFLP	375	Eshed and Zamir (1995, 1996), Alpert et al. (1995), Eshed et al. (1996), Gur and Zamir (2004)
S. pennelli LA0716	Fruit shape	2	1	F_2 from IL2-5 (60)	RFLP	15	Ku et al. (1999)
S. pennellii LA0716	Sensory attributes , aroma volatiles	ns	1	ILS (4)	RFLP	ns	Tadmor et al. (2002)
S. pennellii LA0716	Leaf morphology and size	8	30	ILS (58)	RFLP	375	Holtan and Hake (2003)
S. pennellii LA0716	Fruit color, carotenoids	6	50	ILs (75)	RFLP, CG	637 (614, 23)	Liu et al. (2003)
S. pennellii LA0716 and *S. pimpinellifolium* LA1589	Locule number	2	4	Several F_2	ns	ns	Barrero and Tanksley (2004)
S. pennellii LA0716	Fruit size and composition	9	81	ILs (70)	RFLP, CG	671 (592,79)	Causse et al. (2004)
S. pennellii LA0716	Leaf and flower morphology	22 (18)	36	F_2 (83)	RFLP, SSR, COS	391, (350, 10, 31)	Frary et al. (2004b)

S. pennellii LA0716	Fruit quality, transcriptomic analysis	6	ns	ILs (6)	RFLP	ns	Baxter et al. (2005)
S. pennellii LA0716	Fruit antioxidants	5	20	ILs (76)	RFLP	~600	Rousseaux et al. (2005)
S. pennelli LA0716	Primary metabolites (74), yield related	83	889, 326[g]	ILs (76)	RFLP	~600	Schauer et al. (2006)
S. pennelli LA0716	Yield fitness	35	841	ILs, ILHs (76; 76)	RFLP	~600	Semel et al. (2006)
S. pennelli LA0716	Aroma volatiles (23), organic acids	25 (24)	29	ILs (74)	RFLP	~600	Tieman et al. (2006)
S. pennellii LA0716	Ascorbic acid	1	12	ILs (71)	RFLP	~600	Stevens et al. (2007, 2008)
S. pennelli LA0716	Hybrid incompatibility	4	19	ILs (71)	RFLP	~600	Moyle and Nakazato (2008)
S. pennelli LA0716	Primary metabolites (74)	74	332	ILs, ILHs (68;68)	RFLP	~600	Schauer et al. (2008)
S. pennellii LA1657	Yield, fruit quality, horticultural	25	84	BC_2/BC_2F_1 (175)	RFLP	150	Frary et al. (2004a)
S. pimpinellifolium CIAS27	Fruit quality, horticultural	18	85	F_2 (1,700)	MO, ISO	6, 4	Weller et al. (1988)
S. pimpinellifolium LA1589	Fruit quality, flower morphology, flowering and ripening time	19	54	BC_1 (257)	MO, RAPD, RFLP	120	Grandillo and Tanksley (1996a), Alpert et al. (1995), Grandillo et al. (1996), Ku et al. (2000)
S. pimpinellifolium LA1589	Yield, fruit quality, horticultural	21 (18)	87	$BC_2/BC_2F_1/BC_3$ (~170/170)	MO, RAPD, CAPS, RFLP	121	Tanksley et al. (1996)
S. pimpinellifolium LA0722	Fruit quality, lycopene	7	59	BC_1/BC_1S_1 (119)	RFLP	151	Chen et al. (1999)
S. pimpinellifolium LA1589	Fruit shape	2	2	F_2 (82)	RFLP	82	Ku et al. (1999)
S. pimpinellifolium LA1589	Fruit size and shape	7	30	F_2 (114)	RFLP, CAPS	90	Lippman and Tanskley (2001)
S. pimpinellifolium LA1589	Fruit and ovary shape	2	1	F_2 (100)	RFLP, SNP	108	van der Knaap and Tanksley (2001)
S. pimpinellifolium LA1589	Fruit quality, horticultural	22	71	BC_2F_6 – BILs (196)	RFLP, MO	127	Doganlar et al. (2002b)
S. pimpinellifolium LA1589	Biochemical related to flavor	15	33	$BC_2/BC_2F_1/BC_3$ (~170)	MO, RAPD, CAPS, RFLP	121	Fulton et al. (2002a)
S. pimpinellifolium LA1589	Fruit shape	3	4	F_2 (85)	RFLP	97	van der Knaap et al. (2002)
S. pimpinellifolium LA1589	Fruit shape and size	10	50	F_2 (200)	RFLP	93	van der Knaap and Tanksley (2003)
S. pimpinellifolium LA1589	Fruit shape	15	36, 32, 27[h]	(two) F_2, BC_1 27(99; 130; 100)[h]	RFLP, PCR	111, 111, 108[h]	Brewer et al. (2007)
S. pimpinellifolium LA1589	Fruit shape	14	20, 23, 20[h]	(three) F_2 (130;106; 94)	RFLP, PCR	111, 96, 97	Gonzalo and van der Knaap (2008)

(*continued*)

Table 9.7 (continued)

Wild or donor parent	Main traits analyzed	No. traits evaluated[a]	No. QTL[b]	Mapping population (pop. size)[c]	Marker type[d]	No. Markers	References[e]
S. lycopersicum IVT KTl (breeding line containing S. pimpinellifolium and S. neorickii introgressions)	Fruit size, flowering and ripening time	6	3	F_2, F_3 (292)	RFLP	45	Lindhout et al. (1994c)
S. pimpinellifolium LA1237 (the "selfer") and LA1581 (the "outcrosser")	Flower morphology and number	6 (4)	5	F_2 (147)	RFLP	48	Georgiady et al. (2002)

[a]The number of traits for which QTLs were identified is indicated in parenthesis
[b]An "*" indicates the number of significant markers × traits associations
[c]*ILH* Introgression line hybrid
[d]For markers abbreviations see legends to Tables 9.5 and 9.6
[e]Some of the related and/or follow-up studies are also listed
[f]*ns* not specified
[g]Fruit metabolism and yield-associated traits, respectively
[h]Per population

Type and source of tolerance/resistance[a]	Developmental stage[b]	No. traits or treatments	No. QTL (Q)	Chromosome	Mapping population	Marker type[c]	No. Markers[d]	References
Cold								
S. habrochaites LA1777	Pollen selection	1	2	6, 12	BC_1	ISO	9	Zamir et al. (1982)
S. habrochaites	VG	1	3	6, 7, 12	BC_1	ISO	11	Vallejos and Tanksley (1983)
S. habrochaites LA1778	VG	7	10	1, 3, 5, 6 (three Q), 7, 9, 11, 12	BC_1	RFLP	89	Truco et al. (2000)
S. pimpinellifolium LA0722	SG	1	3–5	1 (two Q), 4	BC_1S_1	RFLP	151	Foolad et al. (1998b)
Drought								
S. pennellii (ns)[e]	VG	1	3	ns[e]	F_3, BC_1S_1	RFLP	17	Martin et al. (1989)
S. pennellii LA0716	VG	1	6	2, 3, 5, 7, 9, 12	ILs, sub-ILs of IL5-4	STS, CAPS, AFLP, SSR	29	Xu et al. (2008)
S. pimpinellifolium LA0722	SG	1	4	1, 8, 9, 12	BC_1S_1	RFLP	119	Foolad et al. (2003)
Salt								
S. galapagenese (L2) and *S. pimpinellifolium* (L1 and L5)	RS	4	31	1–5,7, 9–12	three F_2	RFLP, ISO	19, 3	Monforte et al. (1997a)
S. galapagenese (L2) and *S. pimpinellifolium* (L1 and L5)	RS	2, 4	43	1–5, 7, 9–12	three F_2	RFLP, ISO	19, 3	Monforte et al. (1997b)
S. galapagense L2	VG, RS	6	8	1, 2, 5, 7, 9, 12	F_2	RFLP, ISO	20, 3	Monforte et al. (1999)
S. galapagense (L2) and *S. pimpinellifolium* (L5)	VG, RS	19	12, 23	1–8, 10–12	two F_7-RILs	RFLP, SSR, CG	153, 124	Villalta et al. (2007)
S. galapagense (L2) and *S. pimpinellifolium* (L5)	VG	10	18, 25	1, 3, 5–8, 11, 12	two F_8-RILs	RFLP, SSR, CG	153, 124	Villalta et al. (2008)
S. galapagense (L2) and *S. pimpinellifolium* (L5)[f]	RS	3	8	3, 9, 11	two F_9-RILs	RFLP, SSR, CG	153, 124	Estañ et al. (2009)
S. pennellii LA0716	VG	3	6	1, 2, 4–6, 12	F_2	ISO	15	Zamir and Tal (1987)
S. pennellii LA0716	SG	2	5	1, 3, 7, 8, 12	F_2 (SGe)[g]	ISO	16	Foolad and Jones (1993)
S. pennellii LA0716	SG	1	8	1–3, 7–9 (two Q), 12	F_2 (SGe)	ISO, RFLP	16, 68	Foolad et al. (1997)
S. pennellii LA0716	SG	1	8	1, 3, 5 (two Q), 6, 8, 9, unknown	F_2 (SGe)	RAPD	53	Foolad and Chen (1998)
S. pennellii LA0716	VG	12	125	All	ILs	PCR	122	Frary et al. (2010)
S. pimpinellifolium L1	RS	3	6	1–4, 10, 12	F_2, F_3	ISO, RAPD, RFLP	2, 2, 10	Bretó et al. (1994), Monforte et al. (1996)
S. pimpinellifolium L1	RS	3	12	1–4, 10, 12	F_2	ISO, RFLP	2, 14	Monforte et al. (1996)
S. pimpinellifolium LA0722	SG	1	7	1 (two Q), 2, 5, 7, 9, 12	BC_1S_1	RFLP	151	Foolad et al. (1998a)
S. pimpinellifolium LA0722	VG	1	5	1 (two Q), 3, 5, 9	BC_1S_1	RFLP	151	Foolad and Chen (1999)
S. pimpinellifolium LA0722	VG	1	5	1, 3, 5, 6, 11	BC_1 (SGe)	RFLP	115	Foolad et al. (2001)

[a]*S. galapagense* L2 = it was *Lycopersicon cheesmanii* f. *minor* in the original study by Monforte et al. (1997a)
[b]*RS* reproductive stage, *SG* seed germination, *VG* vegetative growth
[c]For markers abbreviations see legends to Tables 9.5 and 9.6
[d]Per marker type
[e]*ns* accession number, or other data, not specified
[f]Used as a rootstock
[g]*SGe* Selective genotyping

Subsequently, an F_2 population of 83 individuals derived from the cross *S. lycopersicum* (LA0925) × *S.pennellii* (LA0716) was used to construct the first PCR-based reference genetic map covering the entire tomato genome (Frary et al. 2005). The same population has been used to develop a new molecular linkage map using conserved ortholog set (COS) and conserved ortholog set II (COSII) markers derived from a comparison of the tomato expressed sequence tag (EST) database against the entire *Arabidopsis* genome (Fulton et al. 2002b; Wu et al. 2006). These markers have been selected to be single/low copy and to have a highly significant match with a putative orthologous locus in the model species *Arabidopsis thaliana* (L.) Heynh. This map, referred to as Tomato-EXPEN 2000, contains also a subset of RFLP markers from the Tomato-EXPEN 1992 map, a significant number of SSRs identified in ESTs, and other CAPSs, which, as of July 2010, add up to a total of over 2,500 markers (http://solgenomics.net/). Recently, the Tomato-EXPEN 2000 mapping population was used to generate a new linkage map based on SSRs derived from EST (TES) and from genome sequences (TGS) as well as intronic polymorphism markers (TEI) (Table 9.5; Shirasawa et al. 2010). Altogether, this new high-density linkage map includes a total of 2,116 marker loci (1,433 new and 683 existing) covering 1,503.1 cM. The large number of SSR and SNP markers developed in this study provide new useful tools also for molecular breeding in tomato.

Online versions of some of the aforementioned maps are available at the SOL Genomics Network (SGN; http://solgenomics.net/) (Mueller et al. 2005a), the National Center for Biotechnology Information (NCBI) (Wheeler et al. 2004), and http://www.tomatomap.net (Van Deynze et al. 2006) (Table 9.5). Information on the DNA markers developed by Shirasawa et al. (2010) is available at http://www.kazusa.or.jp/tomato/.

The comparison of genetic maps based on interspecific crosses between *S. lycopersicum* and wild tomato species shows a generally conserved gene order (with a few exceptions), suggesting a strong synteny within this plant group (see Sect. 9.2.5), although the total genetic lengths of these published maps can vary (reviewed by Ji and Scott 2007). As was already reported by Rick (1969), recombination is not a process that occurs randomly over the entire genome. Recombination frequencies may vary dramatically in intensity between chromosomal regions and among populations, although it is not yet clear to what extent they might be affected by the phylogenetic relationship between species. Whatever the cause, these phenomena have been exploited to generate dense molecular linkage maps around specific gene(s), sometimes by using combinations of several inter- and/or intraspecific mapping populations (e.g., Balint-Kurti et al. 1994; Ganal and Tanksley 1996; Bonnema et al. 1997). Skewed segregation is another phenomenon that has been reported in many interspecific crosses of tomato, with the extent of skewness generally being greater in wider crosses compared to crosses between closely related species, and also generally greater in F_2 than in BC populations, as well as in F_7 compared to the original F_2 generation (Chen and Foolad 1999; Paran et al. 1995; Villalta et al. 2005).

Comparative genetic mapping studies have been conducted to understand the genetic relationships between the sects. *Lycopersicoides* and *Juglandifolia* species and cultivated tomato, and therefore to evaluate their potential use in breeding programs, as well as their history of evolution and speciation. One such study, based on a *S. lycopersicoides* and *S. sitiens* F_2 population, revealed a high degree of colinearity, except for chromosome 10, where a paracentric inversion was detected (see also Sect. 9.2.5; Pertuzé et al. 2002). More recently, the genetic relationships of the two nightshades *S. ochranthum* and *S. juglandifolium* to tomato and other *Solanum* species were also investigated using comparative genetic linkage maps obtained from a *S. ochranthum* × *S. juglandifolium* F_2 population (Albrecht and Chetelat 2009). This study shows that, in spite of the strong reproductive barriers that isolate these two taxa from the tomatoes (*Solanum* sect. *Lycopersicon*), most regions of the identified twelve linkage groups were co-linear with the tomato reference maps.

9.6.4 Mapping of Genes and Polygenic Clusters

Interspecific crosses have been widely used for genetic analysis in tomato (Stevens and Rick 1986; Kalloo 1991). The reduced polymorphism at the DNA level between cultivated tomato varieties has stimulated the extensive utilization of domesticated-by-wild crosses for mapping studies. Due to the wealth of molecular marker loci available for this crop, progenies deriving from interspecific crosses have also played an important role in gene mapping as well as development of QTL analysis strategies, including map-based cloning approaches (Paterson et al. 1988; Martin et al. 1993; Tanksley 1993; Eshed and Zamir 1995; Tanksley and Nelson 1996; Frary et al. 2000; Fridman et al. 2000). Since the earliest QTL mapping reports based on isozymes and morphological markers, it has been clear that this approach allows more efficient uncovering of "cryptic" genetic variation, and that wild relatives would provide a rich source of favorable alleles for the improvement of elite germplasm, as well as for traits in which the unimproved (wild) species show an inferior phenotype (Tanksley et al. 1982; Weller et al. 1988). Following the demonstration by Paterson et al. (1988) that molecular linkage maps covering the entire genome could be used to resolve quantitative traits into Mendelian factors, QTL mapping studies in tomato, based on interspecific crosses, were extended to hundreds of agronomically important traits involved in plant morphology, adaptation, yield, fruit quality, metabolism, and gene expression. The outcome of these studies has been the identification of thousands of QTLs, many of which are of potential value for tomato breeding, and whose molecular basis is still to be deciphered.

The status of gene and/or QTL mapping in tomato has been the subject of several recent reviews (Labate et al. 2007; S. Grandillo personal communication), and most of the studies have been summarized in Tables 9.6–9.8. Here we will describe some of the main mapping results obtained so far using each species.

9.6.4.1 *Solanum arcanum, Solanum corneliomulleri, Solanum huaylasense, Solanum peruvianum*

These four green-fruited wild species have been segregated from *Solanum peruvianum* s.l. (see Sects. 9.2 and 9.3; Peralta et al. 2005, 2008), formerly considered the most widespread and polymorphic species in *Solanum* sect. *Lycopersicon* (Peralta et al. 2005). *S. arcanum* Peralta and *S. huaylasense* Peralta have been described as new species (Peralta et al. 2005) from Peru, while the other two, *S. peruvianum* s.str. and *S. corneliomulleri* had already been named by Linnaeus (1753) and MacBride (1962), respectively. In some of the reviewed studies that used the old nomenclature, the accession number was not available; therefore, in these cases, genotypes have been referred to as *S. peruvianum* s.l. here. These four species cover a wide diversity of habitats that range from approximately sea level to nearly 3,000 m, thus explaining their adaptation to extreme environments. They also represent a rich reservoir of potentially valuable genes for disease resistance as well as other agronomically important traits. However, mainly due to the hybridization barriers that exist between these species and the cultivated tomato, they have not been thoroughly exploited for breeding purposes (Taylor 1986).

Within *S. arcanum*, the SC accession LA2157, originating from 1,650 m above sea level in northern Peru, is known as source of several beneficial traits for cultivated tomatoes such as cold tolerance, resistance to bacterial and fungal diseases, as well as heat-stable resistance to root nematode caused by *Meloidogyne* spp. (Table 9.6). The cross with *S. lycopersicum* is difficult but possible through in vitro embryo rescue (Brüggemann et al. 1996). Molecular linkage maps have allowed the identification of QTLs underlying the resistance of *S. arcanum* LA2157 to bacterial canker caused by *Clavibacter michiganensis* subsp. *michiganensis* (*Cmm*) (Table 9.6). A first RFLP mapping study was conducted on BC populations derived from the intraspecific cross between *S. arcanum* LA2157 and the susceptible *S. arcanum* accession LA2172, and five QTLs were identified (Sandbrink et al. 1995). Subsequently, Van Heusden et al. (1999) used RFLPs in a F_2 derived from the interspecific *S. lycopersicum* cv. "Solentos" × *S. arcanum* LA2157 cross, and detected three QTLs, which showed a substantial influence on resistance to *Cmm* (Van Heusden et al. 1999).

Recently, a strong source of resistance to an Indonesian isolate of *Alternaria solani*, the causal agent of early blight (EB) was identified in *S. arcanum* LA2157 (Chaerani et al. 2007). Early blight is a devastating fungal disease of tomato worldwide, and most commercial cultivars are susceptible. Classical genetic studies revealed at least two loci with additive and dominance effects and epistatic interactions for resistance to EB symptoms (see references in Chaerani et al. 2007). However, classical breeding approaches have not been successful in developing cultivars with a sufficient level of resistance and adequate commercial quality. Therefore, molecular-based breeding strategies are foreseen as a possible solution to obtain resistant cultivars with early to mid-season maturity and high yield potential. In order to study the genetic basis of this resistance, a QTL analysis was conducted in F_2 and F_3 populations derived from a *S. lycopersicum* cv. "Solentos" × *S. arcanum* LA2157 cross, using AFLP, SSR and SNP markers, which allowed the identification of six QTLs for resistance to EB, some of which also conferred resistance to stem lesions in the field (Chaerani et al. 2007).

The *S. arcanum* accession LA2172 is completely resistant, almost immune, to *Oidium neolycopersici* (previously named *O. lycopersici*) (Kiss et al. 2001), the causal agent of powdery mildew (PM) in tomato (Lindhout et al. 1994a, b). The gene *Ol-4* responsible for this complete resistance was mapped, and subsequently fine-mapped, on tomato chromosome 6 in a pseudo-F_2 population from an interspecific cross between *S. lycopersicum* cv. "Moneymaker" and *S. arcanum* LA2172 (Bai et al. 2004, 2005).

All tomato cultivars with resistance to *Meloidogyne* spp. (*Meloidogyne incognita, M. arenaria* and *M. javanica*) carry the dominant gene *Mi-1* deriving from *S. peruvianum* accession PI 128657, which was mapped on chromosome 6 (Smith 1944; Gilbert and McGuire 1956; Medina-Filho 1980; Klein-Lankhorst et al. 1991; Messeguer et al. 1991). The *Mi-1* gene of tomato was isolated by a positional cloning approach, and it was shown to belong to the NBS-LRR class of *R* genes and to have a dual specificity resistance to the root-knot nematode *M. incognita* and to an unrelated pathogen, the potato aphid *Macrosuphum euphorbiae* (Milligan et al. 1998; Vos et al. 1998). However, this resistance is not active at soil temperatures above 28°C; in contrast, *S. arcanum* LA2157 has been identified as a source for heat-stable resistance, and the gene conferring this resistance, named *Mi-9,* was mapped to the short arm of chromosome 6 in a similar genetic interval as *Mi-1* (Veremis et al. 1999; Veremis and Roberts 2000; Ammiraju et al. 2003). Using virus-induced gene silencing (VIGS) targeted to silence *Mi-1* homologs in *S. arcanum* LA2157, Jablonska et al. (2007) showed that *Mi-9* is likely a homolog of *Mi-1*. Another resistance gene, *Mi-3,* which confers resistance to *Mi-1*-virulent nematode isolates, was mapped to the telomeric region of chromosome 12, using a segregating population of *S. peruvianum* accession PI 126443 clone 1MH (Yaghoobi et al. 1995). Veremis and Roberts (1996a, b) revealed a spectrum of *Meloidogyne* resistance genes in *S. peruvianum* s.str., which are expressed in single dominant gene fashion. They showed the presence of a linked additional gene (*Mi-5*) for heat-stable resistance in the same region of *Mi-3*, and found two weakly linked pairs of genes (*Mi-2* and *Mi-8* in PI 270435 clone 2R2 and *Mi-6* and *Mi-7* in PI 270435 clone 3MH), which seemed to be independent of each other and of the *Mi-1* region on chromosome 6, and also independent from the *Mi-3*/*Mi-5* region on chromosome 12.

Resistances to tobacco mosaic virus (TMV), tomato spotted wilt virus (TSWV), and tomato yellow leaf curl virus (TYLCV) have been studied in *S. peruvianum* s.l. The two allelic genes, *Tm-2* and *Tm2a* (a.k.a.$Tm2^2$), which confer resistance to TMV, were introgressed from *S. peruvianum* PI 128650 into *S. lycopersicum* (Labate et al. 2007 and references there in). The durable $Tm\text{-}2^2$ resistance gene was mapped and fine-mapped to the centromeric region of chromosome 9 (Young et al. 1988; Pillen et al. 1996a). Subsequently, $Tm\text{-}2^2$ was isolated from tomato via transposon tagging, and was shown to be functional in both tomato and tobacco (Lanfermeijer et al. 2003, 2004). The isolation and characterization of the broken *Tm-2* resistance gene showed that the two resistance alleles, $Tm\text{-}2^2$ and *Tm-2,* from tomato differ in four amino acids (Lanfermeijer et al. 2005). CAPS markers have been developed to differentiate the *Tm-2*, $Tm\text{-}2^2$, and *tm-2* (susceptible) alleles (Lanfermeijer et al. 2005).

The single dominant gene (*Sw-5*) originating from *S. peruvianum* s.l. that confers resistance to common strains of TSWV was mapped to the long arm of chromosome 9 (Stevens et al. 1995). The map-based cloning of the *Sw-5* locus showed that it contains a single gene capable of providing resistance to different Tospovirus species and it is a homolog of the root-knot nematode resistance gene *Mi-1* (Brommonschenkel and Tanksley 1997; Brommonschenkel et al. 2000). PCR-based marker systems have been developed that can aid MAS for the *Sw-5* gene (Folkertsma et al. 1999; Garland et al. 2005).

TYLCV is currently considered as one of the most devastating viruses of cultivated tomatoes in tropical and subtropical regions, and resistant cultivars are highly effective in controlling the disease. The breeding line TY172, originating from *Solanum peruvianum* s.l., is highly resistant to TYLCV (Anbinder et al.

2009). QTL analysis showed that TYLCV resistance in TY172 is controlled by a previously unknown major QTL, named *Ty-5*, originating from the resistant line and mapping on chromosome 4, and by four additional minor QTLs, originating either from the resistant or susceptible parents, and mapping on chromosomes 1, 7, 9, and 11 (Anbinder et al. 2009).

Genetic resistance to the soil-borne fungus *Pyrenochaeta lycopersici*, the casual agent of corky root rot, which can cause big losses in tomato production, has been identified in accessions of *S. peruvianum* s.l. and *S. habrochaites* (Hogenboom 1970). Subsequently, a single recessive gene (*pyl*) was shown to control this resistance and was introgressed into *S. lycopersicum* from *S. peruvianum* s.l. (Laterrot 1983). However, breeding programs aimed at transferring this resistance based on phenotypic selection have been hampered by the difficulties associated with greenhouse inoculation and with direct screening of corky root rot resistance. In order to overcome these difficulties, the *pyl* gene was mapped on tomato chromosome 3 using RAPD and RFLP markers, and then codominant CAPS markers were developed to allow a more efficient MAS approach (Doganlar et al. 1998).

S. arcanum has also been used in QTL mapping efforts aimed at exploring the potential value of this wild relative as source of favorable alleles for the improvement of yield, fruit quality, and other horticultural traits (Table 9.7). For this purpose an advanced backcross (AB) population of 200 BC_4 families, derived from the *S. lycopersicum* E6203 × *S. arcanum* LA1708 cross, was analyzed with 205 RFLPs and was evaluated for 35 traits involving yield, processing fruit quality, and plant characteristics. A total of 166 QTLs were identified for 29 of the traits, and, interestingly, for half of the favorable alleles originated from the wild parent (Fulton et al. 1997). The same population was also evaluated for sugars, organic acids, and other biochemical properties possibly contributing to flavor, and 103 QTLs were identified for the 15 analyzed traits (Fulton et al. 2002a). Also in this case, favorable wild QTL alleles were detected for several of the analyzed traits.

9.6.4.2 *Solanum cheesmaniae* and *Solanum galapagense*

The two yellow- to orange-fruited wild species *Solanum cheesmaniae* and *Solanum galapagense* are very closely related to the cultivated tomato and can be reciprocally hybridized with it. Therefore, a number of genes have been transferred from these wild relatives to the cultigen (Rick 1956). *Solanum galapagense* is also particularly valuable as a source for salt tolerance (ST) (Taylor 1986). In contrast, their role as sources of disease and insect resistances has been more limited, probably due to their isolation on the Galápagos islands, which has reduced the exposure of these two taxa to the numerous pests and diseases that instead can be found on the mainland and that attack the other sect. *Lycopersicon* species (Taylor 1986).

One gene that has been introgressed from *S. cheesmaniae* LA0166 into cultivated tomato is the jointless pedicel gene, *jointless-2* (*j-2*); a recessive mutation that completely suppresses the formation of flower and fruit pedicel abscission zone, and which was discovered on the Galápagos Islands of South America by Rick (1956). This gene has been widely used for more than 40 years in the tomato processing industry (Zhang et al. 2000). High resolution genetic and physical mapping have located the *j-2* gene in the centromeric region of tomato chromosome 12 (Zhang et al. 2000; Budiman et al. 2004). Two other genes, the *Beta* (*B*) and the *Beta*-modifier (Mo_B), which control the high concentrations of β-carotene in orange-pigmented tomatoes, were mapped to the long arm of chromosome 6 using segregating populations derived from the two interspecific crosses *S. lycopersicum* × *S. galapagense* LA0317 and *S. lycopersicum* × *S. habrochaites* PI 126445 (Zhang and Stommel 2000). The *B* gene was isolated by map-based cloning approach (Ronen et al. 2000; see also Sect. 9.6.4.9), and codominant SCAR and CAPS markers were developed for use in MAS programs (Zhang and Stommel 2001). Furthermore, interspecific mapping populations derived from crossing between *S. lycopersicum* and *S. galapagense* LA0483 were used to map several genes involved in pigment content and fruit ripening including *high pigment-1* (*hp-1*), *non-ripening* (*nor*), and *ripening-inhibitor* (*rin*) (Giovannoni et al. 1995; Yen et al. 1997; Peters et al. 1998).

Solanum cheesmaniae has been used as source of resistance to blackmold caused by *Alternaria alternata* (Rick 1986a) (Table 9.6). Cassol and St. Clair (1994) showed that resistance in *S. cheesmaniae* LA0422 was quantitatively inherited, and blackmold resistance QTLs were mapped in a progeny derived from a *S. lycopersicum* cv.

"VF145B-7879" × *S. cheesmaniae* LA0422 cross; subsequently, by means of MAS, five resistant QTLs from *S. cheesmaniae* LA0422 were introgressed into cultivated tomato (Robert et al. 2001). The QTL on chromosome 2 had the largest positive effect on black-mold resistance, and was also associated with earliness, a positive horticultural trait.

Solanum galapagense as well as several other wild tomato species including *S. pimpinellifolium*, *S. chilense*, *S. cheesmaniae*, *S. pennellii,* and *S. peruvianum* s.l. represent genetic sources of ST (reviewed by Foolad 2004, 2005). Given the complex nature of ST, most studies have focused on specific developmental stages. In the case of *S. galapagense,* QTL analyses have focused on the vegetative (VG) and/or reproductive (RS) stage of the plant (Table 9.8) (Monforte et al. 1997a, b, 1999; Villalta et al. 2007, 2008). These studies were conducted on F_2 or recombinant inbred line (RIL) populations derived from interspecific crosses between the salt-sensitive *S. lycopersicum* and *S. lycopersicum* "cerasiforme" and the two ST wild species *S. galapagense* and *S. pimpinellifolium* to analyze the effect of salinity on several yield related traits, including fruit weight, fruit number, total fruit weight, as well as Na^+ and K^+ in stems and leaves. Villalta et al. (2007) found that, contrary to the expected, the allele from the wild ST genotype had a favorable effect only at one total fruit yield QTL. These results suggested that alternative approaches need to be pursued in order to improve tomato crop productivity under salinity, and one possibility is by grafting cultivars onto ST wild relatives. Therefore, Estañ et al. (2009) analyzed the rootstock effect on fruit yield of a grafted tomato variety under moderate salinity (75 mM NaCl) using as rootstocks F_9 lines of the two interspecific RIL populations previously used by Villalta et al. (2007, 2008). This study detected at least eight QTLs that contributed to this ST rootstock effect, with the most relevant component being the number of fruits. In addition, Albacete et al. (2009) found that in the *S. galapagense* RIL population rootstock-mediated changes in xylem ionic and hormonal status were correlated with delayed leaf senescence, and increased leaf area and thus crop productivity in salinized tomato.

The *S. galapagense* accession LA0483 has been used in QTL mapping studies aimed at deciphering the genetic basis of fruit quality traits (Table 9.7). Work by Paterson et al. (1991) detailed the identification of 29 putative QTLs for soluble solids content (SSC) measured by brix value, mass per fruit, and pH in a F_2 population derived from a cross between the inbred cultivar "UC204C" and *S. galapagense* LA0483. Subsequently, 97 F_8 RILs were developed from the same cross and were used to identify QTLs for seed weight, fruit weight, SSC, and morphological traits (Goldman et al. 1995; Paran et al. 1997).

9.6.4.3 *Solanum chilense*

Among all wild species of tomato, the green-fruited *Solanum chilense* seems to be one of the most notable as a source of a broad spectrum of disease resistance. In fact, this species shows resistance to several bacterial, fungal, and viral diseases, as well as to root knot nematodes and parasitic plants, such as dodder (*Cuscuta* spp., Convolvulaceae, Rick and Chetelat 1995). Moreover, *S. chilense*, being indigenous to arid and semi-arid environments in South America, has also been considered of interest for its drought tolerance (Rick 1973).

Among viral diseases, cucumber mosaic virus (CMV) is an important disease for tomatoes growing in temperate regions and is the most destructive virus in some areas. Fortunately, several wild tomato species are resistant or tolerant to CMV, including *S. chilense, S. pimpinellifolium, S. peruvianum* s.l., *S. habrochaites*, *S. galapagense*, and *S. lycopersicoides* (Stamova and Chetelat 2000 and references therein). In order to explore the genetic basis of CMV resistance, Stamova and Chetelat (2000) used isozyme and RFLP markers in BILs derived from a *S. lycopersicum* × *S. chilense* LA0458 cross and identified a single dominant resistance gene, *Cmr,* which mapped on chromosome 12 (Table 9.6).

Within *S. chilense*, high levels of resistance to begomoviruses, such as monopartite tomato yellow leaf curl virus (TYLCV) and bipartite tomato mottle virus (ToMoV), transmitted by the whitefly, *Bemisia tabaci*, have been identified in several accessions including LA1969, LA1932, LA2779, and LA1938, which have been useful sources of resistance in tomato breeding programs (Ji et al. 2007b and references therein). The accession LA1969 has been used as source for the TYLCV tolerance locus, *Ty-1*, a partially dominant major gene, which was located on chromosome 6 of tomato using RFLP markers, and subsequently introgressed into cultivated tomato (Table 9.6; Michelson et al. 1994; Zamir et al. 1994). Conventional genetic analysis and QTL mapping conducted in F_2 populations derived from *S. chilense* accessions LA1932, LA1938, and LA2779 revealed three regions on chromosome 6 contributing to resistance to both TYLCV and ToMoV, and RAPD markers linked to each region were identified (Griffiths and Scott 2001; Ji and Scott 2005; Agrama and Scott 2006; Ji et al. 2007a). The first region includes the *Ty-1* locus, while the other two regions flank either side of the *self-pruning* (*sp*) and *potato leaf* (*c*) loci. Two additional TYLCV resistance genes, *Ty-3* and *Ty-4*, were recently discovered in *S. chilense* accessions (LA2779 and LA1932) and mapped to chromosomes 6 and 3, respectively (Ji et al. 2007a, 2009a). The partially dominant gene, *Ty-3,* deriving from *S. chilense* accession LA2779, was mapped on chromosome 6 near the *Ty-1* locus (Ji et al. 2007a). RILs carrying both resistance genes had the highest level of TYLCV resistance (Ji et al. 2009a). PCR-based markers tightly linked to both genes have been developed and used in MAS breeding programs (Jensen et al. 2007; Ji et al. 2007a, b). Finally, TSWV resistance was identified in a breeding line derived from a cross with *S. chilense* LA1938; the same line was previously selected for ToMoV resistance in Florida (Canady et al. 2001).

The *S. chilense* accession LA1969 was identified also as a source of resistance to *Leveillula taurica*, one of the two pathogens responsible for PM in tomato, which has become a serious problem to tomato growers and breeders around the world, but especially in subtropical regions (Chunwongse et al. 1997). A single dominant gene, *Lv*, conferring resistance to this pathogen has been described from *S. chilense* LA1969 and has been introgressed into the cultivated tomato via backcross breeding (Stamova and Yordanov 1990). Subsequently, *Lv* was mapped to a high resolution map position near the centromere of chromosome 12 (Chunwongse et al. 1994, 1997).

Recently, the *S. chilense* accession LA1932 has been used in an AB-QTL mapping study aimed at exploring the potentials of this wild relative as a source of useful QTL alleles for yield-related and fruit quality traits (Termolino et al. 2010; S. D. Tanksley personal communication). Results from this study have demonstrated that, despite its inferior horticultural characteristics, *S. chilense* contains alleles capable of improving several traits of economic importance for processing tomatoes including brix, firmness, and viscosity.

9.6.4.4 *Solanum chmielewskii*

This green-fruited wild species has been studied extensively for its high concentration in soluble solids (SSC or brix; mainly sugars and organic acids), which can reach approximately 10%, almost twice the concentration found in mature fruit of the domestic tomato (Rick 1974). By means of extensive backcrossing and selection, genes for enhanced SSC from *S. chmielewskii* accession LA1028 have been introgressed into *S. lycopersicum* cv. "VF 36" and cv. "VF 145-22-8" resulting in BC_5S_5 lines (including LA1500-1503 and LA1563) with a 40% higher SSC (about 7–8%), but a similar yield, fruit size, and color to the recurrent parent (Rick 1974). Subsequently, the segments introgressed from *S. chmielewskii* were identified using RFLP and isozyme markers and characterized for their effects on SSC, pH, and yield (Table 9.7; Osborn et al. 1987; Tanksley and Hewitt 1988; Azanza et al. 1994). Paterson et al. (1988) conducted a QTL analysis on a *S. lycopersicum* UC82B × *S. chmielewskii* LA1028 BC_1 population using a whole genome RFLP map, and they identified 15 QTLs related to SSC, fruit weight, and pH; some of them were then fine-mapped using a substitution mapping method (Paterson et al. 1990). One of the near isogenic lines (NILs) developed by Paterson et al. (1990), TA1150, contained a 56-cM introgression from *S. chmielewskii* chromosome 1 and had several interesting phenotypic characteristics including fruit with high levels of brix, orange color, thicker pericarp, smaller stem scars, and higher firmness than the control *S. lycopersicum* cv. "E6203" (Frary et al. 2003). The development and field characterization of a set of derived overlapping sub-ILs allowed breaking the undesirable linkage between high brix and orange color (Frary et al. 2003). Moreover, in contrast to *S. lycopersicum*, *S. chmielewskii*, as well as *S. peruvianum* s.l. and *S. habrochaites* fruit, accumulates soluble sugars primarily as sucrose, rather than glucose and fructose (Davies 1966; Yelle et al. 1988). High sucrose accumulation in *S. chmielewskii* and *S. habrochaites* has been suggested to be recessive and monogenic (Yelle et al. 1991), and the gene, denominated *sucr*, was mapped to the pericentromeric region of chromosome

3 using RFLPs (Chetelat et al. 1993). However, after introgressing the *S. chmielewskii* LA1028 *sucr* gene into the genetic background of a hexose-accumulating cultivated tomato, it was observed that associated reduced fertility, due to tightly linked genes or to pleiotropic effects of *sucr*, did not allow a net gain in yield of SSC (Chetelat et al. 1995a, b).

More recently, fruit quality traits and physiological parameters were evaluated on 20 ILs derived from the introgression of *S. chmielewskii* LA1840 into *S. lycopersicum* cv. "Moneyberg" under high (unpruned trusses) and low (trusses pruned to one fruit) fruit load conditions (Table 9.7; Prudent et al. 2009). The results obtained in this study suggested that the relationships between fruit weight and its composition could be mainly related to sink strength through cell division whose intensity was modulated by fruit load (Prudent et al. 2009). Phenotypic analysis of the same *S. chmielewskii* LA1840 IL population revealed three overlapping ILs on chromosome 1 conferring a pink fruit color (Ballester et al. 2010). Genetic mapping, segregation analysis, and VIGS results suggested strongly that the *MYB12* gene is a likely candidate of the locus leading to pink fruit, probably the *Y* locus (Ballester et al. 2010).

Finally, an F_2 population derived from *S. lycopersicum* and its late-flowering wild relative *S. chmielewskii* (line CH6047) was used to study the genetic mechanisms underlying flowering time in tomato (Table 9.7; Jiménez-Gómez et al. 2007). This work allowed the identification of two QTLs affecting days to flowering and six QTLs for leaf number (the number of leaves under the first inflorescence). Interestingly, some of the early flowering QTL alleles were contributed by the *S. chmielewskii* parent, highlighting the usefulness of this wild species for the improvement of flowering time, and in general the importance of exploiting the genetic variation existing among all wild relatives of tomato.

9.6.4.5 *Solanum habrochaites*

This green-fruited wild species is typically found at high elevations, often above 3,000 m, and therefore is expected to be a source of tolerance to low temperatures (Patterson 1988). Moreover, *S. habrochaites* has been typically associated with resistance to a wide range of insect predators and is also a good source of genes for resistance to other pathogens (Rick 1973; Taylor 1986; Farrar and Kennedy 1991; Lukyanenko 1991; Labate et al. 2007). QTL mapping studies conducted with *S. habrochaites* have shown that this wild species is also a valuable source of favorable QTL alleles for numerous other traits including yield and fruit quality, for which the wild phenotype is inferior compared to elite tomato germplasm (Bernacchi et al. 1998a, b; Monforte et al. 2001; S. Grandillo personal communication).

With respect to viral diseases, sources of resistance have been found in *S. habrochaites* accessions. For example, resistances to alfalfa mosaic virus (AMV) have been identified in three accessions of *S. habrochaites* (PI 134417, LA1777, and "Bruinsma") (Parrella et al. 2004). The single dominant gene, *Am*, from *S. habrochaites* PI 134417, which confers resistance to most strains of AMV, was mapped to the short arm of tomato chromosome 6 in the resistance hotspot, which includes the *R*-genes *Mi-1* and *Cf-2*/*Cf-5* and the quantitative resistance factors *Ty-1*, *Ol-1*, and *Bw-5* (Table 9.6; Parrella et al. 2004). A complete resistance to potyviruses (PVY – potato *virus Y*- and TEV – tobacco etch virus) was identified in *S. habrochaites* accession PI 247087, and the recessive gene *pot-1* was mapped to the short arm of tomato chromosome 3 in the vicinity of the recessive *py-1* locus for resistance to corky root rot (Parrella et al. 2002). A comparative genomic approach was used for the molecular characterization of the *pot-1* gene, which was shown to be the tomato ortholog of the pepper *pvr2–elF4E* gene (Ruffel et al. 2005).

The resistance gene, *Tm-l,* to tomato mosaic virus (ToMV), one of the most serious diseases in tomato, originated from *S. habrochaites* by interspecific crossing (Holmes 1957). This gene has been used, either alone or together with one of the other ToMV-resistance genes, *Tm-2* or *Tm-2a* (*a.k.a* $Tm2^2$), to develop resistant varieties. The *Tm-1* gene was mapped near the centromere of chromosome 2, and a number of DNA markers linked to the locus have so far been identified, including RFLPs, RAPDs, SCARs (Table 9.6; Levesque et al. 1990; Tanksley et al. 1992; Ohmori et al. 1996). Due to the reduced frequency of recombination, previous attempts to isolate the *Tm-1* gene using map-based cloning proved unsuccessful; therefore the gene was identified by purifying its inhibitory activity toward ToMV RNA replication in vitro (Ishibashi et al. 2007). *S. habrochaites* is also a source for high resistance to TYLCV, and resistant tomato lines carrying resistance derived from *S. habrochaites* accession B6013 were developed by Kalloo and Banerjee (1990). Later the TYLCV resistance locus, originating from B6013, was mapped to the long arm of chromosome 11, using RFLP markers (Hanson et al. 2000), formally designated *Ty-2*, and further fine-mapped (Hanson et al. 2006; Ji et al. 2007b). PCR markers have been developed, which allow precise monitoring of the introgression of the *Ty-2* gene into elite breeding lines (Ji et al. 2009b).

With respect to bacterial diseases, a source of resistance to *Cmm*, the casual agent of bacterial canker, was identified in *S. habrochaites* accession LA0407, and two major QTLs for resistance to *Cmm* were mapped using a BIL population derived from a *S. lycopersicum* × LA0407 cross (Kabelka et al. 2002). These QTLs were subsequently fine-mapped and an additive-by-additive epistasis between them was confirmed (Coaker and Francis 2004).

Resistances to several fungal diseases have also been identified in *S. habrochaites*. For example, the *S. habrochaites* accession PI 126445 was identified as a source of resistance to EB (*Alternaria solani*) (Nash and Gardner 1988) and was crossed to a susceptible tomato breeding line to generate BC populations suitable for QTL mapping (Foolad et al. 2002; Zhang et al. 2002, 2003b). In total 14 QTLs affecting EB response were detected using different populations and mapping strategies, and four of them, detected as major QTLs in both studies, were considered of potential value for MAS breeding programs (Foolad et al. 2002; Zhang et al. 2002, 2003b). Tomato gray mold (GM) (*Botrytis cinerea*) is a common fungal disease worldwide, which often causes serious production loss by infecting leaves, stems, flowers, and fruits. No modern hybrid tomato cultivars completely resistant to GM are available, although a few cultivars show a certain level of quantitative resistance (ten Have et al. 2007). In contrast, accessions of *S. chilense*, *S. habrochaites*, and *S. neorickii* show marked quantitative resistance to GM, in both leaf and stem segment assays (Egashira et al. 2000; ten Have et al. 2007). Among others, the *S. habrochaites* accession G1.1560 (LYC4) was selected for high level of resistance (ten Have et al. 2007). Finkers et al. (2007a, b) identified three and 10 QTLs for resistance to *B. cinerea*, in an F_2 and an IL population, respectively, both derived from a *S. lycopersicum* cv. "Moneymaker" × *S. habrochaites* LYC4 cross. In similar studies resistance to late blight (LB) (*Phytophthora infestans*) was described for several accessions of *S. habrochaites* (Lobo and Navarro 1987). The multigenic resistance to LB of the highly resistant *S. habrochaites* accession LA1033 was studied using AFLP markers in BC populations derived from an interspecific cross with the cultivated tomato (Lough 2003). QTLs affecting LB response were detected on four to nine linkage groups depending upon the method of analysis used. At least 15 QTLs for quantitative resistance to *P. infestans* have also been identified in reciprocal BC populations derived from a *S. lycopersicum* × *S. habrochaites* LA2099 cross (Brouwer et al. 2004). Three of these QTLs, *lb4, lb5b*, and *lb11b*, were fine-mapped using NILs and sub-NILs (Brouwer and St. Clair 2004).

S. habrochaites has been the source of the gene *Cf-4,* which confers resistance to *Cladosporium fulvum*, the casual agent of tomato leaf mold (Kerr and Bailey 1964; Stevens and Rick 1986). The *Cf-4* gene from *S. habrochaites*, and the gene *Cf-9* derived from *S. pimpinellifolium*, were introgressed into cultivated tomato (Stevens and Rick 1986). A combination of classical and RFLP mapping showed that they are both located on the short arm of tomato chromosome 1 (Jones et al. 1993; Balint-Kurti et al. 1994); subsequently, *Cf-4* was cloned and characterized by Thomas et al. (1997).

Several accessions of *S. habrochaites* (G1.1257, G1.1290, G1.1560, G1.1606 = CPRO742208, LA1775, PI 247087) have been found to be resistant to PM, caused by *Oidium neolicopersici* (Huang et al. 2000b and references therein). The resistance found in the *S. habrochaites* accession Gl.1560 resulted to be largely controlled by an incompletely dominant gene, *Ol-1*, that was mapped by means of RAPD and RFLP markers on the long arm of chromosome 6, near the *Aps-1* locus in the vicinity of the resistance genes *M-1* and *Cf-2*/*Cf-5* to *Meloidogyne* spp. and *C. fulvum*, respectively (Van der Beek et al. 1994; Huang et al. 2000a). Subsequently, the *Ol-1* gene was fine-mapped, and the use of another resistant *S. habrochaites* accession, G1.1290, allowed the identification of a new incompletely dominant gene, designated *Ol-3*, which was also mapped to chromosome 6, in the same chromosome region as *Ol-1* (Huang et al. 2000b; Bai et al. 2005). Another source of resistance to *O. neolycopersici* was identified in the *S. habrochaites* accession PI 247087 and it was shown to be polygenic but with a major gene, *Ol-5*, mapping on the long arm of chromosome 6, about 1 cM proximal of the *Ol-1* locus (Bai et al. 2005).

S. habrochaites is a remarkable source of resistance to many arthropod pests that attack cultivated tomato (Rick 1982; Taylor 1986; Farrar and Kennedy 1991). This resistance is mediated by several factors, including glandular trichome type and density, and presence of particular compounds in trichome glands that possess toxic properties against Lepidoptera or aphids. Great morphological variation and chemical differentiation of trichome secretions can be observed among *S. habrochaites* accessions; in some cases, trichome secretions are predominated by methylketones, often 2-tridecanone (2-TD) and/or 2-undecanone, while in other cases by sesquiterpenoids, often sesquiterpene hydrocarbons (Van der Hoeven et al. 2000; Zhang et al. 2008; Sallaud et al. 2009 and references therein). Moreover, *S. habrochaites* can be immune to insects, suggesting that repellence may be a mechanism of protection (Rick 1982). In this respect, Guo et al. (1993) and Snyder et al. (1993) found that spider mite repellence in trichome secretions on the *S. habrochaites* accessions LA1927 and LA1363 was mainly due to the presence of 2,3-dihydrofarnesoic acid, a sesquiterpene acid. The inheritance of this compound was studied in segregating generations deriving from interspecific crosses between *S. lycopersicum* and *S. habrochaites* LA1363 but no conclusive results were reported (Zhang et al. 2008).

The inheritance of allelochemicals, as well as of other characters related to insect resistance, appears to be complex, and molecular markers have been used to identify QTLs underlying some of these traits. For example, Zamir et al. (1984) reported association of the level of 2-tridecanone in *S. habrochaites* with five isozyme markers mapping on at least four different chromosomes. Nienhuis et al. (1987) found association of 2-TD levels with RFLPs on three linkage groups and of type VI trichome density with one of these marker loci.

Subsequently, RFLP markers were used in an F_2 population derived from an interspecific cross between *S. lycopersicum* cv. "Moneymaker" and *S. habrochaites* to identify QTL for greenhouse whitefly (*Trialeurodes vaporariorum*) resistance (Maliepaard et al. 1995). Two QTLs affecting oviposition rate were mapped to chromosome 1; while two QTLs affecting trichome type IV density and one affecting type VI trichome density were mapped to chromosomes 5, 9, and 1, respectively (Maliepaard et al. 1995). The genetic control of the concentration of 2-TD and 2-undecanone was studied in F_1 and F_2 populations derived from the interspecific cross between *S. lycopersicum* cv. "IPA-6" × *S. habrochaites* PI 134418 (Pereira et al. 2000). Using the ILs resulting from a cross between *S. lycopersicum* and *S. habrochaites* LA1777, Van der Hoeven et al. (2000) showed that the biosynthesis of class I and II sesquiterpene olefins is controlled by two independent loci, *Sst1* and *Sst2*, respectively, mapping on chromosome 6, *Sst1*, and 8 (*Sst2*). By searching into a *S. habrochaites* trichome EST database, Sallaud et al. (2009) identified two candidate genes that are highly and specifically expressed in trichome cells and that mapped to the *Sst2* locus on chromosome 8. These two genes are responsible for the biosynthesis of all chromosome 8-associated class II sesquiterpenes.

A few studies have investigated the genetic basis of chilling tolerance in tomato using interspecific crosses between the cultigen and *S. habrochaites* accessions (Table 9.8). Vallejos and Tanksley (1983) analyzed a BC_1 between *S. lycopersicum* and a *S. habrochaites* cold-tolerant accession with 17 isozyme markers and identified a minimum of three QTLs for growth at low temperatures, two of which had positive effects, and the other negative. In crosses between the same two species, Zamir et al. (1982) conducted pollinations at low temperatures and, using nine isozyme markers, detected two regions of the *S. habrochaites* LA1777 genome on chromosomes 6 and 12, which were highly favored in crosses at low temperature. More recently, Truco et al. (2000) conducted a QTL analysis using RFLPs on a BC_1 between *S. lycopersicum* and *S. habrochaites* LA1778, which allowed the identification of multiple QTLs related with shoot wilting and root ammonium uptake under chilling temperatures. For example, three QTLs were detected for wilting at 2 h, on chromosomes 5, 6, and 9, and the presence of the *S. habrochaites* allele had a favorable effect in decreasing wilting at the two QTLs on chromosomes 5 and 9.

S. habrochaites has also been useful for studying the genetic basis of numerous yield and fruit quality-related traits (Table 9.7). With respect to single genes, for example, Levin et al. (2000) described a locus, *Fgr*, that controls the fructose–glucose ratio in mature fruit, with a *S. habrochaites* LA1777 allele yielding a higher ratio. Later, it was shown that alleles of *S. habrochaites* at two loci interacted to increase this ratio (Levin et al. 2004). The action of the *Beta* (*B*) gene (which increases fruit β-carotene content at the expense of lycopene, resulting in orange-pigmented fruit) was first described in segregants descended from a cross between cultivated tomato and *S. habrochaites* PI 126445 (Lincoln and Porter 1950). Subsequently, studies by Tomes et al. (1954) determined that *B* was dominant but subject to influence by a modifier gene, Mo_B, which segregated independently of *B*. Both genes were mapped to the long arm of chromosome 6 (Zhang and Stommel 2000, 2001), and *B* was cloned (Ronen et al. 2000) (see also Sects. 9.6.4.2 and 9.6.4.9).

Progenies deriving from a *S. lycopersicum* cv. "E6203" × *S. habrochaites* LA1777 cross have also been used in numerous QTL mapping studies (Table 9.7). An AB population (BC_2/BC_3) was analyzed for 19 quantitative traits of agronomic importance in replicated field trials conducted in several locations around the world (Bernacchi et al. 1998a). A total of 121 QTLs were identified for all traits evaluated, and interestingly, for 25 of the QTLs (20%) corresponding to 12 traits (60%), the wild parent allele had a favorable effect on the trait from a horticultural perspective. Favorable wild QTL alleles were identified also for traits for which the wild parent had an inferior phenotype compared to the cultivated parent. For example, wild alleles were associated with increased yield or with improved red color of the fruit, despite the fact that *S. habrochaites* has low yield and produces green fruit that lacks lycopene. The same AB population has been evaluated for traits possibly contributing to flavor, including sugars, organic acids, and other biochemical properties, and 34 QTLs were identified for the 15 analyzed traits (Fulton et al. 2002a). Also in this case, favorable wild QTL alleles were identified for several traits. Starting from the same population, a few cycles of MAS selection allowed the development of improved-processing tomato NILs carrying *S. habrochaites* LA1777 specific QTL alleles (Bernacchi et al. 1998b). The NILs were evaluated for their agronomic performance in five locations worldwide, and for most of them quantitative factors showed the phenotypic improvement predicted by QTL analysis of the BC_3 populations. The same interspecific cross was used to develop a population of 99 NILs, or ILs, and backcross recombinant inbred lines (BCRILs), which were genotyped with 95 RFLP markers (Monforte and Tanksley 2000a). Most of these lines have been evaluated for yield related and fruit quality traits, and several of them showed to carry beneficial wild QTL alleles (Grandillo et al. 2000; S. Grandillo et al. unpublished results). For a few of these lines (e.g., bottom of chromosomes 1 and 4), sub-ILs have been developed and evaluated in order to fine-map the QTLs and find more tightly linked markers, as well as to break undesirable linkages (Monforte and Tanksley 2000b; Monforte et al. 2001; Yates et al. 2004). More recently, the *S. habrochaites* LA1777 NIL/BCRIL population has been used to identify QTLs associated with the emission of fruit volatile compounds associated with flavor and a total of at least 30 QTLs affecting the emission of one or more of 24 volatiles were identified (Mathieu et al. 2009). In a framework of a collaborative project (EU-SOL, funded by the European Commission under FP6, PL 016214–2 EU-SOL) a new set of *S. habrochaites* LA1777 ILs has been produced, anchored to the high density tomato molecular map by means of PCR-based markers (mostly COSIIs), and which will allow a better coverage of the wild parent species genome (Tripodi et al. 2006, 2009; S. Grandillo personal communication).

Another *S. habrochaites* accession, PI 247087, was used to identify QTLs associated with ascorbic acid content (Stevens et al. 2007). A comparison of the results obtained using three different mapping populations (the *S. habrochaites* PI 247087 advanced BC, the *S. pennellii* LA0716 ILs, and the cherry-RILs) allowed the identification of common regions controlling ascorbic acid content on chromosomes 2, 8, 9, 10, and 12; in general, the wild alleles increased ascorbic acid content (Stevens et al. 2007). The same *S. lycopersicum* × *S. habrochaites* LA0407 BIL population, used by Kabelka et al. (2002) to map QTL for resistance to *Cmm*, was also evaluated for fruit color traits. Although no significant fruit color QTL was identified with the favorable allele contributed by the wild parent, the performance of a few lines did suggest some potential of the LA0407 BIL population for the improvement of color (Kabelka et al. 2004).

Gorguet et al. (2008) studied the genetics of parthenocarpy in two different lines, IL5-1 and IVT-line 1, carrying chromosome segments from *S. habrochaites* LYC4 and from an unknown accession, respectively. Four novel parthenocarpy QTLs (on chromosomes 4,

5, and 9) responsible for the seedless fruit development in IL5-1 and IVT-line 1 were identified; moreover, one stigma exsertion locus (*se5.1*) was detected in the line IL5-1.

Progenies deriving from interspecific crosses between the SC tomato *S. lycopersicum* and SI accessions of *S. habrochaites* have been used to explore the genetic basis of the evolution of mating system and of hybrid incompatibility. Bernacchi and Tanksley (1997) used a BC_1 population between *S. lycopersicum* (SC) and *S. habrochaites* LA1777 (SI) for a QTL study of sexual compatibility factors and floral traits. The only QTL for SI identified in this population mapped at the SI locus, *S*, on chromosome 1 (Tanksley and Loaiza-Figueroa 1985; Bernatzky 1993), indicating that the transition from SI to SC that ultimately led to cultivated tomato was mainly the result of mutations that occurred at the *S* locus (Bernacchi and Tanksley 1997). The major QTL controlling unilateral incongruity (UI) also mapped to the *S* locus, which supports the hypothesis that SI and UI are related mechanisms. In addition, the fact that most major QTLs for several floral traits important to pollination biology (e.g., number and size of flowers) were also located at the *S* locus region of chromosome 1, suggested the presence of a gene complex controlling both genetic and morphological mechanisms of reproduction control (deVicente and Tanksley 1993; Bernacchi and Tanksley 1997). In order to guarantee self-fertilization, besides SI, changes in floral morphology are also required. In this respect, one key morphological trait is stigma exsertion, since an exserted stigma promotes outcrossing, while a recessed stigma, below the anthers, promotes self-fertilization (Rick 1979). In the *S. lycopersicum* × *S. habrochaites* LA1777 BC_1 population, Bernacchi and Tanksley (1997) mapped a single major QTL on chromosome 2, called *stigma exsertion 2.1* (*se2.1*) which explained most of the morphological changes that occurred in the evolutionary transition from allogamous to autogamous flowers. The same major QTL was also detected in a *S. lycopersicum* (SC) × *S. arcanum* LA1708 (SI) cross (Fulton et al. 1997). Fine-mapping studies showed that *se2.1* is a complex locus composed of at least five closely linked genes: three controlling stamen length, one controlling style length, and one conditioning anther dehiscence (Chen and Tanksley 2004). The locus controlling style length, named *Style 2.1*, which explained the largest change in stigma exsertion was cloned using map-based cloning method, and the gene resulted to encode a putative transcription factor that regulates cell elongation in developing styles (Chen et al. 2007).

The *S. habrochaites* LA1777 NIL/BCRIL population has also been used to examine the genetic basis of hybrid incompatibility, in terms of traits that potentially contribute to pre-zygotic isolation that can influence pollinator preferences and/or selfing rates (e.g., flower size, flower shape, stigma exsertion, and inflorescence length) and post-zygotic isolation (pollen and seed sterility) between *S. lycopersicum* and *S. habrochaites* (Moyle and Graham 2005; Moyle 2007). The results obtained with the post-zygotic traits showed that hybrid pollen and seed infertility are each based on a relatively small and comparable number of QTLs (Moyle and Graham 2005). Interestingly, similar results were obtained using the *S. pennellii* LA0716 IL population (Moyle and Nakazato 2008). The fact that QTLs for pollen and seed sterility from the two *Solanum* studies colocalized suggested a shared evolutionary history for these QTLs, and also that loci causing sterility are not randomly distributed in the genome.

9.6.4.6 *Solanum lycopersicoides*

This nightshade species possess unique traits, including extreme abiotic stress tolerance and resistance to several insect pests and pathogens that have an impact on the production of tomatoes (Rick 1988; Chetelat et al. 1997). The *S. lycopersicoides* accession LA2951, which was used to develop a population of ILs within the background of *S. lycopersicum* cv. “VF36” (Canady et al. 2005), exhibits high foliar resistance to GM (*Botrytis cinerea*) (Rick 1987; Rick and Chetelat 1995; Chetelat et al. 1997). In order to identify QTLs for resistance to *B. cinerea*, 58 *S. lycopersicoides* LA2951 ILs, which collectively represent more than 96% of the map units in the *S. lycopersicoides* genome, were screened for foliar resistance and susceptibility to *B. cinerea* over a period of more than 2 years (Davis et al. 2009). A total of five putative resistance QTLs were identified, and two for susceptibility, with the major resistance and susceptibility QTL mapping on the long arm of chromosome 1 and on chromosome 11, respectively.

9.6.4.7 *Solanum lycopersicum* "cerasiforme"

The red-fruited cherry tomato, *S. lycopersicum* "cerasiforme," has been postulated as the expected ancestor of the domesticated form, while others more recently have suggested that it is merely a small-fruited form and not necessarily involved in the direct origins of the cultivar (Peralta et al. 2008). In the Andean region, putatively wild and feral forms can be found and *S. lycopersicum* "cerasiforme" is also described as highly invasive (Rick 1976). Recently, a molecular study was conducted to clarify the position of *S. lycopersicum* "cerasiforme" in the evolution of the cultivated tomato (Ranc et al. 2008). The study focused on the red-fruited tomato clade (*S. lycopersicum*, *S. pimpinellifolium*, *S. galapagense,* and *S. cheesmaniae*), and a total of 360 wild, feral, and cultivated accessions (144 of which were cherry tomatoes) were genotyped with 20 SSR markers. The results confirmed the admixture status of *S. lycopersicum* "cerasiforme"; in fact, part of this taxon is genetically close to the cultivated *S. lycopersicum* group and the other part is an admixture of the *S. lycopersicum* and *S. pimpinellifolium* genomes. The molecular data also showed that domesticated and wild tomatoes have evolved as a species complex with intensive level of hybridization; *S. lycopersicum* and *S. pimpinellifolium* have occasionally been classified as conspecific (see Peralta et al. 2008).

Sources of resistance to some diseases have been found in *S. lycopersicum* "cerasiforme." Danesh et al. (1994) used DNA markers to identify regions associated with partial resistance to bacterial wilt (caused by *Pseudomonas solanacearum* a.k.a *Ralstonia solanacearum*) in a F_2 population derived from a cross between a highly resistant line (L285) of cherry tomatoes and a highly susceptible cultivar (Table 9.6). In plants inoculated through roots, genomic regions on chromosomes 6 and 10 were correlated with resistance, while in plants inoculated through shoots significant regions correlated with resistance were identified on chromosomes 6, 7, and 10.

Several different *Cf* genes confer resistance to specific races of *C. fulvum* and have been bred into cultivated tomato to generate NILs (Stevens and Rick 1986; Rivas and Thomas 2005; see also Table 9.6). The gene *Cf-5* was identified in *S. lycopersicum* "cerasiforme" PI 187002 and was mapped to a complex locus on chromosome 6, very closely linked to *Cf-2* (Dickinson et al. 1993; Jones et al. 1993). Dixon et al. (1998) reported the isolation of the *Cf-5* gene and the characterization of the complex locus from three genotypes. Resistance to PM caused by *O. neolicopersici* was identified in the line (LC-95), selected within the LA1230 accession of *S. lycopersicum* "cerasiforme" collected in Ecuador, and an F_2 population obtained by crossing LC-95 and the susceptible cultivar "Super Marmande" was used to study the genetic basis of this resistance. A single recessive gene, named *ol-2*, responsible for a broad-spectrum resistance was mapped around the centromere of chromosome 4 (De Giovanni et al. 2004). Using a candidate gene approach based on comparative genetics, Bai et al. (2008) showed that loss of function of a tomato *Mlo* gene (*SlMlo1*) is responsible for PM resistance conferred by the *ol-2* gene.

S. lycopersicum "cerasiforme" has also been used as a source of favorable alleles for fruit quality traits. In this respect, a population of 144 RILs was developed from a cross between a common *S. lycopersicum* line with large fruit and a common taste, and a cherry tomato line with fruit having very good taste and high aroma intensity (Saliba-Colombani et al. 2000). The cherry-RIL population was used to study the genetic control of several traits involved in the organoleptic quality of tomato including physical and chemical components, and sensory attributes (Table 9.7; Causse et al. 2001; Saliba-Colombani et al. 2001). Eight clusters of QTLs were detected that controlled most of the variation of the organoleptic quality traits, and most of the favorable alleles were conferred by the cherry tomato parent for all of the quality traits (Causse et al. 2002). This allowed the selection of five chromosome regions that showed promise for improving fruit quality. These regions were introgressed into three cultivated tomato lines by means of a marker-assisted BC scheme, and the analysis revealed interactions between QTLs and genetic backgrounds (Lecomte et al. 2004a). Further studies showed that both additivity and epistasis control the genetic variation for fruit quality traits in tomato (Causse et al. 2007). The same cherry-RIL population was used to identify QTLs associated with ascorbic acid content of the fruit (Stevens et al. 2007). Six QTLs were identified, and the cherry allele had a positive effect for the four QTLs expressed in percentage fresh weight.

9.6.4.8 *Solanum neorickii*

Solanum neorickii is a green-fruited wild species, with small fruits and flowers; it can be reciprocally hybridized with the cultivated tomato without having to overcome any major interspecific barriers. However, in spite of the relatively easy of crossability with the cultigen, *S. neorickii* has not been extensively used by plant breeders, partly due to its comparatively recent discovery (Taylor 1986). This can be explained in part by the rather restricted geographic range of *S. neorickii* and the similar *S. chmielewskii* (Taylor 1986; although see Peralta et al. 2008; see Sect. 9.2.2).

The first extensive genetic study conducted on an interspecific tomato cross involving the wild species *S. neorickii* used an AB-QTL mapping strategy to explore the *S. neorickii* LA2133 genome as a potential source of useful QTL alleles for traits of agronomic importance including yield and fruit quality-related characteristics (Fulton et al. 2000) (Tables 9.5 and 9.7). One hundred and seventy BC_2 plants were scored for 131 RFLPs and ~170 BC_3 families were evaluated for 30 horticultural traits, in replicated field trials conducted in three different locations. A total of 199 QTLs were detected for the 30 analyzed traits, and for 19 traits at least one QTL was identified for which the wild allele had a favorable effect, despite the overall inferior phenotype of *S. neorickii*. This AB population was also evaluated for sugars, organic acids, and other biochemical properties possibly contributing to flavor, and 52 QTLs were identified for the 15 analyzed traits (Fulton et al. 2002a). Starting from the same *S. lycopersicum* cv. E6203 × *S. neorickii* LA2133 AB population, a set of 142 BILs (BC_2F_7) has been developed by D. Zamir and collaborators. Within the framework of the EU-SOL project (see above and http://www.eusol.net/), the 142 BILs have been anchored to a common set of COSII markers, and have been evaluated for agronomic traits, including yield, brix, and fruit weight (Tripodi et al. 2010; D. Zamir and S. Grandillo personal communication). Several favorable *S. neorickii* alleles were identified that could be targeted for further marker-assisted introgression into cultivated tomato.

The *S. neorickii* accession G1.1601 has been identified as a source of resistance to PM (caused by *Oidium lycopersici*), and an F_2 mapping population derived from the *S. lycopersicum* cv. "Moneymaker" × *S. neorickii* G1.1601 cross was used for QTL analysis (Bai et al. 2003; Table 9.6). The resistance was found to be controlled by three QTLs: *Ol-qtl1* mapping on chromosome 6, in the same region as the *Ol-1* locus (found in *S. habrochaites*), which is involved in a hypersensitive resistance response to the pathogen, and other two linked QTLs (*Ol-qtl2* and *Ol-qtl3*) that are located on chromosome 12, near the *Lv* locus conferring resistance to the other PM species, *L. taurica* (Bai et al. 2003). Since the *S. neorickii* accession G1.1601 showed also a certain level of resistance to GM (*Botrytis cinerea*) (ten Have et al. 2007), F_3 lines derived from the above-mentioned F_2 population were used to identify QTLs underlying the resistance response to *B. cinerea* using a stem bioassay (Finkers et al. 2008). Three putative QTLs were identified, and for each of them a putative homologous locus had been previously identified in *S. habrochaites* LYC4 (Finkers et al. 2007a, b).

9.6.4.9 *Solanum pennellii*

Solanum pennellii is a green-fruited species that grows at a wide range of elevations along the western slopes of the Andes, and is found on arid slopes and dry washes (see Sect. 9.2.2; Rick 1973). The extreme drought tolerance of this species has motivated numerous studies aimed at transferring its tolerance to the cultivated tomato. Another important characteristic of many *S. pennellii* accessions is their high level of resistance to numerous insects, which has been correlated with high density of type IV glandular trichomes and the presence of high levels of toxic acylsugars in their exudates (Farrar and Kennedy 1991; Labate et al. 2007 and references therein). *S. pennelli* has also been used as a source of disease resistances and, more recently, with the availability of the *S. pennellii* LA0716 IL population, the use of this wild species has greatly increased, and extended to hundreds of different traits of agronomical and biological relevance (Tables 9.6–9.8; see also reviews by Lippman et al. 2007; Grandillo et al. 2008).

With respect to disease resistance, *S. pennellii* LA0716 was found to display an incompatible reaction with race 3 (T3) strains of *Xanthomonas campestris* pv. *vesicatoria,* the casual agent of bacterial spot, indicating the existence of hypersensitive response (HR)-related resistance in this wild species. Using the *S. pennellii* LA0716 IL population a dominant resistance gene, called *Xv4,* was mapped on chromosome 3, and the avirulence gene, *avrXv4,* was isolated (Astua-Monge et al. 2000; Table 9.6).

S. pennellii has been used as a source of resistance to fungal diseases. *Alternaria* stem canker disease in tomato is caused by the necrotrophic fungus *A. alternata* f. sp. *lycopersici*. Genetic analyses showed that high insensitivity to AAL toxins from *S. pennellii* LA0716 is inherited in tomato as a single complete dominant locus, *Asc*, which has been genetically mapped on chromosome 3 of tomato using RFLPs (Van der Biezen et al. 1995). Subsequently, Mesbah et al. (1999) reported the physical analysis of a yeast artificial chromosome (YAC) contig spanning the *Asc* locus. Positional cloning of *Asc* showed that sensitivity is associated with a mutation in the gene that leads to a predicted aberrant ASC protein, a new plant member of the longevity assurance protein family (Brandwagt et al. 2000).

S. pennellii accessions have also been found to be sources of resistance to the soil-borne fungus *Fusarium oxysporum* f.sp. *lycopersici,* the causal agent of *Fusarium* wilt disease. The resistance conferred by genes *I* and *I-2* (both derived from accessions of *S. pimpinellifolium*) was overcome by a new race 3 of the fungus, and therefore a new dominant resistance gene, *I-3*, was identified in two *S. pennellii* accessions, PI 414773 (McGrath et al. 1987) and LA0716 (Scott and Jones 1989), and introgressed into *S. lycopersicum*. The *I-3* gene from LA0716 was mapped to chromosome 7 near the isozyme marker *Got-2* (Bournival et al. 1989, 1990; Sarfatti et al. 1991). The isolation of this resistance gene is being pursued via map-based cloning, and a high resolution genetic and physical map of the *I-3* region has been reported (Hemming et al. 2004; Lim et al. 2008). In addition, a new locus conferring resistance against *F. oxysporum* f. sp. *lycopersici* race 1 was mapped using RFLPs in a BC_1 population derived from a *S. lycopersicum* cv. "Vendor" (susceptible to race 1) × *S. pennellii* LA0716 (resistant) cross (Sarfatti et al. 1991). The locus, called *I-1*, was located on chromosome 7 and was not allelic to *I*, the traditional gene for resistance against the same fungal pathogen that was derived from *S. pimpinellifolium* (Sarfatti et al. 1991). These genes have been introgressed into commercial tomato and their map position further defined (Scott et al. 2004). A genome-wide dissection of *Fusarium* resistance was conducted using the *S. pennellii* LA0716 IL population (Sela-Buurlage et al. 2001). The study allowed the identification of six independent loci; the *I* and *I2* loci, previously introgressed from *S. pimpinellifolium,* were shown to reside on different arms of chromosome 11; three novel loci were identified on chromosomes 2 (loci *I-4* and *I-5*) and 10 (locus *I-6*). The loci *I-5* and *I-6* represented new *S. pennellii* resistance loci with varying degrees of potency; in contrast, the origin of the *I-4* locus was not defined. This study emphasized the complexity of wilt disease resistance revealed at both inter- and intralocus levels.

With respect to insect resistance, the genetic control of acylsugar accumulation in exudates of type IV glandular trichomes has been studied in an interspecific F_2 population derived from the cross *S. lycopersicum* × *S. pennellii* LA0716 (Mutschler et al. 1996). A total of five QTLs were identified, which were subsequently transferred by MAS pyramiding into the cultivated tomato genetic background (Lawson et al. 1997). Although, the obtained multiline accumulated acylsugars, the levels were lower than those of the interspecific F_1 control, suggesting that in order to reach higher level of acylsugar accumulation additional QTLs, still unidentified, might be necessary. Furthermore, the inheritance of acylsugar fatty acid composition was analyzed in an intraspecific F_2 population derived from a cross between *S. pennellii* LA0716 and *S. pennellii* LA1912, and six QTLs were detected for the nine segregating fatty acid constituents (Blauth et al. 1999).

The drought tolerance of *S. pennellii* was found to be related to greater WUE and less negative carbon isotope composition (δ^{13} C) (Martin et al. 1999), compared to the cultivated tomato, due to the ability of its leaves to take up dew (Rick 1973), and also to a rapid closure of stomata upon water deficit stress (Kebede and Martin 1994). Carbon isotope composition (δ^{13} C) is considered an attractive substitute for WUE in research and breeding programs, since in C_3 plants it varies in concert with leaf WUE, and δ^{13} C can be measured with minimal tissue destruction. Therefore, identification and MAS of QTL for WUE by means of δ^{13} C is considered a particularly promising way to break negative pleiotropy between WUE and yield in C_3 species. An RFLP mapping study conducted in F_3 and BC_1S_1 tomato populations derived from an *S. lycopersicum* × *S. pennellii* cross allowed the identification of three genomic regions explaining a large proportion of the genetic variance for δ^{13} C (Martin et al. 1989). More recently, the use of the *S. pennellii* LA0716 IL library allowed the detection of a dominant QTL for δ^{13} C, *QWUE5.1*, in *S. pennellii* IL5-4; at this QTL the wild allele had a favorable effect, since it determined high δ^{13} C (small negative value) (Xu et al. 2008).

S. pennellii LA0716 has also been used to identify QTLs conferring ST during seed germination (SG) (Foolad and Jones 1993; Foolad et al. 1997; Foolad and Chen 1998) or during the vegetative stage (VG) (Zamir and Tal 1987; Frary et al. 2010) (Table 9.8). The studies conducted during SG have shown that ST at this stage in tomato was controlled by a few major QTLs, which act together with a number of smaller effect QTLs. Moreover, some of these QTLs were conserved across species, while other were species specific (Foolad et al. 1997, 1998a; Foolad and Chen 1998). Zamir and Tal (1987) used a *S. lycopersicum* × *S. pennellii* LA0716 F_2 population and 15 isozyme markers to identify QTL that affect Na^+, Cl^-, and K^+ ion contents. The authors detected a minimum of four major loci that affected the contents of both Na^+, Cl^- in leaves, and two other loci influencing K^+ uptake. Recently, the *S. pennellii* LA0716 IL population, along with its parental lines, has been evaluated for growth parameters and for antioxidant paramethers of the leaves, under both control and salt stress (150 mM NaCl) conditions (Frary et al. 2010). The data allowed the identification of 125 QTLs for seven traits related to antioxidant content and to the response of tomato antioxidants to salt stress. It was generally observed that salt stress resulted in higher levels of antioxidant compounds and enzymes in the wild species. However, a direct correlation between antioxidant levels and salinity tolerance could not be definitely shown, and further studies are necessary in order to verify whether higher antioxidant tomato cultivars will show improved ST in the field.

Interspecific mapping populations deriving from crosses between *S. lycopersicum* and *S. pennellii* LA0716 have been used to map several genes involved in pigment content and fruit ripening including *high pigment-2* (*hp-2*) and *jointless* (*j*) loci (Kinzer et al. 1990; Wing et al. 1994; Zhang et al. 1994; van Tuinen et al. 1997; Mustilli et al. 1999; Liu et al. 2003; Rousseaux et al. 2005). In addition, *S. pennellii* LA0716, as well as other green-fruited wild species of tomato, has been the source of the mutation *Delta* (*Del*) that changes fruit color from red to orange as a result of accumulation of δ-carotene at the expense of lycopene. The *Del* gene was located on the RFLP map of tomato chromosome 12, and evidence strongly suggested that the locus *Del* in the fruit-color mutation *Delta* encoded the gene for lycopene ε-cyclase (Ronen et al. 1999). Furthermore, the two *S. pennellii* LA0716 ILs, IL-3-2 and IL3-3, have been used for the positional cloning of the *Beta* (*B*) gene, which encodes a novel type of lycopene β-cyclase, an enzyme that converts lycopene to β-carotene (Ronen et al. 2000).

Progenies derived from the *S. lycopersicum* × *S. pennellii* LA0716 cross have been extensively used to explore the genetic basis of numerous quantitative traits related to yield and fruit quality, and to identify molecular markers linked to favorable wild alleles to be used in MAS breeding programs. The first study, conducted in a *S. lycopersicum* × *S. pennellii* LA0716 BC_1 population, used isozymes to analyze the genetic basis of the four metric traits: leaf ratio, stigma exsertion, fruit weight, and seed weight (Tanksley et al. 1982; Table 9.7). Interestingly, already in this study, it was reported the identification of specific QTL alleles with effects opposite to those expected from the parental phenotypes. A more comprehensive investigation of the genetic basis of wide-cross transgressive segregation was conducted by deVicente and Tanksley (1993) on a large F_2 population derived from the *S. lycopersicum* cv. "Vendor TM2a" × *S. pennellii* LA0716 cross using RFLP markers. A total of 74 significant QTLs were identified for the 11 biological traits evaluated, and 36% of these QTLs had alleles with effects opposite to those predicted by the parental phenotypes, which could be directly related to the appearance of transgressive individuals in the F_2. Another *S. lycopersicum* × *S. pennellii* LA0716 F_2 population was used to study the genetic basis of leaf and flower morphology (Frary et al. 2004b).

In 1994, Eshed and Zamir reported the development of the first generation of the *S. pennellii* LA0716 IL library in the genetic background of *S. lycopersicum* cv. "M82," consisting of 50 ILs, each containing a single RFLP-defined introgression from *S. pennellii* in an otherwise cultivated genomic background. Collectively these lines provide coverage of the entire wild species genome. This new kind of genetic resource, also referred to as "exotic library" (Zamir 2001) was developed with the purpose of improving the efficiency with which wild germplasm could be used in tomato breeding and genetic studies.

The numerous advantages and potentialities of IL populations for the analysis of complex traits have been obvious since the first studies conducted to map and fine-map QTLs underlying horticultural yield and fruit quality traits using this type of genetic resource (Eshed and Zamir 1995, 1996; Eshed et al. 1996; as reviewed by Zamir and Eshed 1998a, b). Since then, the *S. pennellii* IL library, as well as its second generation consisting of 76 ILs and sub-ILs (Liu and Zamir 1999; Pan et al. 2000), has been used to analyze hundreds of traits of agronomical and biological interest including fruit weight, fruit shape, brix, pH, yield, traits related to reproductive fitness (Eshed and Zamir 1995, 1996; Eshed et al. 1996; Monforte et al. 2001; Causse et al. 2004; Baxter et al. 2005; Semel et al. 2006), disease resistance (Astua-Monge et al. 2000; Sela-Buurlage et al. 2001), leaf and flower morphology (Holtan and Hake 2003), locule number (Barrero and Tanksley 2004), carotenoid content in relation to fruit color (Liu et al. 2003), fruit nutritional and antioxidant content (Rousseaux et al. 2005; Stevens et al. 2007, 2008), fruit primary metabolites (Causse et al. 2004; Schauer et al. 2006, 2008), aroma compounds (Tadmor et al. 2002; Tieman et al. 2006), hybrid incompatibility (Moyle and Nakazato 2008), and antioxidant content of the leaves related to salt stress conditions (Frary et al. 2010). All these mapping efforts have allowed the identification of more than 2,800 QTLs (Tables 9.6–9.8; for reviews, see Lippman et al. 2007; Grandillo et al. 2008). Recently, in order to detect the genetic basis of metabolic regulation in tomato fruit, Kamenetzky et al. (2010) constructed a detailed physical map of five genomic regions associated with 104 previously described metabolic QTLs of the *S. pennellii* LA0716 IL population. For this purpose, the genetic and physical maps of *S. pennellii* and *S. lycopersicum* were integrated, providing a large dataset that will constitute a useful tool for QTL fine-mapping and relatively easy screening of target clones in map-based cloning approaches.

Another *S. pennellii* accession, LA1657, has been used in an AB-QTL mapping study aimed at identifying loci for yield, processing, and fruit quality traits (Frary et al. 2004a). A total of 175 BC_2F_1 families derived from the interspecific cross *S. lycopersicum* E6203 × *S. pennellii* LA1657 were grown and phenotyped for 25 traits in three locations, and 84 QTLs were identified. Also in this case a high proportion (26%) of the identified QTLs had *S. pennellii* alleles that enhanced the performance of the elite parent, also for traits for which the wild parent had an inferior phenotype (Frary et al. 2004a).

All these studies have allowed the identification of numerous *S. pennellii* QTL alleles that are of potential interest for breeding. Furthermore, the *S. pennellii* IL library facilitated exploration of the genetic basis of heterosis for "real-world" applications, as shown by the development of a new leading hybrid of processing tomato (Lippman and Zamir 2007; Lippman et al. 2007) (see Sect. 9.7.2.5).

9.6.4.10 *Solanum pimpinellifolium*

This red-fruited wild relative of tomato can be reciprocally hybridized with *S. lycopersicum*, and due to its close relationship with the cultigen and ease of backcrossing it has been extensively used as an attractive source of germplasm for various agriculturally important traits such as disease and insect resistance/tolerance as well as fruit quality traits (Taylor 1986; Peralta and Spooner 2001; Kole et al. 2006; Ashrafi et al. 2009). Additionally, some *S. pimpinellifolium* accessions have been identified as potential sources for abiotic stress tolerance (Foolad 2004, 2005).

With respect to viral diseases, tolerance to TYLCV infection has been reported in *S. pimpinellifolium* accessions, including LA0121, LA0373, and LA0690 (reviewed by Stevens and Rick 1986). Bulk RAPD analyses were performed on F_4 lines segregating for resistance to TYLC derived from *S. pimpinellifolium* "hirsute INRA" (Montfavet, INRA, France), and a major QTL responsible for up to 27.7% of the resistance was identified on chromosome 6 (Table 9.6; Chaguè et al. 1997).

In tomato, resistance to *Pseudomonas syringae* pv. "tomato" strains expressing the avirulence gene *avrPto* requires the presence of at least two host genes, designated *Pto* and *Prf* . The *Pto* gene has been introgressed into a *S. lycopersicum* cultivar from *S. pimpinellifolium* (Pitblado et al. 1984). *Pto* was isolated by a map-based cloning approach and it was shown to be a member of a clustered multigene family, located on the short arm of chromosome 5, with similarity to various proteinserine/threonine kinases (Martin et al. 1991, 1993). Subsequently, the gene *Prf* was identified through a mutational approach and was shown to be tightly linked to *Pto* (Salmeron et al. 1996). Another member of the *Pto* gene cluster termed *Fen* was found to confer sensitivity to fenthion (Loh and Martin 1995).

As discussed in Sect. 9.6.4.9, several studies have been conducted to identify resistances to the soil-borne fungus *F. oxysporum* f. sp. *lycopersici* which causes *Fusarium* wilt of tomato. The first gene (*I*), coferring vertical resistance to race 1 of the pathogen, was found in *S. pimpinellifolium* accession PI 79532 (Bohn and Tucker 1939) and was assigned to chromosome 11 (Paddock 1950). A second dominant gene, *I-2*, for resistance to race 2 was discovered in *S. pimpinellifolium* accession PI 12915 (Stall and Walter 1965; Cirulli and Alexander 1966), and was mapped to chromosome 11 using morphological markers (Laterrot 1976) and later by RFLPs (Sarfatti et al. 1991; Segal et al. 1992). The functional *I-2* resistance gene was isolated by a positional cloning approach and it was shown to be a complex locus (Ori et al. 1997; Simons et al. 1998). More recently, Scott et al. (2004) showed that the race 1 resistance, also present in PI 12915, was controlled by the *I* gene. Both genes have been incorporated into a wide number of commercial tomato cultivars (Bournival et al. 1989).

Resistances to other fungal diseases have also been identified in *S. pimpinellifolium* accessions. For example, the resistance of tomato to gray leaf spot disease caused by four *Stemphylium* species is conferred by a single incompletely dominant gene, *Sm*, which was introgressed into cultivars from *S. pimpinellifolium* accession PI 79532 and was found to be linked to a *Fusarium* race 1 resistance gene, *I*, on chromosome 11 (Dennett 1950). The *Sm* gene was then placed on the RFLP map of tomato using an F_2 population segregating for the resistance (Behare et al. 1991). Numerous cultivars with stable resistance to gray leaf spot have been released (Stevens and Rick 1986). Sources of resistance to GM (*Botrytis cinerea*) have also been identified in *S. pimpinellifolium.* In a study conducted to find new breeding material for resistance to GM, *S. pimpinellifolium* accession LA1246 showed high resistance both in the leaflet and in the stem (Ignatova et al. 2000).

LB caused by the fungal pathogen *P. infestans* is one of the most important diseases of the cultivated tomato and potato (Robertson 1991). Breeding for resistance to LB in tomato has followed two directions: one has been the search for "*R*" genes that confer race-specific or isolate-specific resistance that often exhibit qualitative inheritance, and the other has been the search for quantitative resistance, also referred to as partial resistance, which tends to be multigenic and quantitatively inherited (Wastie 1991; Umaerus and Umaerus 1994). In tomato, three isolate-specific *R* genes have been reported, *Ph-1* (a completely dominant gene), *Ph-2*, and *Ph-3* (both incompletely dominant genes), and *S. pimpinellifolium* was the original source for all of them (Table 9.6; Peirce 1971; Chunwongse et al. 1998; Moreau et al. 1998). The gene *Ph-1* was located on chromosome 7 (Peirce 1971) and *Ph-2* gene, originating from *S. pimpinellifolium* WVa700 was located on the long arm of chromosome 10 by RFLP analysis (Moreau et al. 1998). The *Ph-3* gene was found in an interspecific cross of *S. lycopersicum* and *S. pimpinellifolium* L3708, and mapped to chromosome 9 (Chunwongse et al. 2002). The same interspecific cross was used to study the genetic basis of quantitative resistance to LB in field trials, and two QTLs were identified (Frary et al. 1998). More recently, Kole et al. (2006) mapped another *R*-gene (*Ph-4*) conferred by *S. pimpinellifolium* from a similar cross. Their QTL analysis resulting in significantly high contribution to phenotypic variance also confirmed qualitative nature of inheritance.

Kerr and Bailey (1964) investigated *S. pimpinellifolium* resistance to tomato leaf mold (*C. fulvum*), and identified two genes, *Cf-2* and *Cf-9,* which were later introgressed into commercial tomato (Stevens and Rick 1986). Classical and RFLP mapping allowed more precise positioning of these genes, and revealed the existence of two complex resistance loci in tomato, one on chromosome 6, of which *Cf-2* and *Cf-5* are members, and another on chromosome 1, the *Milky Way* (*MW*) complex locus, of which *Cf-4* and *Cf-9* are members (van der Beek et al. 1992; Dickinson et al. 1993; Jones et al. 1993; Balint-Kurti et al. 1994; reviewed by Rivas and Thomas 2005). *Cf-2* was isolated by positional cloning (Dixon et al. 1996), while the *Cf-9* gene was isolated by transposon tagging (Jones et al. 1994). Functional analysis of a limited number of *S. pimpinellifolium* accessions allowed the identification of novel *Cf* genes (*Cf-ECP1*, *Cf-ECP2*, *Cf-ECP3*, *Cf-ECP4*, and *Cf-ECP5*) that trigger an HR in response to the *C. fulvum* extracellular proteins ECP1, ECP2, ECP3, ECP4, and ECP5 (Laugé et al. 1998a, b, 2000). Genetic mapping showed that *Cf-ECP2* and *Cf-ECP3* defined a new complex locus for *C. fulvum* resistance at *Orion* (*OR*) on the short arm of chromosome 1 (Haanstra et al. 1999a; Yuan et al. 2002) and the mapping of *Cf-ECP5* also defined a new complex locus, located 3 cM proximal to *MW*, which was designated *Aurora* (*AU*) (Haanstra et al. 2000) (see also review by Rivas and Thomas 2005). Soumpourou et al. (2007) showed that both genes, *Cf-ECP1* and *Cf-ECP4*, are located at *MW* complex locus together with *Cf-9* and *Cf-4*.

The *Hero* gene of tomato, a broad spectrum resistance gene that confers a high level of resistance to all pathotypes of the potato cyst nematodes *Globodera rostochiensis* and partial resistance to *G. pallida*, was introgressed into tomato cultivar LA1792 from the wild species *S. pimpinellifolium* LA0121 (Ellis and Maxon-Smith 1971). The gene was mapped to chromosome 4 (Ganal et al. 1995) and subsequently isolated by a map-based cloning approach (Ernst et al. 2002).

With respect to tolerance to abiotic stresses, the *S. pimpinellifolium* accession LA0722 was identified as a source of ST during both SG and VG (Foolad et al. 1998a; Foolad and Chen 1999; Zhang et al. 2003a); in addition it exhibited rapid SG in cold conditions (Foolad et al. 1998b) and under drought stress (Foolad et al. 2003; Table 9.8). QTL analysis of BC_1S_1 families derived from a cross between *S. pimpinellifolium* LA0722 and a moderately salt-sensitive *S. lycopersicum* line (NC84173) allowed the identification of seven QTLs for ST during SG (Foolad et al. 1998a), and of five QTLs for ST during VG in saline solution cultures (Foolad and Chen 1999). The *S. pimpinellifolium* accession had favorable QTLs at six of the seven QTLs identified during SG, and at all five ST QTLs identified during VG. Three of these QTLs for ST during VG were subsequently validated using the selective genotyping approach (Foolad et al. 2001). The same BC_1S_1 families were evaluated for germination at low temperature ($11 \pm 0.5°C$), and two chromosomal locations (3–5 putative QTLs) with significant effects on low temperature germination were identified; the wild species had favorable QTL alleles on chromosomes 1 (Foolad et al. 1998b). Finally, the same population was evaluated for drought tolerance during SG and four QTLs were identified for rate of germination under drought stress. For the two QTLs with larger effect, located on chromosomes 1 and 9, the favorable allele was contributed by *S. pimpinellifolium* donor parent (Foolad et al. 2003).

As described for *S. galapagenense* in Sect. 9.6.4.2, two other *S. pimpinellifolium* accessions (L1 and L5) have been used to identify QTLs for ST during the vegetative and/or reproductive stages (Bretó et al. 1993; Monforte et al. 1996, 1997a, b; Villalta et al. 2007, 2008; Estañ et al. 2009). Also in this case, the *S. lycopersicum* "cerasiforme" × *S. pimpinellifolium* interspecific RILs were used as rootstocks for a commercial hybrid, and were tested under saline conditions (Estañ et al. 2009). The results showed that up to 65% of the rootstock lines raised the fruit yield of the commercial hybrid under saline conditions, and QTLs underlying the ST rootstock effect were identified. Correlation and QTL analyses suggested that rootstock-mediated improvement of fruit yield in the *S. pimpinellifolium* population under salinity was mainly explained by the rootstock's ability to minimize perturbations in scion water status (Asins et al. 2010).

S. pimpinellifolium has been used as a source for a number of plant and fruit desirable traits like earliness, yield, and fruit quality also by means of classical genetic approaches (Kalloo 1991). The first QTL mapping study was conducted by Weller et al. (1988) on a large F_2 population derived from a *S. lycopersicum* × *S. pimpinellifolium* CIAS27 cross using six morphological markers and four isozymes. A total of 85 significant marker by trait combinations were identified for 18 quantitative analyzed traits including brix, fruit weight, fruit shape, and sugar content (Weller et al. 1988). For 14 traits at least one highly significant effect of opposite sign to the one expected based on the parental values was identified.

During the past 15 years, crosses between *S. lycopersicum* and the *S. pimpinellifolium* accession LA1589 have been used for numerous mapping studies. Grandillo and Tanksley (1996a) analyzed a BC_1, population deriving from the above-mentioned cross, for 19 quantitative traits related to fruit quality, flower morphology, flowering and ripening time, and identified 54 QTLs. From the same interspecific cross, an AB population was generated, and approximately 170 BC_2 plants were analyzed with segregating molecular markers covering the entire tomato genome. BC_2F_1 and BC_3 families were evaluated for 21 horticultural traits including yield and fruit quality (Tanksley et al. 1996). A total of 87 QTLs were identified for 18 of the analyzed traits, and, interestingly, trait-enhancing QTL alleles derived from *S. pimpinellifolium* were identified for most traits important in processing tomato production, including traits for which the wild parent had an inferior phenotype. This AB population, along with the ones obtained with *S. arcanum* LA1708, *S. habrochaites* LA1777 and *S. neorickii* LA2133, has been evaluated for sugars, organic acids, and other biochemical properties possibly contributing to flavor, and 33 QTLs were identified for the 15 analyzed traits (Fulton et al. 2002a). Starting from the same interspecific cross, Doganlar et al. (2002b) developed a population of 196 BILs (BC_2F_6), which were genotyped for 127 marker loci covering the entire tomato genome, and were evaluated for 22 quantitative traits, including several fruit quality related traits. In all, 71 significant QTLs were identified and for 48% of them the wild allele was associated with improved agronomic performance.

Other studies have used mapping populations derived from cultivated *S. lycopersicum* × *S. pimpinellifolium* LA1589 crosses to study the genetic basis of extreme fruit size (Lippman and Tanksley 2001), or to map QTL influencing fruit shape (Grandillo et al. 1996; Ku et al. 1999, 2000; Van der Knaap and Tanksley 2001, 2003; Van der Knaap et al. 2002; Brewer et al. 2007; Gonzalo and Van der Knaap 2008). Chen et al. (1999) used the same BC_1S_2 population derived from the cross *S. lycopersicum* fresh-marker breeding line NC84173 × *S. pimpinellifolium* LA0722, used to detect abiotic stress tolerance QTL, to map 59 QTLs related to brix, fruit shape, lycopene content, and pH.

S. pimpinellifolium is a SC species with variation in outcrossing rate correlated with floral morphology, and therefore is an ideal taxon with which to study mating system evolution (Rick et al. 1977). Traits that affect mating behavior (petal, anther, and style lengths) differ greatly between inbreeding and outcrossing populations, whereas other flower parts (sepals, ovaries) show minimal differences. In order to analyze the genetic basis of traits distinguishing outcrossing and self-pollinating forms of *S. pimpinellifolium*, Georgiady et al. (2002) conducted a QTL mapping study on a F_2 population derived from a cross between two accessions with contrasting mating systems; LA1237 the "selfer" and LA1581 the "outcrosser". A total of five QTLs were found to underlie the variation for four of the six morphological traits analyzed. Interestingly, each of these four traits had a QTL of major (>25%) effect on phenotypic variance, which suggests that the genetic basis for these traits follows the pattern of a macromutation with modifiers, as described by Grant (1975).

9.7 Role in Crop Improvement Through Traditional and Advanced Tools

9.7.1 Tomato Domestication and Early Breeding

Wild tomatoes (*Solanum* sect. *Lycopersicon*) are native to western South America, and their natural distribution goes from central Ecuador, through Peru to northern Chile, with two endemic species in the Galápagos Islands (Darwin et al. 2003; Peralta and Spooner 2005). *S. lycopersicum* was domesticated by native Americans, but the original site of this process is still considered an unsolved question (Peralta and Spooner 2007), and two competing hypotheses have been proposed for the original place of domestication, one Peruvian (DeCandolle 1886), and the other Mexican (Jenkins 1948). Very likely, early humans selected for plants with mutations associated with a preferred genotype (e.g., larger fruit), and gradually, enough favorable (e.g.,"large-fruited") mutations accumulated resulting in the domesticated tomato. *S. lycopersicum* "cerasiforme", the cherry tomato, which has fruit weighing only a few grams, was thought to be the putative wild ancestor of the domesticated tomato (Cox 2000); however, recent studies have shown that the plants known as "cerasiforme" are a mixture of wild and cultivated forms rather than being "ancestral" to the cultivated tomatoes (Nesbitt and Tanksley 2002; Ranc et al. 2008).

Severe genetic bottlenecks were associated with tomato domestication as the crop was carried from the Andes to Central America and subsequently to Europe. By the time Europeans arrived to America in the fifteenth century, large fruited types already existed, indicating that tomato domestication was already at a fairly advanced stage (Jenkins 1948; Rick 1995). Further domestication occurred throughout Europe in the 18th and 19th centuries (Sims 1980). In addition, it is possible that the return of tomato from Europe to the New World might have caused further reduction of genetic variation (Rick 1988). During the nineteenth century, tomato cultivars were selected for different purposes, including adaptation to local climate conditions. As a result, by the end of the century, numerous cultivars of tomato were available, which could be considered as landraces and the result of domestication and some early breeding, and most of them required open pollination (Bai and Lindhout 2007).

Similarly to other crops, tomato domestication has resulted in drastic phenotypic changes that can be observed in the wide range of morphological and physiological traits that distinguish domesticated tomato from its wild ancestors. Particularly extreme changes have occurred in the tissues and organs important to humans (for example, seeds, roots, and tubers). Collectively, these changes are referred to as the domestication syndrome, and the exact trait composition varies for each crop (Frary and Doganlar 2003). In tomato, one of the most obvious outcomes of domestication is the enormous increase in fruit size, which has been accompanied by a tremendous variation in fruit shapes; wild and semi-wild forms of tomato bear small, almost invariably round fruit, while fruit of cultivated tomatoes comes in a wide variety of sizes (as much as 1,000 times larger than those of wild progenitors) and shapes including round, oblate, pear-shaped, torpedo-shaped, and bell pepper-shaped (Tanksley 2004). Additionally, domesticated tomatoes produce seeds up to several times larger than their wild relatives (Doganlar et al. 2000; Orsi and Tanksley 2009). However, it is not clear why seed size increased during domestication in crops such as tomato, which are not consumed for their seeds. One explanation might be that, in these species, seed size increased as a result of indirect selection for greater seedling vigor and germination uniformity under field production (Harlan et al. 1973) or as an overall allometric effect.

In tomato, the genetic basis of these domestication syndrome traits has been explored for fruit characters (size, shape, color, morphology, and set) and growth habit (self-pruning, plant height, and earliness) (Pnueli et al. 1998; Grandillo et al. 1999a; Doganlar et al. 2000; Lippman and Tanksley 2001; Frary and Doganlar 2003; Tanksley 2004; Gonzalo and Van der Knaap 2008; see also Sect. 9.6.4). The studies have shown that tomato fruit size and shape are controlled by major and minor QTL loci, and that a relatively small number of genes were involved in the dramatic transition from small-sized fruit of wild progenitors to the extremely large size of some modern cultivars, and these genes control two processes: cell cycle and organ number determination (Lippman and Tanksley 2001; Tanksley 2004). The molecular basis of some of these major QTLs has been deciphered; *FW2.2* and *FAS* control fruit mass by increasing the placenta area and locule number, respectively, and thus affect patterning along the medio-lateral axis (Frary et al. 2000; Cong et al. 2008); the two fruit shape QTLs, *SUN* and *OVATE,* control fruit elongation and therefore affect patterning along the apical–basal axis (Liu et al. 2002; Xiao et al. 2008). Additionally, comparative studies have shown a co-localization of many loci associated with similar characteristics in tomato, pepper, and eggplant, all also members of the family Solanaceae (Doganlar et al. 2002a; Frary and Doganlar 2003).

9.7.2 Role of Wild Species for Tomato Breeding

At the beginning of the twentieth century tomato breeding programs began in public institutes, mainly in the USA, and breeders started introducing disease resistant cultivars, which dominated the US market in the 1920s and 1930s (Bai and Lindhout 2007). Subsequently, the formation of private companies favored the shift from open pollinated cultivars to hybrids, and the first hybrid tomato cultivar "Single Cross" was released in 1946 (Dorst 1946). Eventually, hybrids cultivars ended up dominating the fresh market, as well as an increasing quote of cultivars used for processing market.

Tomato breeding priorities have changed over the years. Until 1950s, cultivars have been developed that assembled several traits useful for both the processing industries and the fresh market. Afterwards, fresh market and processing cultivars started to be reasonably different. In the 1970s the main breeding goal was to increase yield, while in the 1980s the improvement of fruit shelf-life became a priority. Currently, sensorial and nutritional quality has become an important consumer demand (Bai and Lindhout 2007).

Closely related wild species within *Solanum* sect. *Lycopersicon* started to be used in tomato breeding programs in the early 1940s, when they began to be screened for additional disease resistances (Alexander et al 1942). Before that time, breeders had relied entirely on genetic variation in the European sources and their derivatives. This explains the difficulties breeders experienced in achieving most of their breeding objectives, in terms of improved yield, disease resistance, and other important traits (Rick 1988). As a result tomato improvement has been very slow, with very retarded gain in fruit yields until about 1940, when Bohn and Tucker discovered a strong resistance to Fusarium wilt in *S. pimpinellifolium.* Eventually, wild species began to play a significant role in tomato research and breeding. Despite the various difficulties often associated with the use of unadapted germplasm, numerous attributes were transferred from wild species to commercial cultivars, in particular resistance to pathogens, but also tolerance to abiotic stresses, and fruit quality-related traits (Stevens and Rick 1986; Kalloo 1991; Rick and Chetelat 1995; Tanksley and McCouch 1997; Zamir 2001; Bai and Lindhout 2007; Labate et al. 2007; Osborn et al. 2007). However, the potential of wild species in terms of source of valuable alleles for the improvement of cultivated germplasm is far from being fully exploited. During the past two decades, the advent of molecular markers technology has opened new opportunities for a more efficient use of wild germplasm. Molecular mapping studies have demonstrated that favorable alleles in wild relatives can remain cryptic until expressed in an improved background. These results have favored the development of new concepts and approaches aimed at a more efficient use of the genetic variation stored in wild germplasm (Tanksley and Nelson 1996; Tanksley et al. 1996; Tanksley and McCouch 2007; Zamir 2001; McCouch 2004; Lippman et al. 1997; Grandillo et al. 2008).

In this section, we will give an overview of the status of wild tomato species as a source of useful traits for the improvement of cultivated tomato and the main achievements reached in tomato breeding using genes derived from wild species. Moreover, strategies and tools that can facilitate studies on the genetic control of novel traits derived from wild species, the understanding of mechanisms underlying these traits, and their use for tomato improvement will also be discussed.

9.7.2.1 Disease Resistance

Tomato is susceptible to over 200 diseases caused by all types of pathogens, including viruses, bacteria, fungi, and nematodes (Lukyanenko 1991). Since the chemical control of these diseases is often too expensive for growers and in some cases ineffective, the development of resistant cultivars has always been a major breeding objective. Except for a few cases (Table 9.6; e.g., Lukyanenko 1991; Foolad and Sharma 2005; Ji and Scott 2007; Labate et al. 2007; Robertson and Labate 2007), all resistance genes have been derived from tomato wild relatives, with *S. chilense*, *S. peruvianum* s.l., *S. habrochaites,* and *S. pimpinellifolium* being the richest sources. Overall, resistances to over 42 major diseases have been discovered in tomato wild relatives, and at least 20 of them have been bred into tomato cultivars (Rick and Chetelat 1995; Ji and Scott 2007; Robertson and Labate 2007). For example, most commercial tomato hybrids carry different combinations of 15 independently introgressed disease-resistance genes originating from various wild accessions (Laterrot 2000; Zamir 2001; Foolad and Sharma 2005). Generally, they are major resistance genes for diseases such as root-knot nematode, fusarium wilt, verticillium wilt, alternaria stem canker, gray leaf spot, and some bacterial and viral disease (Laterrot 2000; Foolad and Sharma 2005; Ji and Scott 2007; Scott and Gardner 2007). However, in some cases (e.g., for diseases such as early blight, powdery mildew, bacterial canker, and bacterial wilt) horizontal resistance has been transferred since major genes for resistance were not available (Foolad and Sharma 2005). There is no doubt that, so far, the achievements in this area represent the greatest economic contribution of the wild species for the improvement of cultivated tomato germplasm.

Many of these resistance genes have been transferred into tomato cultivars or breeding lines through conventional breeding (see Table 3.2 in Ji and Scott 2007). One of the first examples was the exploitation of *C. fulvum* resistance from *S. pimpinellifolium* in 1934 (Walter 1967). During the last two decades, the use of molecular markers and MAS approaches have facilitated identification, mapping, and transfer of many disease resistance genes and QTLs in tomato (see Sect. 9.6.4) (Foolad and Sharma 2005; Labate et al. 2007). Currently, molecular markers are routinely employed in breeding programs by many seed companies in order to reduce cost and screening time mostly for transferring genes controlling vertical (race-specific) resistance to tomato diseases including bacterial speck, corky toot, fusarium wilt, LB, nematodes, powdery mildew, tobacco/tomato mosaic virus, tomato spotted wilt virus, tomato yellow leaf curl virus, and verticillium wilt (Foolad and Sharma 2005; Labate et al. 2007). Although in most tomato seed companies MAS is not yet employed as a routine approach for manipulating QTLs it has, however, been used to improve quantitative resistance to bacterial canker, bacterial wilt, and TYLCV (Foolad and Sharma 2005). More limited is the application of MAS in public tomato breeding programs; a few examples are given by its use to improve horizontal resistances to blackmold (Robert et al. 2001) and LB (Brouwer and St. Clair 2004) (reviewed by Foolad and Sharma 2005; see also Sect. 9.6).

MAS may not only accelerate the procedure of gene transfer, but, through it, the pyramiding of desirable genes and QTLs for different traits can be also simpler and more effective (Barone and Frusciante 2007). However, many disease resistance genes are clustered in the genome. Therefore, the transfer of multiple resistance genes into single varieties might have to overcome difficulties associated with unfavorable repulsion linkages between clustered resistance loci and unforeseen actions of the resistance genes themselves. In this respect, the use of molecular markers will be a valuable tool for identifying rare recombinants that can be evaluated for improved performance. A solution could be to combine favorable alleles of the target loci in coupling phase linkage; an approach that was applied for the *Mi-1* and *Ty-1* resistance genes located near the centromere of tomato chromosome 6, a region where several other important resistance genes cluster (Hoogstraten and Braun 2005).

Further progress is to be expected in this field in light of the numerous new genetic, genomic, and bioinformatic tools that are becoming available for tomato and other species (Mueller et al. 2009; Sanseverino et al. 2010; see also Sect. 9.8).

9.7.2.2 Insect Resistance

The cultivated tomato is susceptible to a wide array of arthropod pests, some of which can cause severe losses (Farrar and Kennedy 1991; Kennedy 2003). Wild tomato species represent a rich reservoir of resistances to most important insects in tomatoes (Farrar and Kennedy 1991; Kennedy 2007). In particular, *S. habrochaites* is the most significant source of arthropod resistances, carrying resistance to at least 18 pest species (Ji and Scott 2007), followed by *S. pennellii* which shows resistance to at least nine insect species, with one accession, LA0716, being resistant to eight of these pests (Muigai et al. 2003). In addition, some insect resistance has also been found in *S. lycopersicum* "cerasiforme," *S. pimpinellifolium*, *S. cheesmaniae*, *S. chmielewskii*, *S. peruvianum.*, *S. corneliomulleri*, *S. arcanum*, and *S. chilense* (Farrar and Kennedy 1991).

As described in Sects. 9.6.4.5 and 9.6.4.9, several mechanisms can be responsible for tomato resistance to arthropods, including physical and chemical properties of glandular trichomes, and chemical defenses associated with the leaf lamella (Farrar and Kennedy 1991). More specifically, methyl-chetones, such as 2-TD, and sesquiterpenes have been found to be associated with pest resistance in *S. habrochaites,* whereas in many *S. pennellii* accessions high level of resistance to numerous insects, including aphids, whiteflies, tomato fruitworm, beet armyworm, and the agromyzid leafminer is correlated with high density of type IV glandular trichomes and with the presence of high levels of toxic acylsugars in their exudates (references in Labate et al. 2007). QTLs underlying some of these traits have been identified (see Sects. 9.6.4.5 and 9.6.4.9).

Despite the rich source of natural resistance available, partly due to the mobile nature of the organisms involved, breeding for insect-resistance has been more complicated than breeding for disease resistance. As a result, only a few insect-resistant cultivars have been developed so far, and hence advanced molecular-based approaches are foreseen as the tools that might change this trend, although it might be advisable to apply them after having used a combination of breeding and biochemical methods (Mutschler 2006).

9.7.2.3 Abiotic Stress Tolerance

Several environmental stresses, including salinity, drought, excessive moisture, extreme temperature, mineral toxicity, and deficiency as well as pollution can challenge tomato crop, reducing its growth and production. The development of cultivars tolerant to various abiotic stresses is a goal of great economic importance and has been a major practice in tomato breeding (Kalloo 1991; Foolad 2005). Tomato wild relatives represent a rich source of genetic diversity that can be used to improve abiotic stress tolerance of cultivated tomato germplasm. Predicting tolerance to abiotic stresses from observations of habitats of wild species, as proposed by Rick (1973), allowed to identify some useful sources for these traits. For instance, the arid habitats of *S. pennellii* and *S. sitiens* have led the detection of drought tolerance, while the high altitude accessions of *S. habrochaites* have been shown to possess resistance to cold temperatures. Resistance or tolerance to numerous adverse environmental conditions have been transferred in cultivated tomato including cold, heat, drought, excessive moisture conditions, as well as soil salinity and alkalinity (Kalloo 1991). A number of stress tolerant wild species stocks are maintained at TGRC that have been used in breeding programs (Robertson and Labate 2007; http//tgrc.ucdavis.edu/). However, traditional breeding for abiotic stress tolerance has been generally unsatisfactory mainly due to the very complex nature of such traits, except for heat tolerance (Scott et al. 1995).

As described in Sect. 9.6.4, extensive research has been conducted for identifying wild QTL alleles potentially involved in tolerances to different abiotic stresses, and considerable efforts have been invested in mapping research for tomato ST also at the reproductive stage. Several QTLs for drought related traits during important growth stages have been identified from *S. pimpinellifolium* and *S. pennellii*, while *S. habrochaites* has been the source for cold tolerance alleles (Foolad 2005). Moreover, recent QTL mapping studies have provided evidence that in order to be able to fully exploit the genetic potential of wild germplasm for the improvement of tomato crop productivity under salinity alternative approaches might be necessary. For instance, a more effcient utilization of wild germplasm could be via the improvement of rootstocks that confer ST, instead of introgression of beneficial QTL alleles into the genome of the cultivated tomato (Estañ et al. 2009).

In order to improve the effectiveness of these molecular tools, reliable QTLs at all stages of plant development should be identified, which can then be used to enable powerful MAS. In addition, new methodologies that integrate molecular, physiological, and phenotypic data should be explored in order to facilitate the pyramiding of QTLs.

9.7.2.4 Fruit Quality

Breeding objectives for fruit quality vary depending on whether the product is used fresh or processed, and whether we consider the producers', distributors', or consumers' needs. Quality traits important for processing tomato include the content of total soluble solids (SSC or brix; mainly sugars and acids), pH, and paste viscosity; shelf-life and firmness are priorities for distributors and retailers; while nutritional (e.g., antioxidants and vitamins) and sensorial quality play a major role in driving consumers' choices (Causse et al. 2001; Sinesio et al. 2010). Tomato sensorial quality for fresh consumption is a complex character as it relates to visual appearance (size, shape, and color), texture (firmness, mealiness, juiciness), and flavor attributes. The typical flavor of tomato fruit depends on a complex mixture of sugars, acids, amino acids, minerals, and volatile compounds (Baldwin et al. 1991).

Within wild species of tomato, there is a wealth of genetic variability also for fruit quality characters (Sect. 9.6.4; e.g., Stevens and Rick 1986; Rick and Chetelat 1995; Labate et al. 2007; Grandillo et al. 2008). For some of these traits, the value of the wild accession as a source of useful alleles can be assessed on a mere phenotypic basis (e.g., brix, nutritional quality, and in a few cases fruit color), whereas for other traits, such as fruit size, shape, and color, the breeding value depends on cryptic genetic variation that can become manifest once introgressed into cultivated genetic backgrounds, and that can be localized by means of molecular mapping approaches (Tanksley and McCouch 1997; Grandillo et al. 1999a; Zamir 2001; Lippman et al. 2007; Grandillo et al. 2008).

Among others, SSC of tomato fruit is a major concern in both fresh and processed market tomato production (Stevens 1986). This explains why much effort has been invested in trying to improve this quality trait. The SSC of commercial hybrid cultivars generally ranges from 4.5 to 6.0% of the fruit fresh weight, whilst the percentage of some tomato wild species can be much higher (Stevens 1972; Rick 1974; Hewitt and Garvey 1987). For example, *S. pimpinellifolium* and *S. chmielewskii* showed high concentrations (9–15%) of total soluble solids (Rick 1974; Hewitt and Garvey 1987). Generally, the efforts to breed for higher fruit solids have not been very successful because of the negative correlation between yield and SSC. However, Rick (1974) by introgressing *S. chmielewskii* genes into a cultivated tomato variety, developed lines with approximately 40% greater total soluble solids, without any major penalty on yield.

The wild relatives of tomato are also sources of alleles that affect other components of flavor, such as the concentration of specific sugars and organic acids (Fulton et al. 2002a) as well as the accumulation of nutritional compounds, such as lycopene, β-carotene, and ascorbic acid (see Sect. 9.6.4). For example, while fruit of *S. lycopersicum* accumulates primarily reducing sugars (glucose and fructose) and very little sucrose, fruit of *S. chmielewskii*, *S. habrochaites,* and of other green-fruited wild species accumulate high amounts of sucrose, due to the action of the recessive sucrose accumulator gene (*sucr*) (Davies 1966; Yelle et al. 1988; Chetelat et al. 1995a, b). The fructose-to-glucose ratio in the mature tomato fruit was found to be modulated by a major gene (*Fgr*) on chromosome 4, which does not affect total sugar levels (Levin et al. 2000); the incompletely dominant *S. habrochaites* (LA1777) allele at this locus increases the fructose-to-glucose ratio. Firmness of most cultivars has been improved using a *S. pimpinellifolium* background introgressed in the 1940s (Scott 1984).

The red color and the antioxidant activity of tomato fruit is principally determined by their carotenoid pigments content. An important gene that was introduced from several wild tomato species is *Beta* (*B*); the wild allele increases the level of provitamin A (β-carotene) in the fruit by more than 15-fold (as reviewed by Labate et al. 2007). Breakage of the linkage between *B* and *sp* + (the gene for indeterminate growth habit), both located on chromosome 6, allowed the use of *B* for commercial production (Stommel et al. 2005a, b). Another important nutrient of tomato fruit is vitamin C. There is wide range of variation in vitamin C level among tomato and wild tomato species; the concentration may range from 8 to 119 mg per 100 g. Wild tomato accessions are rich in ascorbic acid, a quality that has been lost in many commercial varieties, which contain up to five times less ascorbic acid, although small-fruited varieties are richer in this vitamin than are standard varieties (Stevens 1986). Cultivars with high level of vitamin C have been developed from a cross with *S. peruvianum* s.l., but with little commercial success at that time (Stevens and Rick 1986).

Today, new efforts to explore wild species to obtain new cultivars with high sensorial quality and nutritional value are underway. For example, recent studies have shown that IL libraries, derived from interspecific crosses, provide a very efficient tool to access wide genetic variation also in compositional changes in the fruit, including aroma volatiles (Rousseaux et al. 2005; Tieman et al. 2006; Schauer et al. 2006, 2008; Stevens et al. 2007, 2008; Mathieu et al. 2009; see also Sect. 9.6). To accomplish improvement for these traits, breeding will require clear parameters and efficient methods of analysis. In the future, higher attempts in developing multidisciplinary programs in this research fields are expected.

9.7.2.5 Yield

Improved yield and yield stability has long been recognized as an important objective in plant breeding. The continuous growth of world population, combined with improvements in quality of life and with the on-going reduction of land available for farming, has created an urgent need for greater production of vegetables. There is no doubt that the replacement of inbred varieties with hybrid varieties have significantly contributed to the total genetic gains achieved in yield during the past decades. However, it is difficult to determine which traits, besides yield per se, are responsible for increased crop yields, since adaptive and defensive characters may play a major role in determining the higher yields of modern varieties (Tanksley et al. 1997a, b; Grandillo et al. 1999b).

Recent studies conducted in tomato have highlighted the potential of wild germplasm to affect yield stability in diverse environments, and to be able to lift yield barriers (Gur and Zamir 2004). The authors demonstrated that an exotic library derived from a wild tomato species, with no yield potential, can segregate for a wide array of previously unexplored genetic variation, which is rapidly available to plant breeders for the improvement of crop productivity. More specifically, progress in breeding for increased tomato yield was evaluated using *S. lycopersicum* genotypes carrying a pyramid of three independent yield-promoting genomic regions introgressed from the drought-tolerant green-fruited wild species *S. pennellii* (LA0716). Yield of hybrids obtained by crossing the pyramided genotypes was more than 50% higher than that of a control market leader variety under both wet and dry field conditions that received 10% of the irrigation water. Moreover, the wild introgressions were effective in different cultivated genetic backgrounds, indicating that the cultivated tomato gene pool was missing alleles similar to those of the wild species (Gur and Zamir 2004). The approach of MAS pyramiding beneficial wild species chromosome segments into elite genetic backgrounds provides a new paradigm to revitalize plant breeding (Tanksley and McCouch 1997; Zamir 2001; Morgante and Salamini 2003; Koornneef et al. 2004; Lippman et al. 2007).

The results obtained by Gur and Zamir (2004) using the *S. pennellii* ILs established also a genetic infrastructure to explore the genetic and molecular basis underlying yield heterosis (Semel et al. 2006). This phenomenon has been studied for almost 100 years, and the cumulated research suggests that the genetic basis of hybrid vigor is determined by non-mutually exclusive mechanisms that include dominance complementation, overdominance, and epistasis (Lippman and Zamir 2007; Springer and Stupar 2007). However, the principles that govern heterosis and their molecular basis are still poorly understood. The use of the *S. pennellii* ILs allowed to partition heterosis into defined genomic regions, and, by eliminating a major part of the genome-wide epistasis, it was possible to estimate the importance of loci with overdominant (ODO) effects (Semel et al. 2006). It was shown that classical tomato heterosis is driven predominantly by overdominant QTLs associated with reproductive traits. Recently, Krieger et al. (2010) provided the first example of a single ODO gene for yield. The authors demonstrated that heterozygosity for tomato loss-of-function alleles of *SINGLE FLOWER TRUSS* (*SFT*), which is the genetic originator of the flowering hormone florigen, increases yield by up to 60%. Notably, the effect matched an ODO QTL from the *S. pennellii* LA0716 IL population (Semel et al. 2006). With the coming sequence of the tomato genome it will be easier to isolate those factors that are responsible for the strong ODO effects, and the derived knowledge will surely support further progress in crop breeding.

9.8 Genetic and Genomics Resources

Tomato has long served as a model system for genetic studies in plants, partly due to its importance as a food crop, but also because it has a series of advantageous characteristics including diploid inheritance, self-pollinating nature, ease of seed and clonal propagation, efficient sexual hybridization, easy crossability with most of the wild species, and a relatively short generation time. Tomato is also an excellent species for cytogenetic research, as its 12 chromosomes can be readily identified through analysis of pachytene karyotype, synaptonemal complexes, and chromosome or chromosome arm-specific DNA sequences. Finally, from the perspective of genetic and molecular investigations tomato has the additional advantages of a relatively small genome size among crop species (ca. 950 Mb) (Arumuganathan and Earle 1991). Extensive genetic and genomic tools have been developed in the domesticated tomato (see also reviews by Barone et al. 2008; Moyle 2008). Many of these tools should easily be exportable to tomato wild relatives due to the close relationship between the tomato and the related wild taxa as well as to the ample use of interspecific crosses with the cultigen.

Genetic and genomic resources currently available in tomato include thousands of molecular markers appropriate for use in domesticated and wild species, various molecular linkage maps (see Table 9.5), numerous DNA libraries, including BAC libraries and an advanced physical map, multiple permanent mapping populations, tomato wild species (see also Table 9.4), mutant collections, and Targeting Induced Local Lesions IN Genomes (TILLING) populations. Moreover, well-established genetic transformation protocols, gene-silenced tomato lines, and VIGS libraries (for transient silencing) have been developed, while EST collections are being actively produced worldwide permitting the design of different microarray platforms of which public results are also available. An ongoing genome sequencing initiative is providing insights into the genome structure of tomato with the purpose of generating a reference genome for the family Solanaceae and the Euasterid clade (APG 2009) more broadly. Websites distributed worldwide are providing information about resources for tomato and many of the other members of this plant family, as well as methodologies and bioinformatics tools (Mueller et al. 2005b; Labate et al. 2007). The SOL Genomics Network (SGN) organizes a comprehensive web-based genomics information resource designed to disseminate information for the Solanaceae family and the related families in the Asterid clade (Mueller et al. 2005a, b; http://solgenomics.net/). Besides providing reference information strictly concerning the genome sequencing, such as BAC registry, project statistics, sequence repository, and viewers for the annotated sequence, SGN catalogs and maintains genetic maps and markers of the Solanaceae species (Mueller et al. 2005b). Additionally, it provides links to related sites of interest representing therefore the reference site for the tomato community.

Other web-based resources available for tomato include the TGRC, founded by Charles M. Rick, the central gene bank for wild relatives, and tomato mutant stocks (http://tgrc.ucdavis.edu); the Germplasm Resources Information Network (GRIN), providing germplasm information (http://www.ars-grin.gov/); the tomato core collection from the EU-SOL initiative https://www.eu-sol.wur.nl/, composed of ~7,000 domesticated (*S. lycopersicum*) lines, along with representative wild species, provided by different international sources and from private collections. Tools such as the Tomato Analyzer, a stand-alone piece of software, which performs semi-automated phenotyping of fruit shape (http://www.oardc.ohio-state.edu/vanderknaap/tomato_analyzer.htm), are also available and are flourishing worldwide. Links to these efforts are and will continue to be provided at SGN.

An international consortium of ten countries is sequencing the tomato genome as the cornerstone of the "International Solanaceae Genome Project (SOL): Systems Approach to Diversity and Adaptation" initiative (http://solgenomics.net/solanaceae-project/). The preliminary effort is to produce a high-quality tomato genome sequence starting from the approximately 220 Mb of estimated gene-dense euchromatin (corresponding to less than 25% of the total DNA) (Peterson et al. 1996). Towards this objective, a BAC-by-BAC strategy has been pursued (Mueller et al. 2009), though a whole-genome shotgun approachhas also been undertaken to support the coverage of the entire genome. Currently (July 2010), more than 1,000 BACs are available. Moreover, the first draft of the whole genome sequencing of *S. lycopersicum* cv. "Heinz" is today available at http://solgenomics.net/genomes/Solanum_lycopersicum/. The research groups of D. Ware, W. R. McCombie, and Z. B. Lippman at Cold Spring Harbor Laboratory have released a draft genome sequence of *S. pimpinellifolium* LA1589 (http://solgenomics.net/genomes/Solanum_pimpinellifolium/). This draft sequence provides a relevant added resource of genomic data useful for biological discovery of the processes of plant domestication and evolution, as well as for a better exploitation of the breeding potential of this wild species.

Genome sequences are being released to the GenBank repository (http://www.ncbi.nlm.nih.gov) and are made available at the SGN website (http://solgenomics.net/) as well. The international Tomato Annotation Group (iTAG), a collaborative effort involving several groups from Europe, USA, and Asia, is taking care of the sequence annotation, to provide a high quality, information-enriched, tomato genome.

While waiting for the publication of the annotated tomato genome, preliminary data concerning the available contigs of BACs are made available on the SGN website (http://solgenomics.net/) as well as on cross-linked resources such as ISOLA (http://biosrv.cab.unina.it/isola; Chiusano et al. 2008). Additionally, a number of chromosome specific curated information resources, as well as web-based tools, have been developed in order to allow researchers to access and exploit the emerging genome sequence as it is released by the different participants in the sequencing project (Mueller et al. 2009).

The organization of tomato and other Solanaceae transcript sequence collections is a prerequisite to provide a reliable annotation of the tomato genome consistently supported by experimental evidence. Moreover, this information is relevant for investigation on expression profiles and provides a reference for microarray chip design. Therefore, the genome sequencing initiative has further encouraged the production of EST collections worldwide.

As reference examples, SGN organizes and distributes ESTs sequenced from cDNA libraries from *S. lycopersicum*, *S. pennellii*, *S. habrochaites*, as well as the corresponding assembled consensus sequences; the Tomato Stress EST Database (TSED) contains ESTs from more than ten stress-treated subtractive cDNA libraries from *S. lycopersicum*; the Micro-Tom Database (MiBASE) (Yano et al. 2006) distributes unigenes obtained by assembling ESTs from full-length cDNA libraries of *S. lycopersicum* cv.

"Micro-Tom" and ESTs from other tomato lines; and TomatEST included in SolEST (D'Agostino et al. 2007, 2009) which is a secondary database of EST/cDNA sequences, currently containing 112 libraries from all the tomato species available at dbEST, the NCBI repository of public collections. Other EST databases available for tomato and related species include DFC http://compbio.dfci.harvard.edu/tgi/plant.html and PLANT GDB http://www.plantgdb.org/.

Moreover, several microarray platforms based on the extensive EST collections available in tomato are now available for transcriptional profiling (Barone et al. 2009): Tom1, a cDNA-based microarray containing probes for approximately 8,000 independent genes; Tom2, a long oligonucleotide-based microarray containing probes for approximately 11,000 independent genes (http://ted.bti.cornell.edu/; SOL project, http://www.eu-sol.net); and an Affymetrix Genechip, which contains probe sets for approximately 9,000 independent genes (http://www.affymetrix.com/products/arrays/specific/tomato.affxspecific/tomato.affx).

Results from the different platforms are available from a variety of specific websites such as the Tomato Expression Database (TED) which is a primary database for tomato microarray data (Fei et al. 2006; http://ted.bti.cornell.edu).

Well-established molecular genetic tools are also available for tomato functional analyses. To date, 1,000 monogenic mutant stocks in a variety of genetic backgrounds are publicly available at the TGRC (http://tgrc.ucdavis.edu); seeds from an isogenic tomato "mutation library" consisting of 6,000 EMS-induced and 7,000 fast neutron-induced mutant lines are publicly available for gene function research (Menda et al. 2004; http://zamir.sgn.cornell.edu/mutants/). Insertional mutagenesis systems exploiting exogenous transposon systems have also been described in tomato (reviewed by Barone et al. 2008). Platforms based on TILLING (McCallum et al. 2000) are also under development for tomato in several countries, including the USA, France, Italy, and India, and the EU-SOL project (http://www.eu-sol.net) is coordinating the Franco-Italian effort. Gene silencing approaches have also been widely used as a tool for functional genomics research in tomato. These include early systems of sense and antisense silencing, as well as the more recent technologies of RNA interference (RNAi) and VIGS.

9.9 Conclusions and Future Actions

In this review, we have looked into the plant group *Solanum* sect. *Lycopersicon* (the clade containing the domesticated tomato and its 12 wild relatives) and the four allied species in the immediate outgroups *Solanum* sect. *Lycopersicoides* (*S. lycopersicoides* and *S. sitiens*) and sect. *Juglandifolia* (*S. ochranthum* and *S. juglandifolium*), belonging to the large and diverse family Solanaceae. We have summarized the geographic distribution and morphological characters of these plant groups, describing their evolutionary relationships in the context of a new taxonomic revision at the species level (Peralta et al. 2008). We have shown that cultivated tomato, like many other crops, has a very narrow genetic basis that has limited the breeding potential of this crop for many years. In contrast, wild species are characterized by a wide range of genetic variation, which represents a rich reservoir of valuable alleles that could be used to address present and future breeding challenges. Over the past 60 years, tomato breeders have been at the forefront of establishing new principles for crop breeding based on the use of wild species to improve modern cultivars (Powers 1941; Rick 1974). Although, the most remarkable achievements have been reached in the area of disease resistances, yet exotic germplasm has also been used as a source of useful genes to improve other important traits. The numerous molecular mapping studies conducted using interspecific crosses have clearly demonstrated that the breeding value of exotic germplasm goes much beyond its phenotype. However, in spite of these successful results, it has to be acknowledged that we are still far from having been able to fully exploit the breeding potential of the thousands of accessions stored in seed banks around the world, and that can still be found in natural habitats (Tanksley and McCouch 1997). We need to capitalize on the acquired knowledge and on the evergrowing genetic and "-omics" resources that are becoming available for tomato, to keep developing new concepts and breeding strategies suitable for a more efficient use of the wealth of genetic variation stored in the wild relatives. In this respect, among all model systems, the wild and domesticated species of the tomato clade have pioneered novel population development, such as "exotic libraries" (Zamir

2001; Lippman et al. 2007). The last 15 years of research conducted on the *S. pennellii* LA0716 ILs (the founding population) using cutting edge phenotyping platforms has demonstrated the value of such a resource in fundamental biology, and for exploring and utilizing the hidden breeding potential of wild species for practical use in agriculture.

These results have encouraged the tomato research community to invest in the development of IL populations, or related pre-breds, such as BILs, for a number of other tomato wild species including *S. habrochaites*, *S. arcanum*, *S. pimpinellifolium*, *S. lycopersicoides*, *S. neorickii*, *S. chmielewskii*, and *S. chilense* (see Sect. 9.6.4). Recently, in order to enhance the rate of introgression breeding in tomato, in the framework of a currently running EU project (EU-SOL), "exotic libraries" of tomato from a diverse selection of accessions are being further refined and anchored to a common set of COSII markers (Tripodi et al. 2009; https://www.eu-sol.wur.nl/).

These genetic resources, combined with advances in other fields such as cytogenetics and tissue culture, along with the increasing knowledge deriving from bioinformatics and the many "omics" tools, including the tomato genome sequence (http://sgn.cornell.edu/solanaceae-project), are expected to further improve the efficiency with which wild tomato relatives will contribute to the improvement of this important crop. At the end, breeders will be able to select the best combinations of alleles and to design programs to combine traits in new, superior genotypes following the "breeding by design" concept (Peleman and Van der Voort 2003).

Given the unquestionable value of wild tomato germplasm there is the need to preserve this precious resource for future generations. Therefore, conservation initiatives have to be taken not only for the excellent ex situ collections available worldwide, but also to preserve populations in situ. The appropriate authorities in national governments of the countries of origin – mainly Ecuador, Peru, and Chile – should be helped to take steps to protect their native tomatoes and their habitats from further catastrophic loss. International organizations, such as the CGIAR, are urged to get involved to initiate and/or support such conservation efforts. Without action, the wealth of wild germplasm in the tomato relatives may not be available to future generations.

Acknowledgments Research in the laboratories of S. Grandillo and S. Knapp is supported in part by the European Union (EU) program EU-SOL (contract PL 016214–2 EU-SOL). Research in the laboratories of S. Grandillo is also supported in part by the Italian MIUR project GenoPOM. Research in the laboratories of S. Knapp and D. M. Spooner is supported in part by the National Science Foundation's (NSF) Planetary Biodiversity Inventory program (DEB-0316614 "PBI *Solanum* – a worldwide treatment"). This work was in part supported also by the Italian CNR Short-Term Mobility Program 2009 to S. Grandillo. Contribution nr. 363 from CNR-IGV, Institute of Plant Genetics, Portici.

References

Agrama HA, Scott JW (2006) Quantitative trait loci for Tomato yellow leaf curl virus and Tomato mottle virus resistance in tomato. J Am Soc Hortic Sci 131(2):637–645

Albacete A, Martínez-Andúar C, Ghanem ME, Acosta M, Sánchez-Bravo J, Asins MJ, Cuartero J, Lutts S, Dodd IC, Pérez-Alfocea F (2009) Rootstock-mediated changes in xylem ionic and hormonal status are correlated with delayed leaf senescence, and increased leaf area and crop productivity in salinized tomato. Plant Cell Environ 32:928–938

Albrecht E, Escobar M, Chetelat RT (2010) Genetic diversity and population structure in the tomato-like nightshades *Solanum lycopersicoides* and *S. sitiens.* Ann. Bot. 105:535–554

Albrecht E, Chetelat RT (2009) Comparative genetic linkage map of *Solanum* sect Juglandifolia: evidence of chromosomal rearrangements and overall synteny with the tomatoes and related nightshades. Theor Appl Genet 118:831–847

Alexander L, Lincoln RE, Wright A (1942) A survey of the genus *Lycopersicon* for resistance to the important tomato diseases occurring in Ohio and Indiana. Plant Dis Rep Suppl 136:51–85

Alpert K, Grandillo S, Tanksley SD (1995) *fw2.2*: a major QTL controlling fruit weight is common to both red- and green-fruited tomato species. Theor Appl Genet 91:994–1000

Alvarez AE, van de Wiel CCM, Smulders MJM, Vosman B (2001) Use of microsatellites to evaluate genetic diversity and species relationships in the genus *Lycopersicon*. Theor Appl Genet 103:1283–1292

Ammiraju JSS, Veremis JC, Huang X, Roberts PA, Kaloshian I (2003) The heat-stable root-knot nematode resistance gene *Mi-9* from *Lycopersicon peruvianum* is localized on the short arm of chromosome 6. Theor Appl Genet 106:478–484

Anbinder I, Reuveni M, Azari R, Paran I, Nahon S, Shlomo H, Chen L, Lapidot M, Levin I (2009) Molecular dissection of *Tomato leaf curl virus* resistance in tomato line TY172 derived from *Solanum peruvianum.* Theor Appl Genet 119:519–530

APG III (2009) An update of the Angiosperm Phylogeny Group classification for the orders and families of flowering plants: APG III. Bot J Linn Soc 161:105–121

Arens P, Odinot P, van Heusden AW, Lindhout P, Vosman B (1995) GATA- and GACA-repeats are not evenly distributed throughout the tomato genome. Genome 38(1):84–90

Areshchenkova T, Ganal MW (1999) Long tomato microsatellites are predominantly associated with centromeric regions. Genome 42:536–544

Areshchenkova T, Ganal MW (2002) Comparative analysis of polymorphism and chromosomal location of tomato microsatellite markers isolated from different sources. Theor Appl Genet 104:229–235

Arumuganathan K, Earle ED (1991) Nuclear DNA content of some important plant species. Plant Mol Biol Rep 9:208–218

Ashrafi H, Kinkade M, Foolad MR (2009) A new genetic linkage map of tomato based on a *Solanum lycopersicum* × *S. pimpinellifolium* RIL population displaying locations of candidate pathogen response genes. Genome 52:935–956

Asins MJ, Bolarín MC, Pérez-Alfocea F, Estañ MT, Martínez-Andújar C, Albacete A, Villalta I, Bernet GP, Dodd IC (2010) Genetic analysis of physiological components of salt tolerance conferred by *Solanum* rootstocks. What is the rootstock doing for the scion? Theor Appl Genet 121: 105–115

Astua-Monge G, Minsavage GV, Stall RE, Vallejos E, Davis MJ, Jones JB (2000) *Xv4-vrxv4*: a new gene-for-gene interaction identified between *Xanthomonas campestris* pv. *Vesicatoria* Race T3 and the wild tomato relative *Lycopersicon pennellii*. Mol Plant Microbe Interact 13:1346–1355

Azanza F, Young TE, Kim D, Tanksley SD, Juvik JA (1994) Characterization of the effect of introgressed segments of chromosome 7 and 10 from *Lycopersion chmielewskii* on tomato soluble solids, pH, and yield. Theor Appl Genet 87: 965–972

Bai Y, Lindhout P (2007) Domestication and breeding of tomatoes: what have we gained and what can we gain in the future? Ann Bot 100:1085–1094

Bai Y, Huang C-C, van der Hulst R, Meijer-Dekens F, Bonnema G, Lindhout P (2003) QTLs for tomato powdery mildew resistance (*Oidium lycopersici*) in *Lycopersicon parviflorum* G1.1601 co-localize with two qualitative powdery mildew resistance genes. Mol Plant Microbe Interact 16(2):169–176

Bai Y, van der Hulst R, Huang CC, Wei L, Stam P, Lindhout P (2004) Mapping *Ol-4*, a gene conferring resistance to *Oidium neolycopersici* and originating from *Lycopersicon peruvianum* LA2172, requires mulialleli, single-locus markers. Theor Appl Genet 109:1215–1223

Bai Y, van der Hulst R, Bonnema G, Marcel TC, Meijer-Dekens F, Niks RE, Lindhout P (2005) Tomato defense to *Oidium neolycopersici*: dominant *Ol* genes confer isolate-dependent resistance via a different mechanism than recessive *ol-2*. Mol Plant Microbe Interact 18(4):354–362

Bai Y, Pavan S, Zheng Z, Zappel NF, Reinstädler A, Lotti C, De Giovanni C, Ricciardi L, Pim Lindhout P, Visser R, Theres K, Panstruga R (2008) Naturally occurring broad-spectrum powdery mildew resistance in a Central American tomato accession is caused by loss of *Mlo* function. Mol Plant Microbe Interact 21(1):30–39

Baldwin EA, Nisperos-Carriedo MO, Baker R, Scott JW (1991) Quantitative analysis of flavor parameters in six Florida tomato cultivars. J Agric Food Chem 39:1135–1140

Balint-Kurti PJ, Dixon MS, Jones DA, Norcott KA, Jones JDG (1994) RFLP linkage analysis of the *Cf-4* and *Cf-9* genes for resistance to *Cladosporium fulvum* in tomato. Theor Appl Genet 88:691–700

Balint-Kurti PJ, Jones DA, Jones JDG (1995) Integration of the classical and RFLP linkage maps of the short arm of tomato chromosome 1. Theor Appl Genet 90:17–26

Ballester AR, Molthoff J, de Vos R, Hekkert BL, Orzaez D, Fernández-Moreno JP, Tripodi P, Grandillo S, Martin C, Heldens J, Ykema M, Granell A, Bovy A (2010) Biochemical and molecular analysis of pink tomatoes: deregulated expression of the gene encoding transcription factor SlMYB12 leads to pink tomato fruit color. Plant Physiol 152:71–84

Ballvora A, Pierre M, van den Ackerveken G, Schornack S, Rossier O, Ganal M, Lahaye T, Bonas U (2001) Genetic mapping and functional analysis of the tomato *Bs4* locus governing recognition of the *Xanthomonas campestris* pv. *vesicatoria* AvrBs4 protein. Mol Plant Microbe Interact 14:629–638

Barone A, Frusciante L (2007) Molecular marker-assisted selection for resistance to pathogens in tomato. In: Guimaraes E, Ruane J, Scherf BD, Sonnino A, Dargie JD (eds) Marker-assisted selection: current status and future perspectives in crops, livestock, forestry and fish. FAO, Rome. Agriculture and Consumer Protection Dept, 978-92-5-105717-9, A1120 (ftp://ftp.fao.org/docrep/fao/010/a1120e/a1120e02.pdf), pp 151–164

Barone A, Chiusano ML, Ercolano MR, Giuliano G, Grandillo S, Frusciante L (2008) Structural and functional genomics of tomato. Int J Plant Genom 2008:820274

Barone A, Di Matteo A, Carputo D, Frusciante L (2009) High-throughput genomics enhances tomato breeding efficiency. Curr Genom 10(1):1–9

Barrero LS, Tanksley SD (2004) Evaluating the genetic basis of multiple-locule fruit in a broad cross section of tomato cultivars. Theor Appl Genet 109:669–679

Baxter CJ, Sabar M, Quick WP, Sweetlove LJ (2005) Comparison of changes in fruit gene expression in tomato introgression lines provides evidence of genome-wide transcriptional changes and reveals links to mapped QTLs and described traits. J Exp Bot 56:1591–1604

Behare J, Laterrot H, Sarfatti M, Zamir D (1991) Restriction fragment length polymorphisms mapping of the *Stemphylium* resistance gene in tomato. Mol Plant Microbe Interact 4:489–492

Bennett MD, Smith JD (1976) Nuclear DNA amounts in angiosperms. Philos Trans R Soc Lond Biol Sci 181:109–135

Bentham G, Hooker JD (1873) Solanaceae. Genera Planta 2:882–913

Bernacchi D, Tanksley SD (1997) An interspecific backcross of Lycopersicon esculentum × L. hirsutum: linkage analysis and a QTL study of sexual compatibility factors and floral traits. Genetics 147:861–877

Bernacchi D, Beck-Bunn T, Emmatty D, Eshed Y, Inai S, López J, Petiard V, Sayama H, Uhlig J, Zamir D, Tanksley SD (1998a) Advanced backcross QTL analysis of tomato. II. Evaluation of near-isogenic lines carrying single-donor introgressions for desirable wild QTL-alleles derived from *Lycopersicon hirsutum* and *L. pimpinellifolium*. Theor Appl Genet 97:170–180, erratum 1191–1196

Bernacchi D, Beck-Bunn T, Eshed Y, López J, Petiard V, Uhlig J, Zamir D, Tanksley SD (1998b) Advanced backcross QTL analysis in tomato I. Identification of QTLs for traits of agronomic importance from Lycopersicon hirsutum. Theor Appl Genet 97:381–397

Bernatzky R (1993) Genetic mapping and protein product diversity of the self-incompatibility locus in wild tomato (*Lycopersicon peruvianum*). Biochem Genet 31(3–4):173–184

Bernatzky R, Tanksley S (1986) Toward a saturated linkage map in tomato based on isozymes and random cDNA sequences. Genetics 112:887–898

Blauth SL, Steffens JC, Churchill GA, Mutschler MA (1999) Identification of QTLs controlling acylsugar fatty acid composition in an intraspecific population of *Lycopersicon pennellii* (Corr.) D'Arcy. Theor Appl Genet 99:376–381

Bohn GW, Tucker CM (1939) Immunity to *Fusarium* wilt in the tomato. Science 89:603–604

Bohn GW, Tucker CM (1940) Studies on *Fusarium* wilt of the tomato. I. Immunity in *Lycopersicon pimpinellifolium* Mill. and its inheritance in hybrids. MO Agric Exp Stn Res Bull 311:82

Bohs L (1994) *Cyphomandra* (Solanaceae). FL Neotrop Monogr 63:1–175

Bohs L (1995) Transfer of *Cyphomandra* (Solanaceae) and its species to *Solanum*. Taxon 44:583–587

Bohs L (2005) Major clades in *Solanum* based on *ndhF* sequences. In: Keating RC, Hollowell VC, Croat TB (eds) A Festschrift for William G. D'Arcy: the legacy of a taxonomist. Monographs in Systematic Botany from the Missouri Botanical Garden, vol 104. Missouri Botanical Garden Press, St. Louis, MO, USA, pp 27–49

Bohs L, Olmstead RG (1997) Phylogenetic relationships in *Solanum* (Solanaceae) based on *ndhF* sequences. Syst Bot 22:5–17

Bohs L, Olmstead RG (1999) *Solanum* phylogeny inferred from chloroplast DNA sequence data. In: Nee M, Symon DE, Lester RN, Jessop JP (eds) Solanaceae IV: advances in biology and utilization. Royal Botanic Gardens, Kew, UK, pp 97–110

Bohs L, Olmstead RG (2001) A reassessment of *Normandia* and *Triguera* (*Solanaceae*). Plant Syst Evol 228:33–48

Bonnema G, van Schipper D, Heusden S, Zabel P, Lindhout P (1997) Tomato chromosome 1: High-resolution genetic and physical mapping of the short arm in an interspecific *Lycopersicon esculentum* × *L. peruvianum* cross. Mol Gen Genet 253:455–462

Bournival BL, Scott JW, Vallejos CE (1989) An isozyme marker for resistance to race 3 of *Fusarium oxysporum* f. sp. *lycopersici* in tomato. Theor Appl Genet 78:489–494

Bournival BL, Vallejos CE, Scott JW (1990) Genetic analysis of resistances to races 1 and 2 of *Fusarium oxysporum* f. sp. *lycopersici* from the wild tomato *Lycopersicon pennellii*. Theor Appl Genet 79:641–645

Brandwagt BF, Mesbah LA, Takken FLW, Laurent PL, Kneppers TJA, Hille J, Nijkamp HJJ (2000) A longevity assurance gene homolog of tomato mediates resistance to *Alternaria alternata* f. sp. *lycopersici* toxins and fumonisin B1. Proc Natl Acad Sci USA 97(9):4961–4966

Bretó MP, Asins MJ, Carbonell EA (1993) Genetic variability in *Lycopersicon* species and their genetic relationships. Theor Appl Genet 86:113–120

Bretó MP, Asins MJ, Carbonell EA (1994) Salt tolerance in *Lycopersicon* species III. Detection of quantitative trait loci by means of molecular markers. Theor Appl Genet 88: 395–401

Brewer MT, Moyseenko JB, Monforte AJ, van der Knaap E (2007) Morphological variation in tomato: a comprehensive study of quantitative trait loci controlling fruit shape and development. J Exp Bot 58(6):1339–1349

Brommonschenkel SH, Tanksley SD (1997) Map-based cloning of the tomato genomic region that spans the *Sw-5* tospovirus resistance gene in tomato. Mol Gen Genet 256:121–126

Brommonschenkel SH, Frary A, Tanksley SD (2000) The broadspectrum tospovirus resistance gene *Sw-5* of tomato is a homolog of the root-knot nematode resistance gene *Mi*. Mol Plant Microbe Interact 13:1130–1138

Broun P, Tanksley SD (1996) Characterization and genetic mapping of simple repeat sequences in the tomato genome. Mol Gen Genet 250(1):39–49

Brouwer DJ, St. Clair DA (2004) Fine mapping of three quantitative trait loci for late blight resistance in tomato using near isogenic lines (NILs) and sub-NILs. Theor Appl Genet 108: 628–638

Brouwer DJ, Jones ES, St. Clair DA (2004) QTL analysis of quantitative resistance to *Phytophthora infestans* (late blight) in tomato and comparisons with potato. Genome 47: 475–492

Brüggemann W, Linger P, Wenner A, Koornneef M (1996) Improvement of post-chilling photosynthesis in tomato by sexual hybridisation with a *Lycopersicon peruvianum* line from elevated altitude. Adv Hortic Sci 10:215–218

Budiman MA, Chang S-B, Lee S, Yang TJ, Zhang H-B, de Jong H, Wing RA (2004) Localization of *jointless-2* gene in the centromeric region of tomato chromosome 12 based on high resolution genetic and physical mapping. Theor Appl Genet 108:190–196

Caicedo AL, Schaal BA (2004) Population structure and phylogeography of *Solanum pimpinellifolium* inferred from a nuclear gene. Mol Ecol 13:1871–1882

Canady MA, Stevens MR, Barineau MS, Scott JW (2001) Tomato Spotted Wilt Virus (TSWV) resistance in tomato derived from *Lycopersicon chilense* Dun. LA 1938. Euphytica 117:19–25

Canady MA, Meglic V, Chetelat RT (2005) A library of *Solanum lycopersicoides* introgression lines in cultivated tomato. Genome 48:685–697

Canady MA, Ji Y, Chetelat RT (2006) Homeologous recombination in *Solanum lycopersicoides* introgression lines of cultivated tomato. Genetics 174:1775–1788

Carmeille A, Caranta C, Dintinger J, Prior P, Luisetti J, Besse P (2006) Identification of QTLs for *Ralstonia solanacearum* race 3-phylotype II resistance in tomato. Theor Appl Genet 113:110–121

Cassol T, St. Clair DA (1994) Inheritance of resistance to blackmold (*Alternaria alternata* (Fr.) Keissler) in two interspecific crosses of tomato (*Lycopersicon esculentum* × *L. cheesmanii* f. *typicum*). Theor Appl Genet 88:581–588

Causse M, Saliba-Colombani V, Lesschaeve I, Buret M (2001) Genetic analysis of organoleptic quality in fresh market tomato. 2. Mapping QTLs for sensory attributes. Theor Appl Genet 102:273–283

Causse M, Saliba-Colombani V, Lecomte L, Duffé P, Rousselle P, Buret M (2002) QTL analysis of fruit quality in fresh market tomato: a few chromosome regions control the variation of sensory and instrumental traits. J Exp Bot 53: 2089–2098

Causse M, Duffe P, Gomez MC, Buret M, Damidaux R, Zamir D, Gur A, Chevalier C, Lemaire-Chamley M, Rothan C (2004) A genetic map of candidate genes and QTLs involved in tomato fruit size and composition. J Exp Bot 55(403): 1671–1685

Causse M, Chaïb J, Lecomte L, Buret M, Hospital F (2007) Both additivity and epistasis control the genetic variation for fruit quality traits in tomato. Theor Appl Genet 115(3):429–442

Chaerani R, Smulders MJM, van der Linden CG, Vosman B, Stam P, Voorrips RE (2007) QTL identification for early blight resistance (*Alternaria solani*) in a *Solanum lycopersicum* x *S. arcanum* cross. Theor Appl Genet 114:439–450

Chaguè V, Mercier JC, Guenard M, de Courcel A, Vedel F (1997) Identification of RAPD markers linked to a locus involved in quantitative resistance to TYLCV in tomato by bulked segregant analysis. Theor Appl Genet 95:671–677

Chaïb J, Lecomte L, Buret M, Causse M (2006) Stability over genetic backgrounds, generations and years of quantitative trait locus (QTLs) for organoleptic quality in tomato. Theor Appl Genet 112:934–944

Chen FQ, Foolad MR (1999) A molecular linkage map of tomato based on an interspecific cross between *Lycopersicon esculentum* and *L. pimpinellifolium* and its comparison with other molecular maps of tomato. Genome 42:94–103

Chen KY, Tanksley SD (2004) High-resolution mapping and functional analysis of *se2.1*: a major stigma exsertion quantitative trait locus associated with the evolution from allogamy to autogamy in the genus *Lycopersicon*. Genetics 168: 1563–1573

Chen FQ, Foolad MR, Hyman J, St. Clair DA, Beelaman RB (1999) Mapping of QTLs for lycopene and other fruit traits in a *Lycopersicon esculentum* × *L. pimpinellifolium* cross and comparison of QTLs across tomato species. Mol Breed 5:283–299

Chen KY, Cong B, Wing R, Vrebalov J, Tanksley SD (2007) Changes in regulation of a transcription factor lead to autogamy in cultivated tomatoes. Science 318:643–645

Chetelat RT (2006) Revised list of miscellaneous stocks. Rep Tomato Genet Coop 56:37–56

Chetelat RT (2009) Nuclear DNA content in *Solanum* sect. Juglandifolium and Solanum sect. Lycopersicoides. Tomato Genet Coop Rep 59:11–13

Chetelat RT, Ji Y (2007) Cytogenetics and evolution. In: Razdan MK, Mattoo AK (eds) Genetic improvement of solanaceous crops, vol 2, Tomato. Science, Enfield, NJ, USA, pp 77–112

Chetelat RT, Meglic V (2000) Molecular mapping of chromosome segments introgressed from *Solanum lycopersicoides* into cultivated tomato (*Lycopersicon esculentum*). Theor Appl Genet 100:232–241

Chetelat RT, Klann E, DeVerna JW, Yalle S, Bennett AB (1993) Inheritance and genetic mapping of fruit sucrose accumulation in *Lycopersicon chmielewskii*. Plant J 4:643–650

Chetelat RT, DeVerna JW, Bennett AB (1995a) Introgression into tomato (*Lycopersicon esculentum*) of the *L. chmielewskii* sucrose accumulator gene (*sucr*) controlling fruit sugar composition. Theor Appl Genet 91:327–333

Chetelat RT, DeVerna JW, Bennett AB (1995b) Effects of the *Lycopersicon chmielewskii* sucrose accumulator gene (*sucr*) on fruit yield and quality parameters following introgression into tomato. Theor Appl Genet 91:334–339

Chetelat RT, Cisneros P, Stamoa L, Rick CM (1997) A male-fertile *Lycopersicon esculentum* × *Solanum lycopersicoides* hybrid enables direct backcrossing to tomato at the diploid level. Euphytica 95:99–108

Chetelat RT, Rick CM, Cisneros P, Alpert KB, DeVerna JW (1998) Identification, transmission, and cytological behavior of *Solanum lycopersicoides* Dun. monosomic alien addition lines in tomato (*Lycopersicon esculentum* Mill.). Genome 41:40–50

Chetelat RT, Meglic V, Cisneros P (2000) A genetic map of tomato based on BC1 *Lycopersicon esculentum* × *Solanum lycopersicoides* reveals overall synteny but suppressed recombination between these homeologous genomes. Genetics 154:857–867

Chetelat RT, Pertuzé RA, Faundez L, Graham EB, Jones CM (2009) Distribution, ecology and reproductive biology of wild tomatoes and related nightshades from the Atacama Desert region of northern Chile. Euphytica 167:77–93

Child A (1990) A synopsis of *Solanum* subgenus *Potatoe* (G. Don) D'Arcy *Tuberarium* (Dunal) Bitter (s.l.). Feddes Rep 101:209–235

Chiusano ML, D'Agostino N, Traini A, Licciardello C, Raimondo E, Aversano M, Frusciante L, Monti L (2008) ISOL@: an Italian SOLAnaceae genomics resource. BMC Bioinformat 9(Suppl 2):S7

Chunwongse J, Bunn TB, Crossman C, Jiang J, Tanksley SD (1994) Chromosomal localization and molecular-marker tagging of the powdery mildew resistance gene (*Lv*) in tomato. Theor Appl Genet 89:76–79

Chunwongse S, Doganlar C, Crossman JJ, Tanksley SD (1997) High-resolution genetic map of the *Lv* resistance locus in tomato. Theor Appl Genet 95:220–223

Chunwongse J, Chunwongse C, Black L, Hanson P (1998) Mapping of the *Ph-3* gene for late blight from *L. pimpinellifolium* L 3708. Rep Tomato Genet Coop 48:13–14

Chunwongse J, Chunwongse C, Black L, Hanson P (2002) Molecular mapping of the *Ph-3* gene for late blight resistance in tomato. J Hortic Sci Biotechnol 77(3):281–286

Cirulli M, Alexander LJ (1966) A comparison of pathogenic isolates of *Fusarium oxysporum* f. *lycopersici* and different sources of resistance in tomato. Phytopathology 56: 1301–1304

Coaker GL, Francis DM (2004) Mapping, genetic effects, and epistatic interaction of two bacterial canker resistance QTLs from *Lycopersicon hirsutum*. Theor Appl Genet 108: 1047–1055

Coaker GL, Meulia T, Kabelka EA, Jones AK, Francis DM (2002) A QTL controlling stem morphology and vascular development in *Lycopersicon esculentum* × *Lycopersicon hirsutum* (Solanaceae) crosses is located on chromosome 2. Am J Bot 89:1859–1866

Cong B, Barrero LS, Tanksley SD (2008) Regulatory change in YABBY-like transcription factor led to evolution of extreme fruit size during tomato domestication. Nat Genet 40: 800–804

Correll DS (1958) A new species and some nomenclatural changes in *Solanum* section *Tuberarium*. Madroño 14: 232–236

Correll DS (1962) The potato and its wild relatives. Texas Research Foundation, Renner, TX, USA

Cox S (2000) I Say Tomayto, You Say Tomahto. http://lamar.colostate.edu/~samcox/Tomato.html

D'Arcy WG (1972) Solanaceae studies II: typification of subdivisions of *Solanum*. Ann MO Bot Gard 59:262–278

D'Arcy WG (1987) The circumscription of *Lycopersicon*. Solanaceae Newsl 2:60–61

D'Arcy WG (1991) The Solanaceae since 1976, with a review of its biogeography. In: Hawkes JG, Lester RN, Nee M, Estrada N (eds) Solanaceae III: taxonomy, chemistry, evolution. Royal Botanic Gardens, Kew, UK, pp 75–137

D'Agostino N, Aversano M, Frusciante L, Chiusano ML (2007) TomatEST database: in silico exploitation of EST data to explore expression patterns in tomato species. Nucleic Acids Res 35:D901–D905

D'Agostino N, Traini A, Frusciante L, Chiusano ML (2009) SolEST database: a "one-stop shop" approach to the study of Solanaceae transcriptomes. BMC Plant Biol 9:142

Danesh D, Aarons S, McGill GE, Young ND (1994) Genetic dissection of oligogenic resistance to bacterial wilt in tomato. Mol Plant Microbe Interact 7:464–471

Darwin SC, Knapp S, Peralta IE (2003) Taxonomy of tomatoes in the Galapagos Islands: native and introduced species of *Solanum* section *Lycopersicon* (*Solanaceae*). Syst Biodivers 12:29–53

Davies JN (1966) Occurrence of sucrose in the fruit of some species of *Lycopersicon*. Nature 209(5023):640–641

Davis J, Yu D, Evans W, Gokirmak T, Chetelat RT, Stotz HU (2009) Mapping of loci from *Solanum lycopersicoides* conferring resistance or susceptibility to *Botrytis cinerea* in tomato. Theor Appl Genet 119:305–314

De Giovanni C, Dell'Orco P, Bruno A, Ciccarese F, Lotti C, Ricciardi L (2004) Identification of PCR-based markers (RAPD, AFLP) linked to a novel powdery mildew resistance gene (*ol-2*) in tomato. Plant Sci 166:41–48

DeCandolle A (1886) Origin of cultivated plants. Hafner, New York, USA, 1959 reprint

Dennett RK (1950) The association of resistance to *Fusarium* wilt and *Stemphylium* leaf spot disease in tomato, *Lycopersicon esculentum*. Proc Am Soc Hortic Sci 56:353–357

DeVerna JW, Rick CM, Chetelat RT, Lanini BJ, Alpert KB (1990) Sexual hybridization of *Lycopersicon esculentum* and *Solanum rickii* by means of a sesquidiploid bridging hybrid. Proc Natl Acad Sci USA 87:9490–9496

deVicente MC, Tanksley SD (1991) Genome-wide reduction in recombination of backcross progeny derived from male versus female gametes in an interspecific cross of tomato. Theor Appl Genet 83:173–178

deVicente MC, Tanksley SD (1993) QTL analysis of transgressive segregation in an interspecific tomato cross. Genetics 134:585–596

Dickinson MJ, Jones DA, Jones JD (1993) Close linkage between the *Cf-2*/*Cf-5* and *Mi* resistance loci in tomato. Mol Plant Microbe Interact 6:341–347

Diwan N, Fluhr R, Eshed Y, Zamir D, Tanksley SD (1999) Mapping of Ve in tomato: a gene conferring resistance to the broad-spectrum pathogen, Verticillium dahliae race 1. Theor Appl Genet 98:315–319

Dixon MS, Jones DA, Keddie JS, Thomas CM, Harrison K, Jones JDG (1996) The Tomato *Cf-2* disease resistance locus comprises two functional genes encoding leucine-rich repeat proteins. Cell 84:451–459

Dixon MS, Hatzixanthis K, Jones DA, Harrison K, Jones JD (1998) The tomato *Cf-5* disease resistance gene and six homologs show pronounced allelic variation in leucine-rich repeat copy number. Plant Cell 10(11):1915–1925

Doganlar S, Dodson J, Gabor B, Beck-Bunn T, Crossman C, Tanksley SD (1998) Molecular mapping of the *py-1* gene for resistance to corky root rot (*Pyrenochaeta lycopersici*) in tomato. Theor Appl Genet 97:784–788

Doganlar S, Frary A, Tanksley SD (2000) The genetic basis of seedweight variation: tomato as a model system. Theor Appl Genet 100:1267–1273

Doganlar S, Frary A, Daunay MC, Lester RN, Tanksley SD (2002a) A comparative genetic linkage map of eggplant (*Solanum melongena*) and its implications for genome evolution in the Solanaceae. Genetics 161:1697–1711

Doganlar S, Frary A, Ku HM, Tanksley SD (2002b) Mapping quantitative trait loci in inbred backcross lines of *Lycopersicon pimpinellifolium* (LA1589). Genome 45:1189–1202

Dorst JCEA (1946) Een en twintigste beschrijvende rassenlijst voor landbouwgewassen. Rijkscommissie voor de samenstelling van de rassenlijst voor landbouwgewassen, Wageningen, p 221

Dunal MF (1813) Histoire naturelle, médicale et économique des *Solanum* et des genres qui ont été confundus avec eux. (as cited in Luckwill 1943)

Dunal MF (1852) Solanaceae. In: De Candolle AP (ed) Prodromus systematis naturalis regni vegetabilis. 13:450

Egashira H, Kuwashima A, Ishiguro H, Fukushima K, Kaya T, Imanishi S (2000) Screening of wild accessions resistant to gray mold (*Botrytis cinerea* Pers.) in *Lycopersicon*. Acta Physiol Plant 22:324–326

Ellis PR, Maxon-Smith JW (1971) Inheritance of resistance to potato cyst-eelworm (*Heterodera rostochiensis* Woll.) in the genus *Lycopersicon*. Euphytica 20:93–101

Ernst K, Kumar A, Kriseleit D, Kloos D-U, Phillips MS, Ganal MW (2002) The broad-spectrum potato cyst nematode resistance gene (*Hero*) from tomato is the only member of a large gene family of NBS-LRR genes with an unusual amino acid repeat in the LRR region. Plant J 31(2):127–136

Eshed Y, Zamir D (1994) Introgressions from *Lycopersicon pennellii* can improve the soluble-solids yield of tomato hybrids. Theor Appl Genet 88:891–897

Eshed Y, Zamir D (1995) An introgression line population of *Lycopersicon pennellii* in the cultivated tomato enables the identification and fine mapping of yield-associated QTL. Genetics 141:1147–1162

Eshed Y, Zamir D (1996) Less-than-additive epistatic interactions of quantitative trait loci in tomato. Genetics 143:1807–1817

Eshed Y, Gera G, Zamir D (1996) A genome-wide search for wild species alleles that increase horticultural yield of processing tomatoes. Theor Appl Genet 93:877–886

Estañ MT, Villalta I, Bolarín MC, Carbonell EA, Asins MJ (2009) Identification of fruit yield loci controlling the salt tolerance conferred by *Solanum* rootstocks. Theor Appl Genet 118:305–312

Farrar RR, Kennedy GG (1991) Insect and mite resistance in tomato. In: Kalloo G (ed) Genetic improvement of tomato of monographs on theoretical and applied genetics, vol 14. Springer, Berlin, Germany, pp 122–141

Farris JS, Källersjö M, Kluge AG, Bult S (1995) Testing significance of incongruence. Cladistics 10:315–319

Fei Z, Tang X, Alba R, Giovannoni J (2006) Tomato expression database (TED): a suite of data presentation and analysis tools. Nucleic Acids Res 34:D766–D770

Finkers R, van den Berg P, van Berloo R, ten Have A, van Heusden AW, van Kan JAL, Lindhout P (2007a) Three QTLs for *Botrytis cinerea* resistance in tomato. Theor Appl Genet 114:585–593

Finkers R, van Heusden AW, Meijer-Dekens F, van Kan JAL, Maris P, Lindhout P (2007b) The construction of a *Solanum habrochaites* LYC4 introgression line population and the identification of QTLs for resistance to *Botrytis cinerea*. Theor Appl Genet 114:1071–1080

Finkers R, Bai Y, van den Berg P, van Berloo R, Meijer-Dekens F, ten Have A, van Kan J, Lindhout P, van Heusden AW (2008) Quantitative resistance to *Botrytis cinerea* from *Solanum neorickii*. Euphytica 159:83–92

Folkertsma RT, Spassova MI, Prins M, Stevens MR, Hille J, Goldbach RW (1999) Construction of a bacterial artificial chromosome (BAC) library of *Lycopersicon esculentum* cv. Stevens and its application to physically map the *Sw-5* locus. Mol Breed 5:197–207

Foolad MR (2004) Recent advances in genetics of salt tolerance in tomato. Plant Cell Tiss Organ Cult 76:101–119

Foolad MR (2005) Breeding for abiotic stress tolerances in tomato. In: Ashraf M, Harris PJC (eds) Abiotic stresses: plant resistance through breeding and molecular approaches. Haworth, New York, NY, USA, pp 613–684

Foolad MR, Chen FQ (1998) RAPD markers associated with salt tolerance in an interspecific cross of tomato (*Lycopersicon esculentum* × *L. pennellii*). Plant Cell Rep 17:306–312

Foolad MR, Chen FQ (1999) RFLP mapping of QTLs conferring salt tolerance during the vegetative stage in tomato. Theor Appl Genet 99:235–243

Foolad MR, Jones RA (1993) Mapping salt-tolerance genes in tomato (*Lycopersicon esculentum*) using trait-based marker analysis. Theor Appl Genet 87:184–192

Foolad MR, Sharma A (2005) Molecular markers as selection tools in tomato breeding. Acta Hortic 695:225–240

Foolad MR, Stoltz T, Dervinis C, Rodríguez RL, Jones RA (1997) Mapping QTLs conferring salt tolerance during germination in tomato by selective genotyping. Mol Breed 3: 269–277

Foolad MR, Chen FQ, Lin GY (1998a) RFLP mapping of QTLs conferring salt tolerance during germination in an interspecific cross of tomato. Theor Appl Genet 97:1133–1144

Foolad MR, Chen FQ, Lin GY (1998b) RFLP mapping of QTLs conferring cold tolerance during seed germination in an interspecific cross of tomato. Mol Breed 4:519–529

Foolad MR, Zang LP, Lin GY (2001) Identification and validation of QTLs for salt tolerance during vegetative growth in tomato by selective genotyping. Genome 44:444–454

Foolad MR, Zang LP, Khan AA, Lin GY (2002) Identification of QTLs for early blight (*Alternaria solani*) resistance in tomato using backcross populations of a *Lycopersicon esculentum* × *L. hirsutum* cross. Theor Appl Genet 104: 945–958

Foolad MR, Zang LP, Subbiah P (2003) Genetics of drought tolerance during seed germination in tomato: inheritance and QTL mapping. Genome 46:536–545

Fosberg FR (1987) New nomenclatural combinations for Galapagos plant species. Phytologia 62:181–183

Frary A, Doganlar S (2003) Comparative genetics of crop plant domestication and evolution. Turk J Agric For 27:59–69

Frary A, Graham E, Jacobs J, Chetelat RT, Tanksley SD (1998) Identification of QTL for late blight resistance from *L. pimpinellifolium* L3708. Tomato Genet Coop Rep 48: 19–21

Frary A, Nesbitt TC, Frary A, Grandillo S, van der Knaap E, Cong B, Liu J, Meller J, Elber R, Alpert KB, Tanksley SD (2000) *fw2.2*: a quantitative trait locus key to the evolution of tomato fruit size. Science 289(5476):85–88

Frary A, Doganlar S, Frampton A, Fulton T, Uhlig J, Yates H, Tanksley SD (2003) Fine mapping of quantitative trait loci for improved fruit characteristics from *Lycopersicon chmielewskii* chromosome 1. Genome 46:235–243

Frary A, Fulton TM, Zamir D, Tanksley SD (2004a) Advanced backcross QTL analysis of a *Lycopersicon esculentum* × *L. pennellii* cross and identification of possible orthologs in the Solanaceae. Theor Appl Genet 108:485–496

Frary A, Fritz LA, Tanksley SD (2004b) A comparative study of the genetic bases of natural variation in tomato leaf, sepal, and petal morphology. Theor Appl Genet 109:523–533

Frary A, Xu Y, Liu J, Mitchell S, Tedeschi E, Tanksley SD (2005) Development of a set of PCR-based anchor markers encompassing the tomato genome and evaluation of their usefulness for genetics and breeding experiments. Theor Appl Genet 111:291–312

Frary A, Göl D, Keleş D, Okmen B, Pinar H, Siğva HO, Yemenicioğlu A, Doğanlar (2010) Salt tolerance in Solanum pennellii: antioxidant response and related QTL. BMC Plant Biol 10:58

Fridman E, Pleban T, Zamir D (2000) A recombination hotspot delimits a wild-species quantitative trait locus for tomato sugar content to 484 bp within an invertase gene. Proc Natl Acad Sci USA 97:4718–4723

Fridman E, Carrari F, Liu YS, Fernie AR, Zamir D (2004) Zooming in on a quantitative trait for tomato yield using interspecific introgressions. Science 305:1786–1789

Frodin D (2004) History and concepts of big plant genera. Taxon 53:753–776

Fulton TM, Beck-Bunn T, Emmatty D, Eshed Y, Lopez J, Petiard V, Uhlig J, Zamir D, Tanksley SD (1997) QTL analysis of an advanced backcross of *Lycopersicon peruvianum* to the cultivated tomato and comparisons with QTLs found in other wild species. Theor Appl Genet 95:881–894

Fulton TM, Grandillo S, Beck-Bunn T, Fridman E, Frampton A, López J, Petiard V, Uhlig J, Zamir D, Tanksley SD (2000) Advanced backcross QTL analysis of a *Lycopersicon esculentum* × *Lycopersicon parviflorum* cross. Theor Appl Genet 100:1025–1042

Fulton TM, Bucheli E, Voirol E, López J, Pétiard V, Tanksley SD (2002a) Quantitative trait loci (QTL) affecting sugars, organic acids and other biochemical properties possibly contributing to flavor, identified in four advanced backcross populations of tomato. Euphytica 127:163–177

Fulton TM, van der Hoeven R, Eanetta NT, Tanksley SD (2002b) Identification, analysis, and utilization of conserved ortholog set markers for comparative genomics in higher plants. Plant Cell 14:1457–1467

Ganal MW, Tanksley SD (1996) Recombination around the *TM2a* and *Mi* resistance genes in different crosses of *Lycopersicon peruvianum*. Theor Appl Genet 92:101–108

Ganal MW, Simon R, Brommonschenkel S, Arndt M, Tanksley SD, Kumar A (1995) Genetic mapping of a wide spectrum nematode resistance gene (*Hero*) against *Globodera rostochiensis* in tomato. Mol Plant Microbe Interact 8:886–891

García-Martínez S, Andreani L, Garcia-Gusano M, Geuna F, Ruiz JJ (2006) Evaluation of amplified fragment length polymorphism and simple sequence repeats for tomato germplasm fingerprinting: utility for grouping closely related traditional cultivars. Genome 49:648–656

Garland S, Sharman M, Persley D, McGrath D (2005) The development of an improved PCR-based marker system for *Sw-5*, an important TSWV resistance gene of tomato. Aust J Agric Res 56:285–289

Georgiady MS, Whitkus RW, Lord EM (2002) Genetic analysis of traits distinguishing outcrossing and self-pollinating forms of currant tomato, *Lycopersicon pimpinellifolium* (Jusl.) Mill. Genetics 161:333–344

Gerrior S, Bente L (2002) Nutrient content of the U.S. food supply, 1909–1999: a summary report. USDA, Center for Nutrition Policy and Promotion, Washington, DC, USA

Gidoni D, Fuss E, Burbidge A, Speckmann GJ, James S, Nijkamp D, Mett A, Feiler J, Smoker M, de Vroomen MJ, Leader D, Liharska T, Groenendijk J, Coppoolse E, Smit JJ, Levin I, de Both M, Schuch W, Jones JD, Taylor IB, Theres K, van Haren MJ (2003) Multi-functional T-DNA/Ds tomato lines designed for gene cloning and molecular and physical dissection of the tomato genome. Plant Mol Biol 51:83–98

Gilbert JC, McGuire DC (1956) Inheritance of resistance to several rootknot from *Meloidogyne incognita* in commercial type tomatoes. Proc Am Soc Hortic Sci 68:437–442

Giovannoni JJ, Noensie EN, Ruezinsky DM, Lu X, Tracy SL, Ganal MW, Martin GB, Pillen K, Alpert K, Tanksley SD (1995) Molecular genetic analysis of the *ripening-inhibitor* and *non-ripening* loci of tomato: a first step in genetic map-based cloning of fruit ripening genes. Mol Gen Genet 248:195–206

Goldman IL, Paran I, Zamir D (1995) Quantitative trait locus analysis of a recombinant inbred line population derived from a *Lycopersicon esculentum* × *Lycopersicon cheesmanii* cross. Theor Appl Genet 90:925–932

Gonzalo MJ, van der Knaap E (2008) A comparative analysis into the genetic bases of morphology in tomato varieties exhibiting elongated fruit shape. Theor Appl Genet 116: 647–656

Gorguet B, Eggink PM, Ocaña J, Tiwari A, Schipper D, Finkers R, Visser RGF, van Heusden AW (2008) Mapping characterization of novel parthenocarpy QTLs in tomato. Theor Appl Genet 116:755–767

Graham EB (2005) Genetic diversity and crossing relationships of *Lycopersicon chilense*. PhD Thesis, University of California, Davis, CA, USA, 157 p

Graham EB, Shannon SM, Petersen JP, Chetelat RT (2003) A self-compatible population of *Lycoperisicon peruvianum* collected from N. Chile. Rep Tomato Genet Coop 53:22–24

Grandillo S, Tanksley SD (1996a) QTL analysis of horticultural traits differentiating the cultivated tomato from the closely related species *Lycopersicon pimpinellifolium*. Theor Appl Genet 92:935–951

Grandillo S, Tanksley SD (1996b) Genetic analysis of RFLPs. GATA microsatellites and RAPDs in a cross between L. esculentum and L. pimpinellifolium. Theor Appl Genet 92: 957–965

Grandillo S, Ku HM, Tanksley SD (1996) Characterization of *fs8.1*, a major QTL influencing fruit shape in tomato. Mol Breed 2:251–260

Grandillo S, Ku HM, Tanksley SD (1999a) Identifying loci responsible for natural variation in fruit size and shape in tomato. Theor Appl Genet 99:978–987

Grandillo S, Zamir D, Tanksley SD (1999b) Genetic improvement of processing tomatoes: a 20-year perspective. Euphytica 110:85–97

Grandillo S, Monforte AJ, Fridman E, Zamir D, Tanksley SD (2000) Agronomic characterization of a set of near-isogenic lines derived from a *Lycopersicon esculentum* × *L. hirsutum* cross. In: XLIV Convegno Annuale della Società Italiana di Genetica Agraria (SIGA), Bologna, Italy, 20–23 Sept 2000, p 78

Grandillo S, Tanksley SD, Zamir D (2008) Exploitation of natural biodiversity through genomics. In: Varshney RK, Tuberosa R (eds) Genomics assisted crop improvement, vol 1, Genomics approaches and platforms. Springer, Dordrecht, Netherlands, pp 121–150

Grant V (1975) Genetics of flowering plants. Columbia University Press, New York, USA

Griffiths PD, Scott JW (2001) Inheritance and linkage of Tomato mottle virus resistance genes derived from *Lycopersicon chilense* accession LA 1932. J Am Soc Hortic Sci 126: 462–467

Guo Z, Weston PA, Snyder JC (1993) Repellency to two-spotted spider mite, *Tetranychus urticae* Koch, as related to leaf surface chemistry of *Lycopersicon hirsutum* accessions. J Chem Ecol 19:2965–2979

Gur A, Zamir D (2004) Unused natural variation can lift yield barriers in plant breeding. PLoS Biol 2(10):e245

Haanstra JPW, Laugé R, Meijer-Dekens F, Bonnema G, de Wit PJGM, Lindhout P (1999a) The *Cf-ECP2* gene is linked to, but not part of the *Cf-4*/*Cf-9* cluster on the short arm of chromosome 1 in tomato. Mol Gen Genet 262:839–845

Haanstra JPW, Wye C, Verbakel H, Meijer-Dekens F, van der Berg P, Odinot P, van Heusden AW, Tanksley S, Lindhout P, Peleman J (1999b) An integrated high-density RFLP-AFLP map of tomato based on two *Lycopersicum esculentum* × *L. pennellii* F_2 populations. Theor Appl Genet 99:254–271

Haanstra JPW, Meijer-Dekens F, Laugé R, Seetanah DC, Joosten MHAJ, de Wit PJGM, Lindhout P (2000) Mapping strategy for resistance genes against *Cladosporium fulvum* on the short arm of chromosome 1 of tomato: *Cf-ECP5* near the *Hcr9* milky way cluster. Theor Appl Genet 101:661–668

Hanson PM, Bernacchi D, Green S, Tanksley SD, Muniyappa V, Padmaja AS, Chen H, Kuo G, Fang D, Chen J (2000) Mapping a wild tomato introgression associated with tomato yellow leaf curl virus resistance in a cultivated tomato line. J Am Soc Hortic Sci 15:15–20

Hanson P, Green SK, Kuo G (2006) *Ty-2*, a gene on chromosome 11 conditioning geminivirus resistance in tomato. Tomato Genet Coop Rep 56:17–18

Harlan JR, de Wet JMJ, Price EG (1973) Comparative evolution of cereals. Evolution 27:311–325

Hawkes JG (1990) The potato: evolution biodiversity and genetic resources. Belhaven, London, UK

Hedrick UP, Booth NO (1907) Mendelian characters in tomatoes. Proc Am Soc Hortic Sci 5:19–24

Heine H (1976) Flora de la Nouvelle Caledonie, vol 7. Museum National D'Histoire Naturelle, Paris, France

Helentjaris T, King G, Slocum M, Siedenstrang C, Wegman S (1985) Restriction fragment polymorphisms as probes for plant diversity and their development as tools for applied plant breeding. Plant Mol Biol 5:109–118

Hemming MN, Basuki S, McGrath DJ, Carroll BJ, Jones DA (2004) Fine mapping of the tomato *I-3* gene for fusarium wilt resistance and elimination of a co-segregating resistance gene analogue as a candidate for *I-3*. Theor Appl Genet 109:409–418

Hewitt JD, Garvey TC (1987) Wild sources of high soluble solids in tomato. In: Nevins DJ, Jones RA, Liss AR (eds) Tomato biotechnology, vol 4. Alan R Liss, New York, NY, USA, pp 45–54

Hogenboom NG (1970) Inheritance of resistance to corky root in tomato (*Lycopersicon esculentum* UM Mill.). Euphytica 19:413–425

Holmes FO (1957) True-breeding resistance in tomato to infection by tobacco-mosaic virus. Phytopathology 47:16–17

Holtan HEE, Hake S (2003) Quantitative trait locus analysis of leaf dissection in tomato using *Lycopersicon pennellii* segmental introgression lines. Genetics 165:1541–1550

Hoogstraten JGJ, Braun CJ (2005) Methods for coupling resistance alleles in tomato. U.S. Patent Pending 20050278804. Date published 15 Dec 2005

Huang CC, Cui YY, Weng CR, Zabel P, Lindhout P (2000a) Development of diagnostic PCR markers closely linked to the tomato powdery mildew resistance gene *Ol-1* on chromosome 6 of tomato. Theor Appl Genet 101:918–924

Huang CC, Hoefs-Van De Putte PM, Haanstra-Van Der Meer JG, Meijer-Dekens F, Lindhout P (2000b) Characterization and mapping of resistance to *Oidium lycopersicum* in two *Lycopersicon hirsutum* accessions: evidence for close linkage of two *Ol*-genes on chromosome 6 of tomato. Heredity 85:511–520

Hunziker AT (1979) South American Solanaceae: a synoptic survey. In: Hawkes JG, Lester RN, Skelding AD (eds) The biology and taxonomy of *Solanaceae*. Academic, London, UK, pp 49–85

Hunziker AT (2001) Genera Solanacearum, the genera of *Solanaceae* illustrated arranged according to a new system. ARG Gantner, Ruggell, Germany

Ignatova SI, Gorshkova NS, Tereshonkova TA (2000) Resistance of tomato F_1 hybrids to grey mold. Acta Physiol Plant 22:326–328

Ishibashi K, Masuda K, Naito S, Meshi T, Ishikawa M (2007) An inhibitor of viral RNA replication is encoded by a plant resistance gene. Proc Natl Acad Sci USA 104: 13833–13838

Jablonska B, Ammiraju JSS, Bhattarai KK, Mantelin S, de Ilarduya OM, Roberts PA, Kaloshian I (2007) The *Mi-9* gene from *Solanum arcanum* conferring heat-stable resistance to root-knot nematodes is a homolog of *Mi-1*. Plant Physiol 143:1044–1054

Jenkins JA (1948) The origin of the cultivated tomato. Econ Bot 2:379–392

Jensen KS, Martin CT, Maxwell DP (2007) A CAPS marker, *FER-G8*, for detection of *Ty3* and *Ty3a* alleles associated with *S. chilense* introgressions for begomovirus resistance in tomato breeding lines. University of Wisconsin-Madison. http://www.plantpath.wisc.edu/pp-old/GeminivirusResistantTomatoes/Markers/MAS-Protocols/Ty3a-CAPS.pdf

Ji Y, Chetelat RT (2003) Homoeologous pairing and recombination in *Solanum lycopersicoides* monosomic addition and substitution lines of tomato. Theor Appl Genet 106:979–989

Ji Y, Scott JW (2005) Identification of RAPD markers linked to Lycopersicon chilense derived begomovirus resistant gene on cromosome 6 of tomato. In: Ist international symposium on tomato diseases. Acta Hortic 695:407–416

Ji Y, Scott JW (2007) Tomato. In: Singh RJ (ed) Genetic resources, chromosome engineering, and crop improvement. Ser 4: Vegetable crops. CRC, Boca Raton, FL, USA

Ji Y, Pertuzé R, Chetelat RT (2004) Genome differentiation by GISH in interspecic and intergeneric hybrids of tomato and related nightshades. Chrom Res 12:107–116

Ji Y, Schuster DJ, Scott JW (2007a) *Ty-3*, a begomovirus resistance locus near the *Tomato yellow leaf curl virus* resistance locus *Ty-1* on chromosome 6 of tomato. Mol Breed 20: 271–284

Ji Y, Scott JW, Hanson P, Graham E, Maxwell DP (2007b) Sources of resistance, inheritance, and location of genetic loci conferring resistance to members of the tomato infecting begomoviruses. In: Czosnek H (ed) Tomato Yellow Leaf Curl virus disease: management, molecular biology, breeding for resistance. Kluwer, Dordrecht, Netherlands, pp 343–362

Ji Y, Scott JW, Schuster DJ, Maxwell DP (2009a) Molecular mapping of *Ty-4*, a new tomato yellow leaf curl virus resistance locus on chromosome 3 of tomato. J Am Soc Hortic Sci 134(2):281–288

Ji Y, Scott JW, Schuster DJ (2009b) Toward fine mapping of the tomato yellow leaf curl virus resistance gene *Ty-2* on chromosome 11 of tomato. HortScience 44(3):614–618

Jiménez-Gómez JM, Alonso-Blanco C, Borja A, Anastasio G, Angosto T, Lozano R, Martínez-Zapater M (2007) Quantitative genetic analysis of flowering time in tomato. Genome 50:303–315

Jones DF (1917) Linkage in *Lycopersicon*. Am Nat 51:608–621

Jones DA, Dickinson MJ, Balint-Kurti PJ, Dixon MS, Jones JDG (1993) Two complex resistance loci revealed in tomato by classical and RFLP mapping of the *Cf-2, Cf-4, Cf-5*, and *Cf-9* genes for resistance to *Cladosporium fulvum*. Mol Plant Microbe Interact 6:348–357

Jones DA, Thomas CM, Hammond-Kosack KE, Balint-Kurti PJ, Jones JDG (1994) Isolation of the tomato *Cf-9* gene for resistance to *Cladosporium fulvum* by transposon tagging. Science 266:789–793

Jussieu AL (1789) Genera plantarum. Herissant V & Barrios T, Paris, France

Kabelka E, Franchino B, Francis DM (2002) Two loci from *Lycopersicon hirsutum* LA407 confer resistance to strains of *Clavibacter michiganensis* subsp. *michiganensis*. Phytopathology 92:504–510

Kabelka E, Yang W, Francis DM (2004) Improved tomato fruit color within an inbred backcross line derived from *Lycopersicon esculentum* and *L. hirsutum* involves the interaction of loci. J Am Soc Hortic Sci 129(2):250–257

Kalloo G (1991) Genetic improvement of tomato. Springer, Berlin, Germany

Kalloo G, Banerjee MK (1990) Transfer of tomato leaf curl virus resistance from *Lycopersicon hirsutum* f. *glabratum* to *L. esculentum*. Plant Breed 105:156–159

Kamenetzky L, Asís R, Bassi S, de Godoy F, Bermúdez L, Fernie AR, Van Sluys MA, Vrebalov J, Giovannoni JJ, Rossi M, Carrari F (2010) Genomic analysis of wild tomato introgressions determining metabolism- and yield-associated traits. Plant Physiol 152(4):1772–1786

Kawchuk LM, Hachey J, Lynch DR (1998) Development of sequence characterized DNA markers linked to a dominant *Verticillium* wilt resistance gene in tomato. Genome 41: 91–95

Kawchuk LM, Hachey J, Lynch DR, Kulcsar F, van Rooijen G, Waterer DR, Robertson A, Kokko E, Byers R, Howard RJ, Fischer R, Prufer D (2001) Tomato *Ve* disease resistance genes encode cell surface-like receptors. Proc Natl Acad Sci USA 98(11):6511–6515

Kebede H, Martin B (1994) Leaf anatomy of two *Lycopersicon* species with contrasting gas exchange properties. Crop Sci 34:108–113

Kennedy GG (2003) Tomato, pests, parasitoids, and predators: tritrophic interactions involving the genus *Lycopersicon*. Annu Rev Entomol 48:51–72

Kennedy GG (2007) Resistance in tomato and other *Lycopersicon* species to insect and mite pests. In: Razdan MK, Mattoo AK (eds) Genetic improvement of Solanaceous crops, vol 2, tomato. Science, Enfield, NH, USA, pp 488–519

Kerr EA, Bailey DL (1964) Resistance to *Cladosporium fulvum* cke. obtained from wild species of tomato. Can J Bot 42(11): 1541–1554

Khush GS (1963) Identification key for pachytene chromosomes of *L. esculentum*. Tomato Genet Coop Rep 13:12–13

Khush GS, Rick CM (1963) Meiosis in hybrids between *Lycopersicon esculentum* and *Solanum pennellii*. Genetica 33: 167–183

Khush GS, Rick CM (1966) The origin, identification, and cytogenetic behavior of tomato monosomics. Chromosoma 18:407–420

Khush GS, Rick CM (1968) Cytogenetic analysis of the tomato genome by means of induced deficiencies. Chromosoma 23: 452–484

Kinzer SM, Schwager SJ, Mutschler MA (1990) Mapping of ripening-related or -specific cDNA clones of tomato (*Lycopersicon esculentum*). Theor Appl Genet 79:489–496

Kiss L, Cook RTA, Saenz GS, Cunnington JH, Takamatsu S, Pascoe I, Bardin M, Nicot PC, Sato Y, Rossman AY (2001) Identification of two powdery mildew, *Oidium neolycopersici sp.* nov. and *Oidium lycopersici*, infecting tomato in different parts of the world. Mycol Res 105:684–697

Klein-Lankhorst R, Rietveld P, Machiels B, Verkerk R, Weide R, Gebhardt C, Koornneef M, Zabel P (1991) RFLP markers linked to the root knot nematode resistance gene *Mi* in tomato. Theor Appl Genet 81:661–667

Knapp S (1991) A revision of *Solanum sessile* species group (section *Geminata* pro parte: Solanaceae). Bot J Linn Soc 105:179–210

Knapp S (2000) A revision of *Solanum thelopodium* species group (section *Anthoresis* sensu Sheite, pro parte): Solanaceae. Bull Nat Hist Mus London (Bot) 30:13–30

Knapp S (2002) *Solanum* section *Geminata*. FL Neotrop 84: 1–405

Knapp S, Darwin SC (2007) Proposal to conserve the name *Solanum cheesmaniae* (L. Riley) Fosberg against *S. cheesmanii* Geras. (Solanaceae). Taxon 55:806–807

Kole C, Ashrafi H, Lin G, Foolad M (2006) Identification and molecular mapping of a new R gene, Ph-4, conferring resistance to late blight in tomato. In: Solanaceae conference, University of Wisconsin, Madison, Abstr 449

Koornneef M, Alonso-Blanco C, Vreugdenhil D (2004) Naturally occurring variation in *Arabidopsis thaliana*. Annu Rev Plant Biol 55:141–172

Krieger U, Lippman ZB, Zamir D (2010) The flowering gene *SINGLE FLOWER TRUSS* drives heterosis for yield in tomato. Nat Genet 42(5):459–463

Ku HM, Doganlar S, Chen K-Y, Tanksley SD (1999) The genetic basis of pear-shaped tomato fruit. Theor Appl Genet 9:844–850

Ku HM, Grandillo S, Tanksley SD (2000) *fs8.1*, a major QTL, sets the pattern of tomato carpel shape well before anthesis. Theor Appl Genet 101:873–878

Labate JA, Grandillo S, Fulton T, Muños S, Caicedo AL, Peralta I, Ji Y, Chetelat RT, Scott JW, Gonzalo MJ, Francis D, Yang W, van der Knaap E, Baldo AM, Smith-White B, Mueller LA, Prince JP, Blanchard NE, Storey DB, Stevens MR, Robbins MD, Wang JF, Liedl BE, O'Connell MA, Stommel JR, Aoki K, Iijima Y, Slade AJ, Hurst SR, Loeffler D, Steine MN, Vafeados D, McGuire C, Freeman C, Amen A, Goodstal J, Facciotti D, Van Eck J, Causse M (2007) Tomato. In: Kole C (ed) Genome mapping and molecular breeding in plants, vol 5, Vegetables. Springer, Berlin, Germany, pp 1–96

Lanfermeijer FC, Dijkhuis J, Sturre MJG, de Haan P, Hille J (2003) Cloning and characterization of the durable Tomato mosaic virus resistance gene *Tm-2*^{2} from *Lycopersicon esculentum*. Plant Mol Biol 52:1037–1049

Lanfermeijer FC, Jiang G, Ferwerda MA, Dijkhuis J, de Haan P, Yang R, Hille J (2004) The durable resistance gene *Tm-2*^{2} from tomato confers resistance against ToMV in tobacco and preserves its viral specificity. Plant Sci 167:687–692

Lanfermeijer FC, Warmink J, Hille J (2005) The products of the broken *Tm-2* and the durable *Tm-2*^{2} resistance genes from tomato differ in four amino acids. J Exp Bot 56:2925–2933

Langford AN (1937) The parasitism of *Cladosporium fulvum* Cooke and the genetics of resistance to it. Can J Res 15: 108–128

Laterrot H (1976) Localisation chromosomique de 12 chez la tomate controlant Ia resistance au pathotype 2 de *Fusarium oxysporum* f. *lycopersici*. Ann Amelior Plant 26:485–491

Laterrot H (1983) La lutte genetique contre la maladie des racines liegueses de la tomate. Rev Hortic 238:143–150

Laterrot H (2000) Disease resistance in tomato: practical situation. Acta Physiol Plant 22(3):328–331

Laugé R, Dmitriev AP, Joosten MHAJ, De Wit PJGM (1998a) Additional resistance gene(s) against *Cladosporium fulvum* present on the *Cf-9* introgression segment are associated with strong PR protein accumulation. Mol Plant Microbe Interact 11(4):301–308

Laugé R, Joosten MHAJ, Haanstra JPW, Goodwin PH, Lindhout P, De Wit PJGM (1998b) Successful search for a resistance gene in tomato targeted against a virulence factor of a fungal pathogen. Proc Natl Acad Sci USA 95:9014–9018

Laugé R, Goodwin PH, De Wit PJGM, Joosten MHAJ (2000) Specific HR-associated recognition of secreted proteins from

Cladosporium fulvum occurs in both host and non-host plants. Plant J 23:735–745

Lawson DM, Lunde CF, Mutschler MA (1997) Marker-assisted transfer of acylsugar-mediated pest resistance from the wild tomato, *Lycopersicon pennellii*, to the cultivated tomato, *Lycopersicon esculentum*. Mol Breed 3:307–317

Lecomte L, Duffé P, Buret M, Servin B, Hospital F, Causse M (2004a) Marker-assisted introgression of five QTLs controlling fruit quality traits into three tomato lines revealed interactions between QTLs and genetic backgrounds. Theor Appl Genet 109:658–668

Lecomte L, Saliba-Colombani V, Gautier A, Gomez-Jimenez MC, Duffé P, Buret M, Causse M (2004b) Fine mapping of QTLs of chromosome 2 affecting the fruit architecture and composition of tomato. Mol Breed 13:1–14

Levesque H, Vedel E, Mathieu C, de Coureel AJL (1990) Identification of a short rDNA spacer sequence highly specific of a tomato line containing *Tm-1* gene introgressed from *Lycopersicon hirsutum*. Theor Appl Genet 80:602–608

Levin I, Gilboa N, Teselson E, Shen S, Schaffer AA (2000) *Fgr*, a major locus that modifies fructose to glucose ratio in mature tomato fruits. Theor Appl Genet 100:256–262

Levin I, Lalazar A, Bar M, Schaffer AA (2004) Non GMO fruit factories strategies for modulating metabolic pathways in the tomato fruit. Ind Crops Products 20:29–36

Lim GTT, Wang GP, Hemming MN, McGrath DJ, Jones DA (2008) High resolution genetic and physical mapping of the *I-3* region of tomato chromosome 7 reveals almost continuous microsynteny with grape chromosome 12 but interspersed microsynteny with duplications on Arabidopsis chromosomes 1, 2 and 3. Theor Appl Genet 118(1):57–75

Lincoln RE, Porter JW (1950) Inheritance of beta-carotene in tomatoes. Genetics 35:206–211

Lindhout P, Pet G, Van der Beek H (1994a) Screening wild *Lycopersicon* species for resistance to powdery mildew (*Oidium lycopersicum*). Euphytica 72:43–49

Lindhout P, Van der Beek H, Pet G (1994b) Wild *Lycopersicon* species as sources for resistance to powdery mildew (*Oidium licopersicum*): mapping of the resistance gene *Ol-1* on cromosome 6 of *L. hirsutum*. Acta Hortic 376:387–394

Lindhout P, Heusden S, Pet G, Ooijen JW, Sandbrink H, Verkerk R, Vrielink R, Zabel P (1994c) Perspectives of molecular marker assisted breeding for earliness in tomato. Euphytica 79:279–286

Linkage Committee (1973) Linkage summary. Tomato Genet Coop 23:9–11

Linnaeus C (1753) Species plantarum, 1st edn. L. Salvius, Stockholm, Sweden

Lippman Z, Tanksley SD (2001) Dissecting the genetic pathway to extreme fruit size in tomato using a cross between the small-fruited wild species *Lycopersicon pimpinellifolium* and *L. esculentum* var. Giant Heirloom. Genetics 158:413–422

Lippman ZB, Zamir D (2007) Heterosis: revisiting the magic. Trends Genet 23:60–66

Lippman ZB, Semel Y, Zamir D (2007) An integrated view of quantitative trait variation using tomato interspecific introgression lines. Curr Opin Genet Dev 17:545–552

Liu Y, Zamir D (1999) Second generation *L. pennellii* introgression lines and the concept of bin mapping. Rep Tomato Genet Coop 49:26–30

Liu J, Van Eck J, Cong B, Tanksley SD (2002) A new class of regulatory genes underlying the cause of pear-shaped tomato fruit. Proc Natl Acad Sci USA 99:13302–13306

Liu J, Gur A, Ronen G, Causse M, Damidaux R, Buret M, Hirschberg J, Zamir D (2003) There is more to tomato fruit colour than candidate carotenoid genes. Plant Biotechnol J 1:195–207

Livingstone KD, Lackney VK, Blauth JR, Van Wijk RIK, Kyle Jahn M (1999) Genome mapping in *Capsicum* and the evolution of genome structure in the Solanaceae. Genetics 152: 1183–1202

Lobo MA, Navarro R (1987) Late blight horizontal resistance in *L. esculentum* × *L. hirsutum* hybrids. Tomato Genet Coop Rep 37:52–54

Loh YT, Martin G (1995) The disease-resistance gene *Pto* and the fenthion-sensitivity gene Fen encode closely related functional protein kinases. Proc Natl Acad Sci USA 92:4181–4184

Lough RC (2003) Inheritance of tomato late blight resistance in *Lycopersicon hirsutum* LA1033. Thesis, North Carolina State University, Raleigh, NC, USA. http://www.lib.ncsu.edu/theses/available/etd-04172003-231125/unrestricted/etd.pdf

Luckwill LC (1943) The genus *Lycopersicon*: an historical, biological, and taxonomical survey of the wild and cultivated tomatoes. Aberdeen Univ Stud 120:1–44

Lukyanenko AN (1991) Disease resistance in tomato. In: Kalloo G (ed) Genetic improvement of tomato, vol 14, Monographs on theoretical and applied genetics. Springer, Berlin, Germany, pp 99–119

MacBride JF (1962) Solanaceae. In: Flora of Peru. Field Mus Nat Hist Bot Ser 13:3–267

Maliepaard C, Bas N, Van Heusden S, Kos J, Pet G, Verkerk R, Vrielink R, Zabel P, Lindhout P (1995) Mapping of QTLs for glandular trichome densities and *Trialeurodes vaporariorum* (greenhouse whitefly) resistance in an F_2 from *Lycopersicon esculentum* × *Lycopersicon hirsutum* f. *glabratum*. Heredity 75:425–433

Mangin B, Thoquet P, Olivier J, Grimsley NH (1999) Temporal and multiple quantitative trait loci analyses of resistance to bacterial wilt in tomato permit the resolution of linked loci. Genetics 151:1165–1172

Marshall JA, Knapp S, Davey MR, Power JB, Cocking EC, Bennett MD, Cox AV (2001) Molecular systematics of *Solanum* section *Lycopersicum* (*Lycopersicon*) using the nuclear ITS rDNA region. Theor Appl Genet 103:1216–1222

Martin B, Nienhuis J, King G, Schaefer A (1989) Restriction fragment length polymorphisms associated with water use efficiency in tomato. Science 243(4899):1725–1728

Martin GB, Williams JGK, Tanksley SD (1991) Rapid identification of markers linked to a *Pseudomonas* resistance gene in tomato by using random primers and near-isogenic lines. Proc Natl Acad Sci USA 88:2336–2340

Martin GB, Brommonschenkel S, Chunwongse J, Frary A, Ganal M, Spivey R, Wu T, Earle E, Tanksley S (1993) Map-based cloning of a protein-kinase gene conferring disease resistance in tomato. Science 262:1432–1436

Martin B, Tauer CG, Lin RK (1999) Carbon isotope discrimination as a tool to improve water-use efficiency in tomato. Crop Sci 39:1775–1783

Mathieu S, Dal Cin V, Fei Z, Li H, Bliss P, Taylor MG, Klee HJ, Tieman DM (2009) Flavour compounds in tomato fruits: identification of loci and potential pathways affecting volatile composition. J Exp Bot 60(1):325–337

McCallum CM, Comai L, Greene EA, Henikoff S (2000) Targeting induced local lesions IN genomes (TILLING) for plant functional genomics. Plant Physiol 123(2):439–442

McClean PE, Hanson MR (1986) Mitochondrial DNA sequence divergence among *Lycopersicon* and related *Solanum* species. Genetics 112:649–667

McCouch S (2004) Diversifying selection in plant breeding. PLoS Biol 2(10):e347

McGrath DJ, Gillespie D, Vawdrey L (1987) Inheritance of resistance to *Fusarium oxysporium* f.sp. *lycopersici* races 2 and 3 in *Lycopersicon pennellii*. Aust J Agric Res 38(4): 729–733

Medina-Filho P (1980) Linkage of *Aps-1*, *Mi* and other markers on chromosome 6. Tomato Genet Coop Rep 30:26–28

Menda N, Semel Y, Peled D, Eshed Y, Zamir D (2004) *In silico* screening of a saturated mutation library of tomato. Plant J 38(5):861–872

Menzel MY (1962) Pachytene chromosomes of the intergeneric hybrid *Lycopersicon esculentum* × *Solanum lycopersicoides*. Am J Bot 49:605–615

Mesbah LA, Kneppers TJA, Takken FLW, Laurent P, Hille J, Nijkamp HJJ (1999) Genetic and physical analysis of a YAC contig spanning the fungal disease resistance locus *Asc* of tomato (*Lycopersicon esculentum*). Mol Gen Genet 261: 50–57

Messeguer R, Ganal M, de Vicente MC, Young ND, Bolkan H, Tanksley SD (1991) High resolution RFLP map around the root knot nematode resistance gene (*Mi*) in tomato. Theor Appl Genet 82:529–536

Michelson I, Zamir D, Czosnek H (1994) Accumulation and translocation of tomato yellow leaf curl virus (TYLCV) in a *Lycopersicon esculentum* breeding line containing the *L. chilense* TYLCV tolerance gene *Ty-1*. Phytopathology 84(9):928–933

Miller P (1731) The Gardener's dictionary, 1st edn. John and Francis, Rivington, London, UK

Miller P (1754) The Gardener's dictionary, Abridged 4th edn. John and James. Rivington, London, UK

Miller P (1807) The Gardener's and botanist's dictionary, Posthumous edn. Thomas Martyn, Cambridge, UK

Miller JC, Tanksley SD (1990) RFLP analysis of phylogenetic relationships and genetic variation in the genus *Lycopersicon*. Theor Appl Genet 80:437–448

Milligan SB, Bodeau J, Yaghoobi J, Kaloshian I, Zabel P, Valerie M, Williamson VM (1998) The root knot nematode resistance gene *Mi* from tomato is a member of the leucine zipper, nucleotide binding, leucine-rich repeat family of plant genes. Plant Cell 10:1307–1319

Mittova V, Guy M, Tal M, Volokita M (2004) Salinity upregulates the antioxidative system in root mitochondria and peroxisomes of the wild salt-tolerant tomato species *Lycopersicon pennellii*. J Exp Bot 55:1105–1113

Monforte AJ, Tanksley SD (2000a) Development of a set of near isogenic and backcross recombinant inbred lines containing most of the *Lycopersicon hirsutum* genome in a *L. esculentum* genetic background: a tool for gene mapping and gene discovery. Genome 43:803–813

Monforte AJ, Tanksley SD (2000b) Fine mapping of a quantitative trait locus (QTL) from *Lycopersicon hirsutum* chromosome 1 affecting fruit characteristics and agronomic traits: breaking linkage among QTLs affecting different traits and dissection of heterosis for yield. Theor Appl Genet 100: 471–479

Monforte AJ, Asins AJ, Carbonell EA (1996) Salt tolerance in *Lycopersicon* species IV. Efficiency of marker-assisted selection for salt tolerance improvement. Theor Appl Genet 93:765–772

Monforte AJ, Asins AJ, Carbonell EA (1997a) Salt tolerance in *Lycopersicon* species. V. Does genetic variability at quantitative trait loci affect their analysis? Theor Appl Genet 95: 284–293

Monforte AJ, Asins AJ, Carbonell EA (1997b) Salt tolerance in *Lycopersicon* species VI. Genotype-by-salinity interaction in quantitative trait loci detection: constitutive and response QTLs. Theor Appl Genet 95:706–713

Monforte AJ, Asins AJ, Carbonell EA (1999) Salt tolerance in *Lycopersicon* spp VII. Pleiotropic action of genes controlling earliness on fruit yield. Theor Appl Genet 98: 593–601

Monforte AJ, Friedman E, Zamir D, Tanksley SD (2001) Comparison of a set of allelic QTL-NILs for chromosome 4 of tomato: deductions about natural variation and implications for germplasm utilization. Theor Appl Genet 102:572–590

Moreau P, Thoquet P, Olivier J, Laterrot H, Grimsley N (1998) Genetic mapping of *Ph-2*, a single locus controlling partial resistance to *Phytophthora infestans* in tomato. Mol Plant Microbe Interact 11(4):259–269

Morgante M, Salamini F (2003) From plant genomics to breeding practice. Curr Opin Biotechnol 14:214–219

Moyle LC (2007) Comparative genetics of potential prezygotic and postzygotic isolating barriers in a *Lycopersicon* Species cross. J Hered 98(2):123–135

Moyle LC (2008) Ecological and evolutionary genomics in the wild tomatoes (*Solanum* sect. *Lycopersicon*). Evolution 62(12):2995–3013

Moyle LC, Graham EB (2005) Genetics of hybrid incompatibility between *Lycopersicon esculentum* and *L. hirsutum*. Genetics 169:355–373

Moyle LC, Nakazato T (2008) Comparative genetics of hybrid incompatibility: sterility in two *Solanum* species crosses. Genetics 179:1437–1453

Mueller LA, Solow TH, Taylor N, Skwarecki B, Buels R, Bins J, Lin C, Wright MH, Ahrens R, Wang Y et al (2005a) The SOL genomics network (SGN): a comparative resource for Solanaceous biology and beyond. Plant Physiol 138: 1310–1317

Mueller LA, Tanksley SD, Giovannoni JJ, van Eck J, Stack S, Choi D, Kim BD, Chen M, Cheng Z, Li C, Ling H, Xue Y, Seymour G, Bishop G, Bryan G, Sharma R, Khurana J, Tyagi A, Chattopadhyay D, Singh NK, Stiekema W, Lindhout P, Jesse T, Lankhorst RK, Bouzayen M, Shibata D, Tabata S, Granell A, Botella MA, Giuliano G, Frusciante L, Causse M, Zamir D (2005b) The tomato sequencing project, the first cornerstone of the international Solanaceae project (SOL). Comp Funct Genom 6(3):153–158

Mueller LA, Klein Lankhorst R, Tanksley SD, Giovanonni JJ et al (2009) A snapshot of the emerging tomato genome sequence: the tomato genome sequencing consortium. Plant Genome 2:78–92

Muigai SG, Bassett MJ, Schuster DJ, Scott JW (2003) Greenhouse and field screening of wild *Lycopersicon* germplasm for resistance to the whitefly *Bemisia argentifolii*. Phytoparasitica 31(1):27–38

Müller CH (1940) A revision of the genus *Lycopersicon*. USDA Misc Publ 382:1–28

Mustilli AC, Fenzi F, Ciliento R, Alfano F, Bowler C (1999) Phenotype of the tomato *high pigment-2* mutant is caused by a mutation in the tomato homolog of *DEETIOLATED1*. Plant Cell 11:145–157

Mutschler MA (2006) Combining field and laboratory methods in tomato breeding strategies. Acta Hortic 724:23–27

Mutschler MA, Tanksley SD, Rick CM (1987) Linkage maps of the tomato (*Lycopersicon esculentum*). Rep Tomato Genet Coop 37:5–34

Mutschler MA, Doerge RW, Liu S-C, Kuai JP, Liedl BE, Shapiro JA (1996) QTL analysis of pest resistance in the wild tomato *Lycopersicon pennellii*: QTLs controlling acylsugar level and composition. Theor Appl Genet 92:709–718

Nakazato T, Bogonovich M, Moyle LC (2008) Environmental factors predict adaptive phenotypic differentiation within and between two wild Andean tomatoes. Evolution 62(4): 774–792

Nash AF, Gardner RG (1988) Tomato early blight resistance in a breeding line derived from *Lycopersicon hirsutum* PI 126445. Plant Dis 72:206–209

Nesbitt TC, Tanksley SD (2002) Comparative sequencing in the genus Lycopersicon: implications for the evolution of fruit size in the domestication of cultivated tomatoes. Genetics 162:365–379

Nienhuis J, Helentjarims M, Slocum B, Ruggero B, Schaffer A (1987) Restriction fragment length polymorphism analysis of loci associated with insect resistance in tomato. Crop Sci 27:797–803

Nuez F, Prohens J, Blanca JM (2004) Relationships, origin, and diversity of Galapagos tomatoes: implications for the conservation of natural populations. Am J Bot 91:86–99

Ohmori T, Murata M, Motoyoshi F (1996) Molecular characterization of RAPD and SCAR markers linked to the *Tm-1* locus in tomato. Theor Appl Genet 92:151–156

Olmstead RG, Palmer JD (1997) Implications for phylogeny, classification, and biogeography of *Solanum* from cpDNA restriction site variation. Syst Bot 22:19–29

Olmstead RG, Sweere JA, Spangler RE, Bohs L, Palmer JD (1999) Phylogeny and provisional classification of the *Solanaceae* based on chloroplast DNA. In: Nee M, Symon DE, Lester RN, Jessop JP (eds) Solanaceae IV: advances in biology and utilization. Royal Botanic Gardens, Kew, pp 111–137

Ori N, Eshed Y, Paran I, Presting G, Aviv D, Tanksley S, Zamir D, Fluhr R (1997) The *I2C* family from the wilt disease resistance locus *I2* belongs to the nucleotide binding, leucine-rich repeat superfamily of plant resistance genes. Plant Cell 9:521–532

Orsi CH, Tanksley SD (2009) Natural variation in an ABC transporter gene associated with seed size evolution in tomato species. PLoS Genet 5(1):e1000347

Osborn TC, Alexander DC, Fobes JF (1987) Identification of restriction fragment length polymorphisms linked to genes controlling soluble solids content in tomato fruit. Theor Appl Genet 73:350–356

Osborn TC, Kramer C, Graham E, Braun CJ (2007) Insights and innovations from wide crosses: examples from canola and tomato. Crop Sci 47(S3):S228–S237

Paddock EF (1950) A tentative assignment of *Fusarium-immunity* locus to linkage group 5 in tomato. Genetics 35:683–684

Palmer JD, Zamir D (1982) Chloroplast DNA evolution and phylogenetic relationships in *Lycopersicon*. Proc Natl Acad Sci USA 79:5006–5010

Pan Q, Liu Y-S, Budai-Hadrian O, Sela M, Carmel-Goren L, Zamir D, Fluhr R (2000) Comparative genetics of nucleotide binding site-leucine rich repeat resistance gene homologues in the genomes of two dicotyledons: tomato and *Arabidopsis*. Genetics 155:309–322

Paran I, Goldman I, Tanksley SD, Zamir D (1995) Recombinant inbred lines for genetic mapping in tomato. Theor Appl Genet 90:542–548

Paran I, Goldman I, Zamir D (1997) QTL analysis of morphological traits in a tomato recombinant inbred line population. Genome 40:242–248

Parniske M, Hammond-Kosack KE, Golstein C, Thomas CM, Jones DA, Harrison K, Wulff BBH, Jones JDG (1997) Novel disease resistance specificities result from sequence exchange between tandemly repeated genes at the *Cf- 4/9* locus of tomato. Cell 91:821–832

Parrella G, Ruffel S, Moretti A, Morel C, Palloix A, Caranta C (2002) Recessive resistance genes against potyviruses are localized in colinear genomic regions of the tomato (*Lycopersicon* spp.) and pepper (*Capsicum* spp.) genomes. Theor Appl Genet 105:855–861

Parrella G, Moretti A, Gognalons P, Lesage M-L, Marchoux G, Gebre-Selassie K, Caranta C (2004) The *Am* gene controlling resistance to Alfalfa mosaic virus in tomato is located in the cluster of dominant resistance genes on chromosome 6. Phytopathology 94:345–350

Paterson AH, Lander ES, Hewitt JD, Peterson S, Lincoln SE, Tanksley SD (1988) Resolution of quantitative traits into Mendelian factors by using a complete linkage map of restriction fragment length polymorphisms. Nature 335: 721–726

Paterson AH, DeVerna JW, Lanini B, Tanksley SD (1990) Fine mapping of quantitative trait loci using selected overlapping recombinant chromosomes, in an interspecies cross of tomato. Genetics 124:735–741

Paterson AH, Damon S, Hewitt JD, Zamir D, Rabinowitch HD, Lincoln SE, Lander ES, Tanksley SD (1991) Mendelian factors underlying quantitative traits in tomato: comparison across species, generations, and environments. Genetics 127:181–197

Patterson BD (1988) Genes for cold resistance from wild tomatoes. HortScience 23:794–947

Peirce LC (1971) Linkage test with *Ph* conditioning resistance to race 0 Phytophthora infestans. Rep Tomato Genet Coop 21:30

Peleman JD, van der Voort JR (2003) Breeding by deisgn. Trends Plant Sci 8:330–334

Peralta IE, Spooner DM (2001) Granule-bound starch synthase (GBSSI) gene phylogeny of wild tomatoes (*Solanum* L. section *Lycopersicon* [MILL.] Wettst. Subsection *Lycopersicon*). Am J Bot 88(10):1888–1902

Peralta IE, Spooner DM (2005) Morphological characterization and relationships of wild tomatoes (*Solanum* L. Sect. *Lycopersicon*). American Phytopathology Society (APS), St. Paul, MN, USA, 32 p

Peralta IE, Spooner DM (2007) History, origin and early cultivation of tomato (*Solanaceae*). In: Razdan MK, Mattoo AK (eds) Genetic improvement of Solanaceous crops, vol 2, tomato. Science, Enfield, NH, USA, pp 1–27

Peralta IE, Knapp S, Spooner DM (2005) New species of wild tomatoes (*Solanum* section *Lycopersicon*: Solanaceae) from Northern Peru. Syst Bot 30(2):424–434

Peralta IE, Spooner DM, Knapp S (2008) Taxonomy of wild tomatoes and their relatives (*Solanum* sections *Lycopersicoides*, *Juglandifolia*, *Lycopersicon*; Solanaceae). Syst Bot Monogr 84:1–186

Pereira NE, Leal NR, Pereira MG (2000) Controle genético da concentração de 2-tridecanona e de 2-undecanona em cruzamentos interespecifícos de tomateiro. Bragantia 59(2): 165–172

Pertuzé RA, Ji Y, Chetelat RT (2002) Comparative linkage map of the *Solanum lycopersicoides* and *S. sitiens* genomes and their differentiation from tomato. Genome 45:1003–1012

Pertuzé RA, Ji Y, Chetelat RT (2003) Transmission and recombination of homeologous *Solanum sitiens* chromosomes in tomato. Theor Appl Genet 107:1391–1401

Peters JL, Széll M, Kendrick RE (1998) The expression of light-regulated genes in the *high-pigment-1* mutant of tomato. Plant Physiol 117:797–807

Peterson DG, Price HJ, Johnston JS, Stack SM (1996) DNA content of heterochromatin and euchromatin in tomato (*Lycopersicon esculentum*) pachytene chromosomes. Genome 39:77–82

Pillen K, Ganal MW, Tanksley SD (1996a) Construction of a high-resolution genetic map and YAC-contigs in the tomato *Tm-2a* region. Theor Appl Genet 93:228–233

Pillen K, Pineda O, Lewis CB, Tanksley SD (1996b) Status of genome mapping tools in the taxon Solanaceae. In: Paterson AH (ed) Genome mapping in plants. RG Landes, Austin, TX, pp 282–308

Pitblado RE, MacNeil BH, Kerr EA (1984) Chromosomal identity and linkage relationships of *Pto*, a gene for resistance to *Pseudomonas syringae* pv. *tomato* in tomato. Can J Plant Pathol 6:48–53

Plunknett DL, Smith NJH, Williams JT, Murthi-Anishetti N (1987) Gene banks and the world's food. Princeton University Press, Princeton, NJ, USA

Pnueli L, Carmel-Goren L, Hareven D, Gutfinger T, Alvarez J, Ganal M, Zamir D, Lifschitz E (1998) The *SELF-PRUNING* gene of tomato regulates vegetative to reproductive switching of sympodial meristems and is the ortholog of *CEN* and *TFL1*. Development 125:1979–1989

Powers L (1941) Inheritance of quantitative characters in crosses involving two species of Lycopersicon. J Agric Res 63:149–175

Prudent M, Causse M, Génard M, Tripodi P, Grandillo S, Bertin N (2009) Genetic and physiological analysis of tomato fruit weight and composition: influence of carbon availability on QTL detection. J Exp Bot 60(3):923–937

Ranc N, Muños S, Santoni S, Causse M (2008) A clarified position for *Solanum lycopersicum* var. *cerasiforme* in the evolutionary history of tomatoes (Solanaceae). BMC Plant Biol 8:130

Rick CM (1951) Hybrid between *Lycopersicoides esculentum* Mill. and *Solanum lycopersicoides* Dun. Genetics 37: 741–744

Rick CM (1956) Genetic and systematic studies on accessions of *Lycopersicon* from the Galapagos islands. Am J Bot 43: 687–696

Rick CM (1963) Barriers to interbreeding in *Lycopersicon peruvianum*. Evolution 17:216–232

Rick CM (1969) Controlled introgression of chromosomes of *Solanum pennellii* into *Lycopersicon esculentum*: segregation and recombination. Genetics 62:753–768

Rick CM (1971) Further studies on segregation and recombination in backcross derivatives of a tomato species hybrid. Biol Zentralbl 90:209–220

Rick CM (1973) Potential genetic resources in tomato species: clues from observation in native habitats. In: Srb AM (ed) Genes, enzymes and populations. Plenum, New York, NY, USA, pp 255–269

Rick CM (1974) High soluble-solids content in large-fruited tomato lines derived from a wild green-fruited species. Hilgardia 42:493–510

Rick CM (1975) The tomato. In: King RC (ed) Handbook of genetics, vol 2. Plenum, New York, NY, USA, pp 247–280

Rick CM (1976) Tomato *Lycopersicon esculentum* (*Solanaceae*). In: Simmonds NW (ed) Evolution of crop plants. Longman, London, UK, pp 268–273

Rick CM (1979) Biosystematic studies in *Lycopersicon* and closely related species of *Solanum*. In: Hawkes JG, Lester RN, Skelding AD (eds) The biology and taxonomy of *Solanaceae*, Linn Soc Symp Ser 7. Academic, New York, NY, USA, pp 667–677

Rick CM (1982) The potential of exotic germplasm for tomato improvement. In: Vasil IK, Scowcroft WR, Frey KJ (eds) Plant improvement and somatic cell genetics. Academic, New York, NY, pp 1–28

Rick CM (1986a) Germplasm resources in the wild tomato species. Acta Hortic 190:39–47

Rick CM (1986b) Potential contributions of wide crosses to improvement of processing tomatoes. HortScience 21:881

Rick CM (1986c) Reproductive isolation in the *Lycopersicon peruvianum* complex. In: D'Arcy WG (ed) Solanaceae biology and systematics. Columbia University Press, New York, NY, USA, pp 477–496

Rick CM (1987) Seedling traits of primary trisomics. Rep Tomato Genet Coop 37:60–61

Rick CM (1988) Tomato-like nightshades: affinities, autoecology, and breeders opportunities. Econ Bot 42:145–154

Rick CM (1990) New or otherwise noteworthy accessions of wild tomato species. Tomato Genet Coop Rep 40:30

Rick CM (1995) Tomato. In: Smartt J, Simmonds NW (eds) Evolution of crop plants. Longman, London, pp 452–457

Rick CM, Chetelat RT (1995) Utilization of related wild species for tomato improvement. Acta Hortic 412:21–38

Rick CM, Fobes JF (1975) Allozyme variation in the cultivated tomato and closely related species. Bull Torr Bot Club 102: 376–384

Rick CM, Holle M (1990) Andean *Lycopersicon esculentum* var. *cerasiforme*: genetic variation and its evolutionary significance. Econ Bot 44:69–78

Rick CM, Tanksley SD (1981) Genetic variation in *Solanum pennellii*: comparisons with two other sympatric tomato species. Plant Syst Evol 139:11–45
Rick CM, Yoder JI (1988) Classical and molecular genetics of tomato: highlights and perspectives. Annu Rev Genet 22: 281–300
Rick CM, Kesicki E, Fobes JF, Holle M (1976) Genetic and biosystematic studies on two new sibling species of *Lycopersicon* from interandean Peru. Theor Appl Genet 47: 55–68
Rick CM, Fobes JF, Holle M (1977) Genetic variation in *Lycopersicon pimpinellifolium*: evidence of evolutionary change in mating systems. Plant Syst Evol 127:139–170
Rick CM, Holle M, Thorp RW (1978) Rates of cross-pollination in *Lycopersicon pimpinellifolium*: impact of genetic variation in floral characters. Plant Syst Evol 129:31–44
Rick CM, Fobes JF, Tanksley SD (1979) Evolution of mating systems in *Lycopersicon hirsutum* as deduced from genetic variation in electrophoretic and morphological characters. Plant Syst Evol 132:279–298
Rick CM, Laterrot H, Philouze J (1990) A revised key for the *Lycopersicon* species. Tomato Genet Coop Rep 40:31
Rivas S, Thomas CM (2005) Molecular interactions between tomato and the leaf mold pathogen *Cladosporium fulvum*. Annu Rev Phytopathol 43:395–436
Robert VJM, West MAL, Inai S, Caines A, Arntzen L, Smith JK, St.Clair DA (2001) Marker-assisted introgression of blackmold resistance QTL alleles from wild *Lycopersicon cheesmanii* to cultivated tomato (*L. esculentum*) and evaluation of QTL phenotypic effects. Mol Breed 8:217–233
Robertson NF (1991) The challenge of *Phytophthora infestans*. In: Ingram DS, Williams PH (eds) Advances in plant pathology, vol 10. Academic, London, pp 1–30
Robertson LD, Labate JA (2007) Genetic resources of tomato (*Lycopersicon esculentum* var. *esculentum*) and wild relatives. In: Razdan MK, Mattoo AK (eds.) Genetic improvement of Solanaceous crops, vol 2: Tomato. Science, Enfield, NH, USA, pp 25–75
Ronen GL, Cohen M, Zamir D, Hirschberg J (1999) Regulation of carotenoid biosynthesis during tomato fruit development: expression of the gene for lycopene epsilon-cyclase is down-regulated during ripening and is elevated in the mutant Delta. Plant J 17:341–351
Ronen G, Carmel-Goren L, Zamir D, Hirschberg J (2000) An alternative pathway to β-carotene formation in plant chromoplasts discovered by map-based cloning of *Beta* and *old-gold* color mutations in tomato. Proc Natl Acad Sci USA 97:11102–11107
Rossi M, Goggin FL, Milligan SB, Kaloshian I, Ullman DE, Williamson VM (1998) The nematode resistance gene *Mi* of tomato confers resistance against the potato aphid. Proc Natl Acad Sci USA 95:9750–9754
Rousseaux MC, Jones CM, Adams D, Chetelat R, Bennett A, Powell A (2005) QTL analysis of fruit antioxidants in tomato using *Lycopersicon pennellii* introgression lines. Theor Appl Genet 111:1396–1408
Ruffel S, Gallois JL, Lesage ML, Caranta C (2005) The recessive potyvirus resistance gene *pot-1* is the tomato orthologue of the pepper *pvr2-eIF4E* gene. Mol Genet Genom 274: 346–353
Rush DW, Epstein E (1981) Breeding and selection for salt tolerance by the incorporation of wild germplasm into a domestic tomato. J Am Soc Hortic Sci 106:699–704
Saliba-Colombani V, Causse M, Gervais L, Philouze J (2000) Efficiency of RFLP, RAPD, and AFLP markers for the construction of an intraspecific map of the tomato genome. Genome 43:29–40
Saliba-Colombani V, Causse M, Langlois D, Philouze J, Buret M (2001) Genetic analysis of organoleptic quality in fresh market tomato. 1. Mapping QTLs for physical and chemical traits. Theor Appl Genet 102:259–272
Sallaud C, Rontein D, Onillon S, Jabès F, Duffé P, Giacalone C, Thoraval S, Escoffier C, Herbette G, Leonhardt N, Causse M, Tissiera A (2009) A novel pathway for sesquiterpene biosynthesis from Z, Z-farnesyl pyrophosphate in the wild tomato Solanum habrochaites. Plant Cell 21:301–317
Salmeron J, Oldroyd G, Rommens C, Scofield S, Kim H-S, Lavelle D, Dahlbeck D, Staskawicz B (1996) Tomato *Prf* is a member of the leucine-rich repeat class of plant disease resistance genes and lies embedded within the *Pto* kinase gene cluster. Cell 86:123–133
Sandbrink JM, van Ooijen JW, Purimahua CC, Vrielink M, Verkerk R, Zabel P, Lindhout P (1995) Localization of genes for bacterial canker resistance in *Lycopersicon peruvianum* using RFLPs. Theor Appl Genet 90:444–450
Sanseverino W, Roma G, De Simone M, Faino L, Melito S, Stupka E, Frusciante L, Ercolano MR (2010) PRGdb: a bioinformatics platform for plant resistance gene analysis. Nucleic Acids Res 38:D814–D821
Sarfatti M, Katan J, Fluhr R, Zamir D (1989) An RFLP marker in tomato linked to the *Fusarium oxysporum* resistance gene *I2*. Theor Appl Genet 78:755–759
Sarfatti M, Abu-Abied M, Katan J, Zamir D (1991) RFLP mapping of *I1*, a new locus in tomato conferring resistance against *Fusarium oxysporum* f. sp. *lycopersici* race 1. Theor Appl Genet 82:22–26
Schauer N, Zamir D, Fernie AR (2005) Metabolic profiling of leaves and fruits of wild species tomato: a survey of the *Solanum lycopersicum* complex. J Exp Bot 56:297–307
Schauer N, Semel Y, Roessner U, Gur A, Balbo I, Carrari F, Pleban T, Perez-Melis A, Bruedigam C, Kopka J, Willmitzer L, Zamir D, Fernie AR (2006) Comprehensive metabolic profiling and phenotyping of interspecific introgression lines for tomato improvement. Nat Biotechnol 24:447–454
Schauer N, Semel Y, Balbo I, Steinfath M, Repsilber D, Selbig J, Pleban T, Zamir D, Fernie AR (2008) Mode of inheritance of primary metabolic traits in tomato. Plant Cell 20:509–523
Schornack S, Ballvora A, Gürlebeck D, Peart J, Baulcombe D, Ganal M, Baker B, Bonas U, Lahaye T (2004) The tomato resistance protein Bs4 is a predicted non-nuclear TIR-NB-LRR protein that mediates defense responses to severely truncated derivatives of AvrBs4 and overexpressed AvrBs3. Plant J 37(1):46–60, Erratum Plant J 37(5):787
Schumacher K, Ganal M, Theres K (1995) Genetic and physical mapping of the lateral suppressor (*ls*) locus in tomato. Mol Gen Genet 246:761–766
Scott JW (1984) Genetic source of tomato firmness. In: Proceedings of 4th tomato quality workshop, Miami, FL, USA, 7 Mar 1983, pp 60–67

Scott JW, Gardner RG (2007) Breeding for resistance to fungal pathogens. In: Razdan MK, Mattoo AK (eds) Genetic improvement of Solanaceous crops, vol 2: Tomato. Science, Enfield, NH, pp 421–456

Scott JW, Jones JP (1989) Monogenic resistance in tomato to *Fusarium oxysporum* f. sp. *Lycopersici* race 3. Euphytica 40:49–53

Scott JW, Olson SM, Howe TK, Stoffella PJ, Bartz JA, Bryan HH (1995) Equinox' heat-telerant hybrid tomato. HortScience 30:647–648

Scott JW, Agrama HA, Jones JP (2004) RFLP-based analysis of recombination among resistance genes to *Fusarium* wilt races 1, 2, and 3 in tomato. J Am Soc Hortic Sci 129: 394–400

Seah S, Yaghoobi J, Rossi M, Gleason CA, Williamson VM (2004) The nematode-resistance gene, *Mi-1*, is associated with an inverted chromosomal segment in susceptible compared to resistant tomato. Theor Appl Genet 108: 1635–1642

Segal G, Sarfatti M, Schaffer MA, Ori N, Zamir D, Fluhr R (1992) Correlation of genetic and physical structure in the region surrounding the I_2 *Fusarium oxysporum* locus in tomato. Mol Gen Genet 231:179–185

Seithe A (1962) Die Haararten der Gattung *Solanum* L. und ihre taxonomische Verwertung. Bot Jahrb Syst 81:261–336

Sela-Buurlage MB, Budai-Hadrian O, Pan Q, Carmel-Goren L, Vunsch R, Zamir D, Fluhr R (2001) Genome-wide dissection of *Fusarium* resistance in tomato reveals multiple complex loci. Mol Genet Genom 265:1104–1111

Semel Y, Nissenbaum J, Menda N, Zinder M, Krieger U, Issman N, Pleban T, Lippman Z, Gur A, Zamir D (2006) Overdominant quantitative trait loci for yield and fitness in tomato. Proc Natl Acad Sci USA 103:12981–12986

Sharma A, Zhang L, Niño-Liu D, Ashrafi H, Foolad MR (2008) A *Solanum lycopersicum* × *Solanum pimpinellifolium* linkage map of tomato displaying genomic locations of R-genes, RGAs, and candidate resistance/defense-response ESTs. Int J Plant Genom. doi:10.1155/2008/926090

Sherman JD, Stack SM (1995) Two-dimensional spreads of synaptonemal complexes from solanaceous plants. VI. High resolution recombination nodule map for tomato (*Lycopersicon esculentum*). Genetics 141:683–708

Shirasawa K, Asamizu E, Fukuoka H, Ohyama A, Sato S, Nakamura Y, Tabata S, Sasamoto S, Wada T, Kishida Y, Tsuruoka H, Fujishiro T, Yamada M, Isobe S (2010) An interspecific linkage map of SSR and intronic polymorphism markers in tomato. Theor Appl Genet 121(4):731–739

Simons G, Groenendijk J, Wijbrandi J, Reijans M, Groenen J, Diergaarde P, Van der Lee T, Bleeker M, Onstenk J, de Both M, Haring M, Mes J, Cornelissen B, Zabeau M, Vos P (1998) Dissection of the Fusarium *I2* gene cluster in tomato reveals six homologs and one active gene copy. Plant Cell 10: 1055–1068

Sims WL (1980) History of tomato production for industry around the world. Acta Hortic 100:25–26

Sinesio F, Cammareri M, Moneta E, Navez B, Peparaio M, Causse M, Grandillo S (2010) Sensory quality of fresh French and Dutch market tomatoes: a preference mapping study with Italian consumers. J Food Sci 75:S55–S67

Smith PG (1944) Embryo culture of a tomato species hybrid. Proc Am Soc Hortic Sci 44:413–416

Smith SD, Peralta IE (2002) Ecogeographic surveys as tools for analyzing potential reproductive isolating mechanisms: an example using *Solanum juglandifolium* Dunal, *S. ochranthum* Dunal, *S. lycopersicoides* Dunal, and *S. sitiens* I.M. Johnston. Taxon 51:341–349

Snyder JC, Guo Z, Thacker R, Goodman JP, Pyrek JS (1993) 2,3-Dihydrofarnesoic acid, a unique terpene from trichomes of Lycopersicon hirsutum, repels spider mites. J Chem Ecol 19:2981–2997

Soumpourou E, Iakovidis M, Chartrain L, Lyall V, Thomas CM (2007) The *Solanum pimpinellifolium Cf-ECP1* and *Cf-ECP4* genes for resistance to *Cladosporium fulvum* are located at the *Milky Way* locus on the short arm of chromosome 1. Theor Appl Genet 115:1127–1136

Spooner DM, Anderson GJ, Jansen RK (1993) Chloroplast DNA evidence for the interrelationships of tomatoes, potatoes, and pepinos (Solanaceae). Am J Bot 80:676–688

Spooner DM, van den Berg RG, Rodríguez A, Bamberg J, Hijmans RJ, Lara Cabrera SI (2004) Wild potatoes (*Solanum* section *Petota*; Solanaceae) of North and Central America. Syst Bot Monogr 68:1–209

Spooner DM, Peralta IE, Knapp S (2005) Comparison of AFLPs to other markers for phylogenetic inference in wild tomatoes [*Solanum* L. section *Lycopersicon* (Mill.) Wettst.]. Taxon 54:43–61

Springer NM, Stupar RM (2007) Allelic variation and heterosis in maize: how do two halves make more than a whole? Genome Res 17:264–275

Stack SM, Covey PA, Anderson LK, Bedinger PA (2009) Cytogenetic characterization of species hybrids in the tomato clade. Tomato Genet Coop Rep 59:57–61

Städler T, Roselius K, Stephan W (2005) Genealogical footprints of speciation processes in wild tomatoes: demography and evidence for historical gene flow. Evolution 59:1268–1279

Stall RE, Walter JM (1965) Selection and inheritance of resistance in tomato to isolates of race 1 and 2 of the *Fusarium* wilt organism. Phytopathology 55:1213–1215

Stamova BS, Chetelat RT (2000) Inheritance and genetic mapping of cucumber mosaic virus resistance introgressed from *Lycopersicon chilense* into tomato. Theor Appl Genet 101:527–537

Stamova L, Yordanov M (1990) *Lv* – as a symbol of the gene controlling resistance to *Leveillula taurica*. Tomato Genet Coop Rep 40:36

Stevens MA (1972) Relationships between components contributing to quality variation among tomato lines. J Am Soc Hortic Sci 97:70–73

Stevens MA (1986) Inheritance of tomato fruit quality components. In: Janick J (ed) Plant breeding reviews, vol 4. AVI, Westport, CT, USA, pp 273–312

Stevens MA, Rick CM (1986) Genetics and breeding. In: Atherton JG, Rudich J (eds) The tomato crop: a scientific basis for improvement. Chapman and Hall, London, UK, pp 35–109

Stevens MR, Lamb EM, Rhoads DD (1995) Mapping the *Sw5* locus for tomato spotted wilt virus resistance in tomatoes using RAPD and RFLP analyses. Theor Appl Genet 90: 451–456

Stevens R, Buret M, Duffé F, Garchery C, Baldet P, Rothan C, Causse M (2007) Candidate genes and quantitative trait loci affecting fruit ascorbic acid content in three tomato populations. Plant Physiol 143:1943–1953

Stevens R, Page D, Gouble B, Garchery C, Zamir D, Causse M (2008) Tomato fruit ascorbic acid content is linked with monodehydroascorbate reductase activity and tolerance to chilling stress. Plant Cell Environ 31:1086–1096

Stommel JR (2001) USDA 97L63, 97L66 and 97L97: tomato breeding lines with high fruit beta-carotene content. HortScience 36:387–388

Stommel JR, Zhang Y (1998) Molecular markers linked to quantitative trait loci for anthracnose resistance in tomato. HortScience 33:514

Stommel JR, Zhang Y (2001) Inheritance and QTL analysis of anthracnose resistance in the cultivated tomato (*Lycopersicon esculentum*). Acta Hortic 542:303–310

Stommel JR, Abbott J, Saftner RA, Camp M (2005a) Sensory and objective quality attributes of beta-carotene- and lycopene-rich tomato fruit. J Am Soc Hortic Sci 130:244–251

Stommel JR, Abbott JA, Saftner RA (2005b) USDA 02L1058 and 02L1059: Cherry tomato breeding lines with high fruit beta-carotene content. HortScience 40:1569–1570

Suliman-Pollatschek S, Kashkush K, Shats H, Hillel J, Lavi U (2002) Generation and mapping of AFLP, SSRs and SNPs in *Lycopersicon esculentum*. Cell Mol Biol Lett 7(2A): 583–597

Symon DE (1981) The Solanaceous genera *Browallia, Capsicum, Cestrum, Cyphomandra, Hyoscyamus, Lycopersicon, Nierembergia, Physalis, Petunia, Salpichroa, Withania*, naturalized in Australia. J Adelaide Bot Gard 3:133–166

Symon DE (1985) The *Solanaceae* of New Guinea. J Adelaide Bot Gard 8:1–177

Szinay D (2010) The development of FISH tools for genetic, phylogenetic and breeding studies in tomato (*Solanum lycopersicum*). PhD Thesis, Wagenigen University, Wageningen, Netherlands

Tadmor Y, Fridman E, Gur A, Larkov O, Lastochkin E, Ravid U, Zamir D, Lewinsohn E (2002) Identification of malodorous, a wild species allele affecting tomato aroma that was selected against during domestication. J Agric Food Chem 50(7):2005–2009

Tanksley SD (1993) Mapping polygenes. Annu Rev Genet 27: 205–233

Tanksley SD (2004) The genetic, developmental and molecular bases of fruit size and shape variation in tomato. Plant Cell 16:S181–S189

Tanksley SD, Bernstzky R (1987) Molecular markers for the nuclear genome of tomato. In: Nevins DJ, Jones RA (eds) Plant biology, vol 4, Tomato biotechnology. Alan R. Liss, New York, NY, USA, pp 37–44

Tanksley SD, Costello W (1991) The size of the *L. pennellii* chromosome 7 segment containing the *I-3* gene in tomato breeding lines as measured by RFLP probing. Rep Tomato Genet Coop 41:60–61

Tanksley SD, Hewitt JD (1988) Use of molecular markers in breeding for soluble solids in tomato – a re-examination. Theor Appl Genet 75:811–823

Tanksley SD, Loaiza-Figueroa F (1985) Gametophytic self-incompatibility is controlled by a single major locus on chromosome 1 in *Lycopersicon peruvianum*. Genetics 82: 5093–5096

Tanksley SD, McCouch SR (1997) Seed banks and molecular maps: unlocking genetic potential from the wild. Science 277:1063–1066

Tanksley SD, Nelson JC (1996) Advanced backcross QTL analysis: a method for the simultaneous discovery and transfer of valuable QTLs from unadapted germplasm into elite breeding lines. Theor Appl Genet 92:191–203

Tanksley SD, Rick CM (1980) Isozymic gene linkage map of the tomato: applications in genetics and breeding. Theor Appl Genet 57:161–170

Tanksley SD, Medina-Filho E, Rick CM (1982) Use of naturally-occurring enzyme variation to detect and map genes controlling quantitative traits in an interspecific backcross of tomato. Heredity 49(1):11–25

Tanksley SD, Ganal MW, Prince JP, de Vicente MC, Bonierbale MW, Broun P, Fulton TM, Giovannoni JJ, Grandillo S, Martin GB, Messeguer R, Miller JC, Miller L, Paterson AH, Pineda O, Riider MS, Wing RA, Wu W, Young ND (1992) High density molecular linkage maps of the tomato and potato genomes. Genetics 132:1141–1160

Tanksley SD, Grandillo S, Fulton TM, Zamir D, Eshed Y, Petiard V, Lopez J, Beck-Bunn T (1996) Advanced backcross QTL analysis in a cross between an elite processing line of tomato and its wild relative *L. pimpinellifolium*. Theor Appl Genet 92:213–224

Tanksley SD, Bernacchi D, Fulton TM, Beck-Bunn T, Emmatty D, Eshed Y, Inai S, Lopez J, Petiard V, Sayama H, Uhlig J, Zamir D (1997a) Comparing the performance of a pair of processing lines nearly isogenic for the *I2* gene conferring resistance to *Fusarium oxysporum* race 2. Rep Tomato Genet Coop 47:33–35

Tanksley SD, Bernacchi D, Fulton TM, Beck-Bunn T, Emmatty D, Eshed Y, Inai S, Lopez J, Petiard V, Sayama H, Uhlig J, Zamir D (1997b) Comparing the effects of linkage drag in a set of processing tomato lines nearly isogenic for the *Mi* gene for resistance to root knot nematodes. Rep Tomato Genet Coop 47:35–36

Taylor IB (1986) Biosystematics of the tomato. In: Atherton JG, Rudich J (eds) The tomato crop: a scientific basis for improvement. Chapman and Hall, London, UK, pp 1–34

ten Have A, van Berloo R, Lindhout P, van Kan JAL (2007) Partial stem and leaf resistance against the fungal pathogen *Botrytis cinerea* in wild relatives of tomato. Eur J Plant Pathol 117:153–166

Termolino P, Fulton T, Perez O, Eannetta N, Xu Y, Tanksley SD, Grandillo S (2010) Advanced backcross QTL analysis of a *Solanum lycopersicum* × *Solanum chilense* cross. In: Proceedings of SOL2010 7th Solanaceae conference, Dundee, Scotland, 5–9 Sept 2010 (in press)

Thomas CM, Jones DA, Parniske M, Harrison K, Balint Kurti PJ, Hatzixanthis K, Jones JDG (1997) Characterization of the tomato *Cf-4* gene for resistance to *Cladosporium fulvum* identifies sequences that determine recognitional specificity in *Cf-4* and *Cf-9*. Plant Cell 9:2209–2224

Thoquet P, Olivier J, Sperisen C, Rogowsky P, Laterrot H, Nigel G (1996a) Quantitative trait loci determining resistance to bacterial wilt in tomato cultivar Hawaii7996. Mol Plant Microbe Interact 9(9):826–836

Thoquet P, Olivier J, Sperisen C, Rogowsky P, Priror P, Anaïs G, Mangin B, Bazin B, Nazer N, Nigel G (1996b) Polygenic resistence of tomato plants to bacterial wilt in the West Indies. Mol Plant Microbe Interact 9(9):837–842

Tieman D, Zeigler M, Schmelz EA, Taylor MG, Bliss P, Kirst M, Klee HJ (2006) Identification of loci affecting flavour volatile emissions in tomato fruits. J Exp Bot 57(4):887–896

Tomes ML, Quackenbush FW, McQuistan M (1954) Modification and dominance of the gene governing formation of high concentrations of beta-carotene in the tomato. Genetics 39: 810–817

Tripodi P, Frusciante L, Tanksley SD, Grandillo S (2006) Updates on the development of a whole genome *S. habrochaites* (acc. LA1777) IL population. In: Proceedings of VI international Solanaceae conference on genomics meets biodiversity, Madison, WI, USA, 23–27 July 2006, p 403

Tripodi P, Maurer S, Di Dato F, Al Seekh S, Frusciante L, Van Haaren MJJ, Mohammad A, Tanksley SD, Zamir D, Gebhardt C, Grandillo S (2009) Linking a set of tomato exotic libraries and a potato mapping population with a framework of conserved ortholog set II (COSII) markers. In: Proceedings of plant biology 2009, Honolulu, HI, USA, 18–22 July 2009, pp 194–195

Tripodi P, Brog M, Di Dato F, Zamir D, Grandillo S (2010) QTL analysis in backcross inbred lines of *Solanum neorickii* (LA2133). In: Proceedings of SOL2010 7th Solanaceae conference, Dundee, Scotland, 5–9 Sept 2010 (in press)

Truco MJ, Randall LB, Bloom AJ, St.Clair DA (2000) Detection of QTLs associated with shoot wilting and root ammonium uptake under chilling temperatures in an interspecific backcross population from *Lycopersicon esculentum* × *L. hirsutum*. Theor Appl Genet 101:1082–1092

Umaerus V, Umaerus M (1994) Inheritance of resistance to late blight. In: Bradshaw JE, Mackay GR (eds) Potato genetics. CABI, Wallingford, UK, pp 365–401

Vakalounakis DJ, Laterrot H, Moretti A, Ligoxigasis EK, Smardas K (1997) Linkage between *Frl* (*Fusarium oxysporum* f. sp. *radicis-lycopersici* resistance) and *Tm-2* (tobacco mosaic virus resistance-2) loci in tomato (*Lycopersicon esculentum*). Ann Appl Biol 130:319–323

Vallejos CE, Tanksley SD (1983) Segregation of isozyme markers and cold tolerance in an interspecific backcross of tomato. Theor Appl Genet 66:241–247

Van der Beek JG, Verkerk R, Zabel P, Lindhout P (1992) Mapping strategy for resistance genes in tomato based on RFLPs between cultivars: *Cf9* (resistance to *Cladosporium fulvum*) on chromosome 1. Theor Appl Genet 84:106–112

Van der Beek JG, Pet G, Lindhout P (1994) Resistance to powdery mildew (*Oidium lycopersicum*) in *Lycopersicon hirsutum* is controlled by an incompletely-dominant gene *Ol-1* on chromosome 6. Theor Appl Genet 89:467–473

Van der Biezen EA, Overduin B, Nijkamp HJJ, Hille J (1994) Integrated genetic map of tomato chromosome 3. Tomato Genet Coop Rep 44:8–10

Van der Biezen EA, Glagotskaya T, Overduin B, Nijkamp HJJ, Hille J (1995) Inheritance and genetic mapping of resistance to *Alternaria alternata* f. sp. *lycopersici* in *Lycopersicon pennellii*. Mol Gen Genet 247:453–461

Van der Hoeven RS, Monforte AJ, Breeden D, Tanksley SD, Steffens JC (2000) Genetic control and evolution of sesquiterpene biosynthesis in *Lycopersicon esculentum* and *L. hirsutum*. Plant Cell 12:2283–2294

Van Der Hoeven R, Ronning C, Giovannoni J, Martin G, Tanksley S (2002) Deductions about the number, organization, and evolution of genes in the tomato genome based on analysis of a large expressed sequence tag collection and selective genomic sequencing. Plant Cell 14:1441–1456

Van der Knaap E, Tanksley SD (2001) Identification and characterization of a novel locus controlling early fruit development in tomato. Theor Appl Genet 103:353–358

Van der Knaap E, Tanksley SD (2003) The making of a bell pepper-shaped tomato fruit: identification of loci controlling fruit morphology in Yellow Stuffer tomato. Theor Appl Genet 107:139–147

Van der Knaap E, Lippman ZB, Tanksley SD (2002) Extremely elongated tomato fruit controlled by four quantitative trait loci with epistatic interactions. Theor Appl Genet 104: 241–247

Van der Knaap E, Sanyal A, Jackson SA, Tanksley SD (2004) High-resolution fine mapping and fluorescence in situ hybridization analysis of *sun*, a locus controlling tomato fruit shape, reveals a region of the tomato genome prone to DNA rearrangements. Genetics 168:2127–2140

Van Deynze A, van der Knaap E, Francis D (2006) Development and application of an informative set of anchored markers for tomato breeding. In: Plant animal genome XIV conference, San Diego, CA, USA, P 188

Van Heusden AW, Koornneef M, Voorrips RE, Brüggemann W, Pet G, Vrielink-van Ginkel R, Chen X, Lindhout P (1999) Three QTLs from *Lycopersicon peruvianum* confer a high level of resistance to *Clavibacter michiganensis* ssp. *michiganensis*. Theor Appl Genet 99:1068–1074

Van Ooijen JW, Sandbrink JM, Vrielink M, Verkerk R, Zabel P, Lindhout P (1994) An RFLP linkage map of *Lycopersicon peruvianum*. Theor Appl Genet 89:1007–1013

Van Tuinen A, Cordonnier-Pratt MM, Pratt LH, Verkerk R, Zabel P, Koornneef M (1997) The mapping of phytochrome genes and photomorphogenic mutants of tomato. Theor Appl Genet 94:115–122

VanWordragen MF, Weide RL, Coppoolse E, Koornneef M, Zabel P (1996) Tomato chromosome 6: a high resolution map of the long arm and construction of a composite integrated marker-order map. Theor Appl Genet 92:1065–1072

Veremis JC, Roberts PA (1996a) Relationships between *Meloidogyne incognita* resistance genes in *Lycopersicon peruvianum* differentiated by heat sensitivity and nematode virulence. Theor Appl Genet 93:950–959

Veremis JC, Roberts PA (1996b) Differentiation of *Meloidogyne incognita* and *M. arenaria* novel resistance phenotypes in *Lycopersicon peruvianum* and derived bridge lines. Theor Appl Genet 93:960–967

Veremis JC, Roberts PA (2000) Diversity of heat-stable genotype specific resistance to *Meloidogyne* in Marañon races of *Lycopersicon peruvianum* complex. Euphytica 111:9–16

Veremis JC, van Heusden AW, Roberts PA (1999) Mapping a novel heat-stable resistance to *Meloidogyne* in *Lycopersicon peruvianum*. Theor Appl Genet 98:274–280

Villalta I, Reina-Sánchez A, Cuartero J, Carbonell EA, Asins MJ (2005) Comparative microsatellite linkage analysis and genetic structure of two populations of F_6 lines derived from *Lycopersicon pimpinellifolium* and *L. cheesmanii*. Theor Appl Genet 110:881–894

Villalta I, Bernet GP, Carbonell EA, Asins MJ (2007) Comparative QTL analysis of salinity tolerance in terms of fruit yield using two *Solanum* populations of F_7 lines. Theor Appl Genet 114:1001–1017

Villalta I, Reina-Sánchez A, Bolarín MC, Cuartero J, Belver A, Venema K, Carbonell EA, Asins MJ (2008) Genetic analysis of Na^+ and K^+ concentrations in leaf and stem as physiological components of salt tolerance in tomato. Theor Appl Genet 116:869–880

Villand J, Skroch PW, Lai T, Hanson P, Kuo CG, Nienhuis J (1998) Genetic variation among tomato accessions from primary and secondary centers of diversity. Crop Sci 38: 1339–1347

Vos P, Simons G, Jesse T, Wijbrandi J, Heinen L, Hogers R, Frijters A, Groenendijk J, Diergaarde P, Reijans M, Fierens-Onstenk J, de Both M, Peleman J, Liharska T, Hontelez J, Zabeau M (1998) The tomato *Mi-1* gene confers resistance to both root-knot nematodes and potato aphids. Nat Biotechnol 16:1365–1369

Walter JM (1967) Heredity resistance to disease in tomato. Annu Rev 5:131–160

Wang Y, Tang X, Cheng Z, Mueller L, Giovannoni J, Tanksley SD (2006) Euchromatin and pericentromeric heterochromatin: comparative composition in the tomato genome. Genetics 172:2529–2540

Warnock SJ (1988) A review of taxonomy and phylogeny of the genus *Lycopersicon*. Hortic Sci 23:669–673

Wastie RL (1991) Breeding for resistance. In: Ingram DS, Williams PH (eds) *Phytophthora infestans*, the cause of late blight of potato. Advances in plant pathology, vol 7. Academic, London, UK, pp 193–224

Weese T, Bohs L (2007) A three-gene phylogeny of the genus *Solanum* (Solanaceae). Syst Bot 33:445–463

Weide R, van Wordragen MF, Lankhorst RK, Verkerk R, Hanhart C, Liharska T, Pap E, Stam P, Zabel P, Koorneef M (1993) Integration of the classical and molecular linkage maps of tomato chromosome 6. Genetics 135:1175–1186

Weller JI, Soller M, Brody T (1988) Linkage analysis of quantitative traits in an interspecific cross of tomato (*Lycopersicon esculentum* × *Lycopersicon pimpinellifolium*) by means of genetic markers. Genetics 118:329–339

Whalen MD (1979) Taxonomy of *Solanum* section *Androceras*. Gentes Herb 11:359–426

Whalen MD (1984) Conspectus of species groups in *Solanum* subgenus *Leptostemonum*. Gentes Herb 12:179–282

Wheeler D, Church D, Edgar R, Federhen S, Helmberg W, Madden T, Pontius J, Schuler G, Schriml L, Sequeira E, Suzek T, Tatusova T, Wagner L (2004) Database resources of the National Center for Biotechnology Information: update. Nucleic Acids Res 32:D35–D40

Williams CE, St. Clair DA (1993) Phenetic relationships and levels of variability detected by restriction fragment length polymorphism and random amplified polymorphic DNA analysis of cultivated and wild accessions of *Lycopersicon esculentum*. Genome 36:619–630

Wing RA, Zhang HB, Tanksley SD (1994) Map-based cloning in crop plants. Tomato as a model system: I. Genetic and physical mapping of jointless. Mol Gen Genet 242:681–688

Wu F, Mueller LA, Crouzillat D, Petiard V, Tanksley SD (2006) Combining bioinformatics and phylogenetics to identify large sets of single-copy orthologous genes (COSII) for comparative, evolutionary and systematic studies: a test case in the euasterid plant clade. Genetics 174:1407–1420

Xiao H, Jiang N, Schaffner E, Stockinger EJ, Van der Knaap E (2008) A retrotransposon- mediated gene duplication underlies morphological variation of tomato fruit. Science 319: 1527–1530

Xu X, Martin B, Comstock JP, Vision TJ, Tauer CG, Zhao B, Pausch RC, Knapp S (2008) Fine mapping a QTL for carbon isotope composition in tomato. Theor Appl Genet 117:221–233

Yaghoobi J, Kaloshian I, Wen Y, Williamson VM (1995) Mapping a new nematode resistance locus in *Lycopersicon peruvianum*. Theor Appl Genet 91:457–464

Yang W, Bai X, Kabelka E, Eaton C, Kamoun S, van der Knaap E, Francis D (2004) Discovery of singly nucleotide polymorphisms in *Lycopersicon esculentum* by computer aided analysis of expressed sequence tags. Mol Breed 14(1): 21–34

Yano K, Watanabe M, Yamamoto N, Tsugane T, Aoki K, Sakurai N, Shibata D (2006) MiBASE: a database of a miniature tomato cultivar Micro-Tom. Plant Biotechnol 23: 195–198

Yates HE, Frary A, Doganlar S, Frampton A, Eannetta NT, Uhlig J, Tanksley SD (2004) Comparative fine mapping of fruit quality QTLs on chromosome 4 introgressions derived from two wild tomato species. Euphytica 135:283–296

Yelle S, Hewitt JD, Robinson NL, Damon S, Bennett AB (1988) Sink metabolism in tomato fruit III. Analysis of carbohydrate assimilation in a wild species. Plant Physiol 87: 737–740

Yelle S, Chetelat RT, Dorais M, DeVerna JW, Bennett AB (1991) Sink metabolism in tomato fruit IV Genetic and biochemical analysis of sucrose accumulation. Plant Physiol 95:1026–1035

Yen HC, Shelton BA, Howard LR, Lee S, Vrebalov J, Giovannoni JJ (1997) The tomato high-pigment (*hp*) locus maps to chromosome 2 and influences plastome copy number and fruit quality. Theor Appl Genet 95:1069–1079

Young ND, Zamir D, Ganal MW, Tanksley SD (1988) Use of isogenic lines and simultaneous probing to identify DNA markers tightly linked to the *Tm-2a* gene in tomato. Genetics 120(2):579–585

Yu AT (1972) The genetics and physiology of water usage in *Solanum pennellii* Corr. and its hybrids with *Lycopersicon esculentum* Mill. PhD Dissertation, University of California, Davis, CA, USA

Yu ZH, Wang JF, Stall RE, Vallejos CE (1995) Genomic localization of tomato genes that control a hypersensitive reaction to *Xanthomonas campestris* pv. *vesicatoria* (Doidge) dye. Genetics 141(2):675–682

Yuan YN, Haanstra J, Lindhout P, Bonnema G (2002) The *Cladopsorium fulvum* resistance gene *Cf-ECP3* is part of the Orion cluster on the short arm of chromosome 1. Mol Breed 10:45–50

Zamir D (2001) Improving plant breeding with exotic genetic libraries. Nat Rev Genet 2:983–989

Zamir D, Eshed Y (1998a) Case history in germplasm introgression: tomato genetics and breeding using nearly isogenic introgression lines derived from wild species. In: Paterson A (ed) Molecular dissection of complex traits. CRC, Boca Raton, FL, USA, pp 207–217

Zamir D, Eshed Y (1998b) Tomato genetics and breeding using nearly isogenic introgression lines derived from wild species. In: Paterson AH (ed) Molecular dissection of complex traits. CRC, Boca Raton, FL, USA, pp 207–217

Zamir D, Tal M (1987) Genetic analysis of sodium, potassium and chloride ion content in lycopersicon. Euphytica 36:187–191

Zamir D, Tanksley SD, Jones RA (1982) Haploid selection for low temperature tolerance of tomato pollen. Genetics 101: 129–137

Zamir D, Ben-David T, Rudich J, Juvik J (1984) Frecuency distributions and linkage relationships of 2-tridecanone in interspecific segregrating generations in tomato. Euphytica 33(2):481–482

Zamir D, Ekstein-Michelson I, Zakay I, Navot N, Zaidan N, Sarfatti M, Eshed Y, Harel E, Pleban T, van Oss H, Kedar N, Rabinowitch HD, Czosneck H (1994) Mapping and introgression of a tomato yellow leaf curl virus tolerance gene, *Ty-1*. Theor Appl Genet 88:141–146

Zhang Y, Stommel JR (2000) RAPD and AFLP tagging and mapping of *Beta (B)* and *Beta modifier* (*MoB*), two genes which influence β-carotene accumulation in fruit of tomato (*Lycopersicon esculentum* Mill.). Theor Appl Genet 100: 368–375

Zhang Y, Stommel JR (2001) Development of SCAR and CAPS markers linked to the *Beta* gene in tomato. Crop Sci 41: 1602–1608

Zhang HB, Martin GB, Tanksley SD, Wing RA (1994) Map-based cloning in crop plants: tomato as a model system U Isolation and characterization of a set of overlapping yeast artificial chromosomes encompassing the jointless locus. Mol Gen Genet 244:613–621

Zhang HB, Budiman MA, Wing RA (2000) Genetic mapping of *jointless-2* to tomato chromosome 12 using RFLP and RAPD markers. Theor Appl Genet 100:1183–1189

Zhang LP, Khan A, Niño-Liu D, Foolad MR (2002) A molecular linkage map of tomato displaying chromosomal locations of resistance gene analogs based on a *Lycopersicon esculentum* × *Lycopersicon hirsutum* cross. Genome 45: 133–146

Zhang LP, Lin GY, Foolad MR (2003a) QTL comparison of salt tolerance during seed germination and vegetative growth in a *Lycopersicon esculentum* × *L. pimpinellifolium* RIL population. Acta Hortic 618:59–67

Zhang LP, Lin GY, Niño-Liu D, Foolad MR (2003b) Mapping QTLs conferring early blight (*Alternaria solani*) resistance in a *Lycopersicon esculentum* × *L. hirsutum* cross by selective genotyping. Mol Breed 12:3–19

Zhang X, Thacker RR, Snyder JC (2008) Occurrence of 2,3-dihydrofarnesoic acid, a spidermite repellent, in trichome secretions of *Lycopersicon esculentum* × *L. hirsutum* hybrids. Euphytica 162:1–9

Chapter 10
Momordica

T.K. Behera, K. Joseph John, L.K. Bharathi, and R. Karuppaiyan

10.1 Introduction

The genus name *Momordica* derives from the Latin word "mordeo" (means to bite) in allusion to the jagged seeds as bitten; ironically species such as *Momordica balsamina* L. do not follow this generic character. Generic and species descriptions (along with keys in some cases) are found in various monographic and floristic treatises (Willdenow 1805; Blume 1826; Seringe 1828; Wight and Walker-Arnott 1841; Thwaites 1864; Hooker 1871; Clarke 1879; Keraudren-Aymonin 1975; Jeffrey 1980). It got a prominent mention in van Rheede's Hortus Malabaricus (1678; Vol. 8), the oldest regional flora for any part of the world with descriptions and plates. Many of the provincial (regional) flora also provide a small description of various *Momordica* species. *Momordica charantia* L. (commonly known as bitter gourd, karela or balsam pear or bitter melon) is the most widely cultivated species of *Momordica*. It is grown in India, Sri Lanka, Philippines, Thailand, Malaysia, China, Japan, Australia, tropical Africa, South America, and the Caribbean. Bitter gourd is consumed regularly as part of several Asian cuisines and has been used for centuries in ancient traditional Indian, Chinese, and African pharmacopoeia. It is a common cucurbit in the wild flora of Africa, occurring almost throughout tropical Africa and occasionally collected from the wild as a vegetable or medicinal plant. Other species, apart from their importance as wild relatives of bitter gourd, have direct utility as nutritious vegetables and multipurpose medicinal plants. Species of *Momordica* have been in use in indigenous medical systems in various countries in Asia and Africa. In India, all the *Momordica* species are being grown in wild and/or cultivated forms. Their cultivation is restricted to specialized geographical pockets in different agrogeographical regions mainly by tribals and poor farming communities. The wild species offer great resources for breeding of cultivated bitter gourd for desirable edible/qualitative traits (such as non-bitterness), tolerance to abiotic stresses (e.g., tolerance to drought), and resistance to several insect pests. Besides their use in improvement of bitter gourd, they have great potential to be exploited as alternative crops.

10.2 Basic Botany of the Species

10.2.1 Taxonomy

Momordica belongs to the tribe Joliffieae, family Cucurbitaceae (Jeffrey 1980) and is native to the Paleotropics (Robinson and Decker-Walters 1997). No comprehensive monographs covering its taxonomy and nomenclature are known to exist. Botanists have described over 150 species of *Momordica*. Taxonomic confusion exists because of the widespread use of common names. Schafer (2005) considers the genus *Momordica* to comprise 47 species including eight Asian species, which are all dioecious, and 39 African species of which 20 are dioecious and 19 monoecious. According to de Wilde and Duyfjes (2002), 10 species are reported in Southeast Asia, of which six occur in Malaysia and India (Fig. 10.1), where *M. balsamina* L.,

T.K. Behera (✉)
Division of Vegetable Science, Indian Agricultural Research Institute, New Delhi 110012, India
e-mail: tusar@iari.res.in

C. Kole (ed.), *Wild Crop Relatives: Genomic and Breeding Resources, Vegetables*,
DOI 10.1007/978-3-642-20450-0_10,

Fig. 10.1 Six different species of *Momordica* of Indian occurrence. (**a**) *M. balsamina*, (**b**) *M. charantia*, (**c**) *M. cochinchinensis*, (**d**) *M. dioica*, (**e**) *M. sahyadrica* and (**f**) *M. subangulata* ssp *renigera*

M. charantia L., *M. subangulata* Blume (ssp. *renigera* (G. Don) W. J. de Wilde), and *M. cochinchinensis* (Lour.) Spreng. are common. A few species have been described recently (Thulin 1991; de Wilde and Duyfjes 2002; Jongkind 2002; Joseph John and Antony 2007), and some of the species described by earlier workers were subsequently relegated to sub-specific status or as synonyms. As evident, there is no clarity and consensus on the interspecific taxonomy of the genus, and the botanical names and common names are often used incorrectly or interchangeably (Joseph John et al. 2007).

10.2.2 Morphology

The botanical description of different *Momordica* spp. is not systematic and scanty information is available in the literature. The information related to morphological features and distribution of a few important *Momordica* spp. is described below.

2.1.1. *M. angustisepala* Harms: A large forest climber occurs in Ghana, South Nigeria and Camaroun. The plant is cultivated in Ghana for its stem, which is used for washing sponge.

2.1.2. *M. anigosantha* Hook. f.: A perennial dioecious sparingly hairy climber with pedately five foliate leaves and male flowers in pseudopanicles. Petals are creamish white to apricot orange, the lowermost largest with a horseshoe shaped marking on the claw (male flower). Fruits fleshy 4.5–6.5 × 1.6–3.8 cm, fusiform, beaked bright red ornamented with fleshy tubercles in eight longitudinal rows. It is distributed in tropical Africa.

2.1.3. *M. balsamina* L.: An annual, monoecious herbaceous climber with small lobed leaves and ovoid ellipsoid softly warted fleshy fruits. It is essentially an African species with distribution extending to western India through West Asia. Endowed with medicinal properties, it is a wild gathered vegetable (leaves and tender fruits) and has the potential as an ornamental.

2.1.4. *M. boivinii* Baill.: A monoecious perennial with a tuberous root and annual herbaceous stem. Flowers orange, fruits fleshy, elongated, fusiform cylindrical 2.3–10 × 0.4–0.9 cm longitudinally angled or ribbed. The plant is used as camel fodder, and roots are used in folk medicine. Distributed in Kenya, Tanganyika, and Uganda.

2.1.5. *M. calantha* Gilg: A monoecious perennial climber with tuberous rootstocks and herbaceous stem. Leaf broadly ovate, triangular in outline. Male flowers in umbels; corolla white or tinged pale orange with black centre. Fruits shortly stalked, fleshy, velvety reflexed strap-shaped receptacle lobes. The fruits are used as a vegetable, animal feed, folk medicine, and the plant is used for soil conservation. Distributed in Kenya, Tanganyika, and Uganda.

2.1.6. *M. cardiospermoides* Klotzsch: A monoecious glabrous perennial climber with tuberous rootstock and annual stems. Leaves two ternately compound. Male flowers solitary, flowers 5 cm in diameter, yellow with a purple eye. Fruit ovoid–oblong, acute 10 cm long, fleshy, orange–red when ripe. The leaves are used as a vegetable and in folk medicine. Distributed in Tanzania, Malawi, Zambia, Zimbabwe, Mozambique, and Botswana, down to Swaziland and South Africa.

2.1.7. *M. charantia* L.: An annual monoecious herb with lobed leaves and ovoid–ellipsoid to elongate fruits varying in size from 5 to 500 g. Fruit surface is clothed with crocodile-back-like tubercles and murication. The small-fruited form *M. charantia* var. *muricata* is wild in India and parts of Nepal and also cultivated to some extent in India for its fruits esteemed with medicinal and culinary properties. The large-fruited form M. *charantia* var. *charantia* shows extreme variability in fruit size, shape, and color and is cultivated in whole of South and Southeast Asia.

2.1.8. *M. cissoides* Planch.: A dioecious glabrous extensive climber with three foliately compound leaves and large (2.5–5 cm in diameter) floral bract. Male flowers crowded. Flowers white with black eye. Fruits ovoid, narrowed at both ends densely clothed with spreading pubescent spines. The leaves are used in folk medicine and kernels are used as a food. Distributed in upper Guinea and Sierra Leone.

2.1.9. *M. clarkeana* King: A dioecious perennial climber with simple leaves and ovoid broadly round beaked fruits, 4.5–7 × 3.5–5 cm. Pericarp hard leathery with smooth surface turning orange on ripening. Distributed in peninsular Malaysia.

2.2.10. *M. cochinchinensis* (Lour.) Spreng.: A dioecious, stout perennial with tuberous roots and palmately lobed gland dotted leaves (petiole). Flowers light cream to white with purple–black bull's-eye mark on three inner petals. Fruits very large weighing 350–600 g or more and is used as a vegetable (sweet gourd). It occurs wild in India (Andaman Islands, northeastern states), Philippines, Vietnam, Thailand and is cultivated in Vietnam, Japan, and other Asian countries for its fruits endowed with many medicinal properties.

2.1.11. *M. corymbifera* Hook.f.: A monoecious foetid perennial climber with a large woody tuberous rootstock and slender stem. Leaves palmately three-lobed, pubescent on both sides. Male flowers in 6–8 flowered corymbs, inconspicuously bracteate. Flowers yellow, petals unequal, 2 much large and very concave. Fruit 9 × 5 cm, ellipsoid, shortly rostrate with 16 longitudinal ridges. Distributed in Mozambique.

2.1.12. *M. dioica* Willd.: A dioecious tuberous perennial with ornamental lobed or unlobed small leaves. Flowers are small lemon yellow sweetly scented and open in the evening. Fruits are small, broadly ovoid–oblong clothed with soft spines. It is a wild gathered vegetable (tender fruits) and medicinal plant native of Peninsular India and has the potential for cultivation as a crop.

2.1.13. *M. denticulata* Miq.: A dioecious perennial allied to *M. cochinchinensis*, leaves entire 7–9 × 12–15 cm, not foetid. Male flowers solitary or grouped in racemes. Petals creamy white, elliptic–oblong, 3.0–5.0 cm long. Fruits ovoid ellipsoid, oblong, apex beaked, 8–14 × 5–10 cm, pericarp leathery without ornamentation or with fine sandpaper-like murication. Distributed in North and Central Sumatra, Peninsular Malaysia, and Borneo.

2.1.14. *M. denudata* Thwaites: A perennial, dioecious delicate herb with a tuberous root. Leaves simple, ovate–lanceolate, 4.5–2.5 × 3.5–8.0 cm. Male flowers on branched peduncles, bracts minute, petals yellow, broadly lanceolate, ca. 10 mm long. Fruits broadly ovoid, 3.0–3.5 × 2 cm, beaked. Endemic to Sri Lanka.

2.1.15. *M. foetida* Schumach: A dioecious, perennial climber rooting at nodes and with dark green, flecked stem and simple leaves. Fruit is a long stalked ellipsoid berry with densely soft spiny exocarp, 7 × 5 cm. It is widely distributed in tropical Africa and occasionally cultivated in its native area for leaves, and tender fruits are used as a vegetable and also in traditional medicine.

2.1.16. *M. friesiorum* (Harms) C. Jeffrey: A dioecious perennial climber with a tuberous rootstock and 3–5 foliate, pedately compound leaves. Petals pale yellow with a dark base. Fruits 6 × 2 cm, fusiform, rostrate, longitudinally ridged or winged turning orange and dehiscing in to three valves when ripe. Roots are used in the treatment of Malaria. Distributed in Kenya, Tanganyika, Uganda, Ethiopia, Tanzania, and Malawi.

2.1.17. *M. glabra* A. Zimm.: A monoecious herbaceous climber with simple to shallowly trilobed leaves (5.5–8.5 × 5.0–7.0 cm). Male flowers 5–7 in subumbels, subtended by a sessile broad bract. Corolla zygomorphic; petals 1.3–2.0 × 0.2–0.3 cm. Fruit ovoid, green with eight fleshy longitudinal undulated ridges. Distributed in Tanganyika and Zanzibar.

2.1.18. *M. kirkii* (Hook. f.) C. Jeffrey: A monoecious herbaceous climber with a perennial rootstock. Leaves membranous, orbicular, cordate, obscurely 3–5 angled, lobes entire or margins toothed. Male flowers 1–4 subumbellate subtended by a broad foliar bract. Male flowers larger than female, corolla orange 1.5–2.0 × 0.7–1.0 cm in male and 1.0–1.1 × 0.7–0.8 cm in female. Fruit fleshy, fusiform, pubescent, and beaked, 2.4–3.2 × 0.4–0.6 cm, ribbed, crowned with persistent receptacle. Distributed in Tropical Africa and Mozambique.

2.1.19. *M. leiocarpa* Gilg: A dioecious climber with dark green flecked stem and broadly ovate cordate leaves, 8.5–14 × 7.0–13.0 cm. Male flowers caducous 2–7 fasciculate subtended by a broad bract. Anthesis occurs in the evening. Petals apricot orange, rounded, slightly apiculate 2.8–3.7 × 1.9–3.0 cm. Fruits fleshy, ovoid ellipsoid with eight conspicuous remotely serrate longitudinal wings or ridges. Distributed in Kenya and Tanganyika.

2.1.20. *M. littorea* Thulin: A monoecious climber with a tuberous woody rootstock and hastate leaves and orange yellow flowers. Fruits subterate, 1.7–2.2 × 0.15–0.35 cm, dry when mature one seeded. Distributed in Ethiopia, Kenya, and Somalia.

2.1.21. *M. multiflora* Hook. f.: A monoecious medium climber with cordate simple leaves and male flowers in many flowered pseudopanicles. Fruits cylindrical 11 × 3.5 cm, longitudinally striped, shortly beaked and covered with persistent floral bracts. Essentially a tropical African species distributed in upper and lower Guinea and Angola.

2.1.22. *M. parvifolia* Cogn.: A dioeciuos medium-sized climber with broadly ovate–cordate simple or palmately lobed leaves. Male flowers in many flowered pseudopanicles, flowers white with black nerves below (petals). Fruits cylindrical, longitudinally ribbed, 11 × 3.5 cm. Distributed in Tropical Africa.

2.1.23. *M. peteri* A.Zimm.: A dioecious, robust perennial with biternately pedate 6–9 foliate leaves (5.5–7.0 × 3.2–4.8 cm) and gland dotted petioles. Male flowers subumbellate, subtended by a large bract with numerous stalked obconic reddish disc-glands on its upper surface. Flowers creamy to brownish yellow with a dark eye. Petals 2–2.1 × 0.9–1.0 cm. Fruits fleshy ovate pyriform 7.0–11.2 × 4.2–7.3 cm with eight longitudinal rows of elongate tubercles and smaller ones in the interspace. Distributed in Kenya and Tanganyika.

2.1.24. *M. pterocarpa* Hochst.: A dioecious robust perennial with a tuberous rootstock and pedately 3–5 foliate leaves. Male flowers 4–18, subumbellate, petals yellow. Fruit ellipsoid, beaked with 8 ± 1 longitudinal toothed wings or ridges, fleshy bright orange (when ripe). Leaf sap and dried root tubers are used in folk medicine. Distributed in Tropical Africa – Abyssinia and Nile.

2.1.25. *M. repens* Bremek.: A monoecious climber with a perennial rootstock and deeply cordate, multilobate leaves, 1.3–3.1 × 3.4–5.5 cm. Male flowers 3–5 in subumbels subtended by a foliar bract. Petals 1.4–3.6 cm long, yellow, dark-veined fruit 5–6 × 4–7 cm, ellipsoid to subglobose, rostrate with ten longitudinal ribs and murication on surface. Distributed in South Africa, Botswana, and Zimbabwe.

2.1.26. *M. rostrata* A. Zimm.: A dioecous tuberous perennial with a woody stem and pedately 5–12 foliate leaves. Male flowers in an umbel-like cluster. Fruit is an ovoid berry 3–7 × 1.5–3 cm, beaked rounded or slightly angled bright red with many seeds embedded in yellow pulp. The leaves and fruits are used as vegetables, and the leaves and stem serve as a fodder especially for donkeys and the leaves are also used in traditional medicine. Distributed in Tropical Africa.

2.1.27. *M. rumphii* W.J. de Wilde: A dioecious slender perennial with three foliate leaves, 8–11 cm diameter, male flowers solitary occasionally three per node. Petals elliptic–oblong, 1.3–1.5 × 6 cm fruit broadly ovoid–ellipsoid or subglobose, beaked, 4.5 × 4.0 cm, sparsely muricate on surface. Distributed in West Seram and Ambon.

2.1.28. *M. sahyadrica* Joseph John and Antony: A robust dioecious perennial, rainy season climber endemic to Western Ghats of India. Leaves are triangular-cordate and has large showy yellow flowers and fairy large softly spinicent (50 g) fruits. The fruits are esteemed as a vegetable and have potential for domestication as a high value vegetable.

2.1.29. *M. sessilifolia* Cogn.: Herbaceous monoecious climber with a tuberous woody rootstock. Leaves hastate, amplexicaul. Male flowers solitary occasionally two per node. Fruit 17–22 mm long and 1.5–3.5 mm diameter, dry when mature, subterete, slightly constricted at about one third of its length from the base.

2.1.30. *M. spinosa* (Gilg) Chiov.: A dioecious perennial climber with a woody tuberous rootstock and a pair of blunt spine at each node. Leaves subpentagonal to suborbicular in outline, 7.0–7.2 × 2.3–5.8 cm, shallowly three lobed. Male flowers in many flowered fascicles, petals yellow, rounded above, 1.0–1.1 × 0.6 cm with three inner petals having black basal markings. Fruit ovoid, beaked, yellow–orange, 7.0 × 3.5 cm, ornamented with ten longitudinal ribs. Distributed in Ethiopia, Somalia, and Kenya.

2.1.31. *M. subangulata* Blume: A dioecious rainy season climber, perennating with both taproot and adventitious tubers. It has two subspecies: *M. subangulata* subspp. *subangulata* and *M. subangulata* subspp. *renigera*. The former is more robust and is cultivated and wild in submontane Himalayas, whereas the latter is a delicate herb restricted to Malaysia and Southeast Asia. Both have large showy creamy flowers with a purple bull's-eye blotch on three inner petals. Fruits are ridged prominently in the former and remnant in the latter. Fruits and tender leaves are used as vegetables.

2.1.32. *M.welwitschii* Hook.f.: A monoecious slender climber with membranous leaves 3–5 cm in diameter. Male peduncles slender, minutely bracteate, flowers yellow, 3 cm across. Fruits ovoid, smooth or slightly warted, 2.5–5 cm long. A monoecious slender climber with membranous leaves 3–5 cm in diameter. Male peduncles slender, minutely bracteate, flowers yellow, 3 cm across. Fruits ovoid, smooth or slightly warted, 2.5–5 cm long. Distributed in Tropical Africa, Lower guinea, Angola, and Mozambique.

2.1.33. *Momordica enneaphylla* Cogn.: IUCN (2006) Red Listed as Threatened species from Africa, Gabon, Congo basin, and Cameroon.

10.2.3 Cytology

The monoecious group has $2n = 2x = 22$, whereas the dioecious group has $2n = 2x = 28$. In *M. balsamina*, $n = 11$ was recorded by Whitaker (1933) and McKay (1931), and in *M. charantia*, $2n = 22$ was recorded by McKay (1931) and Bhaduri and Bose (1947). Richharia and Ghosh (1953) reported 14 bivalents at diakinensis and I-metaphase as being distinctly observed in *M. dioica*. In a recent study (Bharathi et al. unpublished), a polyploid chromosome number ($2n = 4x = 56$) has been observed in *M. subangulata* subsp. *renigera* and $2n = 2x = 28$ was observed in *M. sahyadrica* $2n = 2x = 28$.

Yasuhiro Cho et al. (2006) reported that most of the "cultivated kakrol" (most probably *M. subangulata* ssp. *renigera* and not *M. dioica* Roxb. as assigned by the authors) plants in Bangladesh were confirmed to be tetraploid ($2n = 4x = 56$). Polyploidy is normally associated with vigorous growth with large size of vegetative and reproductive plant parts and, in some cases, sterility as well. However, true *M. dioica* has slender and fragile vines (not stout), small-sized floral parts and fruits. The results of previous work on chromosome numbers are summarized in Table 10.1.

Trivedi and Roy (1972) worked out the karyotypes of *M. charantia*, *M. balsamina*, and *M. dioica*.

Table 10.1 Chromosome numbers of some *Momordica* spp.

Species	2*n*	References
Momordica balsamina L.	22	Whitaker (1933), McKay (1931), Roy et al. (1966), Trivedi and Roy (1972)
M. charantia L.	22	McKay (1931), Bhaduri and Bose (1947), Roy et al. (1966), Trivedi and Roy (1972)
M. cohinchinensis (Lour.) Spreng.	28	Jha et al. (1989)
M. denudata C.B. Clarke	28	Beevy and Kuriachan (1996)
M. dioica Wall.	28	Richharia and Ghosh (1953)
M. subangulata Bl.	28	De Sarkar and Majumdar (1993)
M. subangulata ssp. renigera	56	Bharathi et al. (unpublished)
M. sahyadrica	28	Bharathi et al. (unpublished)
M. rostarata	22	Bosch (2004)
M. foetida	44	Mangenot and Mangenot (1957)

They reported that *M. charantia* and *M. balsamina* have almost the same number of median and submedian chromosomes, although the chromosomes of *M. balsamina* are slightly smaller. But the karyotype of *M. dioica* is more asymmetrical than the other above mentioned species. Therefore, it is assumed that *M. dioica* may be the advanced taxon among these three species. The cytological works in other Indian *Momordica* spp. is under progress at the Indian Agricultural Research Institute, New Delhi.

10.3 Conservation Initiatives

Being well distributed and their habitats being intact, the African taxa are reported not to face any imminent threat (Bosch 2004). However, there are a few reports expressing apprehension of *Momordica* spp. as being "endemic," "endangered," or "nearing extinction" in India (Dwivedi 1999; Jha and Ujawane 2002) and neighboring Bangladesh. The only reference to threatened status of *Momordica* is given in 1997 IUCN Red Data Book (Walter and Gillett 1998), where *M. subangulata* Blume from Wynad (Kerala) and South Canara (Karnataka state of India) is accorded "threatened-indeterminate" status. Infact, the species referred to as *M. subangulata* might be *M. sahydrica,* as true *M. subangulata* does not occur in South India (Joseph John and Antony 2007). In Tanzania, *M. pterocarpa* and *M. rostrata* are assessed as vulnerable species (Mitawa et al. 1996).

Field studies in India (Joseph John 2005; Joseph John and Antony 2008c) revealed a grave threat to *M. dioica* in its entire range and *M. sahyadrica* in the Western Ghats of Kerala. Overall, the wild and feral forms of *M. charantia* var. *muricata* face a medium level of threat across its geographic range. Habitat loss and fragmentation brought about by population pressure and developmental activities, poor distribution and low population density coupled with inadequate in situ conservation efforts, and acculturation of the forest dwelling communities are the major factors attributed to their heightened threat status affecting their long-term survival in the wild. In view of their excellent culinary traits and perceived tolerance to biotic and abiotic stresses, these landraces need to be collected and conserved from the whole range of their distribution across India.

Before establishing collection strategy and conservation priorities, the present status of conservation of the genus must be considered. The germplasm holdings in various institutes are presented in Table 10.2.

10.3.1 Collection and Conservation in Ex Situ Gene Banks

Having assessed the genetic erosion status and potential of wild *Momordica* gene pool, different taxa need different level of conservation approach. However, in the absence of any earlier initiative to collect these resources, it is necessary that extensive collection programs be carried out throughout the range of the taxa and the seeds multiplied and stored in ex situ gene banks. The importance of bitter gourd and other *Momordica* species is bound to increase as nutraceuticals (Kole et al. 2010a, b). Fruit fly tolerant lines of *M. charantia* var. *muricata* were abundant in the whole of Peninsular India; some of them still thrive in homesteads as landraces. Rescue collections and rehabilitation in on-farm, thus effecting seed increase leading to gene bank storage is a priority.

10.3.2 In Situ Conservation

Observations of in situ conservation in forest habitats indicate the possibility of setting up genetic reserves for various *Momordica* species in India. In the context of niche requirements, pollinator specificity, and dependence on biotic agents for seed dispersal and possible dormancy break, an ex situ conservation strategy alone may not make much headway. By establishing a few genetic reserves in selected protected areas in Western Ghats, East, North East, and Andaman Islands of India, these species can be afforded in situ protection.

Inventorization is the starting point for in situ conservation. In the absence of inventories, wild species of *Momordica* are not identified or managed as wild crop relatives in the protected areas like any other wild relative. In the context of alien weeds, passive management by according official protection to a forest pocket would not serve to achieve the goals of in situ

Table 10.2 Present status of germplasm holdings in *Momordica* species

S. No	Crop	Number of accessions	Institute	References
1	*M. charantia* var. *charantia*	519	National Gene Bank of NBPGR, New Delhi	Ram and Srivastava (1999)
		1	Institute of Agrobotany, Hungary (ABI)	Horvath (2002)
		15	NI Vavilov Research Institute of Plant Industry (NIR), Russia	Piskunova (2002)
		1	Cukurova University, Turkey	Kucuk et al. (2002)
		95	Kerala Agricultural University, Vellanikkara, India	Raj et al. (1993)
		65	Indian Institute of Horticultural Research, Bangalore, India	Raj et al. (1993)
		219	Indian Institute of Vegetable Research, Varanasi, India	Ghosh and Kalloo (2000)
		30	Vivekananda Parvathiya Krishi Anusandhan Shala, Uttar Pradesh, India	Ghosh and Kalloo (2000)
		2	Aburi Botanic Gardens, Ghana	Harriet (2002)
		281	AVRDC, Taiwan	AVGRIS (2009)
		12	Southern Regional Plant Introduction Station, Georgia, USA	Raj et al. (1993)
		1	National Seed Storage Laboratory, Fort Collins, USA	Raj et al. (1993)
		2	National Institute of Agricultural Sciences, Ibaraki, Japan	Raj et al. (1993)
		72	Institute of Plant Breeding, Laguna, Philippines	Raj et al. (1993)
		7	Division of Plant and Seed control, Pretoria, South Africa	Raj et al. (1993)
		250	Kasetsart University, Bangkok, Thailand	Raj et al. (1993)
	M. charantia var. *muricata*	11	National Genebank of NBPGR, New Delhi	Joseph John and Antony (2008a)
2	*M. cissoides*	1	Aburi Botanic Gardens, Ghana	Harriet (2002)
3	*M. foetida*	Few	New York State Agricultural Experiment Station, USA and National Gene Bank, Kenya	Bosch (2004)
4	*M. cochinchinensis*	6	AVRDC, Taiwan	AVGRIS (2009)
		3	Central Horticultural Expt. Station, Bhubaneswar, India	Bharathi et al. (2006b)
		2	Krishna Chandra Mishra Research Institute of Wild Vegetable Crops	Ghosh and Kalloo (2000)
5	*M. dioica*	60	Central Horticultural Expt Station, Bhubaneswar	Vishalnath et al. (2008)
		8	Indian Institute of Vegetable Research, Varanasi, India	Ghosh and Kalloo (2000)
		2	AVRDC, Taiwan	AVGRIS (2009)
		5	Krishna Chandra Mishra Research Institute of Wild Vegetable Crops	Ghosh and Kalloo (2000)
6	*M. subangulata* ssp. *renigera*	25	Central Horticultural Expt Station, Bhubaneswar, India	Vishalnath et al. (2008)
		< 12	AAU Research Centre, Kahikuchi	Ram et al. (2002)
		2	Krishna Chandra Mishra Research Institute of Wild Vegetable Crops	Ghosh and Kalloo (2000)
7	*M. balsamina*	1	AVRDC, Taiwan	AVGRIS (2009)
		1	NBPGR, New Delhi	Joseph John (2005)
		1	Krishna Chandra Mishra Research Institute of Wild Vegetable Crops	Ghosh and Kalloo (2000)
8	*M. sahyadrica*	10	NBPGR, New Delhi, India	Joseph John (2005)

conservation in the case of *Momordica* species. Artificial seeding and in situ protection in sacred groves especially for *M. dioica* needs consideration in the light of its endangerment especially in coastal lowlands in the Kerala state of India.

10.3.3 In Situ On-Farm Conservation

Cultivation of *M. dioica*, *M. sahyadrica*, and *M. charantia* var. *muricata*, though rare, has been spotted across Western Ghats, India (Joseph John 2005).

Several tribal families across Western Ghats were found to grow various species of wild *Momordica* in their homesteads in a simulated in situ condition. Often in the case of *M. dioica* and *M. sahyadrica*, the planting material (tuber) is collected from the forest. Non-availability of female tubers, poor seed germination, and non-availability of seedlings seem to be the important factors preventing its domestication and spread. *M. charantia* var. *muricata* being exclusively propagated by seeds, domestication attempts have progressed further and the landrace is known by a name, mostly associated with a trait.

With an increasing agronomic input, yield also increased indicating its adaptability to cultivated ecosystem. Even when only female plants are raised, fruit setting is not affected as indicated by higher yield. Being in the forest ecosystem, the natural ecological processes such as pollination and dispersal ensuring establishment of new plants are not hindered. With a little financial support, selected tribal farmers can be persuaded to continue and extend these on-farm conservation attempts. A strong ethnobotanical component will ensure that conservation goes beyond basic authoritarian protective measures. It will help in developing conservation methods that are egalitarian, in harmony with the environment and satisfy the material and cultural needs of the local people.

Hence, home garden adoption within the distribution range must be attempted. In these on-farm attempts, even though the primary aim of the farmer is economic gain, it effects population increase and thereby conservation. The farmer ensures establishment of the tuber uprooted from the forest, and better management care and non-competition leads to production of higher number of fruits and seeds, a certain percentage of which is returned back to nature, even as the mother plant survives as it was in nature.

10.3.4 Ex Situ Home Garden Conservation

An experiment on home garden conservation was taken up on an exploratory basis to assess the prospects of farmer participation in germplasm conservation (Joseph John 2005). Since the taxa being wild or at the most semi-domesticate, not much information on package of practices for the crop was available. Hence, the farmer was at liberty to experiment with his ideas at all levels of execution of the program in his farm. Being at the wild–domestication interphase, on-farm management was considered ideal, as it involves "continued evolution of the taxa in its natural surroundings." Perpetuation of soil seed bank in homesteads indicates operation of basic ecological processes involving pollinators and dispersal agents. These seedlings upon potting and transplanting lead to further spread of the taxa in homesteads, adding to domestication efforts.

In homesteads adopting a tree-based cropping system, the habitat is ideal for seedbank regeneration as in a forest floor. No tillage and low weed growth under partial shade offer ideal habitat for *Momordica* species. Farmers reported abundant seedling populations in *M. charantia* var. *muricata* and *M. sahyadrica* and good tuber production in *M. subangulata* spp. *renigera*. In majority of the cases, the planting materials reached out to their neighbors and friends in the subsequent season. *M. dioica*, *M. sahyadrica,* and selected germplasm of *M. charantia* var. *muricata*, by virtue of their sustainable yield, quality of the fruits, and ease in cultivation or rationability, offer scope for adoption and consequently conservation in homegardens.

Balsam pear, balsam apple, spine gourd, and sweet gourd are treated as ornamentals in Europe and America, where it was grown in glasshouses since Victorian times for their beautiful foliage, pendant orange ripe fruits embedded in green foliage and star-like configuration of bursting fruits (Walters and Decker-Walters 1988; Robinson and Decker-Walters 1997). Miniature-fruited *M. charantia* var. *muricata* and *M. balsamina* have beautiful foliage and orange red fruits. *M. dioica* has musky scented flowers, and *M. sahyadrica* has large showy yellow flowers in profusion; besides, both have ivy-like beautiful foliage and pendant fruits turning orange and bursting in a star-like configuration. All this offers scope for adoption by urban gardeners, thus giving another dimension to on-farm conservation.

10.3.5 Role of Women in Conservation of Genetic Diversity

It has been observed that in the primitive societies, gathering of wild vegetables are usually done by women. Often, they do this while collecting firewood, which is a regular work carried out by tribal women. On-farm conservation is carried out by them intentionally or unknowingly. As it is always the woman who cooks food, it is she who throws out mature or ripe seeds, some of which germinates and develop as new plants. Men also collect wild *Momordica* species either for home consumption or for sale. Mostly, this is carried out along with minor forest produce gathering. In addition, they are a storehouse of information related to various uses and culinary preparations involving *Momordica* species. Hence, any in situ on-farm conservation should center on tribal women in hotspots of diversity.

10.4 Role in Elucidation of Origin and Evolution of Allied Crop Plants

Phylogeny of this group has not been studied adequately. This genus is essentially a native of tropical regions of Asia, Polynesia besides tropical Africa. There is no systematic study on the origin of this group, which made Zeven and Zhukovsky (1975) to regard the genus as one of unidentified origin.

Many workers (Degner 1947; Walters and Decker-Walters 1988) consider the smaller wild variety (var. *muricata* synonymous with *M. charantia* var. *abbreviata* Ser.) as the progenitor of cultivated bitter gourd. The original place of domestication of this flagship species, *M. charantia,* is unknown (Li 1970; Marr et al. 2004). The putative areas proposed by various workers include southern China, eastern India (Walters and Decker-Walters 1988; Raj et al. 1993; Marr et al. 2004) and even southwestern India (Joseph John 2005). It is believed to have taken to rest of tropical Asia and Africa and thence to Brazil and rest of tropical South America with the slave trade (Nguyen and Widodo 1999). A recent study (Marr et al. 2004) based on morphological parameters, isozymes, and nutritional profile of domesticates and wild types from southern China, Southeast Nepal, and northern Laos failed to pinpoint the exact place, though strongly suggested a single place of domestication. Admittedly, the handful of collections from Nepal was not true representative of the wide spectrum of variability encountered in India. Further phylogenetic studies with truly representative wild types from the Northeast and Southwest India may help to resolve the enigma.

The dioecious group is characterized by the non-overlapping distribution in India and rest of South Asia. The floral morphology and pollinators were found to be specific for each species (Joseph John 2005). The morphological distinctness in the wild species is not associated with the evolvement of reproductive barriers except for contrasting anthesis time and consequent pollinator specificity. Here, the species delimitation is based on morphology and geographic isolation. All the three taxa fall under the primary gene pool.

The history of evolution of *M. dioica* and *M. subangulata* ssp. *renigera* (referred to as "wild kakrol" and "cultivated kakrol" by Mondal et al. 2006) is not adequately known. *M. dioica* is indigenous to India and possibly evolved in Central India. Trivedi and Roy (1972) have hypothesized *M. dioica* as having possibly originated from *M. charantia*. Morphological similarity and interspecific crossability suggest the origin of *M. subanulata* ssp. *renigera* from *M. dioica* and/or *M. cochinchinensis*. Character combinations suggest that it may be an amphidiploid between *M. dioica* and *M. cochinchinensis*, arising through hybridization and chromosome doubling (Mondal et al. 2006).

M. sahyadrica having wider pollinator choice has assured fertilization and is more advanced and may have evolved from *M. dioica* in the Western Ghats and may be considered as neoendemic. Fruit and seed morphology has much in common between the two. Seed production following crosses between the taxa and its suspected wild progenitor and the normal growth of the hybrids are evidences to support its ancestry (Ladizinsky 1998). The intermediate behavior of F_1 hybrids of *M. dioica* × *M. sahyadrica* for flower size and anthesis time sheds clear light on the variant forms of *M. dioica* occurring in Southeast Mumbai, for which de Wilde and Duyfjes (2002) assign a separate "taxon of uncertain status, probably of hybrid origin."

M. cochinchinensis must have originated in South Asia, probably in the Cochinchina region of Vietnam (from where originally collected and described) and/or the Andaman Islands (where substantial diversity exists). Pre- and post-zygotic reproductive barriers suggest an origin independent of *M. dioica* (Mondal et al. 2006).

The ongoing phylogenetic studies in the genus, based on nuclear, mitochondrial and chloroplast DNA markers being carried out in University of Munich, Germany (Susane Renner personal communication) and in Asian taxa being carried out in Indian Agricultural Research Institute (L.K. Bharathi personal communication) will hopefully clarify the species delimitation and possible origin of these species. A compiled list of *Momordica* species and their synonyms are listed in Table 10.3.

10.5 Role in Crop Improvement Through Traditional and Advanced Tools

The breeding system depends upon the reproduction system of the plant. Information on floral biology is the basic need before setting up a breeding program. There is very little information about the floral biology and genetic system including number of genes and chromosomes, details of meiosis and pairing, breeding system, sex determination and sex modification, and regulation of gene actions in these species except for bitter gourd (*M. charantia*) and to some extent in *M. dioica*.

10.5.1 Sex Form

M. charantia and *M. balsamina* are monoecious annuals, while the tuberous perennials, *M. dioica*, *M. subangulata* ssp. *renigera*, *M. cochinchinensis*, *M. sahyadrica*, *M. foetida*, and *M. rostrata* are dioecious. However, hermaphrodite flowers in *M. dioica* (Jha and Roy 1989), *M. charantia,* and *M. subangulata* ssp. *renigera* (T.K. Behera unpublished) has been observed. Gynoecious lines originating in India were identified by Behera et al. (2006; lines DBGy-201 and DBGy-202) and Ram et al. (2002; line Gy263B) for use in hybrid development programs. Gynoecism in bitter gourd is under the control of a single, recessive gene (*gy*-1) (Behera et al. 2009) and gynoecious inbred, DBGy-201 showed maximum genetic combining ability (GCA) effect for different yield related traits (Dey et al. 2010). Trivedi and Roy (1973) have reported the appearance of various intermediate sex forms such as andromonoecious, gynoecious, and trimonoecious in colchicine-treated plants of *M. charantia*, but remaining as diploids.

10.5.2 Cytology of Sex Differentiation

Richharia and Ghosh (1953) reported the presence of 14 bivalents in *M. dioica*, with a heteromorphic pair disjoining earlier than the other bivalents. Jha (1990) reported the sexual mechanism in *M. dioica* as an incipient type of sexual dimorphism (an intermediate stage toward X/Y chromosome basis), in which a pair of autosomes is responsible for sexual dimorphism. Sinha et al. (2001) reported the presence of a sex-linked 22 kDa polypeptide (p-22) in the female sex, which was not detected in its male counterpart. Moreover, variation in the intensity of 29 kDa and 32 kDa polypeptides of male and female sex forms suggest that the interplay of these two sex-linked polypeptides may be the contributing factor in controlling sex mechanism of dioecious *M. dioica*. Seshadri and Parthasarathi (2002) considered the differentiation of sex in *M. dioica* to be entirely genic or genetical without any cytological evidence of heterogamety.

In *M. charantia*, Wang et al. (1997) found that initially plants bear hermaphroditic bud primordia, which then yields to the hormone-regulated development of either staminate or pistillate flowers. This process is correlated with RNA and protein synthesis where soluble protein profiles of hermaphrodite flower buds, and male and female flowers differ at three early developmental stages (7, 10, and 13 days after initial bud formation) (Wang and Zeng 1998). Two predominant protein bands, 11 and 30 kDa, are present in pistillate and staminate flowers, respectively, and it is speculated that these proteins may be directly associated with sex expression (Wang and Zeng 1998).

10.5.3 Sex Modification

The principle in sex modification in cucurbits lies in altering the sequence of flowering and sex ratio. Foliar sprays with $AgNO_3$ (400 ppm) at pre-flowering stage could induce 70–90% hermaphrodite flowers in *M. dioica* vines (Rajput et al. 1993). Application of

Table 10.3 List of *Momordica* species

SN	Species	Other names
1	*M. angustisepala* Harms	*Momordica bracteata* Hutch. & Dalz.
2	*M. anigosantha* Hook. f.	*Momordica anigosantha* var. hirtella Cogn.
3	*M. angolensis* R. Fern.	
4	*M.balsamina* L.	*Momordica involucrata* E.Mey. ex Sond., *M. schinzii* Cogn.
5	*M. boivinii* Baill.	*Raphanistrocarpus boivinii* (Baill.) Cogn., *R. asperifolius* Cogn.
6	*M. cabrae* (Cogn.) C. Jeffrey	*Dimorphochlamys mannii* Hook.f., *D. cabrae* Cogn., *D. glomerata* Cogn., *D. crepiniana* Cogn.
7	*M. calantha* Gilg	*Peponia umbellata* Cogn., *Peponium umbellatum* (Cogn.) Engl., *M. umbellata* (Cogn.) A. Zimm.
8	*M. camerounensis* Keraudren	
9	*M. cardiospermoides* Klotzsch	*Momordica clematidea* Sond.
10	*M. charantia* L.	*Momordica thollonii* Cogn., *M. chinensis* Spreng., *M. elegans* Salisb., *M. indica* L., *M. operculata* Vell., *M. sinensis* Spreng., *M. zeylanica* Mill., *M. anthelmintica* Schumach., *M. senegalensis* Lam.
11	*M. cissoides* Planch.	*Momordica gracilis* Cogn.
12	*M. clarkeana* King	
13	*M. cochinchinensis* (Lour.) Spreng.	*Muricia cochinchinensis* Lour., *Momordica mixta* Roxb., *M. meloniflora* Hand.-Mazz.; *M. macrophylla* Gage
14	*M. corymbifera* Hook. f.	*Momordica henriquesii* Cogn.
15	*M. dioica* Willd.	*Momordica roxburghii* G. Don, *M. wallichii* M.J. Roem.
16	*M. dissecta* Baker	
17	*M. denticulata* Miq.	*Momordica racemifera* (Miq.) Cogn., *M. denticulata* var. racemifera Miq. *M. acuminata* Merr.
18	*M. denudata* Thwaites	*Momordica dioica* Willd var. *denudata* Thwaites
19	*M. enneaphylla* Cogn.	*Momordica diplotrimera* Harms
20	*M. foetida* Schumach.	*Momordica mannii* Hook. f., *M. cordifolia* E.Mey. ex Sond., *M. cucullata* Hook. f., *M. schimperiana* Naudin, *M. morkorra* A. Rich., *M. cordata* Cogn.
21	*M. friesiorum* (Harms) C. Jeffrey	*Momordica anigosantha* var. trifoliolata Cogn., *Calpidosicyos friesiorum* Harms
22	*M. glabra* A.Zimm.	
23	*M. gilgiana* Cogn.	*Momordica cogniauxiana* Gilg non De Wild.nom.illegit. (ill.) *M. wildemaniana* Cogn.
24	*M. humilis* (Cogn.) C.Jeffrey	*Raphanocarpus welwitschii* Hook.f. *R. humilis* Cogn. *R. welwitschii* Hook.f. *Momordica welwitchii* Hook. f., *Raphidiocystis welwitschii* Hook.f. *Raphanocarpus humilis* var. *prostratus* Suess
25	*M. jeffreyana* Keraudren	
26	*M. kirkii* (Hook. f.) C. Jeffrey	*Raphanocarpus kirkii* Hook. f.
27	*M. leiocarpa* Gilg	
28	*M. littorea* Thulin	*Raphanocarpus stefaninii* Chiovenda, *Momordica stefaninii* (Chiovenda) Cufodontis
29	*M. macrosperma* (Cogn.) Chiov.	*Momordica bricchettii* Chiov. *Kedrostis macrosperma* Cogn.
30	*M. multiflora* Hook. f.	*M. gaboni* Cogn., *M. laurentii* De Wild., *M.multicrenulata* Cogn., *M. parvifolia* Cogn., *M. affinis* DeWild., *M. gaboni* var. *lobata* Harms, *Coccinia macrocarpa*
31	*M. obtusisepala* Keraudren	
32	*M. parvifolia* Cogn.	*Momordica affinis* De Wild., *M. multicrenulata* Cogn.
33	*M. pauciflora* Cogn. Ex Harms	
34	*M. peteri* A. Zimm.	*Momordica macrocarpa* Jex-Blake

(*continued*)

Table 10.3 (continued)

SN	Species	Other names
35	*M. pterocarpa* Hochst.	*Momordica macrantha* Gilg, *M. grandibracteata* Gilg, *M. runssorica* Gilg, *M. bequaertii* De Wild., *M.rutschuruensis* De Wild.
36	*M. repens* Bremek.	*Momordica marlothii* Harms
37	*M. rostrata* A.Zimm.	*Momordica microphylla* Chiov.
38	*M. rumphii* W.J. de Wilde	*Momordica trifolia* L., *M. trifoliolata* L.
39	*M. sahyadrica* Joseph John and Antony	
40	*M. sessilifolia* Cogn.	*Momordica stephanii* (Chiov.) Cufod. var. membranosa (Chiov.) Cufod., *Raphanocarpus stefaninii* Chiov., *R. stephanii* Chiov. var. membranosus Chiov.
41	*M. silvatica* Jongkind	
42	*M. spinosa* (Gilg) Chiov.	*Kedrostis spinosa* Gilg, *K. brevispinosa* Cogn., *Momordica brevispinosa* (Cogn.) Chiov.
43	*M. subangulata* Blume	
44	*M.welwitschii* Hook. f.	

ethepon to male plants of kakrol (probably *M. subangulata* ssp. *renigera*) did not affect the plants at any level of concentration tested while application of $AgNO_3$ (400 ppm) produced highest number of bisexual flowers per vine (Ali et al. 1991). Spraying of gibberellic acid at 25–100 ppm increases female flower production in *M. charantia*.

Sex expression is affected by environmental conditions under which *M. charantia* seedlings grow (Wang et al. 1997). Short-day cultivars, when grow under short photoperiods, exhibit rapid development and comparatively high gynoecy. To encourage a high frequency of pistillate flowers, such short-day treatments should begin at seedling emergence and proceed to sixth-leaf stage (~20 days post-emergence under growing optimal conditions). While low temperature enhances short-day effects, relatively high temperatures typically delay reproductive growth, weakening short-day responses. Likewise, pistillate flower production under short-days is increased by low temperatures (e.g., 20°C) and nighttime chilling (e.g., 25°C day/15°C night) (Yonemori and Fujieda 1985). Consequently, optimal conditions for *M. charantia* seedling growth are short days and low temperatures (Wang et al. 1997).

The concentration of endogenous growth regulators and polyamines (e.g., spermine, spermidine, cadaverine, and putrescine) in shoot meristems of bitter gourd changes during plant development (Wang and Zeng 1997a). For instance, female flower number increases as IAA and zeatin concentration decreases after anthesis (Wang and Zeng 1997b). Cadaverine content is also higher in staminate and pistillate flowers when compared to vegetative tissues (e.g., leaf and stem) suggesting its possible role in sex detemination (Wang and Zeng 1997a). Likewise, it has been hypothesized that the variation in spermidine content is related to the initiation and development of pistillate flowers, while increases in endogenous putrescine concentrations is related to staminate flower initiation (Wang and Zeng 1997a).

Foliar application of growth regulators can also modify sex expression (Ghosh and Basu 1982). For example, foliar application of gibberellic acid (GA_3) treatment (25–100 mg/l) can dramatically increase gynoecy in bitter gourd, while cycocel (CCC; chlormequat/CCC @ 50–200 mg/l) promotes staminate flower development (Wang and Zeng 1996). Moreover, the appearance of the first staminate flower is delayed and pistillate flower initiation is promoted by relatively low concentrations of GA_3 (0.04–4 mg/l) (Wang and Zeng 1997b). Likewise, foliar application of CCC promotes staminate flower development at 50–200 mg/l, and gynoecy at 500 mg/l. The effects of GA and CCC are sustained for over 80 days, which allows for their use in genetic experiments, the increase of gynoecous lines, and in commercial hybrid production (Wang and Zeng 1996).

Foliar application of ethrel (ethephon), malic hydrazide (MH), GA_3, naphthalene acetic acid (NAA), kinetin, indole acetic acid (IAA), 3-hydroxymethyl oxindole (HMO), morphactin, silver nitrate, and boron when applied at 2- and 4-leaf stage of bitter gourd plants can dramatically affect sex expression (Prakash 1976). Foliar application of silver nitrate (i.e., 250 mg/l at the 5-leaf stage or 400 mg/l at the 3-leaf stage) induces bisexual flower formation, where ovaries and petals are larger than typical pistillate flowers (Iwamoto and Ishida 2005). Likewise, dramatic increases in early pistillate flower appearance can result from foliar application of MH (250 ppm) and ethrel (200 ppm), and staminate flower development can be promoted by application of GA_3 (i.e., 50–75 ppm) (Damodhar et al. 2004). Interestingly, foliar treatment of bitter gourd plants with IAA or HMO at 35 mg/l increases total flower formations, a result that may be due in part to increased ethylene evolution (Damodhar et al. 2004). Regarding such ethylene-dependent sex determination processes, foliar application of ethrel at relatively low concentrations (i.e., 25 mg/l) enhances pistillate flowering, while application of moderately high concentrations (i.e., 100 mg/l) depresses pistillate flower development (Damodhar et al. 2004). Likewise, although exogenous application of GA_3 (i.e., 20–40 mg/l) increases pistillate and staminate flower number, comparatively high concentrations of GA_3 (60 mg/l) increases only pistillate flower number (Ghosh and Basu 1982). Finally, foliar sprays containing 50 ppm NAA stimulate early and abundant pistillate flower development (Shantappa et al. 2005), and boron at 4 ppm enhances pistillate flowers production, and fruit number and weight (Verma et al. 1984).

10.5.4 Parthenocarpy

Singh (1978) has reported the induction of parthenocarpy in *M. dioica* with pollen of related taxa (*M. charantia* and *Lagenaria leucantha*) and mixture of the two pollens. The parthenocarpic fruit setting was higher with the stimulus of extraneous pollen (66% against 36%), compared to natural pollination. The lower fruit setting in natural pollination may be attributed to non-synchronization of anthesis and duration of corolla opening.

10.5.5 Germplasm Development

The importance of germplasm as a basic tool for crop improvement is well recognized. Wild relatives and progenitors of cultivated plants together with semi-domesticates represent a strategic part of germplasm collections. As the genetic base of modern varieties is narrow and variability fast eroding, introgression of genes from wild species can substantially influence the breeding progress. Thus the knowledge of the taxonomic range, morphological similarities, crossing ability and reaction to various biotic and abiotic stresses derived from a germplasm is not only important from a botanical view point, but also it accelerates and increases the potential for its utilization (Dolezalova et al. 2003). Identifying a limited set of accessions likely to be of most interest to specific users from a large collection necessitates the use of a full-fledged descriptor as a tool to provide a generalized search strategy. It helps to differentiate between accessions and to describe the variability in characters of interest. Preliminary characterization and evaluation are prerequisites for successful utilization of plant genetic resources.

There is no published descriptor for bitter gourd, teasle gourd, sweet gourd or spine gourd or any *Momordica* species by IPGRI. The lone reference to a descriptor to *Momordica* was seen in NATP Minimal Descriptor for Vegetable Crops (Srivastava et al. 2001) where bitter gourd, sweet gourd and spine gourd are treated together. These, being evolutionarily divergent groups (bitter gourd on the one hand, sweet gourd and spine gourd on the other hand), should be treated separately as they vary for more than 75% characters by virtue of their breeding behavior and growth forms (Joseph John 2005). Rasul et al. (2004) have proposed a kakrol descriptor, which can be used in characterization of kakrol (*M. dioica*) cultivars and accessions. Joseph John and Antony (2008c) have recently proposed a set of highly discriminating descriptors and descriptor states for the dioecious *Momordica* species of South Asia comprising *M. dioica*, *M. sahyadrica*, *M. subangulata* ssp. *renigera* and *M. cochinchinensis*.

10.5.6 Characterization and Evaluation of Variability

The nature and magnitude of genetic diversity in any crop determines and often limits its utilization in breeding programs. Eleven named landraces of small bitter gourds (*M. charantia* var. *muricata*) have been collected from South Western Ghats and Tamil Nadu plains and their seeds are conserved in the gene bank of National Bureau of Plant Genetic Resources (NBPGR), New Delhi (Joseph John and Antony 2008a).

Random amplified polymorphic DNA (RAPD) marker analysis was found to be a useful tool for genotype identification and estimation of genetic similarity in spine gourd (Rasul et al. 2006). More recently, molecular markers including RAPD (Dey et al. 2006), intersimple sequence repeat (ISSR; Singh et al. 2007), and amplified fragment length polymorphism (AFLP; Gaikwad et al. 2008) have been used to assess the genetic diversity of Indian bitter gourd genotypes including two promising gynoecious lines (DBGy-201 and DBGy-202). A wide range of genetic diversity was detected, indicating that a standard accession reference array for future analyses might include "Pusa Do Mausami-green," "Pusa Do Mausami-white," DBTG-2, Mohanpur Sel-215, and Jaynagar Sel-1. Regardless of the type of marker analysis, however, Mohanpur Sel-125, DBTG-101, and Jaynagar Sel-1 from West Bengal (eastern Indian province) are genetically distinct from other common landrace accessions from North Indian provinces (genetic similarity, GS = 0.57–0.72). Genetic differences between *M. charantia* var. *charantia* and *M. charantia* var. *muricata* accessions are indicative of their use as potential parents for the establishment of narrow- and wide-based mapping populations (Behera et al. 2008). Kole at al. (2010a, b) conducted comprehensive studies on 22 accessions of *M. charantia* var. *charantia* and *M. charantia* var. *muricata* of seven countries. Using 255 AFLP markers they observed wide genetic divergence among the accessions and also detected many AFLPs linked to three fruit quality traits employing association mapping. They also detected wide genetic variability among the accessions with regard to the content of anticancer and antidiabetes phytomedicines in the germplasm. Such exotic populations have been informative for the characterization of qualitative and quantitative traits in other cucurbit species (Serquen et al. 1997; Zalapa et al. 2007).

The entities collected represented the wide range of variability from almost near to wild types, semi-domesticated to cultivated types. In the case of cultivated material, selection criteria might have included fruit size, bitterness, ascribed medicinal quality and other specific traits based on established knowledge and experience of local farmers and/or gatherers.

There is very little information on genetic variation, and no cultivar selection has been reported in *M. cochinchinensis*. Wild and semi-domesticates are maintained on-farm by tribal settlers in Andaman Islands. Two distinct types are found in the wild: the NE specimens having unlobed cordate leaves earlier designated as *M. macrophylla* Gage and the Andaman specimens having trilobed leaves. At the Central Horticultural Experiment Station (CHES, IIHR), Bhubaneswar, India three genotypes from Andaman and Nicobar islands are being maintained in field gene bank (Bharathi et al. 2006b). Individual fruit of these genotypes ranged from 250 to 530 g and days to flowering varied from 140 to 160 days after planting. The β-carotene content and lycopene (from the epicarp of the fruit) was 11.47–16.40 mg/100 g fresh weight and 5.12–8.68 mg/100 g fresh weight, respectively.

Two distinct fruit types are met with in *M. subangulata* ssp. *renigera*: light green oblong (6–10 cm long), weighing up to 80 g ("lambawala") and dark green medium sized (4–6 cm long) round fruits, weighing up to 40 g ("golwala") observed under cultivation in Assam and Andaman Islands. Assam materials grow luxuriantly and are of shy flowering in nature whereas Andaman collections are prolific and come to flower from the 15th node. Reduced vine length, high branching, increased female flower production, early flower initiation and flowering at every node are some of the desirable traits for higher yield/unit area. Like *M. dioica,* it has the unique advantage of tolerance to high rainfall monsoon climate (prevailing in most parts of South Asia), when a very few vegetables grow and produce fruits. At the Central Horticultural Experiment Station (CHES, IIHR), Bhubaneswar, India a high yielding selection from the naturally available variability each in spine gourd (CHSG 28) and teasle gourd (CHTG 2) were identified for commercial cultivation (Vishalnath et al. 2008) and Rasul and Okubo (2002) have evaluated a few teasle gourd genotypes in Bangladesh.

Selection for highly heritable characters such as number of fruits per plant, individual fruit weight, fruit volume are more important for yield improvement in spine gourd (Bharathi et al. 2006a). Rasul et al. (2004) have characterized 29 accessions of *M. dioica* along with one accession of *M. cochinchinensis* collected from different agro-ecological zones of Bangladesh. Considerable morpho-physiological variation was observed among the genotypes (with dissimilarity value ranging from 4.6 to 58.6), though no relationship could be established between genetic divergence and geographical distribution.

10.5.7 Polyploidy Breeding

Polyploids can be produced by treating the seedlings at the cotyledon stage with an emulsion of 0.2% colchicine. Yasuhiro Cho et al. (2006) reported that seed treatment with 0.2, 0.4% colchicine or 0.003% amiprophos-methyl was effective for chromosome doubling, among which the treatment with 0.4% colchicine was most effective. Amiprophos-methyl treatment also produced octoploid plants with high rate of seed germination. Multiple shoot treatments with 0.05% colchicine for 12 and 24 h, and 0.1% colchicine for 24 h also produced octoploid plants. Leaf and guard cell size were bigger, and leaf shape index (leaf length/leaf width) was lower in the octoploid than in tetraploid plants. Leaves of the octoploid plants were uneven on the surface with clear serrations.

Triploid plants of *M. charantia* were obtained by crossing the tetraploid (colchicines induced) and diploid plants (Saito 1957). In seedlings of bitter gourd, colchicine at 0.2% for 18 h to the shoot tip produced tetraploids (Kadir and Zahoor 1965). However, polyploids were inferior to diploids with regard to economic characters.

10.5.8 Heterosis Breeding

As all the species of *Momordica* are cross-pollinated, there is ample scope for exploitation of heterosis. More pronounced hybrid vigor could be observed with the inclusion of diverse parents. Heterosis for earliness, higher number of fruits, and bearing at each flowering node should be exploited. Selection for divergent parent based on number of fruits, fruit weight, fruit length, internodal length, pedicel length and yield will be useful as these characters were the major traits contributing to divergence in *M. dioica* (Bharathi et al. 2005).

Heterosis in *M. charantia* was investigated at the Indian Agricultural Research Institute, New Delhi as early as 1943 (Pal and Singh 1946). Evidence of heterotic effects is supported by genetic analyses that have defined the presence of dominance and complementary gene action for yield as distinguished by its components (Mishra et al. 1998). Heterosis for yield per vine ranges from 27 to 86% depending on genotype (Behera 2004). This heterosis is likely attributable to earliness, first node to bear fruit (first female flowering node), and total increased fruit number (Celine and Sirohi 1998). A few hybrids developed by both private and public sectors are cultivated in Asia including China and India.

10.5.9 Mutation Breeding

In *M. charantia*, progeny (M_1) derived from radiation mutagenesis can possess economically important unique characters, which are controlled by single recessive genes (Raj et al. 1993). One such bitter gourd variety, MDU 1, developed as a result of gamma radiation (seed treatment) of the landrace MC 103 was found to possess improvement for yield (Rajasekharan and Shanmugavelu 1984). Likewise, the white bitter gourd mutant "Pusa Do Mausami" (white-fruited type) was developed through spontaneous mutation from the natural population "Pusa Do Mausmi" (green-fruited type) at the Indian Agriculture Research Institute, India.

10.5.10 Tissue Culture

Nabi et al. (2002) obtained best performance for micropropagation with cotyledon as explant of spine gourd. They also induced the highest number of multiple and vigorous shoots on MS medium fortified with 1.0 mg/l BAP and 0.1 mg/l NAA. Hoque and Rahman (2003) obtained adventitious shoot bud differentiation from the petiolar cut ends of leaf explants cultured on MS medium supplemented with 2.0 mg/l BA and 0.2 mg/l NAA. Callus cultures were also established from leaf, cotyledon and root explants (Bhosle and Paratkar 2005). The excised embryos in case of spine gourd showed a high potential for de novo multiple shoot regeneration and rooting (Meemaduma and Ramanayake 2002). Among different explants and treatments tried auxillary buds in case of explant, MSHP + Ads + BAP + 10 mg/l + IBA at 5 mg/l + gelrite for crop establishment and MSHP + Ads + IBA at 1 mg/l + agar for induction and development of roots survival was observed when plantlets were transferred in sterilized water for 6 h, transferred to soilrite trays, covered with polyethylene and kept open in a polyhouse (Deokar et al. 2003). When MS medium was supplemented with 1.5 mg/l BA and 0.1 NAA, 88% of hypocotyls explant regenerated shoots with an average number of 8.8 shoots per explant (Hoque et al. 1995). Media supplemented with BA at 10 mg/l and IBA at 5 mg/l produced the highest shoot number and optimum establishment and maximum rooting was obtained in media supplemented with 3 mg NAA/l and 0.2% activated charcoal (Pawar et al. 2004).

10.5.11 Interspecific Hybridization in Momordica Species

The identification and incorporation of resistance to economically important pests such as fruit fly and various foliar pathogens is important to bitter gourd production. Genes for resistance to several crop-limiting pests and pathogens are, however, not found in the cultivated *M. charantia* var. *charantia*. Fruit fly is the most serious pest of *M. charantia*. Some of the wild germplasm (var. *muricata*) shows high level of tolerance to fruit fly. It crosses with the cultivated bitter gourd and gene exchange occurs freely within the complex. But *M. charantia* var. *muricata* has dominant trait in F_1 for fruit shape and size, which is not desirable as the size is reduced. *M. charantia* wild and cultivated plants cross readily and there are many intermediate types. Singh (1990) reported 1% fruit set in the cross *M. charantia* × *M. balsamina* and the progeny had a high bivalent frequency with normal meiotic cycle, indicating the close relationship between them and they probably have developed pre-fertilization barrier ensuring their reproductive isolation. Wild African types can be exploited as potential source of disease resistance. *M. foetida* and *M. rostrata* may play a major role in breeding programs of *M. charantia* especially in resistance breeding (Njoroge and van Luijk 2004). Regarding leaf damage due to Aulocophora, *M. balsamina* is highly tolerant. *M. balsamina* may contribute aulocophora tolerance in *Momordica* species.

There have been few attempts at interspecific crosses between *M. charantia* and *M. dioica*, the bitterless, small fruited, tuberous perennial to seek possibilities of transferring desirable attributes of the latter (especially the "bitterless" trait) to the former. Joseph John (2005) reported crossability failure of the dioecious taxa with the two monoecious members (*M. balsamina* and *M. charantia*). Dutt and Pandey (1983) reported that the failure of embryo formation in *M. charantia* × *M. dioica* is due to the abnormal behavior of a number of pollen tubes and heavy deposition of callose at their tips, which obstructed transfer of male gamete and fertilization. However, Vahab and Peter (1993) have reported over 90% success in interspecific crosses between *M. charantia* (cultivated bitter gourd) and *M. dioica* (the dioecious perennial species with bitterless fruits) using pollen stored at 10°C. They attribute the failure in earlier studies to non-synchronization of anthesis in the two species. The study indicated the possibility of utilizing the bitterless, perennial tuberous *M. dioica* in transferring the desirable attributes to the commercially cultivated large fruited bitter gourd.

Ali et al. (1991) have highlighted the scope for transfer of useful traits from the related species to *M. dioica* through interspecific hybridization. There are a few reports of interspecific hybridization between *M. dioica* and *M. cochinchinensis* (Mohanty et al. 1994; Mondal et al. 2006). Joseph John (2005) has reported that both direct and reciprocal crosses were successful within the dioecious group. *M. dioica, M. sahyadrica* and *M. subangulata* ssp. *renigera* are reciprocally interfertile. The highest fruit setting was observed in *M. subangulata* ssp. *renigera* × *M. sahyadrica* crosses. The comparatively low fruit set percentage in crosses involving *M. dioica* as male and female parent may be attributed to the reduction in stigma receptivity and pollen viability of 12–13 h old flowers.

The F_1 plants were intermediate between parents for most of the qualitative traits. Morphologically, the F_1 plants were intermediate between parents. However, expression of maternal character was more dominant especially in the tuber morphology and floral nectar guides. Intermediate behavior in anthesis time was very striking in crosses involving *M. dioica*. When *M. dioica* was used as pollen parent in *M. subangulata* ssp. *renigera*, clutch size was reduced with proportionate reduction in fruit size. When

M. sahyadrica was used as pollen parent in *M. dioica*, days to fruit ripening was extended considerably, beyond that of both the parents. However, the poor germination of F_1 seeds, unsatisfactory growth and flowering in F_1 seedlings, and complete sterility in backcrossing indicate the rather limited potential of hybrids in conventional crop improvement.

10.5.12 Genetic Mapping and Molecular Breeding

Till today, no genetic linkage map has been reported for bitter melon. Kole et al. (2010a, b) have developed a large F_2 population derived from a cross between a horticultural variety, Taiwan White that belongs to the botanical variety *charantia* and produces fruits with higher size and quality but with low content of anti-diabetic and anticancer phytomedicines and a medicinal variety, CBM12 that belongs to the botanical variety *muricata* and bears fruits of inferior size and quality but with high content of phytomedicies (Kole et al. 2009a, b). This population is segregating for fruit size, color, luster, surface and shape. The first genetic linkage map for bitter melon using the above interbotanical variety cross is expected to be developed soon (C. Kole personal communication).

10.6 Quality Attributes

The use of cucurbits, in general, as food plants is not primarily for calorie, mineral or vitamin values, since they are poor or at best only modest sources of these nutrients. However, there are few exceptions including bitter gourds, rich in vitamin C, spine gourd (*M. dioica*), having high protein content, and sweet gourd (*M. cochinchinensis*) containing high carotenoid and lycopene pigments. Local uses of *Momordica* gene pool by the indigenous people in India and rest of SE Asia, and tropical Africa is very extensive.

Despite the use of these plants in such purposes, they have not been given due research attention in terms of their nutritional content, bioavailability and feasibility of their usage as food supplements for major nutritional and antinutritional factors. An attempt is made to bridge the information gap by collating available information on the proximate, minerals and amino acid compositions of these wild vegetables, with hope that it would rekindle the interest in these nature's bounties. Based on the available literature the nutrient composition of some *Momordica* species is given in Table 10.4.

The nutritional value of small bitter gourds is higher or at par for most of the components except phosphorus than that of large bitter gourds. The defatted meals contained 52–61% protein and would be a good source of methionine, compared to the traditional protein-rich legumes. *Momordica* is noted for acids with conjugated double bonds (pumicic acid, alpha-oleo stearic acid). High levels of antioxidant activity (96%) were noticed in *M. balsamina* (Odhav et al. 2007). Teasle gourd is rich in carotene, protein, carbohydrate (Rashid 1993) and vitamin C (154.7 mg/100 g of edible portion) (Bhuiya et al. 1977).

Aberoumand and Deokule (2009) have studied the nutritional values of eight traditional wild edible plants native to Indo-Persian region. In comparison to the other species, *M. dioica* have medium protein content (19.38 mg/g), fat (4.7 mg/g) and phenolic compounds (3.69 mg/g) and have maximum calorie value (4,125.83 kcal/kg). *M. sahyadrica* being recently established taxonomic entity; reliable estimates of its nutritional contents are not available. However, being closely related to *M. dioica*, it can be reasonably assumed to be similar to it.

Sweet gourd (*M. cochinchinensis*) has β-carotene and lycopene at very high levels, with those of lycopene being up to 308 μg/g in the seed membrane, about 10-fold higher than in other lycopene-rich fruits and vegetables (Vuong 2001; Vuong and King 2003; Aoki et al. 2002). Aril tissues contained 2,227 μg total lycopene and 825 μg total carotenoids (718 μg total β-carotene; and 107 μg α-carotene/g FW). Sweet gourd aril contained 22% fatty acids by weight, which are essential for the absorption and transport of β-carotene (Vuong and King 2003; Ishida et al. 2004). Oil extracted from the fruit aril showed a total carotenoid concentration of 5,700 μg/ml, with 2,710 μg of that being β-carotene. This oil also included high levels of vitamin E (Vuong and King 2003). Thus, sweet gourd provides an acceptable source of high levels of valuable antioxidants that have good bioavailability.

Table 10.4 Nutritive value of *Momordica* species

		Moisture (g)	Protein (g)	Fat (g)	Minerals (g)	Fiber (g)	Carbohydrates (g)	Energy (K. cal.)	Calcium (mg)	Magnesium (mg)	Phosphorus (mg)	Iron (mg)	Zinc (mg)	Carotene (mg)	Thiamine (mg)	Riboflavin (mg)	Niacin (mg)	Vitamin C (mg)
1	*M. charantia* var. *muricata*[a]	83.2	2.1	1.0	1.4	1.7	10.6	60.0	23.0	–	38.0	2.0	–	0.13	0.07	0.6	0.4	96
2	*M. charantia* var. *charantia*[a]	92.4	1.6	0.2	0.8	0.8	4.2	25.0	20.0	–	70.0	1.8	–	0.13	0.07	0.9	0.5	88.0
3	*M. balsamina*[b] (Leaves)	89.4	3.0	0.1	–	0.9	3.6	–	340	87.1	27.7	12.7	0.9	–	0.01	0.09	0.7	0.4
4	*M. balsamina*[b]	89.4	2.0	0.1	–	1.8	5.1	–	35.9	41.2	35.8	2.6	1.0	–	0.04	0.06	–	0.5
5	*M. balsamina*[c] (Leaves)	85.0	5.0	0.5	–	2.75	6.82	–	2688	613	356	23	12	–	–	–	–	–
8	*M. dioica* (fruits)[a]	84.1	3.1	1.0	1.1	3.0	7.7	52.0	33.0	–	42.0	4.6	–	1620	0.05	0.1	0.6	–
9	*M. subangulata* ssp. renigera[a]	90.4	0.6	0.1	0.9	1.6	6.4	29.0	27.0	–	38.0	–	–	–	–	–	–	–
10	*M. cochinchinensis*[a] (fruits)	88.6	1.5	0.1	1.1	1.1	7.6	37.0	64.0	–	89.0	–	–		–	–	–	–
9	*M. cochinchinensis*[d] (fruits)	93	0.94	–	–	1.03	–	–	23	–	–	0.34	–	91	–	–	–	0.04
10	*M. foetida*[e]	–	3.3	–	–	3.2	–	22	1.1	–	–	3.4	0.4	5.4	–	–	–	20.6

[a]Gopalan et al. (1982)
[b]Arnold et al. (1985)
[c]Odhav et al. (2007)
[d]AVRDC (2002)
[e]Nesamvuni et al. (2001)

10.7 Genomics Resources Developed

Online search of public domain databases like that of National Centre for Biotechnology Information (NCBI) have revealed that genomic resources in *Momordica* is rather limited. *M. charantia* is by far the most studied species with nucleotide sequence (63), expressed sequence tag (EST) records (7), protein sequence (127) and three-dimensional macromolecular structures (21). In *M. cochinchinensis,* five nucleotide sequences, ten protein sequences, and five three-dimensional macromolecular structures are available. In *M. balsamina,* single record of nucleotide sequence and two protein sequences are only known.

10.8 Scope for Domestication and Commercialization

10.8.1 Traditional Medicinal Uses

Usefulness of various *Momordica* species in SE Asia as anthelmintic, vermifuge, cathartic, hypoglycemic, aphrodisiac, antipyretic, antiulcerogenic and hepatoprotective and in the treatment of burns, bilious disorders, diabetes, cataract, hypertension, leprosy, jaundice, snake bite, haemorrhoids and piles have been mentioned by Van Rheede (1678), Kirtikar and Basu (1933), Walters and Decker-Walters (1988), Yang and Walters (1992), Fernandopulle and Ratnasooriya (1996), Decker-Walters (1999), Dwivedi (1999), Jeffrey (2001), and Deshmukh and Rothe (2003). Traditional knowledge related to use of these species by indigenous tribes is not yet fully documented in published literature. Some of these ethnobotanical claims are validated by phytochemical/animal studies in recent years.

10.8.1.1 *M. charantia*

Whole plant, leaves and especially fruits are used in the folk medicine to treat diabetes in Asia (Perry 1980; Khajuria and Thomas 1993; Platel and Srinivasan 1995; Fernandopulle and Ratnasooriya 1996; Decker-Walters 1999), West Africa (Burkill 1935), and even in the New World (Coe and Anderson 1996; Marr et al. 2004). Lira and Caballero (2002) have reported the use of the feral wild type as an aphrodisiac in Mexico.

10.8.1.2 *Momordica balsamina*

Common and widespread uses are as anthelminthic, against fever and extreme uterine bleeding, to treat syphilis, rheumatism, hepatitis and skin disorders, stomach and intestinal complaints (Hutchings et al. 1996). Other uses include abortifacient, lactogenic including veterinary (Geidam et al. 2004) and hypoglycaemic (Hutchings et al. 1996).

10.8.1.3 *M. dioica*

It was reported to possess hypoglycemic, hepatoprotective, gastroprotective and ulcer healing activities, analgesic, expectorant, post coital antifertility, nematocidal, antiallergic, antimalarial, antifeedant, antibacterial and antifungal activity (Fernandopulle and Ratnasooriya 1996).

10.8.1.4 *M. shyadrica*

Tender fruits are consumed as health food for asthmatic and intestinal ulcer patients. The medicinal uses for mastitis, hydrocele, breast swelling and pain in the early days after child birth and painful eruptions underlie its anti-inflammatory properties, which however needs to be scientifically evaluated. Use of tuber paste as detergent and toilet soap hold promise in the cosmetic and health care industry.

10.8.1.5 *M. cochinchinensis*

The fruits are esteemed as the fruit from Heaven for its ability to promote longevity, vitality and health. It is traditionally used for wound healing, to improve eye health and to promote normal growth in children. The seeds are known in traditional Chinese medicine as "Mubiezi," reported to have resolvent and cooling properties, are used for treating liver and spleen disorders, chest complaints, abdominal pains and dysentery, wounds, hemorrhoids, bruises, swelling, and pus (Voung 2001).

10.8.1.6 *M. foetida*

The juice of crushed leaves is used to relieve cough, stomach-ache, intestinal disorders, headache, earache, toothache and as an antidote for snakebites. Leaves are also used to ameliorate the effects/scars caused by smallpox, boils, spitting cobra poison and malaria. The other uses are as emmenagog, ecbolic, aphrodisiac and abortifacient. The roots, said to be poisonous, and the crushed seeds are used in East Africa to cure constipation (Njoroge and Newton 2002). The fruit pulp is said to be poisonous to weevils, moths and ants, and is used as an insect repellent in Tanzania (Watt and Breyer-Brandwijk 1962).

10.8.2 Scientific Documentation/ Validation of Health Benefits

Even today almost 80% of the human population in developing countries is dependent on plant resources for healthcare (Farnsworth et al. 1985). Indigenous healthcare practices provide low cost alternatives in situation, where modern healthcare services are not available, too expensive and/or ineffective. Documentation of traditional knowledge especially on the medicinal uses of plants has lead to the discovery of many important drugs of the modern day (Cox and Ballick 1994; Fabricant and Farnsworth 2001). Natural products are the basis of many standard drugs used in modern medicine and at least 50 plant derived drugs are developed from ethnobotanical leads (Cox and Ballick 1994). Besides, plants seem to have served as models for drug development also (Fabricant and Farnsworth 2001).

10.8.2.1 Antidiabetic Properties

Over a hundred research papers have been published on pharmacological especially hypoglycaemic properties, and phytochemical characterization. A number of patents have been submitted on actives and processes of *Momordica* spp. for insulin-type properties. In vitro and in vivo studies in animals have shown that karela acts to inhibit glucose absorption, acts as an insulin secretogog, and exerts insulinomimetic effects, though not all of these effects have been supported by in vivo studies. Antifertility side effects and hepatotoxicity have been reported in animals administered with karela, with pregnant animals being especially susceptible. However, no such adverse effects in humans seem to have reported, despite its widespread use medicinally and as a vegetable.

In small-scale clinical trials with limited patients, karela juice/powder have been shown to significantly improve glucose tolerance without increasing blood insulin levels and to improve fasting blood glucose levels (Aslam and Stockley 1979; Leatherdale et al. 1981; Akhtar 1982). Oral administration of powdered fruit (Srivastava et al. 1993) and seeds (Grover and Gupta 1990) have lead to a fall in blood and urine sugar levels and post prandial blood glucose levels also fell.

Studies in *M. balsamina* have confirmed the acclaimed hypoglycaemic property of the stem–bark extract of the plant that could be used in the management of hypoglycaemic conditions (Geidam et al. 2007). Olaniyi (1975) has isolated the hypoglycaemic principle foetidin, consisting of equal parts of beta glucoside and 5,25-stigmatastadien-3 beta-ol glucoside (similar to charantin from *M. charantia*) from whole plants and unripe fruits of *M. foetida*. Foetidin was shown to lower blood glucose levels in normal rats, but it had no significant effect in diabetic animals except at 18 h samples (Marquis et al. 1977; Raman and Lau 1996; Bosch 2004).

10.8.2.2 Anticancer Properties

Antitumour activity of crude extracts of *M. charantia* and *M. cochinchinensis* have been demonstrated in vivo (West et al. 1971; Jilka et al. 1983).

10.8.2.3 Antiviral Activity

The leaf extracts of *M. charantia* has been demonstrated to inhibit the growth of Herpes simplex virus 1 (Foa-Tomasi et al. 1982) and human immuno deficiency virus 1 (Lifson et al. 1988; Lee et al. 1990; Zhang 1992b). The proteins α- and β-momorcharin and MAP are believed to be the active antiviral components of *Momordica charantia* (Zhang 1992b). *M. balsamina* fruit pulp was reported to be commonly used for its antiviral efficacy in poultry and even claimed to be having healing effects for human AIDS in northern Nigeria. An anti-HIV property of the fruit pulp extract (aqueous) was studied for in vitro (in humans) by Bot et al. (2007). The results showed that the plant extract treatment significantly ($p < 0.05$) increased CD4+ lymphocytes count when compared to the untreated peripheral blood mononuclear cells (PBMC), indicating its ameliorative role.

10.8.2.4 Analgesic and Anti-inflammatory Effects

A methanolic extract of the seeds from unripe fruit of *M. charantia* has been shown to produce a marked dose-dependent analgesic effect in mice (Biswas et al. 1991). Karumi et al. (2003) has established the analgesic and anti-inflammatory effect of the aqueous leaf extract of *M. balsamina* in mice.

10.8.2.5 Antifertility Effects

Several components with abortifacient properties (α-,β-momorcharins) have recently been isolated from *M. charantia*. Aqueous and ethanol extracts from the roots were found to be most effective in causing significant abortifacient activity besides showing moderate estrogenic activity in female rats (Shreedhar et al. 2001). A protein isolated from fresh tubers of *M. cochinchinensis* was found capable of inducing mid-term abortion in mice.

10.8.2.6 Hepatotoxicity

Ng et al. (1994) have found that α- and β-momorcharins can induce cytoplasmic blebs and other morphological changes in rat hepatocyte in vitro. Secretion of various enzyme markers of cell damage is also raised. However, in a recent study (Semiz and Sen 2007), *M. charantia* fruit extract exhibited hepatoprotective effects in $CCl4$-intoxicated rats. The results suggest that the *M. charantia* fruit extract possess the protective activities, antioxidant effects besides having hypoglycemic and antidiabetic effects in rats. *M. dioica* leaves (Jain et al. 2008) and fruits (Thirupathi et al. 2006) have potent hepatoprotective action and ethanolic extract was found more potent.

10.8.2.7 Toxicity in Humans

Although toxicity has been observed in some animal studies (Sharma et al. 1960; Zhang 1992a), there are no published reports of fatal or serious effects in humans at normal doses (50 ml, given orally). Karumi et al. (2006) and Geidam et al. (2007) have conducted toxicity studies in *M. balsamina*. Graded doses of aqueous leaf extract were administered orally and intraperitonially to separate groups of rats to determine the acute toxicities and it was concluded that *M. balsamina* at low dosages is safe.

10.8.3 Phytochemistry and Active Ingredients Isolated

M. charantia fruit contains steroids, charantin, momordicosides (G, F1, F2, I, K, L), acyl glucosyl sterols, linolenoyl glucopyranosyl elenosterol, amino acids, fatty acids, and phenolic compounds. The seeds contain galactose-binding lectins, vicine, amino acids, fatty acids, terpenoids, and momordicosides (A, B, C, D and E). The phytochemicals isolated from the whole plant, vines or leaves include saponins, sterols, steroidal glycosides, alkaloids, amino acids and proteins (Raman and Lau 1996).

Phytochemical screening of *M. balsamina* leaves revealed the presence of tannins, saponins and lectins (Akinniyi et al. 1983). The alkaloids include momordicin (Watt and Breyer-Brandwijk 1962), momordocins (Karumi et al. 2004) and cucurbitacins, which impart bitter taste along with saponins. Phytochemicals of pharmaceutical importance (anti-inflamatory, antiviral, antibacterial and antioxidant activities) like momordin II (ribosome inactivating protein) and rosmaric acid (caffeic acid ester) have been isolated from *M. balsamina* (Bosch 2004).

Phytochemical investigations in *M. dioica* have revealed the presence of traces of alkaloids and ascorbic acid in fruits. Lectins, β-sitosterol, saponin glycosides, triterpenes of ursolic acid, hederagenin, oleanolic acid, α-spiranosterol, stearic acid, gypsogenin, momodicaursenol and some aliphatic constituents were isolated from different parts of this plant (Ghosh et al. 1981; Sadyojatha and Vaidya 1996; Ali and Srivastava 1998; Luo et al. 1998).

Trypsin inhibitors are known to produce a cancer-preventive effect in humans and to confer in crops resistance against insects. Hernandez et al. (2000) have separated three trypsin inhibitors from *M. cochinchinensis* seeds. Five trypsin isoinhibitors (TI-1 to TI-5 of the squash-type) differing in molecular mass, specific trypsin-inhibitory activity, and N-terminal amino acid sequence were isolated from *M. cochinchinensis* seeds (Wong et al. 2004). Tien et al. (2005) have investigated in vivo and in vitro the antitumor activity of the crude water extract from Gac fruit (*M. cochinchinensis*). The antitumor component was confirmed as a protein with molecular weight of 35 kDa, retained in the water-soluble high molecular weight fraction as distinct from lycopene, another compound with potential antitumor activity.

Cochinin B, a novel ribosome-inactivating protein (RIP) with a molecular weight of 28 kDa, was purified from the seeds of *M. cochinchinensis* (Chuethong et al. 2007). The purified Cochinin B displayed a strong inhibitory activity on protein synthesis and strong antitumor activities. RIPs have been linked to defense by antiviral, antifungal and insecticidal properties demonstrated in vitro and in transgenic plants (Nielsen and Boston 2001).

Mulholland et al. (1997) have isolated several cucurbitane triterpenoids from the leaves of *M. foetida* including a few novel compounds. Momordicines and foetidin (identical to charantin) were reported from fruits and leaves of *M. foetida*. Momordicines have been found to be both bacteriostatic and insecticidal; foetidin has slight antispasmodic and anticholinergic effects.

10.8.4 Use as Dietary Supplements

Inequitable food availability and inadequate food intake among population at the lower socio-economic strata and especially among children, pregnant and lactating mothers is a grave problem in the developing countries (Andersen et al. 2003; Seena et al. 2005). The major food crops being roots and tubers besides cereals, the diets in these parts are predominantly starchy. Thus, nutrient deficiency especially of protein, macro- and micro-elements are prevalent. A long-term sustainable solution to alleviate the problem is to enlarge the food basket by exploiting under-exploited and lesser-known wild plants as sources of nutrient supplements. In this direction, many researchers have reported the nutritional composition of various types of edible wild plants including those of *Momordica* genus in use in the developing worlds.

10.8.5 Protein and Mineral Supplements

Communities in Africa have a long history of using traditional indigenous leafy vegetables to supplement their diets (Odhav et al. 2007; Rensburg et al. 2007). Introduction and extensive cultivation of more remunerative and aggressive crops, first from Asia and subsequently from Europe and South America during colonization, have marginalized several of these traditional native species. Presently, urbanization and the influence of urban lifestyle on the rural African population have resulted in replacing the traditional vegetables with introduced modern vegetables. Besides, the indigenous traditional knowledge related to the cultivation and uses of these traditional vegetables are also at risk of getting lost at household and community level. Considering their potential nutritional value, these traditional vegetables could contribute in a major way to the food security and balanced diets of rural households and possibly urban households as well. However, further research on other aspects like nutritional profile and bioavailability, genetic improvement and better cultivation practices are warranted.

The leaves of *M. balsamina* are a popular vegetable, consumed regularly in the eastern parts of South Africa (Fox and Norwood Young 1982; Van Wyk and Gericke 2000; Hart and Vorster 2006; Rensburg et al. 2007; Odhav et al. 2007). The study by Hassan and Umar (2006) has detected 17 amino acids with glutamic acid, leucine and aspartic acid being the predominant amino acids. Isoleucine, leucine, valine and aromatic acids were found to be higher than WHO/FAO/UNU (1985) requirement pattern for children, while sulphur containing amino acids are the only limiting amino acids for adults. Comparing the leaves' mineral contents with RDA values, the results indicated that the *M. balsamina* leaves could be good supplement for some mineral elements particularly K, Ca, Mg, Fe, Cu and Mn.

Compared with cabbage, lettuce and spinach, these wild Momordica vegetables contain more protein and fat, while the fiber content is less. Among the minerals analyzed, the leaves of *M. balsamina* had higher values than those reported for the exotic vegetables, except for sodium. The wild vegetables like *M. balsamina* leaves could be promoted as a protein supplement for cereal-based diets in poor rural communities, while its high potassium content could be utilized for the management of hypertension and other cardiovascular conditions. The relatively high concentrations of zinc, iron and manganese could contribute toward combating the problem of micronutrient deficiencies (Flyman and Afolayan 2007).

Odhav et al. (2007) have presented preliminary nutritional data for 20 traditional leafy vegetables including *M. balsamina*. The results of this study provide evidence that these local traditional vegetables, which do not require formal cultivation, could be important contributors to improve the nutritional content of rural and urban people. From this study, it was determined that most of the leafy vegetables provide mineral concentrations exceeding 1% of plant dry weight and are much higher than typical mineral concentrations in conventional edible leafy vegetables; they are thus recommended for future commercial cultivation. High levels of antioxidant activity (96%) were also noticed in *M. balsamina*.

From the results of these analyses, it can be seen that *M. balsamina* leaves could be important green leafy vegetable as a source of nutrients to supplement other major sources. However, chemical analysis alone should not be the sole criteria for judging the nutritional importance of a plant's parts. Thus, it becomes imperative to consider other aspects such as presence of antinutritional and toxicological factors and biological evaluation of nutrients' form, content, availability and utilization. However, research efforts in these areas are rather meager.

The use of leaves in human nutrition is not considered in general with the importance it deserves in some low protein diets. When limiting amino acid of main foods is lysine (as in the case of cereal based diets) the protein of leaves may be of a great importance in its supplementation. The leaves and fruits of *M. balsamina* are used as spinach along with groundnut meal in a traditional cooking called "cacana" in Mozambique. Oliveira and De Carvalho (1975) have studied its nutritional value from the standpoint as a protein supplement of common diets. Along with *Amaranthus spinosus* and *Colocasia antiquorum*, *M. balsamina* leaves have high protein content and scores high in all indices of nutritional value of leaf proteins. However, the use of *M. balsamina* in a traditional dish with groundnut meal is of no interest, since S-amino acids are limiting in both cases. The substitution with/addition of maize meal would be a corrective procedure to be envisaged.

10.8.6 Vitamin Supplements

Malnutrition is an ugly specter in most parts of Africa and Asia and chronic vitamin A deficiency stands out as one of the most persistent nutritional problems though in most cases a food source of retinol and provitamin A carotenoids is plentiful. Sustainable solutions to micronutrient deficiencies that capitalize upon indigenous resources and foodstuffs offer a long-term mechanism for elevating the health status of disadvantaged people. The gac fruit (*M. cochinchinnensis*) is an excellent source of β-carotene (17–35 mg/100 g of edible part). This fruit is familiar to the people within its distribution range and is easy to grow.

Nutritional supplementation trials in Vietnam have shown that children fed with "xoi gac" (rice cooked with fruit pulp of *M. cochinchinensis*, popularly called gac) have significantly higher plasma β-carotene, compared to those who received synthetic β-carotene powder or none (control). Increases in plasma retinol, α-carotene, zeaxanthin, and lycopene levels were also significantly greater in children given gac (Vuong et al. 2002). It is likely that the fatty acids in gac are what make its β-carotene more bioavailable than that of the synthetic form (Vuong et al. 2002).

Voung (2001) and Vuong et al. (2002) have described an exemplary case study in Vietnam where the use of the traditionally cultivated gac fruit (*M. cochinchinensis*) was demonstrated to be an ideal tool in managing chronic cases of vitamin A deficiency in children. This is an example of a highly successful long-term and sustainable strategy using the indigenously available food resources. Extended cultivation coupled with better pre- and post-harvest technologies can make its availability throughout the year. Research efforts toward breaking seed dormancy, development of improved cultivars, and better package of cultivation practices can contribute toward higher productivity and production. More importantly, better efforts are needed to educate the local population about the health benefits of this nature's bounty, "the fruit from heaven" as this fruit is known in Vietnam.

10.9 Prospects as Alternative Crops

Indigenous vegetables have great potential in poverty and malnutrition alleviation and diversification of agricultural environment (Engle and Faustino 2007). They are relatively neglected crops, being left out from research agenda and marginalized from development schemes. They are being increasingly replaced by a handful of cultivated species and their improved varieties. The indigenous knowledge associated with their collection/cultivation, sustainable utilization, and conservation is also under serious threat of being lost in the long run.

Though the current exploitation of these indigenous perennial vegetables by the rural and tribal populations is restricted in the areas of their distribution, some of these have shown equal acceptability among the urban population as well (Chadha and Patel 2007). The dioecious perennial bitterless species native to Asia, i.e., *M. dioica*, *M. sahyadrica*, *M. subangulata* ssp. *renigera* and *M. cochinchinesis* have tremendous potential as cultivated crops on their own. There is better consumer preference for "bitterless bitter gourds," as vegetables, though much of the traditional medicinal properties are ascribed to the bitter constituents. Though rich in minerals, vitamins, and antioxidants such as carotene, lycopene and flavonoids, their potential health benefits are still unknown to the large section of society. Being perennial hardy species propagated by tubers, cultivation of these species is less demanding and ideal for homestead/on-farm cultivation.

M. sahyadrica has come up very well as a natural component in cardomom and coffee estates in the high ranges of Western Ghats of India. In fact, all these species can be very well integrated into any cropping pattern suitable or prevalent in their geographic range. These plants can perenniate under natural conditions for up to 5 years and are highly adapted to organic farming. Being high-value vegetables with wide acceptability, their cultivation will lead to diversification in diet besides nutritional security. Besides, they are an ideal component for home gardens and marginal and subsistence farming.

In spite of being promising fruit vegetables, there has been not much research thrust paid on kakrol's genetic improvement, though there have been few efforts on addressing the problems related to propagation and agronomic aspects (Ali et al. 1991; Mishra and Sahu 1983). Though reported to have *Ctenoplectra* bees as pollinators in South China (Schafer 2005), and the floral characteristics suggest the possibility of such a plant–pollinator mutualism, no such pollinators are reported from India or rest of Asia. The natural fruit set is rather poor and requires hand pollination in the cultivated kakrol (*M. subangulata* ssp. *renigera*).

The dioecious species are shown to have varying degrees of seed dormancy. Hence, tubers serve as better propagules because of their vigorous growth, precocious flowering and fruiting, besides the great advantage of plants of desired sex (10:1 is ideal). Several workers have pointed out the inadequate availability of female tubers in dioecious species as the single most limiting factor in their large-scale cultivation. Recently, Joseph John et al. (2008) have reported that longitudinal splitting of tubers into 2–4 segments (with a portion of apical meristem) in *M. dioica* and *M. sahyadrica*, whereas longitudinal/cross-sectional cutting into several pieces in *M. subangulata* ssp. *renigera* is efficient in multiplication of propagules. However, studies on tuber morphology, germination behavior, and tuber multiplication have yet to be conducted in the other three species.

Though these crops are traditionally raised as rainfed crop, pure rainfed crop cultivation especially in dry zones cannot be practiced at present due to the erratic nature of rainfall prevailing in most parts of South Asia. Navaratne and Kodithuwakku (2006) have explored the possibility of enhancing indigenous vegetable production in Sri Lanka through low-cost microirrigation approach. The yield obtained from the crop grown under designed irrigation system is two-times higher than the yield of crop grown under rainfed condition. Irrigation water consumption by crops under subsurface irrigation system was 1.6-times less when compared to manual irrigation. Meerabai et al. (2006) advocate biofarming in bitter gourd, incorporating the use of organic manures and biofertilizers. This enables the production of superior quality produce devoid of toxic residues and preferred for their flavor, taste, nutritive value, and extended shelf-life. A similar approach incorporating the essentials of traditional farming practices, organic cultivation, and low-cost irrigation and other production technologies can make cultivation of these crops profitable for the farmers, thereby ensuring a sustainable system.

M. balsamina leaves, an important component of the traditional African leafy vegetables, are a good source of supplements for protein and micronutrients to the cereal-based African traditional diets. The same is true in Asian taxa, though its use as leafy vegetable is limited to a handful of ethnic/linguistic tribes. The phytochemical analyses show that these locally adapted indigenous vegetables are rich in proteins (including most of essential amino acids), minerals, and antioxidants, much better than many of the modern vegetable crops. In fact, proximate analysis show leaves are better source of nutrients than fruits with the exception of carotenoids. However, bioavailability of these nutritional components and conflicting reports on toxicity in humans need to be further investigated. Collection of diverse types from wide ranging areas and further evaluation and selection can lead to less bitter and non-toxic types. Research efforts in these aspects can help in promotion of these traditional wild vegetables for the benefit of a much larger population.

Besides their diversified uses, the underexploited indigenous vegetables such as *M. balsamina* have a great innate capacity of giving good returns under scanty and erratic monsoon conditions, prevailing in arid regions of India and a larger part of Africa.

These vegetables provide the nutritional security to the people particularly during the drought year. In drought years, only these vegetables can give good yield when most other vegetables fail. However, very less attempts have so far been made to improve the existing landraces of these indigenous vegetables (Maurya et al. 2006; Rai et al. 2006).

10.10 Utilization as Breeding Resources

The genetic variability present in the wild and underutilized vegetables is of importance for the successful breeding of improved cultivars with desirable qualitative traits and for biotic stress tolerance. The wild species offer great resources for breeding of cultivated bitter gourd for desirable edible/qualitative traits (such as non-bitterness), abiotic stresses (tolerance to drought) and resistance for several insect pests. There are several genotypes of wild bitter gourds (var. *muricata*) that have been collected from southwestern India and reported to have high fruit fly tolerance (Joseph John and Antony 2008b). *M. balsamina* is highly tolerant to most of the typical cucurbit diseases and pests including ladybird beetle (*Epilacna septima*), red pumpkin beetle (*Aulocophora fevicoli*), pumpkin caterpillar (*Margaronia indica*), gall fly (*Lasioptera falcata*), root-knot nematode (*Meladogyne incognita*), cucurbit yellow mosaic, and little leaf disease (Joseph John and Antony 2008b). Small-fruited wild types of bitter gourd (*M. charantia* var. *muricata*) are, in general, tolerant to pumpkin caterpillar and root-knot nematode, whereas some specific accessions with fruit fly tolerance have also been reported from Western Ghats of India. *M. dioica* has been found to be tolerant to pumpkin caterpillar, gall fly, and root-knot nematode, whereas *M. sahyadrica* is highly tolerant to pumpkin caterpillar and root-knot nematode, and most importantly to fruit fly incidence. *M. subangulata* ssp. *renigera* is highly susceptible to most of the diseases and pests, most notably root-knot nematode, whereas it is resistant to cucurbit yellow mosaic and little leaf diseases (Joseph John 2005). The African taxa, especially *M. foetida* and *M. rostrata*, are reputed for their insect-repellant properties (Bosch 2004) and offer great promise in resistance breeding in *M. charantia*.

References

Aberoumand A, Deokule SS (2009) Studies on nutritional values of some wild edible plants from Iran and India. Pak J Nutr 8:26–31

Agarwal PK, Roy RP (1976) Natural polyploids in Cucurbitaceae I. Cytogenetical studies in triploid Momordica dioica Roxb. Caryologia 29:7–13

Akhtar MS (1982) Trial of *Momordica charantia* Linn (Karela) powder in patients with maturity-onset diabetes. J Pak Med Assoc 32:106–107

Akinniyi JA, Uvais M, Bawa S (1983) Glossary of Kanuri names of plants with botanical names, distribution and uses in animals of Borno. Ann Borno 1:85–98

Ali M, Srivastava V (1998) Characterization of phytoconstituents of the fruits of *Momordica dioica*. J Pharm Sci 60:287–289

Ali M, Okubo H, Fujii T, Fujiedan K (1991) Techniques for propagation and breeding of kakrol (*Momordica dioica* Roxb.). Sci Hortic 47:335–343

Andersen LT, Thilsted SH, Nielsen BB, Rangasamy S (2003) Food and nutrient intakes among pregnant women in rural Tamil Nadu, South India. Public Health Nutr 6:131–137

Aoki H, Kieu NTM, Kuze N, Tomisaka K, Chuyen NV (2002) Carotenoid pigments in GAC Fruits (*Momordica cochinchinensis* Spreng). Biosci Biotechnol Biochem 66:2479–2482

Arnold TH, Wells MJ, Wehmeyer AS (1985) Khoisan food plants: taxa with potential for future economic exploitation. In: Wickens GE, Goodin JR, Field DV (eds) Plants for arid lands. Proceedings of Kew international conference on economic plants for arid lands. Allen & Unwin, London, pp 69–86

Aslam M, Stockley IH (1979) Interaction between curry ingredient (karela) and drug (chlorpropamide). Lancet 1:607

AVGRIS (2009) AVRDC vegetable genetic resources information system. http://203.64.245.173/avgris/. Accessed 11 Feb 2009

AVRDC (2002) AVRDC progress report 2002. Asian Vegetable Research and Development Center, Shanhua, Tainan, Taiwan, p 122

Beevy SS, Kuriachan P (1996) Chromosome numbers of south Indian Cucurbitaceae and a note on the cytological evolution in the family. J Cytol Genet 31:65–71

Behera TK (2004) Heterosis in bitter gourd. In: Singh PK, Dasgupta SK, Tripathi SK (eds) Hybrid vegetable development. Haworth, Binghamton, NY, USA, pp 217–221

Behera TK, Dey SS, Sirohi PS (2006) DBGy-201 and DBGy-202: two gynoecious lines in bitter gourd (*Momordica charantia* L.) isolated from indigenous source. Indian J Genet 66:61–62

Behera TK, Singh AK, Staub JE (2008) Comparative analysis of genetic diversity of Indian bitter gourd (*Momordica charantia* L.) using RAPD and ISSR markers for developing crop improvement strategies. Sci Hortic 115:209–217

Behera TK, Dey SS, Munshi AD, Gaikwad AB, Pal A, Singh I (2009) Sex inheritance and development of gynoecious hybrids in bitter gourd (*Momordica charantia* L.). Sci Hortic 120:130–133

Bhaduri PN, Bose PC (1947) Cytogenetical investigation in some cucurbits with special reference to fragmentation of chromosomes as a physical basis of speciation. J Genet 48:237–256

Bharathi LK, Naik G, Dora DK (2005) Genetic divergence in spine gourd. Veg Sci 32:179–181

Bharathi LK, Naik G, Dora DK (2006a) Studies on genetic variability in spine gourd. Indian J Hortic 63:96–97

Bharathi LK, Naik G, Vishalnath (2006b) An exquisite vegetable variety. ICAR News 12(4):14–15

Bhosle DS, Paratkar GT (2005) Callus cultures from *Momordica dioica* (Roxb.). J Cell Tiss Res 5:431–434

Bhuiya MRH, Habib AKMA, Rashid MM (1977) Content of loss of vitamin C in vegetables during storage and cooking. Bangladesh Hortic 5:1–16

Biswas AR, Ramaswamy S, Bapna JS (1991) Analgesic effect of *Momordica charantia* seed extract in mice and rats. J Ethnopharmacol 31:115–118

Blume CL (1826) Bijdragen tot de flora van Nederlandsch Indie. Batavia, Ter Lands Drukkerij, pp 927–940

Bosch CH (2004) *Momordica balsamina* L. In: Grubben GJH, Denton OA (eds) PROTA 2: Vegetables /Legumes [CD-Rom]. PROTA, Wageningen, Netherlands, pp 384–392

Bot YS, Mgbojikwe LO, Nwosu C, Abimiku A, Dadik J, Damshak D (2007) Screening of the fruit pulp extract of *Momordica balsamina* for anti-HIV property. Afr J Biotechnol 6:047–052

Burkill HM (1935) The useful plants of West Tropical Africa, vol 1: Families A–D, 2nd edn. Royal Botanic Gardens, Kew, UK, 960 p

Celine VA, Sirohi PS (1998) Generation mean analysis for earliness and yield in bitter gourd (*Momordica charantia* L.). Veg Sci 25:51–54

Chadha KL, Patel VB (2007) Prospect of indigenous perennial plants as source of vegetable. Acta Hortic 752:49–54

Cho Y, Ozaki Y, Okubo H, Matsuda S (2006) Ploidies of kakrol (*M. dioica* Roxb.) cultivated in Bangladesh. Sci Bull Fac Agric Kyushu Univ 61:49–53

Chuethong J, Oda K, Sakurai H, Saiki I, Leelamanit W (2007) Cochinin B, a novel ribosome-inactivating protein from the seeds of *Momordica cochinchinensis*. Biol Pharm Bull 30:428–432

Clarke CB (1879) Cucurbitaceae. In: Hooker JD (ed) Flora of British India. Reeve, London, UK, pp 616–619

Coe FG, Anderson GJ (1996) Ethnobotany of the Garifuna of Eastern Nicaragua. Econ Bot 50:71–107

Cox PA, Ballick MJ (1994) The ethnobotanical approach to drug discovery. Sci Am 270:82–87

Damodhar VP, Ghode PB, Nawghare PD, Sontakke MB, Pawar PM (2004) Studies on after effects of foliar application of PGR on sex-expression and sex-ratio in bitter gourd (*Momordica charantia* L.) cv. Hirkani. Karnataka J Hortic 1:86–88

De Sarkar D, Majumdar T (1993) Cytological and palynological investigations in *Momordica subangulata* (Cucurbitaceae). J Econ Tax Bot 17:151–153

De Wilde WJJO, Duyfjes BEE (2002) Synopsis of *Momordica* (Cucurbitaceae) in SE-Asia and Malesia. Bot Z 87:132–148

Decker-Walters DS (1999) Cucurbits, sanskrit, and the Indo-Aryas. Econ Bot 53:98–112

Degner (1947) Flora Hawaiiensis, Book 5. Privately Published, Honolulu, HI, USA

Deokar PL, Panchabhai DM, Tagade UG (2003) Tissue culture studies in spine gourd (*Momordica dioica* (L.) Roxb). Ann Plant Physiol 17:64–69

Deshmukh SP, Rothe SP (2003) Ethno-medicinal study from Melghat tribal region of Amaravathi district, Maharashtra. J Econ Taxon Bot 27:582–584

Dey SS, Singh AK, Chandel D, Behera TK (2006) Genetic diversity of bitter gourd (*Momordica charantia* L.) genotypes revealed by RAPD markers and agronomic traits. Sci Hortic 109:21–28

Dey SS, Behera TK, Munshi AD, Pal A (2010) Gynoecious inbred with better combining ability improves yield and earliness in bitter gourd (*Momordica charantia* L.). Euphytica 173:37–47

Dolezalova I, Kristkova E, Lebeda A, Vinter V, Astley D, Boukema IW (2003) Basic morphological descriptors for genetic resources of wild *Lactuca* spp. PGR Newsl 134:1–9

Dutt B, Pandey C (1983) Callose deposition in relation to incompatibility in *Momordica* species. In: Sinha RD, Sinha U (eds) Current approaches in cytogenetics. Spectrum, Patna, Bihar, pp 201–205

Dwivedi SN (1999) Traditional health care among tribals of Rewa district of Madhya Pradesh with special reference to conservation of endangered and vulnerable species. J Econ Taxon Bot 2:315–320

Engle LM, Faustino FC (2007) Conserving the indigenous vegetable germplasm of Southeast Asia. Acta Hortic 752:55–60

Fabricant DS, Farnsworth NR (2001) The value of plants used in traditional medicine for drug discovery. Environ Health Perspect 109:69–75

FAO (1985) Food composition table for use in Africa. Calculations to derive nutrients. http://www.fao.org/docrep/003/x6877e/X6877E20.htms

Farnsworth NR, Akerele O, Bingel AS (1985) Medicinal plants in therapy. Bull World Health Org 63:965–981

Fernandopulle BMR, Ratnasooriya WD (1996) Evaluation of two Cucurbits (genus *Momordica*) for gastroprotective and ulcer healing activity in rats. Med Sci Res 24:85–88

Flyman MV, Afolayan AJ (2007) Proximate and mineral composition of the leaves of *Momordica balsamina* L.: an underutilized wild vegetable in Botswana. Int J Food Sci Nutr 58:419–423

Foa-Tomasi L, Campadelli-Fiume G, Barbieri L, Stirpe E (1982) Effect of ribosome inactivating proteins on virus infected cells. Inhibition of virus multiplication and of protein synthesis. Arch Virol 71:323–332

Fox FW, Norwood Young ME (1982) Food from the veld: edible wild plants of Southern Africa. Delta Books, Johannasberg, South Africa, 400 p

Gaikwad AB, Behera TK, Singh AK, Chandel D, Karihaloo JL, Staub JE (2008) AFLP analysis provides strategies for improvement of bitter gourd (*Momordica charantia* L.). HortScience 43:127–133

Geidam MA, Pakmam I, Laminu H (2004) Effects of aqueous stem bark extract of *Momordica balsamina* L. on serum electrolytes and some haematological parameters in normal and alcohol fed rats. Pak J Biol Sci 7:1430–1432

Geidam MA, Dauda E, Hamza HG (2007) Effects of aqueous stem-bark extract of *Momordica balsamina* Linn. on some serum enzymes in normal and ethanol fed rats. J Biol Sci 7:397–400

Ghosh S, Basu PS (1982) Effect of some growth regulators on sex expression of *Momordica charantia*. Sci Hortic 17:107–112

Ghosh SP, Kalloo G (2000) Genetic resources of indigenous vegetables and their uses in South Asia. Tech Bull No 4, IIVR, Varanasi, UP, India, pp 36–37

Ghosh PN, Dasgupta B, Sircar PK (1981) Purification of lectin from a tropical plant *Momordica dioica* Roxb. Indian J Exp Biol 19:253–255

Gopalan C, Ram Sastri BV, Balasubramanian SC (1982) Nutritive values of Indian foods. National Institute of Nutrition (ICMR), Hyderabad, AP, India, 161 p

Grover JK, Gupta SR (1990) Hypoglycaemic activity of seeds of *Momordica charantia*. Eur J Pharmacol 183:1026–1027

Harriet Gillett (2002) Conservation and sustainable use of medicinal plants in Ghana Medicinal Plant Accession Data. Aburi Botanic Garden, Botanic Gardens Conservation International, 53 p

Hart GB, Vorster HJ (2006) Indigenous knowledge on the South African landscape –potentials for agricultural development. Urban, rural and economic development programme. Occasional Paper No 1. HSRC, Cape Town, South Africa

Hassan LG, Umar KJ (2006) Nutritional value of Balsam apple (*M.balsamina* L.) leaves. Pak J Nutr 5:522–529

Hernandez JF, Gagnon J, Chiche L, Nguyen TM (2000) Squash trypsin inhibitors from *Momordica cochinchinensis* exhibit an atypical macrocyclic structure. Biochemistry 39: 5722–5730

Hooker J (1871) Momordica L. In: Oliver D (ed) Flora of tropical Africa, vol 2. Reeve, London, UK, pp 534–540

Hoque A, Rahman SM (2003) Adventitious shoot regeneration from leaf explants of kakrol cultured *in vitro*. Pak J Bot 35:13–16

Hoque A, Islam R, Joarder OL (1995) *In vitro* Plantlets differentiation in Kakrol (*Momordica dioica* Roxb.). Plant Tiss Cult 5:119–124

Horvath L (2002) Status of the national cucurbit collection in Hungary. In: Díez MJ, Pico B, Nuez F (eds) First *ad hoc* meeting on cucurbit genetic resources, Adana, Turkey, 30 p

Hutchings A, Scott AH, Lewis G, Cunningham AB (1996) Zulu medicinal plants, an inventory. University of Natal Press, Pietermaritzburg, South Africa

Ishida BK, Turner C, Chapman MH, Mckeon TA (2004) Fatty acids and carotenoid composition in gac (*Momordica cochinchinensis* Spreng.). Fruit J Agric Food Chem 52:274–279

Iwamoto E, Ishida T (2005) Bisexual flower induction by the application of silver nitrate in gynoecious balsam pear (*Momordica charantia* L.). Hortic Res 4:391–395

Jain A, Manish S, Lokesh D, Anurekha J, Rout SP, Gupta VB, Krishna KL (2008) Antioxidant and hepatoprotective activity of ethanolic and aqueous extracts of *Momordica dioica* Roxb. leaves. J Ethnopharmacol 115:61–66

Jeffrey C (1980) A review of the Cucurbitaceae. Bot J Linn Soc 81:233–247

Jeffrey C (2001) Cucurbitaceae. In: Hanelt P (ed) Encyclopedia of agricultural and horticultural crops, vol 3. Springer, Berlin, Germany, pp 1510–1557

Jha UC (1990) Autosomal chromosomes carrying sex genes in *Momordica dioica* Roxb. Curr Sci 59:606–607

Jha UC, Roy RP (1989) Hermaphrodite flowers in dioecious *Momordica dioica* Roxb. Curr Sci 58:1249–1250

Jha UC, Ujawane RG (2002) Collection, evaluation and utilization of *Momordica* species. Contributed paper in the international conference on vegetables, Bangalore, India, 11–14 Nov 2002, p 342

Jha UC, Dutt B, Roy RP (1989) Mitotic studies in *Momordica cochinchinensis* (Lour.) a new report. Cell Chrom Res 12:55–56

Jilka C, Strifler B, Fortner GW, Hays EE, Takemoto DJ (1983) *In vivo* antitumor activity of the bitter melon (*Momordica charantia*). Cancer Res 43:5151–5155

Jongkind CCH (2002) A new species of *Momordica* (Cucurbitaceae) from West Africa. Blumea 47:343–345

Joseph John K (2005) Studies on ecogeography and genetic diversity of the genus *Momordica* L. in India. PhD Thesis, Mahatma Gandhi University, Kottayam, Kerala, India

Joseph John K, Antony VT (2007) *Momordica sahyadrica* sp. nov. (Cucucrbitaceae), an endemic species of Western Ghats of India. Nord J Bot 24:539–542

Joseph John K, Antony VT (2008a) Ethnobotanical investigations in the genus *Momordica* L. in the Southern Western Ghats of India. Genet Resour Crop Evol 55:713–721

Joseph John K, Antony VT (2008b) Occurrence, distribution and ex situ regeneration of Balsam pear (*Momordica balsamina* L.) in India. Indian J Plant Genet Resour 21(1):51–54

Joseph John K, Antony VT (2008c) Characterization and Evaluation Descriptors for dioecious *Momordica* species. Indian J Plant Genet Resour (in press)

Joseph John K, Antony VT, Roy YC (2007) On the occurrence, distribution and taxonomy of *Momordica subangulata* Blume ssp. *renigera* (G. Don) de Wilde in India. Genet Resour Crop Evol 54:1327–1332

Joseph John K, Antony VT, Jose M, Karuppaiyan R (2008) Tuber morphology, germination behaviour and propagation efficiency in three wild edible *Momordica* species of India. Genet Resour Crop Evol. doi:10.1007/s10722-009-9407-5

Kadir ZBA, Zahoor M (1965) Colchiploidy in *Momordica charantia* L. Sind Univ Res J 1:53

Karumi YPA, Onyeyili PA, Ogugbuaja VO (2003) Anti-inflammatory and antinociceptive (analgesic) properties of *M. balsamina* Linn. (Balsam apple) leaves in rats. Pak J Biol Sci 6:1515–1518

Karumi YPA, Onyeyili PA, Ogugbuaja VO (2004) Identification of active principles of *M. balsamina* (Balsam apple) leaf extract. J Med Sci 4:179–182

Karumi YPA, Onyeyili PA, Ogugbuaja VO (2006) Toxicity studies and effects of *Momordica balsamina* (Balsam apple) aqueous extract on serum electrolytes and plasma trace elements. Sahel J Vet Sci 5:13–19

Keraudren-Aymonin M (1975) Cucurbitacees. In: Aubreville A, Leroy J-F (eds) Flore du Cambodge, du Laos et du Vietnam. Muséum National D'Histoire Naturelle, Paris, France, pp 36–44

Khajuria S, Thomas J (1993) Traditional Indian beliefs about the dietary management of diabetes: an exploratory study of the implications for the management of Gujarati diabetics in Britain. Hum Nutr Dietet 5:311–321

Kirtikar KR, Basu BD (1933) Indian medicinal plants, vols I–IV. Lalit Mohan Basu, Allahabad, India

Kole C, Olukolu B, Kole P, Abbott AG (2010a) Towards phytomedomics with bitter melon (*Momordica charantia* L.) as a model. In: International Conference on the Status of Plant & Animal Genome Research, January 9–13, 2010, San Diego, CA, USA, P164

Kole C, Olukolu B, Kole P, Abbott AG (2010b) Association mapping of fruit traits and phytomedicine contents in a structured population of bitter melon (*Momordica charantia* L.). In: Thies JA, Kousik CS, Levi A (eds) Proc Cucurbitaceae 2010. Section Breeding and Genetics. ASHS, Alexandria, Va, pp 1-4

Kucuk A, Abak, K, Sari N (2002) Cucurbit genetic resources collections in Turkey. In: Diez MJ, Pico B, Nuez F (eds) First ad hoc Meeting on Cucurbit genetic resources, Adana, Turkey, p 46

Ladizinsky G (1998) Plant evolution under domestication. Kluwer, London, UK, p 146

Leatherdale BA, Panesar RK, Singh G, Atkins TW, Bailey CJ, Bignell AH (1981) Improvement in glucose tolerance due to *Momordica charantia* (karela). Br Med J (Clin Res Ed) 282:1823–1824

Lee HS, Huang PL, Nara PL, Chen HC, Kung HF, Huang P, Huang HI, Huang PL (1990) MAP 30: a new inhibitor of HIV-1 infection and replication. FEBS Lett 272:12–18

Li HL (1970) The origin of cultivated plants in Southeast Asia. Econ Bot 24:3–19

Lifson JD, McGrath MS, Yeung HW, Hwang K (1988) Int Patent No W088/0912

Lira R, Caballero J (2002) Ethnobotany of the wild Mexican Cucurbitaceae. Econ Bot 56:380–398

Luo L, Li Z, Zhang Y, Huang R (1998) Triterpenes and steroidal compounds from *Momordica dioica*. Yaoxue Xuebao 33:839–842

Mangenot S, Mangenot G (1957) Nombres chromosomique nouveaux chez diverses dicotylŽdones et monocotylŽdones d'Afrique occidentale. Bull Jard Bot Etat 27:639

Marquis VO, Adanlawo TA, Olaniyi AA (1977) The effect of Foetidin from *Momordica foetida* on blood glucose level of albino rats. Planta Med 31:367–374

Marr KL, Xia YM, Bhattarai NK (2004) Allozyme, morphological and nutritional analysis bearing on the domestication of *Momordica charantia* L. (Cucurbitaceae). Econ Bot 58:435–455

Maurya IB, Kavita SSK, Rajesh J (2006) Status of indigenous vegetables in southern part of Rajasthan. Acta Hortic 752:193–196

McKay JW (1931) Chromosome studies in the cucurbitaceae. Univ CA Publ Bot 16:339

Meemaduma VN, Ramanayake SMSD (2002) *De novo* shoot regeneration in excised embryos of *Momordica dioica* in the presence of Thidiazuron. In: Proceedings of 58th annual session of the sri lanka association for the advancement of science, Part-1, Abstract of presentations, 2–7 Dec 2002, Colombo, Sri Lanka

Meerabai M, Jayachandran BK, Asha KR (2006) Biofarming in bittergourd (*Momordica charantia* L.). Acta Hortic 752:349–352

Mishra KC, Sahu RP (1983) Large scale cultivation of small biter gourd, problems and possibilities. Indian Hortic 28:5–8

Mishra HN, Mishra RS, Parhi G, Mishra SN (1998) Diallel analysis for variability in bitter gourd (*Momordica charantia*). Indian J Agric Sci 68:18–20

Mitawa GM, Marandu WYF, Dar-es-Salaam (1996) Tanzania. In: Country Report to the FAO international technical conference on plant genetic resources, Liepzig, Germany, 17–23 June 1996, p 80

Mohanty CR, Maharana T, Tripathy P, Senapati N (1994) Interspecific hybridization in *Momordica* species. Mysore J Agric Sci 28:151–156

Mondal A, Ghosh GP, Zuberi MI (2006) Phylogenetic relationship of different kakrol collections of Bangladesh. Pak J Biol Sci 9:1516–1524

Mulholland DA, Vikash S, Roy O, Karl HP, Joseph D (1997) Connolly cucurbitane triterpenoids from the leaves of *Momordica foetida*. Phytochemistry 45:391–395

Nabi SA, Rashid MM, Al-Amin M, Rasul MG (2002) Organogenesis in Teasle gourd (*Momordica dioica* Roxb.). Plant Tissue Cult 12:173–180

Navaratne CM, Kodithuwakku W (2006) Improvement of indigenous vegetable production in Sri Lanka through low-cost micro irrigation approach. Acta Hortic 752:291–295

Nesamvuni C, Steyn NP, Potgieter MJ (2001) Nutritional value of wild, leafy plants consumed by the Vhavenda. S Afr J Sci 97:51–54

Ng TB, Liu WK, Tsao SW, Yueng HW (1994) Effect of trichosanthin and momorcharins on isolated rat hepatocytes. J Ethnopharmacol 43:81–87
Nguyen HH, Widodo SH (1999) *Momordica* L. In: de Padua LS, Bunyapraphatsara N, Lemmens RHMJ (eds) Plant resources of South-East Asia No 12(1): Medicinal and poisonous plants 1. Backhuys, Leiden, Netherlands, pp 353–359
Nielsen K, Boston RS (2001) Ribosome-inactivating proteins: a plant perspective. Annu Rev Plant Physiol Plant Mol Biol 52:785–816
Njoroge GN, Newton LE (2002) Ethnobotany and distribution of wild genetic resources of the family Cucurbitaceae in the Central highlands of Kenya. PGR Newsl 132:10–16
Njoroge GN, van Luijk MN (2004) *Momordica charantia* L. In: Grubben GJH, Denton OA (eds) PROTA 2: vegetables. PROTA, Wageningen, Netherlands, pp 385–390
Odhav B, Beekrum S, Akula U, Baijnath H (2007) Preliminary assessment of nutritional value of traditional leafy vegetables in KwaZulu-Natal. S Afr J Food Compos Anal 20:430–435
Olaniyi AA (1975) A neutral constituent of *Momordica foetida*. Lloydia 38:361–362
Oliveira JS, De Carvalho MF (1975) Nutritional value of some edible leaves used in Mozambique. Econ Bot 29:255–263
Pal BP, Singh H (1946) Studies in hybrid vigour II. Notes on the manifestation of hybrid vigour in the brinjal and bitter gourd. Indian J Genet 6:19–33
Pawar SV, Patil SC, Jambhale VM, Mehetre SS (2004) Micropropagation studies in Kartoli (*Momordica dioica* Roxb). Adv Plant Sci 17:275–278
Perry LM (1980) Medicinal plants of East and South-East Asia. MIT Press, Cambridge, MA, USA
Piskunova T (2002) Status of the cucurbit collections in Russia. In: Díez MJ, Pico B, Nuez F (eds) First *ad hoc* meeting on cucurbit genetic resources, Adana, Turkey, 19 Jan 2002, p 37
Platel K, Srinivasan K (1995) Effect of dietary intake of freeze dried bitter gourd (*Momordica charantia*) in streptozotocin induced diabetic rats. Narhung 39:3977–3986
Prakash G (1976) Effect of plant growth substances and vernalization on sex expression in *Momordica charantia* L. Indian J Exp Biol 14:360–362
Rai M, Pandey S, Ram D, Rai N, Pandey AK, Yadav DS (2006) Plant genetic resources of legumes and under-utilized vegetable crops in India. Acta Hortic 752:225–230
Raj NM, Prasanna KP, Peter KV (1993) Bitter gourd *Momordica* spp. In: Kalloo G, Bergh BO (eds) Genetic improvement of vegetable plants. Pergamon, Oxford, UK, pp 239–246
Rajasekharan KR, Shanmugavelu KG (1984) MDU-1 bitter gourd. S Indian Hortic 32:47
Rajput JC, Parulekar YR, Sawant SS, Jamadagni BM (1993) Sex modification in kartoli (*Momordica dioica* Roxb.) by foliar sprays of silver nitrate ($AgNo_3$). Curr Sci 66:779
Ram D, Srivastava U (1999) Some lesser known minor cucurbitaceous vegetables: their distribution, diversity and use. Indian J Plant Genet Resour 12:307–316
Ram D, Kumar S, Banerjee MK, Singh B, Singh S (2002) Developing bitter gourd *(Momordica charantia* L.) populations with very high proportion of pistillate flowers. Cucurbit Genetic Coop Rep 25:65–66
Raman A, Lau C (1996) Anti-diabetic properties and phytochemistry of *Momordica charantia* L. (Cucurbitaceae). Phytomedicine 2:349–362
Rashid MM (1993) Sabjee Biggan (in Bengali). Bangla Academy, Dhaka, Bangladesh
Rasul MG, Okubo H (2002) Genetic Diversity in teasle gourd (*M. dioica* Roxb.) Bangladesh. J Plant Breed Genet 15:7–15
Rasul MG, Hiramatsu M, Okubo H (2004) Morphological and physiological variation in kakrol (*M. dioica* Roxb.). J Fac Agric Kyushu Univ 49:1–11
Rasul MG, Hiramatsu M, Okubo H (2006) Genetic relatedness (diversity) and cultivar identification by randomly amplified polymorphic DNA (RAPD) markers in teasle gourd (*Momordica dioica* Roxb.). Sci Hortic 111:271–279
Rensburg WSJ, van Averbeke W, Slabbert R, Faber M, van Jaarsveld P, van Heerden I, Wenhold F, Oelofse A (2007) African leafy vegetables in South Africa. Water SA 33(3) (Spl edn):317–325. http://www.wrc.org.za
Richharia RH, Ghosh PN (1953) Meiosis in *Momordica dioica* Roxb. Curr Sci 22:17–18
Robinson RW, Decker-Walters DS (1997) Cucurbits. CABI, Wallingford, Oxford, pp 97–101
Roy RP, Thakur V, Trivedi RN (1966) Cytogenetical studies in *Momordica* L. J Cytol Genet 1:30–40
Sadyojatha AM, Vaidya VP (1996) Chemical constituents of the roots of *Momordica dioica* Roxb. Indian Drugs 330:473–475
Saito K (1957) Studies on the induction and utilization of polyploidy in some cucurbits. II. On polyploidy plants of *Momordica charantia*. Jpn J Breed 6:217
Schafer H (2005) The biogeography of *Momordica*. Cucurbit Network News 12:5
Seena S, Sridhar KR, Jung K (2005) Nutritional and antinutritional evaluation of raw and processed seeds of a wild legume, *Canavalia cathartica* of coastal sand dunes of India. Food Chem 92:465–472
Semiz A, Sen A (2007) Antioxidant and chemoprotective properties of *Momordica charantia* L. (bitter melon) fruit extract. Afr J Biotechnol 6:273–277
Seringe NC (1828) Cucurbitaceae. In: De Candolle AP (ed) Prodromus Systematis Naturalis Regni Vegetabilis, vol 3. Treuttel & Wurtz, Paris, France, pp 311–312
Serquen FC, Bacher J, Staub JE (1997) Mapping and QTL analysis of a narrow cross in cucumber (*Cucumis sativus* L.) using random amplified polymorphic DNA markers. Mol Breed 3:257–268
Seshadri VS, Parthasarathi VA (2002) Cucurbits. In: Bose TK, Som MG (eds) Vegetable crops in India. Naya Prakash, Calcutta, India, pp 91–164
Shantappa T, Gouda MS, Reddy BS, Adiga JD, Kukanoor L (2005) Effect of growth regulators and stages of spray on growth and seed yield in bitter gourd (*Momordica charantia* L.). Karnataka J Hortic 1:55–62
Sharma VN, Sogani RK, Arora RB (1960) Some observations on hypoglycaemic activity of *Momordica charantia*. Indian J Med Res 48:471–477
Shreedhar CS, Pai KSR, Vaidya VP (2001) Postcoital antifertility activity of the root of *Momordica dioica* roxb. Indian J Pharm Sci 63:528–531
Singh H (1978) Parthenocarpy in *Trichosanthes dioica* Roxb. and *Momordica dioica* Roxb. Curr Sci 47:735

Singh AK (1990) Cytogenetics and evolution in the cucurbitaceae In: Bates DM, Robinson RW, Jeffrey C (eds) Biology and utilization of Cucurbitaceae. Comstock Public Association, Cornell University Press, Ithaca, NY, USA, pp 10–28

Singh AK, Behera TK, Chandel D, Sharma P, Singh NK (2007) Assessing genetic relationships among bitter gourd (*Momordica charantia* L.) accessions using inter simple sequence repeat (ISSR) markers. J Hortic Sci Biotechnol 82:217–222

Sinha S, Debnath B, Guha A, Sinha RK (2001) Sex Linked Polypeptides in Dioecious *Momordica dioica*. Cytologia 66 (1):55–58

Srivastava Y, Bhatt HV, Verma Y, Venkiah K (1993) Antidiabetic and adaptogenic properties of *Momordica charantia* extract: an experimental and clinical evaluation. Phytother Res 7:285–289

Srivastava U, Mahajan RK, Gangopadyay KK, Singh M, Dhillon BS (2001) Minimal descriptors of agri-horticultural crops vegetable crops. Part-II. NBPGR, New Delhi, India, pp 61–66

Thirupathi K, Sathesh Kumar S, Govardhan P, Ravi Kumar B, Rama Krishna D, Krishna Mohan G (2006) Protective effect of *Momordica dioica* against hepatic damage caused by carbon tetrachloride in rats. Acta Pharm Sci 48:213–222

Thulin M (1991) A new species of *Momordica* (Cucurbitaceae) from tropical Africa. Nord J Bot 11:425–427

Thwaites GHK (1864) Enumeratio Plantarum Zeylaniae (An enumeration of Ceylon Plants). Dulau, London, UK, 126 p

Tien PG, Kayama F, Konishi F, Tamemoto H, Kasono K, Hung NT, Kuroki M, Ishikawa SE, Van CN, Kawakami M (2005) Inhibition of tumor growth and angiogenesis by water extract of Gac fruit (*Momordica cochinchinensis* Spreng.). Int J Oncol 26:881–889

Trivedi RN, Roy RP (1972) Cytological studies in some species of *Momordica*. Genetica 43:282–291

Trivedi RN, Roy RP (1973) Cytogenetics of *Momordica* charantia *and its* polyploids. Cytologia 38:317–325

Vahab MA, Peter KV (1993) Crossability between *Momordica charantia* and *Momordica dioica*. Cucurbit Genet Coop Rep 16:84

Van Rheede HA (1678–1693) Hortus Malabaricus, vols 1–12. VS Joannis and DV Joannis, Amsterdam, Netherlands (repr edn 2003)

Van Wyk BE, Gericke N (2000) People's plants, a guide to useful plants of Southern Africa. Briza, Pretoria, South Africa, 351 p

Verma VK, Sirohi PS, Choudhry B (1984) Chemical sex modification and its effect on yield in bitter gourd (*Momordica charantia* L.). Prog Hortic 16:52–54

Vishalnath MS, Bharathi LK, Naik G, Singh HS (2008) CHTG 2 (Neelachal Gaurav): a soft-seeded teasel-gourd variety. ICAR News 14:15

Voung LT (2001) A fruit from heaven. Vietnam J Oct 2001. http://www.vietnamjournal.org/article.phpsid=5. Accessed 12 Feb 2009

Vuong LT, King JC (2003) A method of preserving and testing the acceptability of gac fruit oil, a good source of beta-carotene and essential fatty acids. Food Nutr Bull 24:224–230

Vuong LT, Stephen RD, Suzanne PM (2002) Plasma B-carotene and retinol concentrations of children increase after a 30-d supplementation with the fruit Momordica cochinchinensis (gac). Am J Clin Nutr 75:872–879

Walter KS, Gillett HJ (1998) 1997 IUCN red list of threatened plants. Compiled by the World Conservation Monitoring Centre. IUCN – The World Conservation Union, Gland, Switzerland, 862 p

Walters TW, Decker-Walters DS (1988) Balsampear (*Momordica charantia*, Cucurbitaceae). Econ Bot 42:286–288

Wang QM, Zeng GW (1996) Effects of gibberellic acid and Cycocel on sex expression of *Momordica charantia*. J Zhejiang Agric Univ 22:541–546

Wang QM, Zeng GW (1997a) The effect of phytohormones and polyamines on sexual differentiation of *Momordica charantia*. Acta Hortic Sin 24:48–52

Wang QM, Zeng GW (1997b) Hormonal regulation of sex differentiation on *Momordica charantia* L. J Zhejiang Agric Univ 23:551–556

Wang QM, Zeng GW (1998) Study of specific protein on sex differentiation of *Momordica charantia*. Acta Bot Sin 40:241–246

Wang QM, Zeng GW, Jiang YT (1997) Effects of temperature and photoperiod on sex expression of *Momordica charantia*. China Veg 1:1–4

Watt JM, Breyer-Brandwijk MG (1962) Medicinal and poisonous plants of southern and eastern Africa, edn 2. Livingstone, Edinburgh, London, UK, p 1457

West ME, Sidrak GH, Street SPW (1971) The anti-growth properties of extract from *Momordica charantia* L. West Indian J Med 20:25–34

Whitaker TW (1933) Cytological and phylogenetic studies in the Cucurbitaceae. Bot Gaz 94:780–790

Wight R, Walker-Arnott GA (1841) Prodromus Florae Peninsulae India Orientalis. Parbury, Allen, London, UK, 348 p

Willdenow CL (1805) Linnaei species plantarum, vol 4. GC Nauk, Berolini, Vienna, pp 601–605

Wong CHR, Fong WP, Ng TB (2004) Multiple trypsin inhibitors from *Momordica cochinchinensis* seeds, the Chinese drug mubiezhi. Peptides 25:163–169

Yang SL, Walters TW (1992) Ethnobotany and the economic role of the Cucurbitaceae of China. Econ Bot 46:349–367

Yonemori S, Fujieda K (1985) Sex expression in *Momordica charantia* L. Sci Bull Coll Agric Univ Ryukyus Okinawa 32:183–187

Zalapa J, Staub JE, Chung SM, Cuevas H, McCreight JD (2007) Mapping and QTL analysis of plant architecture and fruit yield in melon. Theor Appl Genet 114:1185–1201

Zeven AC, Zhukovsky PM (1975) Dictionary of cultivated plants and their centres of diversity. Centre for Agricultural Publication and Documentation, Wageningen, Netherlands

Zhang QC (1992a) Bitter melon: a herb warranting a closer look. PWA Coalition Newsline 81:48–49

Zhang QC (1992b) Preliminary report on the use of *Momordica charantia* extract by HIV patients. J Naturpath Med 3:65–69

Chapter 11
Raphanus

Yukio Kaneko, Sang Woo Bang, and Yasuo Matsuzawa

11.1 Introduction

Members of *Raphanus* are annual and outcrossing plant species of the Brassicaceae family. They are believed to have originated in the regions from Mediterranean to Black Sea. The genome was described as RR and RrRr for *Raphanus sativus* and *R. raphanistrum*, respectively (Baranger et al. 1995; Chevre et al. 1996; Darmency et al. 1998). The somatic chromosome number of *Raphanus* species is $2n = 18$ (Banga 1976; Harberd 1976; Prakash et al. 1999). Genome size of *R. sativus* was reported to be ca. 526 Mbp (Arumuganathan and Earle 1991) and 573 Mbp of 0.583 pg of genome content (Johnston et al. 2005). These values are intermediate between 529 Mbp (0.539 pg) for *Brassica rapa* and 696 Mbp (0.710 pg) for *B. oleracea* (Johnston et al. 2005), although genome size of *R. raphanistrum* is not yet clear.

Seven species compose the genus *Raphanus* classified under two sections of *Raphanis* DC. and *Hesperidopsis* Boiss. Section *Raphanis* comprises six species including *R. sativus*, *R. raphanistrum*, *R. microcarpus*, *R. rostras*, *R. maritimus*, and *R. landra*, and section *Hesperidopsis* includes only one species, *R. aucheri* (Kitamura 1958). On the other hand, Fujii (1977) and Hida (1990) supposed six species except for *R. aucheri*, and even eight species were proposed by Cheam and Code (1995). Banga (1976), George (1985), and Hida (1990) independently presumed that four wild species, *R. raphanistrum*, *R. maritimus*, *R. landra*, and *R. rostras*, might be involved in the development of cultivated radish. These wild radishes are growing as the dominant weedy plants around the coastal areas of the Sea of Marmara and the Bosphorus Straits of Mediterranean (Dixon 2007). Harberd (1976) proposed that the genus *Raphanus* could be included into a cytodeme based on the somatic chromosome number, chromosome configuration at MI in pollen mother cells (PMCs) and fertility supported by Prakash et al. (1999) and Dixon (2007). Tsunoda (1979) and Hinata (1995) proposed that these wild species were included in *R. raphanistrum* and the genus *Raphanus* was classified into two species, *R. sativus* and *R. raphanistrum*.

11.2 Application of Wild Species for Radish Breeding

Potential of wild radishes and their hybrid lines was evaluated by Matsuzawa et al. (1996), but the genetic resources of wild species have not been exploited to achieve divergent taste, tolerances to new disease and insect damage, effective seed production system, and so on. The agricultural traits of the genus *Raphanus* were summarized by Warwick (1993). These include cytoplasmic male-sterility (CMS) (Ogura 1968; Yamagishi and Terachi 1996), resistance to pod shattering (Agnihotri et al. 1991), tolerance to saline soils (Warwick 1993), and resistance to blackleg (Salisbury 1987).

11.3 Domestic and Wild Radish

Brassica juncea and *B. rapa* grow in a natural state at the banks and roadsides. *B. napus* was also growing in the semi-wild at the Meiji Period. A wild radish

Y. Kaneko (✉)
Laboratory of Plant Breeding, Faculty of Agriculture, Utsunomiya University, 350 Minemachi, Utsunomiya 321–8505, Utsunomiya, Japan
e-mail: kaneko@cc.utsunomiya-u.ac.jp

C. Kole (ed.), *Wild Crop Relatives: Genomic and Breeding Resources, Vegetables*,
DOI 10.1007/978-3-642-20450-0_11, © Springer-Verlag Berlin Heidelberg 2011

Fig. 11.1 Characteristics of Japanese wild radish. (**a**) Koubou-daikon population at Yonezawa, Yamagata Pref., (**b**) Morphology of Koubou-daikon, (**c**) Pods of Koubou-daikon (*right*) and cultivated radish (*left*), (**d**) Dormancy of Hama-daikon in about 1 month after sown

(*R. sativus* L. var. *hortensis* f. *raphanistroides* Makino) growing in the coastal area of Japan, Korea, and South China is known as Hama-daikon (Furusato and Miyazawa 1958; Kitamura 1958; Ohnishi 1999). This species have been considered to be involved in the development of domestic varieties such as Moriguchi, Hatano, and Hosone (Fujii 1977). The other wild radish, known as Nora-daikon or No-daikon, was grown at the restricted inland areas in Tohoku district (Aoba 1967, 1988), and was suggested to be the island type of Hama-daikon by Ohnishi (1999). No-daikon has been grown around the areas 70 km away from the coast being isolated by the mountainous region. This radish was called Koubou-daikon, since Saint Koubou had introduced it to this area about 1,200 years ago. In the 1950s, large colonies were identified in Yamagata and Fukushima Prefectures (Aoba 1981). The plants of this wild radish grow naturally in the field where other crops such as buckwheat was harvested and plowed over (Fig. 11.1a). Their leafy rosettes vigorously grow covering the field and cannot be harvested by hand for development of lateral roots (Fig. 11.1b).

No-daikon was assumed to be relative to Kozena that was locally cultivated at the Kozena Area in the Onoda Village, Miyagi Pref. Kozena is one of the important resources for CMS (Namai 1991; Sakai and Imamura 1994, 2003), although people do not take the plumped root but harvest the fresh leaf of this radish (Furusato and Miyazawa 1958; Sasaki 1994).

Both Hama-daikon and No-daikon are sharply pungent and develop solid roots, and constricted pods yielding a seed (Fig. 11.1c). The 100 seeds weights are ca. 1.306 g, 2.463 g and 2.823 g for Hama-daikon, Minowase, and Horyo, respectively. Wild radishes showed seed dormancy as shown in Tables 11.1–11.3. Most of the seeds did not germinate immediately after sowing (Table 11.1; Fig. 11.1d). Within a month after sowing, 13% of seeds germinated and about 50% of them germinated within 45 days after sowing (Table 11.2). Seed dormancy was broken by chilling treatment of more than 12 h (Table 11.3). The seed dormancy was also observed for wild radish, Koubou-daikon (Aoba 1967) and *R. raphanistrum* examined at Victoria in Australia (Cheam and Code 1995).

Recent studies on wild radishes are focused on analysis of origin of cultivated radishes and evolution of wild radish, model system of competition and suppression between wild radish and their hybrid groups in California, and gene flow from transgenic rapeseed (*B. napus*) to wild radishes or other relatives. For the progress of works on genetics, breeding, and ecology of radishes, it might be important to analyze the genomic feature of wild radishes including Japanese wild ones.

11.4 Researches in Origin of Radish Using Gene Markers

The origin of cultivated radish is not so clear even though the molecular biology studies have been performed.

Table 11.1 Dormancy of Japanese wild radish (Hama-daikon)

No. of strain observed	Date sown	
	1-Jul	31-Jul
3[a]	0	13.3
Cultivated radishes		
Minowase	46.7	93.3
Horyo	63.3	100

[a]Seeds were harvested from Chiba Pref. on 25 June 1993

Table 11.2 Seed germination rate in Japanese wild radish

No. of strain observed	Date of observation[b]				Total germination rates (%)
	27 Jul	11 Aug	22 Aug	02 Sept	
3[a]	10	11.7	20	6.7	48.4
Cultivated radishes					
Minowase	100				
Horyo	100				

[a]Seeds were harvested from Chiba Pref. on 25 June 1993
[b]Seeds were sown on 18 July 1993

Table 11.3 Effects of chilling treatment on dormancy in Japanese wild radish

No. of strain observed[a]	Hours of chilling	Date of observation[b]		Total germination rates (%)
		10 Aug	25 Aug	
3	12	47	29.1	76.1
3	24	51.7	20.8	72.5
3	72	38.1	15.4	53.5

[a]Seeds were harvested from Chiba Pref. on 25 June 1993
[b]Seeds were sown on 01 Aug 1993

For this purpose, Yamagishi and Terachi (1994, 1996) studied on the *orf138* gene of mitochondrial DNA (mtDNA) in cytoplasm (Ogura type; Ogura 1968) that induced CMS. Distribution of *orf138* gene was surveyed for radish cultivars and wild radishes (Hama-daikon) were collected in Japan. However, *orf138* gene could not be detected in the cultivars except for a few ones (MS Gensuke, Kozena, Nou-No.1 bred in Okinawa Pref., Sabaga). On the other hand, this gene was widely distributed in more than 40% of wild radish populations growing in East Asia including Japan and could be identified in both *R. raphanistrum* and *R. maritimus* (Yamagishi and Terachi 1997, 2003). It was assumed that the Ogura cytoplasm carrying the *orf138* gene might be originated in the course of cytoplasmic differentiation to wild radish after the genus *Raphanus* had evolved (Yamagishi 2006).

The nucleotide sequence of an entire coding region of *orf138* was examined for 107 plants of Japanese wild radish, 29 cultivars, and seven strains of *R. raphanistrum* (Yamagishi and Terachi 2001). Six sites of nucleotide change, and one single insertion/deletion (39 bp) were observed in the *orf138* region from wild and cultivated radishes. By the analysis of these seven mutations, the *orf138* sequences of 143 plants were classified into nine types (Type A to Type I) and two types (B and C), eight types except for H type and five types (A, B, E, F and H) were identified in *R. raphanistrum*, Japanese wild radishes, and cultivars, respectively. The type B was detected in 41% of Japanese wild radishes, and found in 13 out of 19 populations growing in Japan. In addition, this type was more frequent (86%) in *R. raphanistrum* suggesting that Type B is one of the ancestral types of *orf138*. Giancola et al. (2007) recently found the mutated *orf138* gene in three populations from France and one from South England, although it was of low frequency. The Ogura-type cytoplasm carrying Type B and Type C sequences might have evolved in *R. raphanistrum*. Furthermore, these cytoplasms might introgress into Japanese wild radish along with the nuclear *Rf* gene(s). By the aid of the *Rf* gene, Ogura-type cytoplasm was transmitted into wide range of radishes in Japan. Several independent mutations occurred in *orf138* of Japanese wild radish, and the variants including Type A have been maintained in populations. Ogura-type cytoplasms with specific types of *orf138* have also been introgressed into some native varieties (Yamagishi and Terachi 2001; Yamagishi 2006). Iwabuchi et al. (1999) suggested *orf125* to be CMS-inducing gene for domestic variety Kozena grown in Miyagi Pref. of Tohoku district. On the other hand, Yamagishi (2006) described that *orf125* might be equivalent to Type F of Japanese wild radish growing in Yamagata and Aomori prefectures of Tohoku district.

Out of nine types of *orf138,* five types – A, B, D, F and H – were ascertained to induce male-sterility. On the other hand, it is important to survey the nuclear restorer gene (*Rf* gene; Brown et al. 2003) inhibiting the expression of *orf138* for pollen fertility. *R. raphanistrum* and most of the Japanese wild radishes had the restorer gene, although a half of the cultivated radishes were carrying it (Yamagishi 1998). Twenty-three out of 28 Chinese and European varieties possessed the restorer gene, although 20 out of 28 Japanese varieties did not carry it. Koizuka et al. (2003) isolated the *Rf* gene, *orf687*, from the Chinese radish var. Yuan hong, and Giancola et al. (2007) also found *Rf* gene in *R. raphanistrum* from France. In addition, Yasumoto et al. (2008) investigated the distribution of the *Rf* gene in Japanese wild radish including 226 individuals collected from 15 regions in Japan and two regions of Korea. The examined 95 plants (42%) of the Japanese wild radish possessed *orf138*, 207 (91.6%) had the *Rf* gene in both Japanese wild radish and F_1 hybrids bred by CMS variety with *orf138*. By polymerase chain reaction-restriction fragment length polymorphism (PCR-RFLP) analysis of the nucleotide sequence of between the *Rf* gene and *orf687*, it was clarified that plants with an identical RFLP pattern of *orf687* were restrictively found in three populations collected from southern region of Japan including Iriomote, Zanpazaki, and Imajuku.

From these studies, it was confirmed that the *orf138* gene (Ogura-CMS gene; Ogura 1968) identified in Japanese native variety is now widely distributed among *R. raphanistrum* in Europe and Japanese wild radish. Moreover, the origin of cultivated radish could be demonstrated by the molecular analysis of the wild radish species.

11.5 Origin of Japanese Wild Radish

Huh and Ohnishi (2001, 2006) investigated the genetic diversity of Japanese wild radish growing around East Asia using allozyme and amplified fragment length polymorphism (AFLP) marker analysis. The Korean populations were of smaller size being isolated like dots, but they maintained a high level of genetic diversity. The average percentage of polymorphic loci, the mean number of alleles per locus, and the average heterozygosity for Korean populations were 63.1%, 2.27, and 0.281, respectively, whereas these parameters for Japanese populations were 53.3%, 2.26, and 0.278, respectively. These results might show the higher genetic diversity in the Japanese wild radishes. These parameters for *R. raphanistrum* in Europe were 94%, 3.25, and 0.45, respectively. In AFLP analysis, both Japanese and Korean populations of Japanese wild radishes showed a wide range of variation where the average percentage of polymorphic loci and the average heterozygosity were 66.3% and 0.128, respectively. *R. raphanistrum* and two lines of *R. sativus* from Kazakhstan were confirmed to be closely related. Therefore, *R. raphanistrum* might have an involvement in the origin of *R. sativus*.

As has been mentioned, plants with cytoplasm carrying *orf138* and nuclear *Rf* gene are widely distributed in Japanese wild radishes, but most of the Japanese cultivars have not incorporated these genes. The European and Chinese varieties had the *Rf* gene, but many varieties except for those in Taiwan have not yet carried the *orf138* gene. Therefore, Yamagishi (2006) suggested that Japanese wild radish might be the descendants from *R. raphanistrum*. Terachi et al. (2001) investigated the nucleotide sequence of the entire coding and flanking regions of *orfB* using *R. raphanistrum*, Japanese wild radishes and cultivated ones. The plants with Ogura male-sterile (MS) cytoplasm carried only the type 1 sequence, whereas the plants with normal cytoplasm had either type 2 or type 3. As regards to the distribution of these three types among varieties and populations, 78% of cultivated radishes exclusively possessed type 2 and 92% of Japanese wild radishes had two or three types (Yamagishi 2004). These facts might suggest that the Japanese wild radishes were not the escape from cultivated radish but the descendants from *R. raphanistrum* (Yamagishi 2006).

11.6 Roots of the Cultivated Radish

Yamagishi and Terachi (2003) analyzed the structural variations in the mitochondrial *cox*I and *orf*B regions by PCR using three wild species (*R. raphanistrum*, *R. landra*, *R. maritimus*) and 43 cultivated radishes. Five (I–V) and four (I–IV) types of mtDNA were identified for *R. raphanistrum* and *R. maritimus*, respectively. They considered that the cytoplasmic differentiation of these two species might be too small to discriminate them as independent species. All of the European cultivars were type II and 60% of Chinese ones were type III. Japanese cultivars classified into type I (42%) and type III (42%) were common to the northern Chinese cultivars. Yamagishi et al. (2005) also analyzed the sequences of *trn*L to *psb*G region in the chloroplast genome by PCR-RFLPs using 118 plants of 22 strains in three wild species (*R. landra*, *R. maritimus*, and *R. raphanistrum*), 219 plants of 13 populations in Japanese wild radishes and 80 plants of 48 varieties in the cultivated radishes. They identified eight haplotypes, although type 1 and type 2 could not be classified. All the seven types were identified in *R. raphanistrum* and five types in *R. maritimus* showed the highest rate (40.6%) of type 3 in the two species. Japanese wild radish contained 39.3% of type 8, 33.6% of type 7 and 27.4% of type 4. In Japanese wild radishes, the prominent genetic diversity clearly showed that 11 populations out of 13 observed maintained more than two types. The cultivated radishes were classified into four types including 1/2, 4, 7, and 8, and a large number of European varieties was of type 4 (66.7%), the Chinese and Japanese varieties were of type 7 (42.4%), type 4 (36.4%), type 8 (18.2%), and 1/2 type (3%), respectively. On the other hand, Yamane et al. (2005) investigated the nucleotide sequence of the 5′-*mat*K region of chloroplast DNA and five regions (*apt*A–*apt*I, *ndh*D–*ndh*H, *rpo*A–*rps*3, *rpo*C3–*trn*C and *psb*B–*psb*H) by PCR-RFLP analysis for 17 accessions including the cultivated radish, *R. raphanistrum*, and East Asian wild radish. They suggested that *R. raphanistrum* was not the maternal ancestor of the cultivated radish, and the East Asian wild radish has contributed to the establishment of the East Asian cultivated radish. Furthermore, Yamane et al. (2009) also studied the origin of cultivated radish by single sequence repeat (cpSS) variation in chloroplast DNA using 59 cultivated radishes and 23 strains of three wild *Raphanus* species. Seven of the 25 cpSSRs studied were polymorphic (two to four alleles), and a total of 20 haplotypes were designated as A to T. Haplotype E is widely distributed in Europe and East Asia. Haplotype N is geographically restricted in Asia. A total of 13 haplotyes were detected in wild *Raphanus*, and seven of them (A, B, G, J, Q, S, and T) were distributed only in them. By minimum-spanning network (MSN) analysis, it was found that there were at least three independent domestication events, relating black Spanish radish and those grouped with two distinct cpSSR haplotypes. Yamane et al. (2009) presumed that the Asian cultivated radish was not originated from the diffused descendants of European cultivated radish, but probably originated from a wild species that is distinct from the wild ancestor of European cultivated radish.

As a result, the cultivated radishes were suggested to be developed through multiple origin, a part of which cytoplasm might inherit from wild relatives of the genus *Raphanus*, plants maintaining differentiated cytoplasm were independently domesticated at the districtive area. A few types of cytoplasm from *R. raphanistrum* were transmitted to the Japanese wild radish. The cultivated radish developed in the Eurasian continent might have produced a large number of varieties around the world. In Japan, the ancestors with the distinct two types of cytoplasm might be firstly introduced, and then new domestic cultigens have evolved through hybridization between types. The wild radishes distributed widely in the Eurasian continent, Japanese wild radish originated the wild radishes adapted to East Asia, and then, the native varieties have been bred by the hybridization among Japanese wild radish (Yamagishi 2006).

Genetic approaches for wild and cultivated radish of Eurasian Continent would provide some potential information to understand the evolutional relationships of radishes around the world.

11.7 Characteristics of *R. raphanistrum* as a Weed

R. raphanistrum grows vigorously as a weed in all continents except for Antarctica (Hinata 1995; Holm et al. 1997). Two species of *R. raphanistrum* and *R. sativus* grow in the wild in North America (Warwick 1993), and the authors observed both these species in Paraguay and South America also (Fig. 11.2). In Australia, three species have been identified:

Fig. 11.2 The wild radishes in Paraguay, South America

R. raphanistrum, *R. maritimus*, and *R. sativus* (Cheam and Code 1995). Lefol et al. (1997) indicated that *R. raphanistrum* was a major weed in Weed Seeds Order of the Canadian Seed Act. This species is distributed in the provinces on the Atlantic seaboard, the Canadian Pacific coast, but is infrequent in the prairie region of western Canada.

Wild radish not only has colonized new regions, but also it has caused yield losses in a variety of crops (Sahli et al. 2008). In Australia, the weedy *R. raphanistrum* is estimated to bring about an average yield loss of 2 ton ha^{-1} and A$ 30 ha^{-1} for herbicide in rainfall areas (Streibig et al. 1989). At Rutherglen and Victoria, 10% and 50% of yield losses were confirmed in the field with only seven wild plants m^{-2} and 200 wild plants m^{-2}, respectively (Cheam and Code 1995). Webster and MacDonald (2001) surveyed 83 troublesome weeds in Georgia and wild radish was found to be the most troublesome one both in the fields of cereals and vegetables. Wild radish is one of the most troublesome weed for the production of spring-sown wheat and barley in Sweden (Bostrom 2003).

Lehtila and Brann (2007) performed a selection for flower size of *R. raphanistrum* concerning to both reproductive and vegetative traits through two generations. The lines having large flowers produced smaller seeds and flowered later than the lines with small flowers. The lines selected for large flower size had more flowers and a larger plant size than those selected for small flowers. The authors also showed that flower size had a positive genetic correlation with flowering and plant height. As to the pollination systems and pollinators for propagation of weedy *R. raphanistrum*, Conner et al. (1995) surveyed the effect of flower morphology of wild radish on pollination using four pollinators: honey bees, small native bees, butterflies, and syrphid flies. The nectar-feeding butterflies had higher potentials for pollination than both the nectar- and pollen-feeding honey bees. Flowers with intermediate stigma exertion had the highest number of pollen grains deposited on their stigma by butterflies, but stigma exertion had no effect on the deposition by honey bees. For both butterflies and honey bees, pollen deposition on the recipient flower increased with the amount of pollen removed from the donor. Furthermore, Conner and Rush (1996) also observed the effects of flower number and the size on visitation of syrphid flies and small bees to wild radish for 3 years could clarify the positive correlation between both flower number and corolla size and the number of flower-visiting syrphid flies. Increase in flower size might cause a little increase for small bee. Additionally, they observed that the 12 small bees of three families (Anthophoridae, Colletidae, and Halictidae) and syrphid flies of nine genera visited to *R. raphanistrum* in 2 years. Sahli and Conner (2007) inspected the pollinators to *R. raphanistrum* and showed that sweat bees of the genus *Dialictus* visited more frequently than the syrphid flies *Toxomerus*, *Syritta*, and *Eristalis*. These four genera accounted for 81% of visitors to *R. raphanistrum,* although 14 visitors of 15 genera were effective pollinators. On the other hand, effects on seed setting widely varied from 0.10 seeds a visit of *Sphaeorphoria* to 1.66 seeds a visit of *Apis*. Larger pollinators were more effective than smaller ones for seed setting. Concerning to the differentiation of quantitative traits in *R. raphanistrum*, they concluded that the no-rosette and the early flowering traits might be the significant adaptations enabling *R. raphanistrum* to be a major agricultural weed.

11.8 Introgression Between the Cultivated Radish and Wild Radish in California

Both cultivated radish and *R. raphanistrum* were introduced to the San Francisco Bay area from Europe in the middle of 1800s. *R. raphanistrum* might be introduced as a weed contaminant together with cereals (Robbins 1940).

From a study on the wild *R. sativus* and *R. raphanistrum* by Panetsos and Baker (1967), the differentiation of two species was confirmed. The wild *R. sativus* has the distinctive traits including the white or partially purple flower on a white background, and tender and slightly thick pod made of spongy parenchyma. On the other hand, *R. raphanistrum* has yellow flower, and slender and hardy pod breaking into pieces when ripened. In California, the above two species were growing with their intermediate types that would be bred from natural hybridization between them. The F_1 plants induced by artificial crossing between the two species showed intermediate characteristics, chromosome configurations of 1IV + 7II at first metaphase (MI) of PMCs and moderate fertility of 50% in both pollen formation and seed setting. From these facts, it was suggested that the F_1 plants were carrying some reciprocal translocations in one pair of chromosomes. When the wild *R. sativus* type was naturally backcrossed to cultivated radish, some progenies did not form the quadrivalent chromosomes, and had showed high seed setting. These facts suggest that the introgression might occur from *R. raphanistrum* to cultivated radishes in the course of acclimatization of radishes. Using eight morphological traits and 10 allozyme loci, Hegde et al. (2006) analyzed the genetic features of more than 50 wild radish populations collected from Cannon Beach, Oregon, through California and south of Santo Tomas, northern Baja California, Mexico, including both the coastal and inland valley areas. They concluded that California wild radishes were intermediate in specific combination of traits, and confirmed the introgression among populations in California. Today, the genetically pure populations of *R. raphanistrum* indicated by Panetsos and Baker (1967) have been replaced by the hybrid populations in these 40 years.

Campbell et al. (2006) investigated the genetic changes using four *R. raphanistrum* collected from Michigan and hybrid populations (*R. raphanistrum* × *R. sativus*). Hybrid and wild populations showed similar growth for four generations, and the pollen fertility of hybrid ones was enhanced progressively. The advanced progenies showed lower fertility than *R. raphanistrum* exhibiting ca. 270% greater lifetime fecundity and ca. 22% greater survival rate than California wild radish. They proposed a hypothesis that crop–wild hybridization might generate new genotypes with potentials enough to surpass the parental species under new environments.

Ridley et al. (2008) studied the evolutionary processes in four cultivars of radish, six populations of European *R. raphanistrum* and 11 populations of Californian wild radish analyzing the *trn*L–*rpl*132 intergenic region of chloroplast DNA. Eight haplotypes were identified in cpDNA of the genus *Raphanus*. Cultivated radishes had B and D types and *R. raphanistrum* had five haplotypes (A, C, E, F, and H). No haplotype was commonly shared between *R. raphanistrum* and cultivated radishes. On the other hand, the Californian wild radish possessed six haplotypes, B and D of which were similar to *R. sativus*, and A, E and F were to *R. raphanistrum*, respectively. Only haplotype G was independently unique. In the geographical distribution, little genic inclination was observed either on North–South axis or West–East (coastal–inland) axis. Genetic diversity among populations of Californian wild radish was 0.412, whereas that of *R. raphanistrum* was 0.742. It was confirmed that the Californian wild radishes originated through hybridizations between *R. sativus* and *R. raphanistrum*. Ridley and Ellstrand (2009) investigated on the Californian wild radish populations using the reproductive traits including days to flowering, length of the longest leaf at flowering, root crown diameter at flowering, number of pods per plant, average number of seeds per pod, and total seeds per plant. Survival rates for both Californian wild radish and *R. raphanistrum* were higher over years and growing sites. Californian wild radish showed intermediate phenotypes to its progenitors in many traits at the beginning of flowering. For potentials of pod and seed production, a significant correlation could be identified between genotype and environment. Californian wild radish has become more fertile than parental progenitors through ca. 150 generations since the original interspecific hybridization. Ridley and Ellstrand (2009) suggested that the progressive hybridization might contribute to enhancement of fertility contributing to higher survival rate and reproductive potential.

Eber et al. (1994, 1998) also confirmed the introgression from *R. raphanistrum* into cultivated radish by the analysis of hybrids between the native *R. raphanistrum* and cultivated radish in France. On the other hand, Kato and Fukuyama (1982) observed normal chromosome configuration of 9II in meiosis and high seed setting in F_1 hybrids between the cultivated radish and *R. raphanistrum* L. ssp. *landra*. From these results, it could be assumed that the chromosomal reconstruction might occur even in *R. raphanistrum*.

11.9 Intergeneric Hybridization Between Wild Radishes and *B. napus* and Its Implications on Gene Flow from Transgenic Rapeseeds

The rapeseed (*B. napus*, $2n = 38$, AACC) is known as incompletely self-fertilized crop with the outcrossing rate of 30% (Rahow and Woods 1987; Rieger et al.

2001). It is feared that the transgene of transgenic rapeseeds would be introduced into wild relatives through hybridization, and the risk of gene dispersal into crops was assumed to be high. The successive steps to assess the probability of transgene transfer into wild relatives are (1) the crops and wild relatives grow side by side and come simultaneously into flower, (2) pollen scatter and seed dispersal, (3) production of viable and fertile F_1 hybrids, (4) stable introgression of the transgene into wild relatives, and (5) persistence of introgressed gene in natural populations (Prieto et al. 2005; Devos et al. 2007).

As to the crossability to *B. napus*, Devos et al. (2007) pointed out that *R. raphanistrum* in Belgium was a member of rank 3. *B. rapa* and *Hirschfeldia incana* were classified to rank 1 and rank 2 in hybridizing potentials for introgression. Kelran et al. (1992) investigated pollen germination and hybrid production in vitro for risk assessment of outcrossing between transgenic rapeseed and *R. raphanistrum*. In the case of *B. napus* × *R. raphanistrum*, no pollen tube penetrated into the style of *B. napus*. In the reciprocal cross, on the other hand, the germination percentage of rapeseed pollen grains was significantly lower (12%) than the former, although the pollen tubes of rapeseed did not penetrate into stigma of *R. raphanistrum*. As a result, they obtained a few hybrids ($2n = 28$, ACR and RAC) in reciprocal crossings by ovary culture.

Prieto et al. (2005) developed cross-specific transposable element (SINE) markers for screening of introgression of rapeseed genes into *R. raphanistrum*. In 47 markers observed, 30 revealed polymorphism in which 17 markers were dominant and 13 were codominant. Polymorphic markers were mapped on 10 linkage groups of *B. napus*. They presumed that the SINE markers might provide efficient tools to analyze the introgression from transgenic rapeseeds to *R. raphanistrum*.

Based on these information, we could survey the transgenic hybridization in reciprocal crossings between *B. napus* and *R. raphanistrum* under natural conditions as shown in Table 11.4.

Eber et al. (1994) analyzed 7,018 seeds obtained from the crossing between *B. napus* and *R. raphanistrum*. All plants obtained from large seeds (longer than 2 mm in diameter) were not hybrid. On the other hand, 3,734 small seeds (shorter than 1.6 mm) developed into 188 hybrids were $2n = 28$ with ACRr genome. One plant was ascertained to be amphidiploid ($2n = 56$, AACCRrRr). The $2n = 28$ plants formed trivalents and quadrivalents in 9.2% of the PMCs, suggesting the possibility of crossing-over between AC and Rr. In the isolated field where the intergeneric hybrid plants were surrounded with *R. raphanistrum*, 8.5% of the 281 F_1 plants were confirmed to produce pods. From 234 pods obtained, 20 seeds were larger than the F_1 seeds. In the case of rapeseed field, 17 plants out of 20 F_1 plants developed pods, and 567 seeds were obtained. From this result, they assumed that fertile progenies could be maintained by backcross with parental donor. Baranger et al. (1995) examined the spontaneous outcrossing of six male-sterile (CMS) rapeseed lines and *R. raphanistrum*. The 42.4% of 651,400 seeds were large (longer than 1.6 mm in diameter) and 57.6% were small (less than 1.6). All the 244 plants obtained from large seeds were true rapeseed. Among 240 plants from small seeds, however, many triploid hybrids ($2n = 28$, ACRr) were bred, in addition to AACCRrRr amphidiploids with $2n = 56$, diploidal AACC ($2n = 38$) and haploid AC ($2n = 19$) plants. It was suggested that the frequent genomic reconstitution was proceeding in those plants with $2n = 28$ from small seeds. Confirming the production of BC_1 plants, they suggested that close relative between Rr genome and the A or C genome might explain the wide hybridization with *B. napus*. Chevre et al. (1996) obtained from 2.8 seeds per 100 pollinated flowers to 23.8 seeds in the pollination of CMS rapeseeds with *R. raphanistrum* under field conditions for 3 years. By the cytogenetical analysis of the 427 F_1 plants, 423 were ACRr with $2n = 28$ and four plants were AACCRrRr with $2n = 56$, respectively. The BC_1 seeds were produced from F_1 hybrids with ACRr genome through open pollination. They observed multivalents at MI of PMCs of BC_1 plants with $2n = 37$ (AACCRrRr) and $2n = 56$. From these facts, they suggested that the chromosome rearrangement between transgenic rapeseeds and *R. raphanistrum* might occur in their following progenies. Lefol et al. (1997) performed hybridizations of *R. raphanistrum* with *B. napus*, *B. rapa* and *B. juncea* in the reciprocal crossings. Only two seeds (0.4 seeds per 100 pollinated flowers) were obtained from the crossing of *B. napus* with wild radish, three and 13 seeds were obtained from *B. rapa* and *B. juncea*, respectively. In these crossings, only two plants generated in the cross between *B. napus* and wild radish were confirmed to be true hybrid. The F_1 plants produced four seeds by the backcrossing with *B. napus*. In the

Table 11.4 Production of F_1 hybrid and their progenies between *B. napus* and wild radish under natural pollinations

Combination	Researchers and years published	Production of F_1 hybrid[a]	Production of progeny from F_1 hybrid[b]
B. napus × wild radish	Eber et al. (1994)	○	○
	Baranger et al. (1995)	○	○
	Chevre et al. (1996)	○	○
	Lefol et al. (1997)	○	○
	Darmency et al. (1998)	○	○
	Chevre et al. (2000)	○	?
	Rieger et al. (2001)	○	○
	Chevre et al. (2007)	○	○
Wild radish × *B. napus*	Lefol et al. (1997)	×	–
	Darmency et al. (1998)	○	○
	Chevre et al. (2000)	○	?
	Rieger et al. (2001)	×	–
	Thalmann et al. (2001)	×	–
	Warwick et al. (2003)	○	?
	Halfhill et al. (2004)	×	–
	Daniels et al. (2005)	×	–

[a]○ More than one hybrid was obtained successful in hybridization, × no hybrid, ? Unknown
[b]○ Production of progenies by backcrossing with parents or by self-pollination, ? Unknown

reciprocal cross, no hybrid plant could be obtained in the crosses between wild radish and *Brassica* species. Darmency et al. (1998) investigated the spontaneous hybridization between rapeseed and wild radish by growing them in cages and field. When CMS rapeseed and wild radish were grown in the field in a ratio of 1:1 and 1:2, 40% and 80% of the germinated seeds developed into true hybrid, respectively. On the other hand, only two spontaneous hybrid plants were obtained from 147,671 seeds when plant materials were grown in cage. The chromosome number of these plants were $2n = 28$ suggesting to be the genome constitution of RrAC. Furthermore, hybrids were fertile in the following progenies. Chevre et al. (2000) obtained 23 hybrids from 73,847 seedlings in the cross between CMS rapeseeds and *R. raphanistrum*, hybridization ratios of which ranged from 2×10^{-5} to 5×10^{-4}. The chromosome number of hybrids were shown to be $2n = 28$, $2n = 37$ and $2n = 56$ for 18 triploids (ACRr), one tetraploid (ACRrRr), and four amphidiploids (AACCRrRr), respectively. In contrary, only one hybrid (RrRrAC, $2n = 37$) was developed from 189,084 seedlings in the reciprocal cross. Most of the hybrids were obtained from small seeds shorter than 1.6 mm in diameter. The chromosome configuration at MI of PMCs in ACRr and RrRrAC hybrids showed trivalents and/or quadrivalents suggesting the allosyndetic chromosome pairing between Rr genome and A or C genome. Rieger et al. (2001) tried the natural hybridization in vivo, where one or four plants of *R. raphanistrum* were planted randomly in each 100 m^2 field of rapeseeds. Two hybrids with $2n = 56$ (AACCRrRr) were obtained from 52,000,000 seedlings via rapeseeds in 2 years. These hybrids were fertile in selfing and backcrossing with *R. raphanistrum*. On the other hand, no hybrid was produced from 25,000 seedlings via *R. raphanistrum*. Thalmann et al. (2001) studied on the spontaneous hybridization between *R. raphanistrum* collected from three regions on Swiss Plateau and rapeseeds under agricultural conditions. A total of 754 plants developed from *R. raphanistrum* were confirmed not to be hybrid not only by the flow cytometry but also by the RAPD analysis. These facts could suggest that higher hybridization barrier might operate between *R. raphanistrum* as recipient and rapeseeds as pollen donor, and gene flow may be extremely rare. Warwick et al. (2003) studied on the natural cross of *R. raphanistrum* with rapeseeds using seven Canadian populations and six European ones in greenhouse and using three Canadian populations and one population from France under field conditions. No hybrid developed from the experiment in the former. On the other hand, in field conditions, only one hybrid was found from 32,821 individuals of four populations, showing 0.003% of hybridization ratios. The chromosome number of this hybrid was $2n = 34$ or 35 with genomic constitution of RrRrAC. From this and previous data (Chevre et al. 1998, 2000), Warwick et al. (2003)

concluded that hybridization between *R. raphanistrum* and *B. napus* was extremely rare. Halfhill et al. (2004) studied on hybridization between 11 transformed rapeseed lines and *R. raphanistrum* under a higher crop ratio (600:1) and a lower one (180:1) to wild radish. No natural hybridization was confirmed in 19,274 seedlings screened with green fluorescent protein (GFP) phenotype. By contrast, 9.7% of hybridization rate was observed between *B. rapa* and *B. napus* in the same condition. Daniels et al. (2005) observed the intergeneric hybridization between *R. raphanistrum* or *R. sativus* and six related species in the cultivated conditions. Seedlings grown from 1,793 seeds via *R. raphanistrum* in 3 years and 117 seedlings via *R. sativus* in 1 year developed no herbicide tolerant plants. Chevre et al. (2007) studied on the evolution of hybrid genome during the successive generations followed by the formation of intergeneric hybrid of *B. napus* × *R. raphanistrum*, in order to set up a framework for gene flow modeling between these two species. They used seven CMS genetically modified (GM)-rapeseed lines that had different insertion sites of the same transgene carrying the same genetic background. More than 500 plants were the true intergeneric F_1 hybrids of ACRr genomic construction with $2n = 28$. A majority of BC_1 plants were expected to have $2n = 37$ chromosomes, but the chromosome number showed a wide range from 24 to 80. When the BC_1 plants with $2n = 28$ to 64 were successively backcrossed with *R. raphanistrum*, the chromosome number decreased progressively from BC_2 to BC_4 generations.

As described above, using the CMS rapeseeds as recipient in hybridization production between *R. raphanistrum* and *B. napus*, a large number of plants really showed hybridity. *B. napus* forming normal pollen was not bred by hybridization but by self-pollination within themselves. As a seed parent, *R. raphanistrum* is so incompatible to the related species that the risk of gene introgression from GM rapeseed to *R. raphanistrum* was extremely low.

11.10 Prospects for Wild Radish

Wild radish has not provided the agronomic traits except for CMS for hybrid seed production system. However, it may offer useful information for genetics and breeding of cultivated radishes in near future.

References

Agnihotri A, Shivanna KR, Raina SN, Lakshmikumaran M, Prakash S, Jagannathan V (1991) Production of *Brassica napus* × *Raphanobrassica* hybrids by embryo rescue: an attempt to introduce shattering resistance into *B. napus*. Plant Breed 105:292–299

Aoba T (1967) On wild radish, Nora-daikon (*Raphanus* spp.). Bull Fac Agri Yamagata Univ 24:7–12 (in Japanese)

Aoba T (1981) Geographical distribution of domestic varieties in radish. In: Vegetables. Publication Department, Housei University, Tokyo, Japan, pp 232–251 (in Japanese)

Aoba T (1988) Pedigree and variation of wild radish in Japan. Technol Agric 12:94–114 (in Japanese)

Arumuganathan K, Earle ED (1991) Nuclear DNA content of some important plant. Plant Mol Biol Rep 9:208–218

Banga O (1976) Radish, *Raphanus sativus* (Cruciferae). In: Simmonds NW (ed) Evolution of crop plants. Longman, London, UK, pp 60–62

Baranger A, Chevre AM, Eber F, Renard M (1995) Effect of oilseed rape genotype on the spontaneous hybridization rate with a weedy species: an assessment of transgene dispersal. Theor Appl Genet 91:956–963

Bostrom U (2003) Yield loss in spring-sown cereals related to the weed flora in the spring. Weed Sci 51:418–424

Brown GG, Formanova N, Jin H, Wargachuk R, Dendy C, Patil P, Laforest M, Zhang J, Cheung WY, Landry BS (2003) The radish *Rfo* restorer gene of Ogura cytoplasmic male sterility encodes a protein with multiple pentatricopeptide repeats. Plant J 35:262–272

Campbell LG, Snow AA, Ridley CE (2006) Weed evolution after crop gene introgression: greater survival and fecundity of hybrids in a new environment. Ecol Lett 9:1198–1209

Cheam AH, Code GR (1995) The biology of Australian weeds 24. *Raphanus raphanistrum* L. Plant Protect Quart 10:2–13

Chevre AM, Eber F, Baranger A, Kerlan MC, Barret P, Festoc G, Vallee P, Renard M (1996) Interspecific gene flow as a component of risk assessment for transgenic *Brassicas*. Acta Hortic 407:169–179

Chevre AM, Eber F, Baranger A, Hureau G, Barret P, Picault H, Renard M (1998) Chracterization of backcross generations obtained under field conditions from oilseed rape-wild radish F_1 interspecific hybrids: an assessment of transgene dispersal. Theor Appl Genet 97:90–98

Chevre AM, Eber F, Darmency H, Fleury A, Picault H, Letanneur JC, Renard M (2000) Assessment of interspecific hybridization between transgenic oilseed rape and wild radish under normal agronomic conditions. Theor Appl Genet 100:1233–1239

Chevre AM, Adamczyk K, Eber F, Huteau V, Coriton O, Letanneur JC, Laredo C, Jenczewski E, Monod H (2007) Modelling gene flow between oilseed rape and wild radish. I. Evolution of chromosome structure. Theor Appl Genet 114:209–221

Conner JK, Rush S (1996) Effects of flower size and number on pollinator visitation to wild radish, *Raphanus raphanistrum*. Oecologia 105:509–516

Conner JK, Davis R, Rush S (1995) The effect of wild radish floral morphology on pollination efficiency by four taxa of pollinators. Oecolgia 104:234–245

Daniels R, Boffey C, Mogg R, Bond J, Clarke R (2005) The potential for dispersal of herbicide tolerance genes from genetically-modified, herbicide-tolerant oilseed rape crops to wild relatives. Final Report to Defra, pp 3–23

Darmency H, Lefol E, Fleury A (1998) Spontaneous hybridizations between oilseed rape and wild radish. Mol Ecol 7:1467–1473

Devos Y, Schrijver ADe, Reheul D (2007) Using an oilseed rape × wild/weedy relative gene flow index for the monitoring of transgenic oilseed rape. J Verbr Lebensm 2(Suppl 1): 88–89

Dixon GR (2007) Origin and diversity of *Brassica* and its relatives. In: Atherton J, Rees A (Ser Eds) Crop production science in horticulture 14. Vegetable Brassicas and Related Crucifers. CABI, Wallingford, UK, pp 1–33

Eber F, Chevre AM, Baranger A, Vallee P, Tanguy X, Renard M (1994) Spontaneous hybridization between a male-sterile oilseed rape and two weeds. Theor Appl Genet 88:362–368

Eber F, Boucherie R, Broucqsault LM, Bouchet Y, Chevre AM (1998) Spontaneous hybridization between vegetable crops and weeds. 1: Garden radish (*Raphanus sativus* L.) and wild mustard (*Sinapis arvensis* L.). Agronomie 18:489–497

Fujii T (1977) Radish asahi-Encyclopedia. World Plant 61:1403–1415 (in Japanese)

Furusato K, Miyazawa A (1958) Varieties of Japanese radish viewed from horticulture. In: Nishiyama I (ed.) Japanese Radish. Jpn Soc From Sci, Tokyo, pp 138–161 (in Japanese)

George RAT (1985) Radish: *Raphanus sativus* L. In: vegetable seed production. Longman, London, UK, pp 148–151

Giancola S, Rao Y, Chaillou S, Hiard S, Martin-Canadell A, Pelletier G, Budar F (2007) Cytoplasmic suppression of Ogura cytoplasmic male sterility in European natural populations of *Raphanus raphanistrum*. Theor Appl Genet 114:1333–1343

Halfhill MD, Zhu B, Warwick SI, Raymer PL, Millwood RJ, Weissinger AK, Stewart CN Jr (2004) Hybridization and backcrossing between transgenic oilseed rape and two related weed species under field conditions. Environ Biosaf Res 3:73–81

Harberd DJ (1976) Cytotaxonomic studies of *Brassica* and related genera. In: Vaughan JG, MacLeod AJ, Jones BMG (eds) The biology and chemistry of the Cruciferae. Academic, London, UK, pp 47–68

Hegde SG, Nason JD, Clegg JM, Ellstrand N (2006) The evolution of California's wild radish has resulted in the extinction of its progenitors. Evolution 60:1187–1197

Hida K (1990) Root crops: 'Radish'. In: Matsuo T (ed.) Collected data of plant genetic resources, vol 2. Koudansha, Tokyo, Japan, pp 823–834 (in Japanese)

Hinata K (1995) Radish asahi-Encyclopedia. World Plant 67:206–207 (in Japanese)

Holm L, Doll J, Holm E, Pancho J, Herberger J (1997) World weed. Natural histories and distribution. Wiley, New York, USA

Huh MK, Ohnishi O (2001) Allozyme diversity and population structure of Japanese and Korean populations of wild radish, *Raphanus sativus* var. *hortensis* f. *raphanistroides* (Brassicaceae). Genes Genet Syst 76:15–23

Huh MK, Ohnishi O (2006) Genetic diversity and genetic relationships of East Asian natural populations of wild radishes revealed by AFLP. Breed Sci 52:79–88

Iwabuchi M, Koizuka N, Fujimoto H, Sakai T, Imamura J (1999) Identification and expression of the kosena radish (*Raphanus sativus* cv. Kosena) homologue of the ogura radish CMS-associated gene, *orf138*. Plant Mol Biol 39:183–188

Johnston JS, Pepper AP, Hall AE, Chen ZJ, Hodnett G, Drabek J, Lopez R, Price HJ (2005) Evolution of genome size in Brassicaceae. Ann Bot 95:229–235

Kato M, Fukuyama T (1982) Production of *Raphanus sativus* 'Aokubi-miyashige' with *R. raphanistrum* 'Seiyo-nodaikon' cytoplasm. I. Process nucleus substitution and selection in the 4th-6th generations. Bull Exp Farm Coll Agric EhimeUniv 4:29–37

Kelran MC, Chevre AM, Eber F, Baranger A, Renard M (1992) Risk assessment of outcrossing of transgenic rapeseed to related species: I. Interspecific hybrid production under optimal conditions with emphasis on pollination and fertilization. Euphytica 62:145–153

Kitamura S (1958) Varieties and transitions of radish. In: Nishiyama I (ed) Japanese radish. Japanese Science Society, Tokyo, Japan, pp 1–19 (in Japanese)

Koizuka N, Imai R, Fujimoto H, Hayakawa T, Kimura Y, Kohno-Murase J, Sakai T, Kawasaki S, Imamura J (2003) Genetic characterization of a pentatricopeptide repeat protein gene, *orf687*, that restores fertility in the cytoplasmic male-sterile Kosena radish. Plant J 34:407–415

Lefol E, Seguin-Swartz G, Downey RK (1997) Sexual hybridization in crosses of cultivated *Brassica* species with the crucifers *Erucastrum gallicum* and *Raphanus raphanistrum*: potential for gene introgression. Euphytica 95:127–139

Lehtila K, Brann KH (2007) Correlated effects of selection for flower size in *Raphanus raphanistrum*. Can J Bot 85:160–166

Matsuzawa Y, Kaneko Y, Bang SW (1996) Prospects of the wide cross for genetics and plant breeding in *Brassiceae*. Bull Coll Agric Utsunomiya Univ 16:5–10

Namai H (1991) Breeding using cytoplasmic male sterility (CMS) in radish. Jpn Soc Seed Prod Breed Eng 1991: 165–181 (in Japanese)

Ogura H (1968) Studies on the new male-sterility in Japanese radish, with special reference to the utilization of this sterility towards the practical raising of hybrid seeds. Mem Fac Agric Kagoshima Univ 6:39–78

Ohnishi O (1999) Chlorophyll-deficient and several morphological genes concealed in Japanese natural populations of wild radish *Raphanus sativus* var. hortensis f. raphanistroides. Genes Genet Syst 74:1–7

Panetsos CA, Baker HG (1967) The origin of variation in wild Raphanus sativus (Cruciferae) in California. Genetica 38:243–274

Prakash S, Takahata Y, Kirti PB, Chopra VL (1999) Cytogenetics. In: Gomez-Campo C (ed) Biology of *Brassica* Coenospecies. Elsevier Science, Amsterdam, pp 59–106

Prieto JL, Pouilly N, Jenczewski E, Deragon JM, Chevre AM (2005) Development of crop-specific transposable element (SINE) markers for studying gene flow from oilseed rape to wild radish. Theor Appl Genet 111:446–455

Rahow G, Woods D (1987) Outcrossing in rape and mustard under Saskatchewan prairie conditions. Can J Plant Sci 67:147–151

Ridley CE, Ellstrand NC (2009) Evolution of enhanced reproduction in the hybrid-derived invasive, California wild radish (*Raphanus sativus*). Biol Invas 11:2251–2264

Ridley CE, Kim S-C, Ellstrand NC (2008) Bidirectional history of hybridization in California wild radish, *Raphanus sativus* (Brassicaceae), as revealed by chloroplast DNA. Am J Bot 95:1437–1442

Rieger MA, Potter TD, Preston C, Powles SB (2001) Hybridization between *Brassica napus* L. and *Raphanus raphanistrum* L. under agronomic field conditions. Theor Appl Genet 103:555–560

Robbins WW (1940) Alien plants growing without cultivation in California. University of California Agricultural Experiment Station Bulletin 637. University of California, Berkeley, USA

Sahli HF, Conner JK (2007) Visitation, effectiveness, and efficiency of 15 genera of visitors to wild radish, *Raphanus raphanistrum* (Brassicaceae). Am J Bot 94: 203–209

Sahli HF, Conner JK, Shaw FH, Howe S, Lale A (2008) Adaptive differentiation of quantitative traits in the globally distributed weed, wild radish (*Raphanus raphanistrum*). Genetics 180:945–955

Sakai T, Imamura J (1994) Somatic hybridization between radish (*Raphanus sativus*) and rapeseed (*Brassica napus*). In: Bajaj YPS (ed) Biotechnology in agriculture and forestry, vol 27; Somatic hybridization in crop improvement. Springer, Berlin, pp 320–333

Sakai T, Imamura J (2003) Development and utilization of hybrid-seed production system in rapeseeds (*Brassica napus* L.). Breed Res 5:93–102

Salisbury PA (1987) Blackleg resistance in weedy crucifers. Cruciferae Newsl 12:90

Sasaki H (1994) Encyclopedia for wholly enjoyment of radish. Nobunkyou, Tokyo, Japan, pp 1–112 (in Japanese)

Streibig JC, Combellack JH, Pritchard GH, Richardson RG (1989) Estimation of thresholds for weed control in Australian cereals. Weed Res 29:117–126

Terachi T, Yamaguchi K, Yamagishi H (2001) Sequence analysis on the mitochondrial *orfB* locus in normal and Ogura male sterile cytoplasm from wild and cultivated radishes. Curr Genet 40:276–281

Thalmann C, Guadagnuolo R, Felber F (2001) Seach for spontaneous hybridization between oilseed rape (*Brassica napus* L.) and wild radish (*Raphanus raphanistrum* L.) in agricultural zones and evaluation of the genetic diversity of the wild species. Bot Helv 111(2):107–119

Tsunoda S (1979) Ecology of wild species of the tribe *Brassiceae* and origin of cultivation. In: Recent advances in breeding. Jpn Soc Breed 20:41–45 (in Japanese)

Warwick SI (1993) Guide to the wild germplasm of *Brassica* and allied crops. Part IV. Wild species in the tribe *Brassiceae* (*Cruciferae*) as sources of agronomic traits. Tech Bull 1993-17E:1–19

Warwick SI, Simard M-J, Legere A, Beckie HJ, Braun L, Zhu B, Mason P, Seguin-Swartz G, Stewart CN Jr (2003) Hybridization between transgenic *Brassica napus* L. and its wild relatives: *Brassica rapa* L., *Raphanus raphanistrum* L., *Sinapis arvensis* L., and *Erucastrum gallicum* (Willd.) O. E. Schulz. Theor Appl Genet 107:528–539

Webster TM, MacDonald GE (2001) A survey of weeds in various crops in Georgia. Weed Technol 15:771–790

Yamagishi H (1998) Distribution and allelism of restores genes for Ogura cytoplasmic male sterility in wild and cultivated radishes. Genes Genet Syst 73:79–83

Yamagishi H (2004) Assessment of cytoplasmic polymorphisms by PCR-RFLP of the mitochondrial *orfB* region in wild and cultivated radishes (*Raphanus*). Plant Breed 123:141–144

Yamagishi H (2006) Phylogeny of wild and cultivated *Raphanus* and the origin of Ogura male-sterile cytoplasm. Breed Res 8:107–112 (in Japanese)

Yamagishi H, Terachi T (1994) Molecular and biological studies on male sterile cytoplasm in Cruciferae. II. The origin of Ogura male sterile cytoplasm inferred from the segregation pattern of male sterility in the F_1 progeny of wild and cultivated radishes (*Raphanus sativus* L.). Euphytica 80:201–206

Yamagishi H, Terachi T (1996) Molecular and biological studies on male-sterile cytoplasm in the Cruciferae. III. Distribution of Ogura-type cytoplasm among Japanese wild radishes and Asian radish cultivars. Theor Appl Genet 93: 325–332

Yamagishi H, Terachi T (1997) Molecular and biological studies on male-sterile cytoplasm in *Cruciferae*. IV. Ogura-type cytoplasm found in the wild radish, *Raphanus raphanistrum*. Plant Breed 116:323–329

Yamagishi H, Terachi T (2001) Intra- and inter-specific variations in the mitochondrial gene *orf138* of Ogura-type male-sterile cytoplasm from *Raphanus sativus* and *Raphanus raphanistrum*. Theor Appl Genet 103:725–732

Yamagishi H, Terachi T (2003) Multiple origins of cultivated radishes as evidenced by a comparison of the structural variations in mitochondrial DNA of *Raphanus*. Genome 46:89–94

Yamagishi H, Tateishi M, Terachi T, Murayama S (1998) Genetic relationships among Japanese wild radishes (*Raphanus sativus* f. *raphanistroides* Makino), cultivated radishes and *R. raphanistrum* revealed by RAPD. J Jpn Soc Hortic Sci 67:526–531

Yamagishi H, Iida T, Ishibashi A, Ozaki A (2005) Origin of cultivated radishes inferred by PCR-RFLP based on the sequences in chloroplast genome. Breed Res (Suppl 1–2): 308 (in Japanese)

Yamane K, Na Lu, Ohnishi O (2005) Chloroplast DNA variations of cultivated radish and its wild relatives. Plant Sci 168:627–634

Yamane K, Na Lu, Ohnishi O (2009) Multiple origins and high genetic diversity of cultivated radish inferred from polymorphism in chloroplast simple sequence repeats. Breed Sci 59:55–65

Yasumoto K, Matsumoto Y, Terachi T, Yamagishi H (2008) Restricted distribution of *orf687* as the pollen fertility restore gene for Ogura male sterility in Japanese wild radish. Breed Sci 58:177–182

Chapter 12
Solanum

Gavin Ramsay and Glenn Bryan

12.1 Basic Botany of the Genus *Solanum*

Solanum is a large genus of around 1,500 species with a near-cosmopolitan distribution, and although only one genus of around 90 genera in the Solanaceae family includes almost half of the species in the family. The Solanaceae family sits in the Euasterid I group alongside the Boraginaceae, the Lamiales (e.g., the Lamiaceae and the Oleaceae), and the Gentianales (e.g., Gentianaceae) (Angiosperm Phylogeny Group 2003; Haston et al. 2007). The centers of diversity for the genus are largely tropical, but the adaptability of the group is demonstrated by the occurrence of representatives in cold-temperate environments and at the altitudinal maximum for vegetation around 4,500 m in parts of the Andes for some tuber-bearing *Solanum* species (Hawkes 1990). Life forms in the genus vary from small, high-altitude, rosette-forming species to rambling lianes, arboreal epiphytes, and bushy and woody perennial forms. South America hosts around one third of the worldwide total species in the genus (Hunziker 1979). In a recent molecular phylogenetic study using three chloroplast genes, the established infrageneric structure was found to be inaccurate. It is now suggested that the genus comprises two large clades with around ten subclades, each with informal names pending further investigation (Weese and Bohs 2007).

Solanum species are generally pollinated by bees. Flowers are pentamerous and usually radially symmetrical but sometimes zygomorphic and without floral nectaries, thus providing only a pollen reward for pollinators. Although most members of the genus release pollen through a terminal pore, there are also longitudinal-dehiscent and intermediate types within the genus (García et al. 2008). Limited access to pollen in enclosed anthers is a feature of flowers adapted for buzz pollination by bee species capable of the appropriate behavior. In such flowers, bees access the pollen by vibration of the anther cone and the collection of the resultant flow of pollen from the pores. The robust anther cone associated with such a pollination syndrome appears, from study of the microstructure, to have evolved more than once in the genus (Glover et al. 2004). Corolla color varies from white to purple and occasionally yellow, often with prominent floral guides. Although white flowers are often associated with pollinators flying at night, in a study in a subtropical moist forest in Belize, several white-flowered species opened their flowers only in daylight hours, whereas some species with white or pigmented corollas flowered intermittently at any time (Smith and Knapp 2002).

The genus *Solanum* has contributed several crop species of global standing. The fourth crop in terms of global yield and the most important non-cereal food crop is the potato, *Solanum tuberosum* L. Additional species grown by man include the tomato (*S. lycopersicum* L., reviewed in a separate chapter), the eggplant, aubergine, or brinjal, (*Solanum melongena* L.), the nakati or Ethiopian eggplant (*S. aethiopicum* L.), the naranjilla (*S. quitoense* Lam.), and the invasive tropical shrub the Turkey berry or devil's fig (*Solanum torvum* Sw.).

The potato, as a staple foodstuff, is a crop that supported civilizations in South America over several millennia, rose to a similar status in the western fringes of Europe in the eighteenth and nineteenth centuries,

G. Ramsay (✉)
The James Hutton Institute, Invergowrie, Dundee DD2 5DA, UK
e-mail: gavin.ramsay@hutton.ac.uk

C. Kole (ed.), *Wild Crop Relatives: Genomic and Breeding Resources, Vegetables*,
DOI 10.1007/978-3-642-20450-0_12,

and is now lauded as an answer to food insecurity and a means of meeting the Millennium Development Goals in a wide range of developing countries. The localized production and lack of trade as a global commodity insulates consumers of potato from the volatility of global markets, and instead makes price largely determined by local factors (Prakash 2008). Although the majority of production is now processed in North America and some European countries (Kirkman 2007), globally only around 15% of the crop is processed. In the last 20 years, production in the traditional main areas in Europe has suffered a gentle decline, but there has been a large increase in demand and production in Latin America, Africa and Asia, where China and India now produce more than one-third of the global harvest (FAOSTAT 2008).

The cultivated potato was domesticated by the tribes that preceded the Huari and Inca peoples around 8,000 years ago in South-Central Peru. The initial domestication was from species in the *Solanum bukasovii* Juz. complex (Spooner et al. 2005: see Sect. 12.3 for further details) suggesting that cultivation began at around 3,000–4,500 m asl. According to Hawkes (1990) the potato belongs to a group of over 220 tuber-bearing species from section Petota within the genus *Solanum*. Since its initial domestication, the potato has migrated both north and south in the Andes, moved to lower altitudes in the Andes, and found its way to coastal Chile where forms developed which were more suited to temperate agriculture. The initial introductions (of Andean stock) into Europe were made in the 1570s first reaching the Canary Isles. Subsequent spread took the potato as far as China in 1609 and finally the USA in 1719. Potatoes are now grown in 149 countries from latitudes 65°N to 50°S and at altitudes from sea level to over 4,000 m (Hijmans 2001).

The tomato (now *S. lycopersicum*) is also a major world crop with an origin in the Andes of South America (Jenkins 1948; Peralta and Spooner 2007). This crop and its wild relatives will be discussed elsewhere in this book series.

In the Old World, the genus *Solanum* has provided one major crop, the eggplant, aubergine or brinjal, *S. melongena*. This species grows as a perennial shrub a meter or so high in tropical regions, but as a crop in temperate zones it is grown as an annual. China produces more than half of the world harvest of eggplant ahead of India, Egypt and Turkey as other major producers. The main center of diversity of the eggplant is the Indian subcontinent and possibly other parts of Southeast Asia where an exceptionally wide range of fruit types can be found. Vavilov (1928) also considered India as the center of origin for the eggplant, with China as a later secondary center of diversity. Without archeological specimens to date the domestication of the crop, other means of dating domestication have been sought. Recent studies have suggested that the eggplant has been under selection in China since at least 59 BC (Wang et al. 2008). However *Sanskrit* documents take the recorded history of the crop back to at least 800 BC (Daunay and Janick 2007) by which time it was already a popular vegetable fruit in the Indian subcontinent. Most studies indicate that *S. incanum* L. or *S. insanum* L. is the wild progenitor of *S. melongena* (Behera et al. 2006) although an ultimate origin from *S. macrocarpon* L. in Africa cannot be ruled out. The history of the eggplant is further confused by the related African species *Solanum aethiopicum*, known as the Ethiopian eggplant or the nakati. This is a related fruit-bearing species, also used for leaf production to be cooked in a manner similar to spinach, and is thought to have originated from *Solanum anguivi* Lam. (Lester and Niakan 1986). There is a form of the species known as Gilo, now generally considered to be a cultivar group of *S. aethiopicum*. With additional molecular studies of the group, additional clarity on the interrelationships of these species may be forthcoming.

An additional New World *Solanum* species exploited by man is the naranjilla, *Solanum quitoense* Lam. This species is grown for its fruits and is found in Peru, southern Colombia and Ecuador and reaches 2.5 m high, forming a woody shrub when mature. It is found at altitudes between 1,200 and 1,800 m where it prefers humid environments and light shade. As with eggplant, the cultivated form has a greatly reduced complement of spines compared to the wild forms. A further shrubby fruit-bearing species originating in the Andes is *S. betaceum* Cav., until recently placed in the genus *Cyphomandra* (Bohs 1995). This species has entered international trade and is grown for export in Portugal and New Zealand, where it has recently been given the name the tamarillo for marketing purposes.

The basic chromosome number of $x = 12$ for the family is found in almost all *Solanum* species although a reduction to $x = 11$ has been reported for *S. mammosum* L. and *S. platense* Diekmann (Chiarini and

Bernardello 2006). Most species remain diploid, but there are both autopolyploid and allopolyploid species, including the most common form of the cultivated potato, *Solanum tuberosum* L. ($2n = 4x = 48$). Haploid genome sizes in the two major New World crops in the genus, potato and tomato, are 850 and 1,000 Mbp respectively, and the main Old World crop, the aubergine or egg plant, has a genome size of 1,100 Mbp (Arumuganathan and Earle 1991). These values are between six- and eight-fold greater than the *Arabidopsis* genome.

12.2 Conservation Initiatives

With most climate change scenarios predicting major disruption of food production in most regions in the middle decades of the twenty-first century, adapting crops to future climates and other changing pressures has never been more important. The store of diversity available for breeders in the genus *Solanum* is very extensive and represents a source of genes which could make a marked improvement to productivity in different environments and to the ability to resist the pressures from the rapidly changing ranges of pests and pathogens predicted under a changing climate.

A pressing threat to this adaptive and potentially valuable diversity in the wild relatives of *Solanum* crops comes from the habitat loss associated with the change of land use arising from population pressures. The IUCN Red Data list of species (IUCN 2009) includes 45 species from the genus *Solanum*, 11 of them already considered to be endangered or critically endangered. Although only one of the Red Data List *Solanum* species was a tuber-bearing relative of the potato, and it was categorized as of least concern, this masks a much broader concern about the sustainability of many wild potato populations. For example, Coca Morante et al. (2007) surveyed one of the great centers of diversity for tuber-bearing wild potato species, the Department of La Paz, Bolivia (Knapp 2002). In the five provinces in the north of the department visited during the study, habitat loss by encroaching agriculture and direct damage by pests and pathogens from that agriculture such as *Premnotrypes* sp. and *Phytophthora infestans* were threatening populations of wild species and it was considered that the local endemic wild potato species *S. achacachense* Cárd. was now in danger of extinction. Many of the high altitude tuber-bearing *Solanum* species live in scattered populations and little is known of their current status and vulnerabilities.

In some parts of the range of wild potatoes there has been, in the last decade or more, a wholesale conversion of pristine habitat into land with intensive agricultural production. This change in landscape use is on a much broader scale than that threatening *S. achacachense* in the Bolivian highlands and can be readily seen from satellite images and aerial photographs using tools such as Google Earth. For example, CPC 3757 *S. microdontum* Bitt. (http://www.hutton.ac.uk/cpc) is one accession from a locality that is now under intensive agriculture.

Habitat loss in the Central Andes may, to some extent, be curbed by the extensive protected areas established by national governments. Large areas of the highly diverse eastern Andes have already been deforested, with Yungas forest lost to the extent of 38%, 25% and 15% in three areas of Peru, and deforestation in eastern Bolivia accelerating to about 2,900 km^{-2} per year. The vegetation of the higher Andes is a little more secure with 54% of the area of high Andean vegetation in Peru and Bolivia within protected areas (Young 2007), but even protected and putatively protected areas are not immune to this loss.

In the Old World, there are also wild crop relatives vulnerable to extinction. For example, Prohens et al. (2007) reported on the genetic structure of two endangered eggplant wild relatives found in the Canary Isles, and suggested that measures were established to aid their survival in the wild.

The greatest threat to wild species of *Solanum* may come from the pressures associated with changing climates. The high altitude species of wild potato are under threat just as are many species internationally. Although in some cases they may be able to migrate to an equivalent climatic zone at high altitude, the migration depends on the existence of a suitable habitat as well as an ability to move during the relatively short period that may be available before the original habitat becomes unsuitable. However, given the topography and the habitat fragmentation induced by man, such migrations will, in many cases, not be possible. The likely rates of loss of wild relatives of three crops, groundnut, potato, and *Vigna* bean, were modeled by Jarvis et al. (2008) using standard projections for climate change. Potato wild relatives were found to be at intermediate levels of threat and the modeling suggested that about 7–14% of species (with no or unlimited migration permitted) will go extinct simply due to climate shifts by 2055 under the conditions specified. The modeling also suggested that their range sizes would decrease by between 38% and 69%. It should be stressed that additional pressures of the conversion of land for other use, and the changes in range of pests and pathogens expected from a changing climate, will add further stresses affecting the survival of wild species. Jarvis et al. (2008) suggest that to safeguard important biodiversity for mankind, increased effort should be made on the collection of material in the gene banks.

By the first half of the nineteenth century, wild potatoes had been collected and were established in botanical gardens in Europe. *S. demissum* Lindl. was described in 1848 by Lindley from material in the Royal Horticultural Society gardens in England, sent to him by Udhe in Mexico. One spontaneous hybrid, *S.* × *edinense* Berth., formed at the Royal Botanic Garden in Edinburgh between cultivated potato, *S. tuberosum*, and a Mexican hexaploid species, *S. demissum* and was described by Berthault in Paris in 1911 (Hawkes 1990). The impact of the discovery of late blight resistance in *S. demissum* on potato breeding was great, stimulating the breeding work of Salaman (1931), Black (1943) and others who used *S. demissum* as a source of genes for the breeding of varieties resistant to late blight. The realization of the value of wild species for breeding gave a further impetus to the collection of wild species and fostered a series of collecting expeditions in the early decades of the twentieth century. The first collecting expedition specifically to collect potato germplasm was that of Bukasov in 1925 and again with Juzepczuk, Vavilov, Lekhnovitch, and Kesselbrenner in 1926–1932 (Correll 1962). The Vavilov collection of wild and landrace potatoes has, unfortunately, largely been lost. German, Swedish, and American expeditions were made in the early 1930s. Although the German and Swedish expeditions were relatively unproductive in terms of living material collected, due largely to the unsuitable season, the American expeditions by Russell, Erlanson, and others to Mexico and Chile brought back around 600 accessions, which were used by breeders although not studied systematically at the time. In 1938, EK Balls and JG Hawkes began their collecting expeditions on behalf of the Commonwealth Bureau of Plant Breeding and Genetics in England. In the first expedition, they collected around 1,200 accessions, giving a major boost to the availability of tuber-bearing *Solanum* to breeders and scientists worldwide. Hawkes (2004) published a detailed account of his early collecting expeditions, recounting his preparatory visit to Russia to learn from the prominent taxonomists of the time, and his meeting with Vavilov during his stay. Particularly during the next four decades, many collecting missions visited Central and South America, establishing the main potato wild relative collections in existence today.

The main collections of potato wild species are now found at CIP in Lima, Peru; Sturgeon Bay in Wisconsin, USA; Wageningen, the Netherlands; Dundee, UK; Braunschweig, Germany; and Chile, Argentina, Bolivia, and Peru in South America. A loose grouping of international gene banks was formed under the banner of the Association of Potato Intergenebank Collaboration and has produced a basic combined database built on those accessions in the collections still retaining a single original collector number (Huamán et al. 2000). The extant accessions in the world's potato gene banks numbered 11,819 wild species accessions, 7,112 of which were not duplicates and 5,306 of which retained their original collector's number. Others had lost that identity or were the results of crossing between accessions losing viability due to bottlenecking. Around 30% of the accessions were from Argentina, whereas the major centers of diversity, Peru and Bolivia, contributed only 24% and 20%, respectively. Of the 188 taxa in the collections, 72 were represented by five or fewer accessions and about 70 taxa were not represented in any gene bank.

The subset of this world assembly of wild potato accessions that is held in Europe is now accessible via a combined database available from the CGN gene bank Web site at Wageningen. Again, the original collector's number is used to provide a unique reference point for combining evaluation data from each collection sharing the accession. More detailed additional information is available from individual gene banks such as the Sturgeon Bay gene bank available through the GRIN system, and the Germinate database of the CPC available at the James Hutton Institute.

Most gene banks preserve material as true seed in cold, dehumidified stores according to protocols close to IBPGR standards (Hanson 1985). In some rare cases, particularly when individual clones have been well characterized, it would be beneficial to maintain clonal material in a safe form, but so far cryopreservation and the use of slow growth medium has been successfully applied only to clonal material of cultivated varieties, breeder's lines and mapping populations as the benefit for this type of material is clear (Foroughi-Wehr et al. 1977; Schäfer-Menuhr et al. 1996). However, true seed of *Solanum* species generally retains a high viability when stored both dried and cold, and practical experience has shown that seed stored in a reasonable condition retains its viability for at least 30 years.

The in situ conservation route through the development of protected areas and by implication the species within the habits that exist there has already been mentioned. In Peru, there are also specific initiatives focused on the potato, the most important of which is the Potato Park in Pisac, Peru. This groundbreaking initiative by local community groups sought to repatriate genetic resources and maintain them for the benefit of local communities. Although the main focus is on the native cultivated biodiversity, the park sits in the center of the area of the original domestication of the potato and includes the wealth of diversity available in the wild potatoes of the area. The sympathetic management of the area of 9,300 ha by the participants in the initiative will afford protection for some of the wild potato species found there.

Eggplant and its wild relatives do not have the same well-developed international network charged with the preservation of genetic resources. However, an EU project on eggplant germplasm resources built a Europe-wide database of 3,500 accessions of eggplant and its wild relatives (van der Weerden and Barendse 2007), which has since grown to around 6,000 accessions under the auspices of the ECP/GR program (http://www.bgard.science.ru.nl/WWW-IPGRI/eggplant.html). About one quarter of the total European holdings of eggplant are thought to be wild relatives of the eggplant species. A further major repository is the National Bureau for Genetic Resources in Pusa, Delhi, where around 2,500 accessions of eggplant are held, many of them indigenous cultivated landraces. The USDA network of gene banks in the GRIN system holds about 760 eggplant accessions and 860 taxa in the genus *Solanum*, although the numbers of unique species held will be somewhat lower.

12.3 Role of Wild *Solanum* Species in Elucidation of Origin and Evolution of Allied Crop Plants

Competing claims to be the source of the cultivated potato have been made by Chile, Bolivia and Peru, encouraged by a confused literature on the topic. There is no simple answer to the question of the origin of the cultivated potato, and to a degree all three countries mentioned above and others can claim a part in the story. One remarkable aspect of the story of the cultivated potato was the discovery by Hosaka (2003, 2004) that the Chilean potato carried the chloroplast type found today only in an Argentinean and Bolivian white starry-flowered species with a covering of glandular hairs, *S. tarijense* Hawkes. Subsequent work has shown that the landraces of potato in the Andes mostly carry the chloroplast signature of the species of the *S. bukasovii* group from Peru, with which all potatoes have a strong affinity at the nuclear DNA level, and also a Bolivian group of species once thought to be better candidates as ultimate ancestors than the Peruvian species (Sukhotu et al. 2006). The signature of *S. tarijense* has not been detected in the nuclear DNA of cultivated potato, yet the clear implication of this work is that an Andean stock of potato hybridized with this species, which has a natural range today in Bolivia and northern Argentina, and that derivatives of that event, backcrossed to Andean stock, made the journey south to coastal Chile where Chilean potatoes became adapted to conditions very different from their home in the Andes. Ames and Spooner (2008) have used the chloroplast DNA deletion characteristic of Chilean potato to survey European potatoes and confirm that potatoes in Europe largely carry Chilean cytoplasm, though prior to the late blight epidemic of the mid-1840s a wider diversity of chloroplast types were found.

Hawkes (1990), without the benefit of the detailed molecular data that was to follow over a decade later, considered that the origin of the potato could be somewhere in the center of diversity of the wild potato, from central Peru to the Lake Titicaca region and in particular northern Bolivia. Although earlier Russian researchers were inclined to place the origin of the potato in Peru, as Hawkes had been earlier in his career, he considered that as the most wild-looking biotypes of landrace Stenotomum potatoes (then given specific rank but now considered a group within *S. tuberosum*) were to be found in Bolivia then it was likely that potatoes were descended from Bolivian ancestors. *S. leptophyes* Bitt. seemed to fit most closely due to its occurrence in the same phytogeographical region and at a similar altitude to Stenotomum potatoes, the living representatives of the earliest domesticated types. This was at variance with some of the early molecular data appearing in print at the time, including the study of Debener et al. (1990), which placed Stenotomum and Andigena accessions close to *S. canasense* Hawkes and not the Bolivian and Argentinean group of species, which included *S. gourlayi* Hawkes, a close ally of *S. leptophyes.* This arrangement of species agrees with the position of Bukasov (1966) who placed the origin of the cultivated potato in the Peruvian group to which *S. canasense* (part of *S. bukasovii sensu lato*) belongs.

Spooner et al. (2005) published a large-scale study using amplified fragment length polymorphism (AFLP) markers. A total of 438 AFLP polymorphisms were assayed on 362 wild and 98 landrace accessions of potato from the Commonwealth Potato Collection and the USDA Potato Genebank at Sturgeon Bay. Cladograms were presented using Wagner parsimony methods, and these demonstrated a clustering of the majority of the accessions in the study into two clades representing the northern and southern *S. brevicaule* complexes, the main groups of species found in Peru and Bolivia plus Argentina, respectively. All the landrace accessions in the study clustered with the northern group and in particular a group of species from southern Peru. This group from the northern Brevicaule complex is regarded by some as one broad species with the earliest name *S. bukasovii*, and by others as several distinct species *S. ambosinum* Ochoa, *S. bukasovii* Juz. *sensu stricto, S. candolleanum* Berth., *S. canasense, S. marinasense* Vargas, and *S. multidissectum* Hawkes. Later exploration of the data using principal coordinate analysis suggested that within this group the accessions labeled *S. canasense* demonstrated the greatest overlap with cultivated landraces, although other species are also candidates for involvement in the ancestry of the cultivated potato.

Together, these studies indicate that the original domestication event took place in central or southern Peru at an altitude of around 3,000–4,300 m, that the original domesticates formed hybrids with other Andean wild species after migration under domestication, and that backcrossing has yielded new combinations of mostly Peruvian nuclear DNA with a range of different cytoplasms, including a relatively distant event to generate the nuclear–cytoplasmic combination found in most Chilean and European potatoes. There is also no indication of a hybridization event with *S. sparsipilum* (Bitt.) Juz. et Buk. to form tetraploid landraces as suggested by Hawkes (1990), but instead the move to tetraploidy seems more likely to have been directly from diploid landraces.

No comparable study exists which clarifies the origin of the eggplant using DNA techniques. With a number of possible contributors to the diversity found in cultivated species, and the ability shown by other members of the genus to achieve introgression from a number of related species, the story of the origin of the eggplant may be complex.

12.4 Role in Development of Cytogenetic and Genetic Stocks

Cultivated potato, *S. tuberosum*, is most commonly grown in the tetraploid form, although diploid forms are also frequent, particularly in the Andes. Also in the Andes are forms of cultivated potato, which have arisen by interspecific hybridization with *S. megistacrolobum* Bitt. or *S. acaule* Bitt. These forms, *S.* × *ajanhuiri* Juz. et Buk. and *S.* × *juzepczukii* Buk., respectively, give frost tolerance to domesticated potato at the expense of reintroducing high levels of tuber glycoalkaloids bred out by the original people domesticating the potato. These high-altitude forms are grown for the production of the ancient food, chuño, a dried product produced with a long soaking period, which de-bitters previously freeze-dried whole small tubers.

With no appreciable differentiation at the DNA level between the traditional diploid and tetraploid taxa of domesticated potato, there is a general acceptance that the material previously given specific rank, for example in Hawkes (1990), should be classified as groups arising under domestication. Dodds (1962) described a system of classification entirely consistent with these modern ideas on the limits of *S. tuberosum*, and his classification is used here. He placed five groups under *S. tuberosum*: Stenotomum, Phureja (both diploid), Andigena, Tuberosum (both tetraploid), and Chaucha (crosses between a diploid and Andigena). In addition, *S.* × *juzepczukii* (*S. tuberosum* group Stenotomum × *S. acaule*, 3×) and *S.* × *curtilobum* Juz. et Buk. (*S. tuberosum* group Andigena *S.* × *juzepczukii*) retained specific rank thanks to their origin as interspecific hybrids, and the position of *S.* × *ajanhuiri* as an interspecific hybrid between *S. tuberosum* Group Stenotomum and *S. megistacrolobum* was clarified by Johns et al. (1987). Some recent authors have used the more recent system proposed by Huamán and Spooner (2002), but we consider that the similar system of Dodds (1962), retained specific epithets for known interspecific hybrids, to be preferable.

There has been a great deal of research performed on the means of moving from one ploidy level in potato to another. Some of this work has used wild relatives of potato, for example to introduce the embryo spot marker to permit the identification of true hybrids rather than the desired dihaploids (Hermsen and Verdeniu 1973). The two processes involved in breeding across ploidy level are the spontaneous doubling of the chromosome complement during gamete formation (Ross and Langton 1974) (or the artificial doubling of chromosome number with chemical agents), and the reduction of chromosome number through a failed pollination, often using the Phureja lines with an introgressed gene for nodal pigmentation (e.g., Uijtewaal et al. 1987), or anther culture. To be able to move between ploidy levels opens up new possibilities for approaches to breeding and in particular allows the breeder the possibility of the benefit of selection at the diploid level and the recreation of productive tetraploid genotypes from such diploid lines (Peloquin et al. 1989). Clulow and Rousselle-Bourgeois (1997) challenged the assumption of parthenogenesis following fertilization and provided evidence for direct introgression of Phureja DNA into the derived dihaploids from incomplete fertilization or from the loss of chromatin soon after fertilization. Although this may provide a means of more rapidly introgressing DNA from wild species into cultivated potato, its use for this purpose has yet to be demonstrated. Furthermore, recent studies have questioned the evidence for direct introgression from Phureja, suggesting that dihaploid induction does generally operate via parthenogenesis (Samitsu and Hosaka 2002; Straadt and Rasmussen 2003).

The ability to inbreed germplasm opens up possibilities to expose recessive alleles, prepare for hybrid true seed production, provide an alternative to monoploids for near-homozygous DNA isolation, and develop material to investigate heterosis. Much of this is best achieved at the diploid level. Although in general self-compatibility is lost when moving from tetraploidy to diploidy in cultivated potato, there are rare genotypes that maintain self-compatibility in the diploid state. Wild species have contributed to the development of inbred lines. Cappadocia et al. (1986) described self-compatibility arising from anther culture doubled haploids of the wild potato *S. chacoense* Bitt. Hosaka and Hanneman (1998) described a gene from *S. chacoense, Sli*, which suppresses the self-incompatibility system normally functioning in diploid potatoes, that this has now been exploited to generate inbred lines as far as the S_5 (Phumichai et al. 2005).

Aneuploids have been recorded several times during the introgression of traits from wild species to the cultivated potato. Observations of aneuploidy in derivatives of *S. demissum* × *S. tuberosum* were made by Cooper and Howard (1952). More recently Wilkinson (1992) observed aneuploidy in aerial parts of cv. Torridon and aneusomaty in the roots. However, such aneuploidy has been recorded as a product of interspecific breeding but not explicitly exploited as a genetic tool in its own right.

12.5 Role in Classical and Molecular Genetic Studies

As an outbreeding tetraploid, potato has not been an attractive subject for classical genetic studies. Such studies in potato have burgeoned since the advent of quantitative trait loci (QTL) mapping based on molecular marker maps, often using parents bearing wild species introgressions (Meyer et al. 2005; Bradshaw et al. 2008; Ritter et al. 2009). As a direct result of this mapping effort potato has become a model for the development of autotetraploid mapping (Luo et al. 2001; Hackett et al. 2003). Some of these gene mapping studies have been performed in hybrids between wild species prior to transfer of the genes to cultivated stock (Danan et al. 2009). However, unlike its close relative tomato, most of the effort has been directed to existing cultivated germplasm rather than using wild relatives directly, although that cultivated stock usually carries uncharacterized introgressions from wild species. These uncharacterized wild species introgressions have often been from species donating pest and disease resistance into cultivated breeding lines. As such they will have encountered few rounds of meiotic recombination since their introduction and may suffer inherently less frequent recombination in any case. The relatively large distances over which linkage disequilibrium occurs in the first published studies of association genetics in potato (Gebhardt et al. 2004; Simko 2004; Simko et al. 2006) may be due to this special nature of introgressed segments originating from wild potato species.

12.6 Role in Crop Improvement Through Traditional and Advanced Tools

Potato, unlike the other major bulk food crops as the cereals, can and has benefited from the extensive use of wild relatives in breeding. The value contributed globally by the wild relatives of the potato is likely to exceed than that of those of any other crop. Prakash (2008) estimated the global value of international trade in potato at US$6 billion and, as much potato production is not traded across international boundaries, the true value must be several-fold higher. As most potato varieties have genetic contributions from deliberate breeding using wild species, the contribution to the global economy from the protection offered by these genes can be seen to be immense.

Potato late blight was the first major crop disease tackled by the exploitation of resistance in wild relatives. The promise of resistance in *S. demissum* and its hybrid derivatives was noted by Salaman in 1909 (Hawkes 1958), and the outputs of breeding with these lines were reported by Salaman (1931) and Black (1943). Today, resistance genes, many of them now overcome by new strains of the pathogen, are found in most modern cultivars of the potato. This unprecedented scale of penetration of wild germplasm into the cultivated types still does not mean that the pathogen has been defeated. The disease still poses a threat to crop production and food security across the world, and new forms of the disease repeatedly appear and pose a new threat to the food security of the region (Drenth et al. 1994; Montarry et al. 2010). In response, the breeders and geneticists have resorted to explore the diversity of resistance genes available in the wild relatives of potato and have reported high frequencies of useful resistance in species from the northern and southern parts of the range of tuber-bearing species, with very few naturally blight resistant species from Peru (Colon and Budding 1988; Ramsay et al. 1999).

One source of promising new resistance genes thought to confer broad-spectrum and possibly more durable resistance to late blight is *S. bulbocastanum*. As this Mexican species does not readily cross with cultivated potato, three different approaches have been taken to overcome this difficulty – bridging crosses (Hermsen and Ramanna 1973), protoplast fusion (Helgeson et al. 1998), and gene isolation for genetic transformation (Song et al. 2003). Both the bridging cross approach and the genetic modification (GM) route have yielded commercializable varieties. This work has stimulated further searches for novel R-genes, using approaches that target *avr* genes in the pathogen (Hein et al. 2007; Vleeshouwers et al. 2008).

Further exploitation of wild tuber-bearing species has focused on the next most important global pathogen of the potato, the potato cyst nematode (PCN). At Dundee, the wild species *S. vernei* Bitt. et Wittm. has been used for PCN resistance, taking 28–30 years to achieve the 5–6 backcrossing cycles required to introgress resistance genes but remove undesirable traits from *S. vernei* (Bradshaw et al. 1995). The successful introduction of pest and disease resistance traits into potato cultivars comes from a relatively small number of wild species. *S. demissum* and *S. stoloniferum* Schltdl. et Bché. have donated late blight resistance and resistance to viruses, additional virus resistance has come from *S. chacoense* and *S. acaule*, whereas both *S. vernei* and *S. spegazzinii* Bitt. have donated resistance to PCN (see for review Bradshaw et al. 2005).

By contrast, eggplant has benefited on only a small scale from the use of wild species in breeding due to difficulties in making fertile interspecific crosses (Sękara et al. 2007) and the wealth of diversity in the crop species. However, wild species are seen as potential sources of pest and disease resistance traits, and various biotechnological approaches have been made to access this variation (Kashyap et al. 2003).

12.7 Genomics Resources Developed

For the entire *Solanum* genus, entries lodged in the CBI databases for nucleotides and expressed sequence tags (ESTs) at the end of 2009 amounted to about 144k nucleotide database sequences and 727k EST sequences. The majority of these sequences were for tomato and the immediate wild relatives of tomato. For the wild relatives of potato, eggplant, and other minor *Solanum* species, the nucleotide entry counts were between 200 and 400 for *S. demissum*, *S. bulbocastanum* Dun., *S. melongena*, and *S. quitoense*, and around 100 for *S. chacoense* and *S. nigrum* L. The only wild *Solanum* species outside the tomato group with any significant number of EST sequences was *S. chacoense* with 7.7k sequences.

This resource is meager compared to the full genome sequence of the potato, which was made available in draft form in September 2009. By the end of 2010, it is intended to have the sequence fully aligned and annotated. The specific goals of the project include the presentation of the genome sequence with >95% of genes together with the regulatory regions, to have >95% of known potato ESTs located on the genome and to anchor at least 50% of the genome and 80% of the genes to chromosomes. The consortium is also committed to public access to the data, and to as full annotation as is practicable (Visser et al. 2009).

Although at this stage no sequencing of wild potatoes is envisaged, the current genome scaffolds and the expected full sequence provide an excellent starting point for comparative sequencing of the genomes of related species, and this is expected to be a major result of the publication of the sequence.

Unlike in tomato and its wild relatives, potato, eggplant, and their wild relatives do not have organized lists and collections of mutants, although some provisional mutagensis has been attempted in the self-compatible diploid potato relative, *S. verrucosum*.

12.8 Scope for Domestication and Commercialization

Solanum species provide a huge reservoir of bioactive compounds. Ethnobotanical reports suggest widespread use of members of the genus for a diversity of therapeutic and other purposes. Applications include the use of a decoction taken as a drink of *S. americanum* Mill. used to treat hypertension by traditional healers in Côte-d'Ivoire (N'guessan et al. 2009). In Mexico, *S. chrysotrichum* Schltdl. is used to control skin fungal infections (Lozoya et al. 1992). The roots of *S. virginianum* L. are used in a mixture of extracts to treat tuberculosis in Bangladesh (Mahabub Nawaz et al. 2009). The juice of crushed fruits of *S. erianthum* D. Don is applied to exposed skin by the Paliyar tribes, an indigenous people of Tamil Nadu, India, to repel leeches during visits to the forest (Ignacimuthu et al. 2006). Other studies show a wide diversity of uses; for example, *S. nigrum* and *S. surattense* Burm. are used for treating heart disease, rheumatism, fever, and chest complaints in northern Pakistan (Wali Khan and Khatoon 2008), and a review of all uses of *Solanum* extracts in India has been published by Amir and Kumar (2004).

Among the most promising uses of *Solanum* extracts is its use for the treatment of cancer. *S. aculeastrum* Dunal is used to make a decoction to treat cancer and is the most commonly used plant for this purpose in South Africa's Eastern Cape, administered daily until symptoms disappear (Koduru et al. 2007). Glycoalkaloids from *S. linnaenum* Hepper & P.-M.L. Jaeger (previously *S. sodomaeum*) have been identified to have antitumor effects (Cham et al. 1987; Cham and Meares 1987). These solasodine rhamnosyl glycosides specifically bind cancer cell receptors and induce cell death so effectively that farmers in Australia apply crushed leaves to retard the progress of ocular squamous cell carcinomas in cattle. The mix of solasodine glycosides from this plant has now been shown to be effective against a range of cancers and is now available in the form of a topical cream for the treatment of malignant and non-malignant skin cancers (Cham 2007).

12.9 Some Dark Sides and Their Addressing

There are several *Solanum* species considered as noxious weeds in some parts of their range. Some are considered to be useful species in one situation and harmful in another. For example, *S. torvum* is widely planted in some countries for its fruit and is used as a bacterial wilt resistant rootstock for the grafting of eggplant (Rahman et al. 2002). In Hawaii and Papua New Guinea, it is an invasive species, particularly in pasture.

Solanum elaeagnifolium Cav., the silverleaf nightshade, is one of several species widely regarded as a noxious weed. It is a native of northeast Mexico and the southwest of USA but has spread to many US states, Morocco, around the Mediterranean basin, in South Africa, and in Australia. Its aggressive growth from deep rootstocks makes it difficult to control, and long-dormant seeds make it liable to recur even when cleared from sites (Anonymous 2007).

12.10 Recommendations for Future Actions

There are a number of crucial issues for the future effective conservation and use of wild *Solanum* germplasm. It is very clear that even in potato, which has a long history of the use of wild germplasm for improvement of the crop, the useful germplasm used in breeding is a very small fraction of the available diversity. To be able to use that diversity effectively to help safeguard food production in the future, the following requirements exist:

- Appropriate understanding of the diversity available and the uses which can be made of that diversity
- Appropriate access to that biodiversity, bearing in mind that countries donating germplasm to collections now rightly expect to share in the benefits of its exploitation
- Conservation of that diversity in situ and ex situ so that it is readily available to researchers now and preserved for uses in the future
- Means of identifying the genes responsible for traits within a realistic timescale
- Societal permission to use technologies, including GM, that can ensure the transfer of traits without the need for very many rounds of recombination to recover properly adapted mixes of genes

Technology and politics combine to intervene in many of these areas. For example, genetic techniques to identify and understand useful genes have been continually improving, making it possible to isolate the full gene sequence for many traits by a variety of techniques. The availability of the full genome sequence of potato will make this task much easier in the future, although for complex traits such as resistance to abiotic stress the lack of a full understanding of the biology of the trait will hinder the identification of the relevant genes. In some parts of the world, significant resistance remains to moving these genes into cultivars by the most efficient route, and so much longer traditional sexual introgression may be required. There is clearly much yet to be done to reduce the public's mistrust of such technologies and to gain acceptance of their use.

The International Treaty for Plant Genetic Resources for Food and Agriculture (http://www.planttreaty.org/) provides a framework for the sustainable use of plant genetic resources and aims to provide a mechanism that returns benefit to the countries donating useful germplasm to the world community. For this to succeed, it is necessary that the mechanisms function well and that such benefit is seen as commensurate with the benefits obtained by the countries using the germplasm. At present, not all countries holding significant *Solanum* genetic resources are permitting internationally based collecting of genetic resources in their countries. However, as much of the wealth of genetic resources found in the wild are coming under increasing risk from various pressures including the rapid shifting of climate zones, there is a greater need than ever to safeguard useful crop relative biodiversity for the future of mankind.

To summarize, the actions required include the continual improvement of the understanding of the important traits for the sustainability of crop production, the integration of this work with genome science, actions to reduce societal reluctance to embrace new technology and enhance global cooperation relating to genetic resources, and an increased effort to bring invaluable crop-related genetic resources into well-managed facilities to make them available to the international community.

References

Ames M, Spooner DM (2008) DNA from herbarium specimens settles a controversy about origins of the European potato. Am J Bot 95:252–257

Amir M, Kumar S (2004) Possible industrial applications of genus *Solanum* in the twenty-first century – a review. J Sci Ind Res 63:116–124

Angiosperm Phylogeny Group (2003) An update of the Angiosperm Phylogeny Group classification for the orders and families of flowering plants: APG II. Bot J Linn Soc 141:399–436

Anonymous (2007) Data sheet on quarantine pests: *Solanum elaeagnifolium*. OEPP/EPPO Bull 37:236–245

Arumuganathan K, Earle ED (1991) Nuclear DNA content of some important plant species. J Plant Mol Biol Rep 9:208–218

Behera TK, Sharma P, Singh BK, Kumar G, Kumar R, Mohapatra T, Singh NK (2006) Assessment of genetic diversity and species relationships in eggplant (*Solanum melongena* L.) using STMS markers. Sci Hortic 107:352–357

Black W (1943) Inheritance of resistance to two strains of blight (*Phytophthora infestans* de Bary) in potatoes. Trans R Soc Edinb 61:137–147

Bohs L (1995) Transfer of *Cyphomandra* (Solanaceae) and its species to *Solanum*. Taxon 44:583–587

Bradshaw JE, Dale MFB, Phillips MS (1995) Breeding potatoes at SCRI for resistance to potato cyst nematodes. SCRI Annu Rep 1995:30–34

Bradshaw JE, Bryan GJ, Ramsay G (2005) Genetic resources (including wild and cultivated *Solanum* species) and progress in their utilisation in potato breeding. Pot Res 49:49–65. doi:10.1007/s11540-006-9002-5

Bradshaw JE, Hackett CA, Pande B, Waugh R, Bryan GJ (2008) QTL mapping of yield, agronomic and quality traits in tetraploid potato (*Solanum tuberosum* subsp *tuberosum*). Theor Appl Genet 116:193–211

Bukasov SM (1966) Die Kulturkartoffel und ihre wildwachsenden Vorfahren. Z Pflanzenzuecht 55:139–164

Cappadocia M, Cheng DSK, Ludlum-Simonette R (1986) Self-compatibility in doubled haploids and their F1 hybrids, regenerated via anther culture in self-incompatible *Solanum chacoense* Bitt. Theor Appl Genet 72:66–69

Cham BE (2007) Solasodine rhamnosyl glycosides specifically bind cancer cell receptors and induce apoptosis and necrosis. Treatment for skin cancer and hope for internal cancers. Res J Biol Sci 2:503–514

Cham BE, Meares MM (1987) Glycoalkaloids from *Solanum sodomaeum* L. are effective in the treatment of skin cancers in man. Cancer Lett 36:111–118

Cham BE, Gilliver M, Wilson L (1987) Antitumour effects of glycoalkaloids isolated from *Solanum sodomaeum* L. Planta Med 53:34–36

Chiarini F, Bernardello G (2006) Karyotype studies in South American species of *Solanum* subgen. *Leptostemonum* (Solanaceae). Plant Biol 8:486–493

Clulow SA, Rousselle-Bourgeois F (1997) Widespread introgression of *Solanum phureja* DNA in potato (*S. tuberosum*) dihaploids. Plant Breed 116:347–351

Coca Morante M, Ticona VH, Castillo Plata W, Tordoya IT (2007) Distribution of wild potato species in the north of the Department of La Paz, Bolivia. Spn J Agric Res 5:326–332

Colon LT, Budding DJ (1988) Resistance to late blight (*Phytophthora infestans*) in ten wild *Solanum* species. Euphytica 5:77–86

Cooper JP, Howard HW (1952) The chromosome numbers of seedlings from the cross *Solanum demissum* × *tuberosum* backcrossed by *S. tuberosum*. J Genet 50:511–521

Correll DS (1962) The potato and its wild relatives. Section Tuberarium of the genus Solanum. Texas Research Foundation, Renner, TX, USA

Danan S, Chauvin JE, Caromel B, Moal JD, Pelle R, Lefebvre V (2009) Major-effect QTLs for stem and foliage resistance to late blight in the wild potato relatives *Solanum sparsipilum* and *S. spegazzinii* are mapped to chromosome X. Theor Appl Genet 119:705–719

Daunay M-C, Janick J (2007) History and iconography of eggplant. Chron Hortic 47:16–22

Debener T, Salamini F, Gebhardt C (1990) Phylogeny of wild and cultivated *Solanum* species based on nuclear restriction fragment length polymorphism. Theor Appl Genet 79:360–368

Dodds KN (1962) Classification of cultivated potatoes. In: Correll DS (ed) The potato and its wild relatives. Contributions from Texas Research Foundation. Bot Stud 4:517–539

Drenth A, Tas ICQ, Govers F (1994) DNA fingerprinting uncovers a new sexually reproducing population of *Phytophthora infestans* in the Netherlands. Eur J Plant Pathol 100:97–107

FAOSTAT (2008) http://faostat.fao.org

Foroughi-Wehr B, Wilson HM, Mix G, Gaul H (1977) Monohaploid plants from anthers of a dihaploid genotype of *Solanum tuberosum* L. Euphytica 26:361–367

García CG, Matesevach M, Barboza G (2008) Features related to anther opening in *Solanum* species (Solanaceae). Bot J Linn Soc 158:344–354

Gebhardt C, Ballvora A, Walkemeier B, Oberhagemann P, Schuler K (2004) Assessing genetic potential in germplasm collections of crop plants by marker-trait association: a case study for potatoes with quantitative variation of resistance to late blight and maturity type. Mol Breed 13:93–102

Glover BJ, Bunnewell S, Martin C (2004) Convergent evolution within the genus *Solanum*: the specialised anther cone develops through alternative pathways. Gene 331:1–7

Hackett CA, Pande B, Bryan GJ (2003) Constructing linkage maps in autotetraploid species using simulated annealing. Theor Appl Genet 106:1107–1115

Hanson J (1985) Procedures for handling seeds in genebanks. IBPGR Practical Manuals for Genebanks no 1. IBPGR Secretariat, Rome, Italy

Haston E, Richardson JE, Stevens PF, Chase MW, Harris DJ (2007) A linear sequence of angiosperm phylogeny group II families. Taxon 56:7–12

Hawkes JG (1958) Significance of wild species and primitive forms for potato breeding. Euphytica 7:257–270

Hawkes JG (1990) The potato: evolution, biodiversity and genetic resources. Belhaven, London, UK

Hawkes JG (2004) Hunting the wild potato in the South American Andes. Memories of the British Empire potato collecting expedition to South America 1938–1939. Botanical and Experimental Garden University of Nijmegen, Nijmegen, Netherlands

Hein I, Squires J, Birch P, Bryan GJ (2007) Screening wild potato accessions for resistance to the virulent allele of the *Phytophthora infestans* effector Avr3a. In: Proceedings of 13th international congress on molecular plant-microbe interactactions, Sorrento, Italy, p 259

Helgeson JP, Pohlman JD, Austin S, Haberlach JT, Wielgus SM, Ronis D, Zambolim L, Tooley P, McGrath JM, James RV,

Stevenson WR (1998) Somatic hybrids between *Solanum bulbocastanum* and potato: a new source of resistance to late blight. Theor Appl Genet 96:738–742

Hermsen JGT, Ramanna MS (1973) Double-bridge hybrids of *Solanum bulbocastanum* and cultivars of *Solanum tuberosum*. Euphytica 22:457–466

Hermsen JGT, Verdeniu J (1973) Selection from *Solanum tuberosum* Group Phureja of genotypes combining high-frequency haploid induction with homozygosity for embryo-spot. Euphytica 22:244–259

Hijmans RJ (2001) Global distribution of the potato crop. Am J Pot Res 78:403–412

Hosaka K (2003) T-type chloroplast DNA in *Solanum tuberosum* L. ssp *tuberosum* was conferred from some populations of *S. tarijense* Hawkes. Am J Pot Res 80:21–32

Hosaka K (2004) Evolutionary pathway of T-type chloroplast DNA in potato. Am J Pot Res 81:153–158

Hosaka K, Hanneman RE Jr (1998) Genetics of self-compatibility in a self-incompatible wild diploid potato species *Solanum chacoense*. I. Detection of an S locus inhibitor (*Sli*) gene. Euphytica 99:191–197

Huamán Z, Spooner DM (2002) Reclassification of landrace populations of cultivated potatoes (*Solanum* sect. Petota). Am J Bot 89:947–965

Huamán Z, Hoekstra R, Bamaberg JB (2000) The inter-genebank potato database and the dimensions of available wild potato germplasm. Am J Pot Res 77:353–362

Hunziker AT (1979) South American Solanaceae. In: Hawkes JG, Lester RN, Skelding AD (eds) The biology and taxonomy of the Solanaceae, vol 7. Academic, London, UK, pp 49–85

Ignacimuthu S, Ayyanar M, Sankara Sivaraman K (2006) Ethnobotanical investigations among tribes in Madurai District of Tamil Nadu (India). J Ethnobiol Ethnomed 2:25. doi:10.1186/1746-4269-2-25

IUCN (2009) IUCN red list of threatened species. Version 2009.1. http://www.iucnredlist.org. Accessed 01 Oct 2009

Jarvis A, Lane A, Hijmans RJ (2008) The effect of climate change on crop wild relatives. Agric Ecosyst Environ 126:13–23

Jenkins JA (1948) The origin of the cultivated tomato. Econ Bot 2:379–392

Johns T, Huamán Z, Ochoa C, Schmiediche P (1987) Relationships among wild, weed, and cultivated potatoes in the *Solanum* × *ajanhuiri* complex. Syst Bot 12:541–552

Kashyap V, Vinod Kumar S, Collonnier C, Fusari F, Haicour R, Rotino G, Sihachakr D, Rajam MV (2003) Biotechnology of eggplant. Sci Hortic 97:1–25

Kirkman MA (2007) Global markets for processed potato products. In: Vreugdenhil D (ed) Potato biology and biotechnology: advances and perspectives. Elsevier, Oxford, UK, pp 27–44

Knapp S (2002) Assessing patterns of plant endemism in Neotropical uplands. Bot Rev 68:22–37

Koduru S, Grierson DS, Afolayan AJ (2007) Ethnobotanical information of medicinal plants used for treatment of cancer in the Eastern Cape Province, South Africa. Curr Sci 92:906–908

Lester RN, Niakan L (1986) Origin and domestication of the scarlet eggplant *Solanum aethiopicum* from *Solanum anguivi* in Africa. In: D'Arcy WG (ed) Solanaceae: biology and systematics. 2nd International symposium, St. Louis, Missouri, Columbia University Press, New York, USA, pp 433–456

Lozoya X, Navarroa V, García M, Zurita M (1992) *Solanum chrysotrichum* (Schldl.) a plant used in Mexico for the treatment of skin mycosis. J Ethnopharmacol 36:127–132. doi:10.1016/0378-8741(92)90011-F

Luo ZW, Hackett CA, Bradshaw JE, McNicol JW, Milbourne D (2001) Construction of a genetic linkage map in tetraploid species using molecular markers. Genetics 157:1369–1385

Mahabub Nawaz AHMD, Hossain M, Karim M, Khan M, Jahan R, Rahmatullah M (2009) An ethnobotanical survey of Jessore District in Khulna Division, Bangladesh. Am-Euras J Sustain Agric 3:195–201

Meyer S, Nagel A, Gebhardt C (2005) PoMaMo – a comprehensive database for potato genome data. Nucleic Acids Res 33: D666–D670

Montarry J, Andrivon D, Glais I, Corbiere R, Mialdea G, Delmotte F (2010) Microsatellite markers reveal two admixed genetic groups and an ongoing displacement within the French population of the invasive plant pathogen *Phytophthora infestans*. Mol Ecol 19:1965–1977

N'guessan K, Tiébré M-S, Aké-Assi E, Zirihi GN (2009) Ethnobotanical study of plants used to treat arterial hypertension, in traditional medicine, by Abbey and Krobou populations of Agboville (Côte-d'Ivoire). Eur J Sci Res 35:85–98

Peloquin SJ, Yerk GL, Werner JE, Darmo E (1989) Potato breeding with haploids and 2n gametes. Genome 31: 1000–1004

Peralta IE, Spooner DM (2007) History, origin and early cultivation of tomato (Solanaceae). In: Razdan MK, Mattoo AK (eds) Genetic improvement of Solanaceous crops, vol 2. Science, Enfield, USA, pp 1–27

Phumichai C, Mori M, Kobayashi A, Kamijima O, Hosaka K (2005) Toward the development of highly homozygous diploid potato lines using the self-compatibility controlling *Sli* gene. Genome 48:977–984. doi:10.1139/G05-066

Prakash A (2008) Global potato economy. FAO International Year of the Potato factsheet. http://www.potato2008.org/en/potato/economy.html. Accessed 26 Oct 2009

Prohens J, Anderson GJ, Herraiz FJ, Bernardello G, Santos-Guerra A, Crawford D, Nuez F (2007) Genetic diversity and conservation of two endangered eggplant relatives (*Solanum vespertilio* Aiton and *Solanum lidii* Sunding) endemic to the Canary Islands. Genet Resour Crop Evol 54:451–464

Rahman MA, Rashid MA, Hossain MM, Salam MA, Masum ASMH (2002) Grafting compatibility of cultivated eggplant varieties with wild *Solanum* species. Pak J Biol Sci 5:755–757. doi:10.3923/pjbs.2002.755.757

Ramsay G, Stewart H, de Jong WS, Bradshaw JE, Mackay GR (1999) Introgresssion of late blight resistance into *S. tuberosum*. In: Scarascia Mugnozza GT, Porceddu E, Pagnotta MA (eds) Genetics and breeding for crop quality and resistance. Proceedings of XV EUCARPIA Congress, Viterbo, Italy

Ritter E, Ruiz I, de Galarreta J, Hernandez M, Plata G, Barandalla L, Lopez R, Sanchez I, Gabriel J (2009) Utilization of SSR and cDNA markers for screening known QTLs for late blight (*Phytophthora infestans*) resistance in potato. Euphytica 170:77–86

Ross H, Langton FA (1974) Origin of unreduced embryo sacs in diploid potatoes. Nature 247:378–379. doi:10.1038/247378a0

Salaman RN (1931) Récent progrès dans la création de varietés de pommes de terre résistant au Mildiou 'Phytophthora infestans'. In: 2nd Congr Int Pathol Comparée, Paris, CR et Comm, pp 435–437

Samitsu Y, Hosaka K (2002) Molecular marker analysis of 24-and 25-chromosome plants obtained from *Solanum tuberosum* L. subsp *andigena* (2n = 4x = 48) pollinated with a *Solanum phureja* haploid inducer. Genome 45: 577–583

Schäfer-Menuhr A, Müller E, Mix-Wagner G (1996) Cryopreservation: an alternative for the long-term storage of old potato varieties. Potato Res 39:507–513

Sękara A, Cebula S, Kunicki E (2007) Cultivated eggplants – origin, breeding objectives and genetic resources, a review. Folia Hortic 19:97–114

Simko I (2004) One potato, two potato: haplotype association mapping in autotetraploids. Trends Plant Sci 9:441–448

Simko I, Haynes KG, Jones RW (2006) Assessment of linkage disequilibrium in potato genome with single nucleotide polymorphism markers. Genetics 173:2237–2245

Smith SD, Knapp S (2002) The natural history of reproduction in *Solanum* and *Lycianthes* (Solanaceae) in a subtropical moist forest. Bull Nat Hist Mus Bot 32:125–136. doi:10.1017/S0968044602000051

Song J, Bradeen JM, Naess SK, Raasch JA, Wielgus SM, Haberlach GT, Liu J, Kuang H, Austin-Phillips S, Buell CR, Helgeson JP, Jiang J (2003) Gene RB cloned from *Solanum bulbocastanum* confers broad spectrum resistance to potato late blight. Proc Natl Acad Sci USA 100: 9128–9133

Spooner DM, McLean K, Ramsay G, Waugh R, Bryan GJ (2005) A single domestication for potato based on multilocus amplified fragment length polymorphism genotyping. Proc Natl Acad Sci USA 102:14694–14699. doi:10.1073/pnas.0507400102

Straadt IK, Rasmussen O (2003) AFLP analysis of *Solanum phureja* DNA introgressed into potato dihaploids. Plant Breed 122:352–356

Sukhotu T, Kamijima O, Hosaka K (2006) Chloroplast DNA variation in the most primitive cultivated diploid potato species *Solanum stenotomum* Juz. et Buk. and its putative wild ancestral species using high-resolution markers. Genet Resour Crop Evol 53:53–63

Uijtewaal BA, Huigen DJ, Hermsen JGT (1987) Production of potato monohaploids (2n = x = 12) through prickle pollination. Theor Appl Genet 73:751–758

van der Weerden GM, Barendse GWM (2007) A web-based searchable database developed for the EGGNET project and applied to the Radboud University Solanaceae database. Solanaceae VI: Genomics Meets Biodiversity. Proc VIth Int Solanaceae Conf, Madison, WI, USA, 23–27 July, 2006. Acta Hortic 745:503–506

Vavilov NI (1928) Geographical centres of our cultivated plants. In: Proceedings of V international congress on genetics, New York, USA, pp 342–369

Visser RGF, Bachem CWB, de Boer JM, Bryan GJ, Chakrabati SK, Feingold S, Gromadka R, van Ham RCHJ, Huang S, Jacobs JME, Kuznetsov B, de Melo PE, Milbourne D, Orjeda G, Sagredo B, Tang X (2009) Sequencing the potato genome: outline and first results to come from the elucidation of the sequence of the world's third most important food crop. Am J Pot Res 86:417–429. doi:10.1007/s12230-009-9097-8

Vleeshouwers VGAA, Rietman H, Krenek P, Champouret N, Young C, Oh S-K, Wang M, Bouwmeester K, Vosman B, Visser RGF, Jacobsen E, Govers F, Kamoun S, van der Vossen EAG (2008) Effector genomics accelerates discovery and functional profiling of potato disease resistance and *Phytophthora infestans* avirulence genes. PLoS ONE 3:e2875

Wali Khan S, Khatoon S (2008) Ethnobotanical studies on some useful herbs of Haramosh and Bugrote valleys in Gilgit, northern areas of Pakistan. Pak J Bot 40:43–58

Wang J-X, Gao T-G, Knapp S (2008) Ancient Chinese literature reveals pathways of eggplant domestication. Ann Bot 102:891–897. doi:10.1093/aob/mcn179

Weese TL, Bohs L (2007) A three-gene phylogeny of the genus Solanum (Solanaceae). Syst Bot 32:445–463. doi:10.1600/036364407781179671

Wilkinson MJ (1992) The partial stability of additional chromosomes in *Solanum tuberosum* cv Torridon. Euphytica 60:115–122

Young BE (ed) (2007) Endemic species distributions on the East Slope of the Andes in Bolivia and Peru. Nature Serv, Arlington, VA, USA

Chapter 13
Spinacia

Sven B. Andersen and Anna Maria Torp

13.1 Basic Botany of the Species

The cultivated spinach, *Spinacia oleracea* L., belongs to the family Chenopodiaceae with two known wild relative species *S. turkestanica* iljin and *S. tetrandra* Stev. A study of taxonomic position of *Spinacia* with its two closely related genera *Beta* and *Chenopodium* and their taxonomic position in Caryophyllales based on chloroplast restriction polymorphism can be found in Downie and Palmer (1994). The two wild species are found distributed over western parts of Asia, *S. turkistanica* in Turkmenistan, Uzbekistan, and Kazakstan and *S. tetrandra* in the Caucasus area, in Armenia and Kurdistan between Iran, Iraq, and Turkey. The two wild species have some tendency to become weedy, but remain geographically restricted (Astley and Ford-Lloyd 1981; Gabrielian and Fragman-Sapir 2008). Cultivated spinach is diploid with $2n = 2x = 12$ chromosomes (Bemis and Wilson 1953). It is widely cultivated as a leaf vegetable all over the world with a worldwide production in 2007 of 14 MT fresh weight, most of these (12 MT) produced in China (FAO 2007).

13.2 Conservation Initiatives

There are quite large collections worldwide of cultivated spinach, while activities concerning the wild species seem sporadic. The largest organized conservation initiative for the wild species seems to be the Crop Wild Relative initiative with reference to 48 accessions of *S. tetrandra* registered in the Armenian database in vivo (Crop Wild Relatives 2009). Most gene banks such as the International Spinach Data Base of The Netherlands (12 *S. turkestanica* and 14 *S. tetrandra*) (International Spinach Database 2009), the National Plant Germplasm System USDA/ARS, GRIN (4 *S. turkestanica* and 2 *S. tetrandra*) (National Plant Germplasm System 2009), and the Plant Genetics Resources of Germany (2 *S. turkestanica* and 3 *S. tetrandra*) (Plant Genetic Resources 2009) have few accessions of the wild species. It has been argued that for reasons of species conservation at least in the Russian area the wild *Spinacia* does not need conservation in gene banks because they are not threatened (Ul'janova 1977); however, for use in the improvement of cultivated spinach, larger seed bank collections would be highly desirable.

13.3 Role in Elucidation of Origin and Evolution of Spinach

The exact origin of the cultivated spinach is not known, and the species is not found as wild types. The geographical distribution of the close relatives *S. turkestanica* and *S. tetrandra* is the main reason for the general assumption that cultivated spinach originated from the same geographical area of West Asia. The general high sexual compatibility between cultivated *S. oleracea* and the two wild species supports the idea that cultivated spinach originated through domestication of one or both of the wild species. The cultivated spinach seems to have spread to China during the seventh century and to Europe

S.B. Andersen (✉)
Department of Agriculture and Ecology, Faculty of Life Sciences, Copenhagen University, Thorvaldsensvej 40, 1871 Frederiksberg, Denmark
e-mail: sba@LIFE.ku.dk

C. Kole (ed.), *Wild Crop Relatives: Genomic and Breeding Resources, Vegetables*,
DOI 10.1007/978-3-642-20450-0_13,

during the twelfth century (Sneep 1982; Sauer 1993; Rubatzky and Yamaguchi 1997).

13.4 Role of Development of Cytogenetic Stocks and Their Utility

The wild *S. tetrandra* has been used as a source of heteromorphic chromosomes to study the genetics of dioecious sex expression in cultivated spinach (Iizuka and Janick 1971; Ramanna 1976).

13.5 Role in Classical and Molecular Genetic Studies

Spinacia turkestanica has been used as a parent for construction of genetically broad segregating offspring populations for construction of a genetic map and for mapping of the genetic factor determining dioecious sex expression in spinach (Khattak et al. 2006). *S. tetrandra* has been used in a genetic diversity study of spinach using target region amplification markers (TRAP), and the single accession of *S. tetrandra* did not cluster much distant from the 47 *S. oleracea* accessions of the study. This indicates that the genetic differences between cultivated and wild species are small and that a rather large genetic variation is still maintained in the cultivated material (Hu et al. 2007). Large genetic variation maintained within cultivated *S. oleracea* is also concluded from a diversity study using genic microsatellite markers (Khattak et al. 2007).

13.6 Role in Spinach Improvement Through Traditional and Advanced Tools

A major problem in production of cultivated spinach is the frequent attack by specialized races of downy mildew (*Peronospora farinose* f.sp. *spinaciae*). These fungal epidemics pose a threat to spinach production without widespread chemical sprays. Specific highly efficient genes for resistance against at least seven different races of downy mildew have been identified and used in breeding, but the duration of effective resistance of such genes become increasingly shorter as spinach production intensifies. At least one such race-specific gene against the race D3 is known to be transferred to cultivated spinach from a line of *S. turkestanica*, and another specific gene conferring resistance against *Peronospora* race LN4 has also been transferred from *S. turkestanica* (Hanke et al. 2000). Other diseases such as leaf spot caused by the fungus *Stemphylium botryosum* (Mou et al. 2008) as well as several pests including leaf miners *Liriomzyza* spp. (Mou 2008a) cause problems in spinach production, and more resistant plant types may be found in future in the wild material.

13.7 Genomics Resources Developed

The complete annotated chloroplast genome is available for spinach (Schmitz-Linneweber et al. 2001), but further published sequence information in *Spinacia* is still limited. The NCBI database (NCBI 2009) contained only 363 genomic sequences, 284 mRNA sequences, and 16 rRNA sequences. The sequences in the database was used by Groben and Wricke (1998) and Khattak et al. (2007) to develop a set of 35 primer pairs for amplification of microsatellites also known as simple sequence repeats (SSRs) of spinach, 13 of which were used for a diversity study of the cultivated spinach.

13.8 Scope for Domestication and Commercialization

There seems to be little reason to domesticate or commercialize the two wild relatives of spinach, since they do not seem very different from the already cultivated material. The two wild species, however, may be a valuable source of genes for resistance to common diseases of cultivated spinach. If such variation is conserved in the wild material, it may be used to introgress the resistance genes into the cultivated spinach and, thus, to maintain a sustainable low chemical input for vegetable production worldwide. Apparently, so far, the wide use of genes from the wild relatives has not been necessary, since enough resistance has been available within the cultivated material; however, in the future the resource of the wild gene pool may be valuable. An undesirable feature of cultivated spinach in addition to disease and pest susceptibility is the tendency to accumulate high amounts of oxalates, which may induce nutritional problems. Screening for low oxalate genotypes have been performed with mainly the cultivated types of spinach, with some progress (Mou 2008b). Screening of larger collections of wild types of spinach may provide further useful genes for this type of breeding.

13.9 Some Dark Sides and Their Addressing

The wild and cultivated types of spinach are reported to be only slightly weedy and apparently not very invasive, since they are still only found wild in limited geographic areas. They are, however, strong wind-pollinated outbreeders whose pollen can spread over long distances. Thus, they will be expected to be difficult to contain in case of transgenic cultivated types being marketed.

13.10 Recommendations for Future Actions

The wild species and their genetic variation have already been shown to be valuable as sources of disease resistance genes for cultivated types of spinach. Further international support for local conservation both in situ and ex situ collections should be encouraged. With new generations of sequencing technology, it has become efficient to establish much better maps and synteny studies, which should be used to ease future introgression of important traits from wild to cultivated spinach. Furthermore, the new marker techniques are efficient tools to characterize diversity of wild species for efficient stratification of the material during conservation initiatives.

References

Astley D, Ford-Lloyd BV (1981) The evolutionary significance of multigermicity in the genus *Spinacia* (*Chenopodiaceae*). Plant Syst Evol 137:57–61

Bemis WP, Wilson GB (1953) A new hypothesis explaining the genetics of sex determination in *Spinacea oleracea* L. J Hered 44:91–95

Crop wild Relatives (2009) Bioversity International. Rome, Italy. http://www.cropwildrelatives.org/. Accessed 20 May 2009

Downie SR, Palmer JD (1994) A chloroplast DNA phylogeny of the Caryophyllales based on structure and inverted repeat restriction site variation. Syst Bot 19:236–252

FAO (2007) Food and Agriculture Organisation of the United Nations. FAOSTAT database, Rome, Italy

Gabrielian E, Fragman-Sapir O (2008) Flowers of the Transcaucasus and adjacent areas. Koeltz Scientific Books, Koenigstein, Germany

Groben R, Wricke G (1998) Occurrence of microsatellites in spinach sequences from computer databases and development of polymorphic SSR markers. Plant Breed 117:271–274

Hanke S, Seehous C, Radies M (2000) Detection of a linkage of the four dominant mildew resistance genes “M1M2M3M4” in spinach from the wild type *Spinacia turkestanica*. Gartenbauwissenschaft 65:73–78

Hu J, Mou B, Bick BA (2007) Genetic diversity of 38 spinach (*Spinacia oleracea* L.) germplasm accessions and 10 commercial hybrids assessed by TRAP markers. Genet Resour Crop Evol 54:1667–1674

Iizuka M, Janick J (1971) Sex chromosome variation in *Spinacia oleracea* L. J Hered 62:349–352

International Spinach DataBase (2009) Centre for Genetic Resources. Wageningen, Netherlands. http://www.cgn.wur.nl/NL. Accessed 20 May 2009

Khattak JZK, Torp AM, Andersen SB (2006) A genetic linkage map of *Spinacia oleracea* and localization of a sex determination locus. Euphytica 148:311–318

Khattak JZK, Christiansen JL, Torp AM, Andersen SB (2007) Genic microsatellite markers for discrimination of spinach cultivars. Plant Breed 126:454–456

Mou BQ (2008a) Leafminer resistance in spinach. HortScience 43:1716–1719

Mou BQ (2008b) Evaluation of oxalate concentration in the U.S. spinach germplasm collection. HortScience 43:1690–1693

Mou BQ, Koike ST, du Toit LJ (2008) Screening for resistance to leaf spot diseases of spinach. HortScience 43:1706–1710

National Plant Germplasm System (2009) Agricultural Research System, Washington, DC, USA. http://www.ars-grin.gov/npgs. Accessed 20 May 2009

NCBI (2009) National Institute of Health. Rockville, MA, USA. http://www.ncbi.nlm.nih.gov. Accessed 20 May 2009

Plant Genetic Resources (2009) Bundesanstalt für Landwirtschaft und Ernährung. Bonn, Germany. http://www.genres.de. Accessed 20 May 2009

Ramanna MS (1976) Are there heteromorphic sex chromosomes in spinach (*Spinacia oleracea* L.)? Euphytica 25:277–284

Rubatzky VE, Yamaguchi M (1997) World vegetables: principles, production and nutritive values, 2nd edn. Chapman & Hall, New York, USA
Sauer JD (1993) Historical geography of crop plants: a select roster. CRC, Boca Raton, FL, USA
Schmitz-Linneweber C, Maier RM, Alcaraz J-P, Cottet A, herrmann RG, Mache R (2001) The plastid chromosome of spinach (*Spinacia oleracea*): complete nucleotide sequence and gene organization. Plant Mol Biol 45:307–315
Sneep J (1982) The domestication of spinach and the breeding history of its varieties. Euphytica (Suppl 2):27
Ul'janova TN (1977) Segetal and ruderal relatives of cultivated plants and the problems of their conservation and use: a Russian perspective. Genet Resour Crop Evol 44:5–8

Index

C. Kole (ed.), *Wild Crop Relatives: Genomic and Breeding Resources, Vegetables*,
DOI 10.1007/978-3-642-20450-0, © Springer-Verlag Berlin Heidelberg 2011